This is a collection of surveys and primarily expository articles focusing on recent developments concerning various quantitative aspects of “thin groups.” There are discrete subgroups of semisimple Lie groups that are both big (Zariski dense) and small (of infinite covolume). This dual nature leads to many intricate questions. Over the past few years, many new ideas and techniques, arising in particular from arithmetic combinatorics, have been involved in the study of such groups, leading, for instance, to far-reaching generalizations of the strong approximation theorem in which congruence quotients are shown to exhibit a spectral gap, referred to as superstrong approximation. This book provides a broad panorama of a very active field of mathematics at the boundary between geometry, dynamical systems, number theory, and combinatorics. It arose from the MSRI Hot Topics workshop of the same name in February 2012.

Mathematical Sciences Research Institute
Publications

61

Thin Groups and Superstrong Approximation

Mathematical Sciences Research Institute Publications

Thin Groups and Superstrong Approximation

Edited by

Emmanuel Breuillard
Université Paris-Sud 11

Hee Oh
Yale University

CAMBRIDGE
UNIVERSITY PRESS

Emmanuel Breuillard
Université Paris-Sud 11
emmanuel.breuillard@math.u-psud.fr

Hee Oh
Yale University
hee.oh@yale.edu

Silvio Levy (*Series Editor*)
Mathematical Sciences Research Institute
Berkeley, CA 94720
levy@msri.org

The Mathematical Sciences Research Institute wishes to acknowledge support by the National Science Foundation and the *Pacific Journal of Mathematics* for the publication of this series.

CAMBRIDGE
UNIVERSITY PRESS

32 Avenue of the Americas, New York NY 10013-2473, USA

Cambridge University Press is part of the University of Cambridge.

It furthers the University's mission by disseminating knowledge in the pursuit of education, learning and research at the highest international levels of excellence.

www.cambridge.org
Information on this title: www.cambridge.org/9781107036857

First published 2014

A catalogue record for this publication is available from the British Library

ISBN 978-1-107-03685-7 Hardback

Thin groups and superstrong approximation
MSRI Publications
Volume **61**, 2013

Contents

Thin groups and superstrong approximation
MSRI Publications
Volume **61**, 2013

Preface

The workshop "Thin Groups and Super-strong Approximation" was held at MSRI from February 6th through 10th, 2012. The organizing committee consisted of Emmanuel Breuillard (Orsay), Jordan Ellenberg (Madison), Alex Gamburd (New York), Emmanuel Kowalski (Zurich), Hee Oh (New Haven), and, last but not least, Peter Sarnak (Princeton), whose initiative it was to organize this workshop in the first place as part of the MSRI Hot Topics series.

The present volume contains expanded versions of most of the invited lectures and represents quite faithfully the range of topics discussed at the workshop. The articles in this volume are almost entirely expository. The goal was to put together in a single volume a series of state-of-the-art surveys on the subject of superstrong approximation and its recent developments. Videos of the talks given at the workshop are available on the MSRI website and can constitute a valuable resource complementing the articles in this volume.

Let us explain briefly the two phrases in the title: *thin groups* and *superstrong approximation.* Discrete subgroups of Lie groups with finite covolume, also known as *lattices* (for example, $SL(2, \mathbb{Z})$ inside $SL(2, \mathbb{R})$), are central objects of mathematics, and they have been extensively studied in the second half of the twentieth century.

Let $\mathbf{G}$ be a connected semisimple real algebraic group. A celebrated result of Borel and Harish-Chandra in 1962 asserts that if $\mathbf{G}$ is defined over $\mathbb{Q}$, then its subgroup $\mathbf{G}(\mathbb{Z})$ of integer points, called *an arithmetic subgroup*, is a lattice in the Lie group $\mathbf{G}(\mathbb{R})$, providing many important examples of lattices.

By Borel's density theorem, any lattice Γ in $\mathbf{G}(\mathbb{R})$ is Zariski-dense, unless $\mathbf{G}(\mathbb{R})$ has a non-trivial compact connected normal subgroup. A discrete subgroup Γ of $\mathbf{G}(\mathbb{R})$ is called *thin* if it is Zariski-dense in $\mathbf{G}$ and yet $\mathbf{G}(\mathbb{R})/\,\Gamma$ is of infinite volume.

Although less obviously important than lattices, such groups do appear naturally in mathematics. And it is one of the themes of this volume to present a variety of situations where these groups are the natural objects to study. For instance, the symmetries of an Apollonian circle packing form a thin subgroup of the orthogonal group $O(3, 1)$. Many classical properties of lattices continue to hold for thin subgroups. Non-abelian harmonic analysis and ergodic theory played a key role in the study of lattices, and not surprisingly they are again on the front line as far as thin groups are concerned. Yet many new ideas and techniques had to be developed. For example, group combinatorics and the study of *approximate groups* have played a major role, as is reflected in this volume.

Arithmetically defined thin groups, i.e., subgroups of an arithmetic subgroup $\mathbf{G}(\mathbb{Z})$ of infinite index, also share many of the known properties of arithmetic subgroups. For example, it is well known and easy to prove that the reduction mod p homomorphism from $SL_d(\mathbb{Z})$ to $SL_d(\mathbb{Z}/p\mathbb{Z})$ is surjective for every prime p. This is called *strong approximation* and holds for any arithmetic subgroup $\mathbf{G}(\mathbb{Z})$ as long as $\mathbf{G}$ is simply connected.

Remarkably, the strong approximation phenomenon continues to hold for arithmetically defined thin subgroups of $\mathbf{G}$. More precisely, the reduction mod p map remains surjective, for all but finitely many primes p, in restriction to any given thin subgroup Γ of $\mathbf{G}(\mathbb{Z})$, as proved by Matthews, Vaserstein, and Weisfeiler in the early 1980s. In other words, for a fixed finite generating subset S of Γ, the Cayley graph $\mathcal{C}(\mathbf{G}(\mathbb{Z}/p\mathbb{Z}), S)$ induced by the reduction mod p map is connected for almost all p. Due to the collective work of a number of mathematicians, pioneered by Helfgott and by Bourgain and Gamburd, we now know that these Cayley graphs are not only connected, but also form an expander family; we say that Γ has the *superstrong approximation* property. This result is a major achievement of recent years and the current volume is centered around it and its many applications. The property of being an expander is equivalent to the combinatorial spectral gap property on Cayley graphs that the smallest positive eigenvalue of the combinatorial Laplacian on these graphs is bigger than some positive number, which is independent of p.

Though the articles are organized alphabetically by last name, many readers will wish to turn first to the last contribution, by Peter Sarnak. It gives a panoramic overview of recent developments in connection with superstrong approximation for thin groups, and is therefore a good entry point for anyone willing to discover what this is all about.

Andrei Rapinchuk's article is a thorough introduction to strong-approximation for arithmetic groups as well as for their Zariski-dense subgroups.

The surveys by Alireza Salehi Golsefidy and Emmanuel Kowalski describe a number of recent applications of superstrong approximation to sieving and counting primes in orbits, applications that were one of the main motivations for establishing superstrong approximation in the first place.

The articles by Emmanuel Breuillard and by Laci Pyber and Endre Szabó focus on recent combinatorial advances which are key to the proof of superstrong approximation and concern approximate groups and the growth of finite subsets of large finite simple groups.

The article by Amir Mohammadi reports on his recent work with Oh on the ergodicity criterion for unipotent flows with respect to the Burger-Roblin measure on $\Gamma\backslash SO(n, 1)$ for geometrically finite subgroups Γ; this measure plays a key role in counting problems on orbits of Γ.

Michael Larsen's article is a survey of recent results regarding word maps in finite simple groups, their image and distribution.

The article by Hee Oh describes how to use ergodic theory and harmonic analysis to extend to thin subgroups many of the counting results known for lattices and its application to Apollonian circle packings and more general sphere packings.

The article by Gopal Prasad and Andrei Rapinchuk surveys their joint work, spread over the past few years and regarding arithmetic properties of Zariski-dense subgroups of semisimple algebraic groups, the notion of weak commensurability, the existence of regular elements with special properties, etc.

Jordan Ellenberg's paper introduces monodromy groups as a chief source of examples of thin groups and describes his recent work with Hall and Kowalski on how to apply superstrong approximation to prove gonality estimates and height bounds in certain situations arising in arithmetic geometry.

Elena Fuchs reports on recent investigations regarding a very explicit class of monodromy groups arising naturally from differential equations and various attempts to determine whether the given groups are thin or not.

Alan Reid's contribution, coauthored by Darren Long, focuses on various constructions of arithmetically defined subgroups of arithmetic groups (such as $SL_3(\mathbb{Z})$) which are isomorphic to the fundamental groups of closed surfaces.

Igor Rivin surveys some interesting results and questions regarding generic elements (in various senses) of geometrically interesting groups.

The articles by Jean Bourgain and Alex Kontorovich describe part of their work on various problems connected to local-global principles for thin (semi-)groups arising from Apollonian circle packings and the Zaremba conjecture.

We hope the reader will enjoy browsing and reading this volume. It is our pleasure to thank all the people involved in the making of this book for helping us assemble this nice collection of articles in a speedy manner: the referees for their prompt and professional advice, Silvio Levy and the editorial staff of Cambridge University Press for their support, and the authors for their contributions.

Emmanuel Breuillard
Université Paris-Sud, Orsay, France

Hee Oh
Yale University, New Haven, USA

Thin groups and superstrong approximation
MSRI Publications
Volume **61**, 2013

Some Diophantine applications of the theory of group expansion

JEAN BOURGAIN

1. Introduction

The recent years have seen considerable progress in the theory of expansion and spectral gaps for so-called thin groups (as opposed to the classical theory of arithmetic groups.) These developments rely in part on methods and results from the rather novel research area of "arithmetic combinatorics" and underlying are general principles such as the "sum-product theorem" in finite fields and "product theorems" in linear groups. These advances turned out to be of interest well beyond group theory and have applications to geometry, number theory, theoretical computer science and even mathematical physics. At this point, many aspects of the extensive story were already accounted for in several survey papers, such as [Bourgain 2010], the Bourbaki exposé of E. Kowalski [2012] and those of B. Green [2009] and A. Lubotzky [2012] based on AMS lectures. A discussion of the "ubiquity" of thin groups from a broader perspective appears in [Sarnak 2014] in the present volume.

We will focus here on two specific number-theoretic applications. The first relates to integral Apollonian circle packings (ACP for short) and the problem of a local/global principle for the curvatures, as proposed in [Graham et al. 2003] and [Sarnak 2011]. The other concerns progress towards Zaremba's conjecture [1972] on continued fraction expansions of rationals and we will briefly review the paper [Bourgain and Kontorovich 2011]. Another exciting application of the theory of group expansion to finiteness in arithmetic geometry may be found in the work of J. Ellenberg, C. Hall and E. Kowalski [Ellenberg et al. 2012] but will not be discussed here. Neither will we get into the role of expansion to sieving theory (originating from [Bourgain et al. 2010a]) which triggered many of the later developments.

Our reference list is far from complete and strictly serves this exposé.

2. Background on expansion in Cayley graphs induced by thin groups

We start by recalling the notion of graph expansion and expander families. The reader is referred to the excellent survey paper [Hoory et al. 2006] for a detailed discussion of this theory. Let $\mathscr{G}$ be a k-regular graph on a finite vertex set V, with $|V| = n$. Here one should see k as fixed whereas $n \to \infty$. The Busemann–Cheeger constant is then defined as

$$h(\mathscr{G}) = \min_{|S| \le n/2} \frac{|\partial S|}{|S|}, \tag{2-1}$$

where the minimum is taken over all subsets S of V and ∂S refers to the set of edges from S to $V \backslash S$. Having fixed k, a collection of k-regular graphs $\mathscr{G}_1, \mathscr{G}_2, \mathscr{G}_3, \ldots$ on vertex sets $V_1, V_2, V_3, \ldots$ with $|V_i| \to \infty$ is called an expander family provided

$$h(\mathscr{G}_i) > c \quad \text{for all } i, \tag{2-2}$$

for some $c > 0$.

Expansion has a well-known spectral interpretation on the level of the adjacency matrix $A(\mathscr{G})$ of $\mathscr{G}$, defined by

$$A_{i,j} = \begin{cases} 1 \text{ if } (i,j) \in \mathscr{G}, \\ 0 \text{ otherwise.} \end{cases} \tag{2-3}$$

Since $\mathscr{G}$ is undirected, A is symmetric and $k = \lambda_0 \ge \lambda_1 \ge \cdots \ge \lambda_{n-1}$ since we assumed $\mathscr{G}$ k-regular. The following inequalities relate then the Cheeger constant to the "spectral gap" $\lambda_0 - \lambda_1$

$$\tfrac{1}{2}(k - \lambda_1) \le h(\mathscr{G}) \le \sqrt{2k(k - \lambda_1)}. \tag{2-4}$$

Do expander families exist? It was proven by Pinsker [1973] that given $k \ge 3$, a random (= typical) k-regular graph on n vertices ($n \to \infty$) is an expander graph. Around the same time, Margulis [1973] came up with explicit constructions based on Cayley graphs of groups. Recall that if $V = \langle S \rangle$ is a group generated by a finite set S of elements, the Cayley graphs $\mathscr{G}(V, S)$ consists of the edges $(x, y), x, y \in V$, for which $xy^{-1} \in S \cup S^{-1}$. Of particular importance to our discussion is Selberg's theorem on the congruence graphs induced by $\mathrm{SL}_2(\mathbb{Z})$.

Theorem 1. *Assume $\langle S \rangle$ a finite index subgroup of $\mathrm{SL}_2(\mathbb{Z})$ and denote by π_q : $\mathrm{SL}_2(\mathbb{Z}) \to \mathrm{SL}_2(q)$ the $(\bmod\, q)$ reduction. There is an integer q_0 such that the family of Cayley graphs*

$$\mathscr{G}\big(\mathrm{SL}_2(q), \pi_q(S)\big), \quad \textit{where } q \in \mathbb{Z}_+,\ (q, q_0) = 1, \tag{2-5}$$

is an expander family.

The assumption that $\langle S \rangle$ is of finite index is quite restrictive. What happens with "thin" subgroups? Recall that if G is an algebraic group, a subgroup $\Gamma \subset G(\mathbb{Z})$ is called "thin" provided $[G(\mathbb{Z}) : \Gamma] = \infty$. Assuming $\langle S \rangle \subset \mathrm{SL}_2(\mathbb{Z})$ contains the free group F_2 on 2 generators (equivalently, $\langle S \rangle$ is nonelementary meaning that $\langle S\rangle$ does not contain a solvable subgroup of finite index), the Zariski closure of $\langle S\rangle$ equals SL_2. Citing the strong approximation property due to Matthews, Vaserstein, and Weisfeiler [1984], recall that if Γ is a subgroup of $SL_d(\mathbb{Z})$ which is Zariski dense in SL_d, there is some $q_0 \in \mathbb{Z}_+$ such that $\pi_q(\Gamma) = SL_d(\mathbb{Z}/q\mathbb{Z})$ for $(q, q_0) = 1$. Thus, if $\langle S\rangle \subset SL_2(\mathbb{Z})$ is nonelementary, the Cayley graphs $\mathcal{G}(SL_2(q), \pi_q(S))$ for $q \in \mathbb{Z}_+$, $(q, q_0) = 1$ are connected. Moreover Theorem 1 generalizes.

Theorem 2. *Assume $\langle S\rangle$ a nonelementary subgroup of $SL_2(\mathbb{Z})$. There is $q_0 \in \mathbb{Z}$ such that $\mathcal{G}(SL_2(q), \pi_q(S))$, $(q, q_0) = 1$, forms an expander family.*

Under the assumption that the limit set of $\Gamma = \langle S\rangle$ has dimension $\delta_\Gamma > \frac{5}{6}$, Theorem 2 is due to A. Gamburd [2002]. His result left unanswered Lubotzky's 1-2-3 problem [1994] (a folklore question in the subject), noting that Selberg's theorem applies to

$$S_1 = \left\{\begin{pmatrix}1 & 1\\ 0 & 1\end{pmatrix}, \begin{pmatrix}1 & 0\\ 1 & 1\end{pmatrix}\right\} \quad \text{and} \quad S_2 = \left\{\begin{pmatrix}1 & 2\\ 0 & 1\end{pmatrix}, \begin{pmatrix}1 & 0\\ 2 & 1\end{pmatrix}\right\},$$

but not to

$$S_3 = \left\{\begin{pmatrix}1 & 3\\ 0 & 1\end{pmatrix}, \begin{pmatrix}1 & 0\\ 3 & 1\end{pmatrix}\right\}.$$

The latter case was settled in [Bourgain and Gamburd 2008], which implies Theorem 2 for q prime. A crucial ingredient is Helfgott's product theorem in $SL_2(p)$ [Helfgott 2008]. Theorem 2 for q square free was proven in [Bourgain et al. 2010a], motivated by applications to prime number sieving, and the case of unrestricted modulus follows from [Bourgain and Varjú 2012]. Meanwhile, based on deep work due to Helfgott [2011], Green, Breuillard, and Tao [2011] and Pyber and Szabó [2010] on the combinatorial side, Theorem 2 has been vastly generalized, at least for prime and square-free moduli (see [Varjú 2012; Salehi Golsefidy and Varjú 2012]). Although the work described later in this exposé is SL_2-based ($SL_2(\mathbb{Z})$ and the Gauss-integer extension $SL_2(\mathbb{Z} + i\mathbb{Z})$), we record one of the most general results obtained in this context.

Theorem 3 [Salehi Golsefidy and Varjú 2012]. *Let k be a number field and $\Gamma \subset SL_d(k)$, $\Gamma = \langle S\rangle$ with S a finite symmetric set. Assume the Zariski-closure of Γ is semisimple. Then the Cayley graphs $\mathcal{G}(\pi_q(\Gamma), \pi_q(S))$ form a family of expanders when q ranges over square-free ideals of the integers $\mathcal{O}$ of k with large prime factors.*

Returning to Lubotzky's problem, let us also mention the Lubotzky–Weiss conjecture, stating that expansion in $SL_2(p)$ or, more generally, $SL_n(p)$-Cayley graphs, is in fact a group property:

Conjecture 4. *There is an absolute constant $c = c_k > 0$ such that*

$$h\big(\mathcal{G}(\mathrm{SL}_2(p), S)\big) > c, \tag{2-6}$$

whenever $S \subset \mathrm{SL}_2(p)$, $|S| = k$, is generating and p prime.

A positive answer would be conceptually very pleasing.

Evidence of its truth is the result of Breuillard and Gamburd [2010] establishing the conjecture for p outside a small exceptional set of the primes. Lubotzky–Weiss' problem will not be essential in our subsequent discussion as the Cayley graphs will be induced by a fixed set of elements in $\mathrm{SL}_2(\mathbb{Z})$ or $\mathrm{SL}_2(\mathbb{Z} + i\mathbb{Z})$. On the other hand, what is essential for us is to have unrestricted modulus q.

3. Hyperbolic lattice point counting

The spectral gap in congruence Cayley graphs described above in conjunction with Brun's combinatorial sieve, have been used to carry out prime and pseudo-prime sieving in the orbits of thin groups (see [Bourgain et al. 2010a; Sarnak 2008; Salehi Golsefidy and Sarnak 2011] for the ultimate generalization of this theory.) The "balls" involved in the sieving process are defined in terms of the word metric on the generators.

Other number-theoretic applications as presented in the next sections are based on different analytical tools (more specifically, the Hardy–Littlewood circle method) and require precise counting in Archimedean balls in the congruence subgroups. Such information may be obtained by hyperbolic lattice point counting and we briefly review the results for thin groups obtained in [Bourgain et al. 2011]. See the same paper also for related references.

Denote by $\mathbb{H} = \mathbb{H}^2 = \{x + iy \in \mathbb{C} : y > 0\}$ the hyperbolic plane on which $\mathrm{SL}_2(\mathbb{R})$ acts by Moebius transformation

$$gz = \frac{az+b}{cz+d}, \quad \text{where } g = \begin{pmatrix} a & b \\ c & d \end{pmatrix} \in \mathrm{SL}_2(\mathbb{R}). \tag{3-1}$$

Note that the theory discussed below has a counterpart for $\mathbb{H}^3$ and the action of $\mathrm{SL}_2(\mathbb{C})$, which is relevant for the application to the Apollonian group discussed in the next section. With g as in (3-1), we have

$$\|g\|^2 = a^2 + b^2 + c^2 + d^2 = 4u(gi, i) + 2, \tag{3-2}$$

where

$$u(z, w) = \frac{|z-w|^2}{4 \operatorname{Im} z \operatorname{Im} w} \quad \text{and} \quad \cosh d_{\mathbb{H}}(z, w) = 1 + 2u(z, w).$$

This clarifies the significance of hyperbolic space to the Archimedean counting problem.

Next, let Γ be a finitely generated subgroup of $\mathrm{SL}_2(\mathbb{Z})$, $L = L(\Gamma) \subset \mathbb{R}$ its limit set and $\delta = \delta_\Gamma$ the Hausdorff dimension of L. The assumption that Γ is nonelementary is equivalent to $0 < \delta \leq 1$. Thus our goal is Archimedean counting in the orbits of Γ and its congruence subgroups. We distinguish two cases.

• $\delta > \frac{1}{2}$: In this case, the hyperbolic surface $\mathbb{H}/\Gamma$ (which has infinite volume if $\delta < 1$) has an L^2-spectral theory and we rely on Lax–Phillips' theory based on automorphic methods and the wave equation.

• $\delta > 0$: If $0 < \delta \leq \frac{1}{2}$, there is no L^2-spectral theory and instead we use the thermodynamical approach based on symbolic dynamics and Ruelle's transfer operator, as developed by Lalley, Dolgopyat, Naud and others. This method is quite flexible and applies also in the semigroup setting (relevant in the application to Zaremba's problem in Section 5.)

We first discuss the spectral approach, assuming $\delta(L) > \frac{1}{2}$. The spectrum

$$0 \leq \lambda_0 < \lambda_1 \leq \cdots \leq \lambda_{\max} < \tfrac{1}{4} \quad \xrightarrow{\text{continuous spectrum}} \tag{3-3}$$

of the Laplace operator on $\mathbb{H}/\Gamma$ has lowest eigenvalue $\lambda_0 = \delta(1-\delta)$ and $\lambda_1 > \lambda_0$. Defining δ_j by $\lambda_j = \delta_j(1-\delta_j)$ and denoting by $\{\varphi_j\}$ the corresponding eigenfunctions, we state the theorem of Lax and Phillips [1982].

Theorem 5. *With the above notation, one has for $w_0, w \in \mathbb{H}$*

$$\bigl|\{\gamma \in \Gamma : d_{\mathbb{H}}(w, \gamma w_0) \leq s\}\bigr| = \sum_{j \geq 0} C_j \varphi_j(w) \varphi_j(w_0) e^{\delta_j s} + O\bigl(e^{\frac{1}{3}(1+\delta_0)s}\bigr). \tag{3-4}$$

What happens in congruence subgroups $\Gamma(q) = \{\gamma \in \Gamma : \gamma = \left(\begin{smallmatrix}1&0\\0&1\end{smallmatrix}\right) (\operatorname{mod} q)\}$?

Considering the spectrum of $\mathbb{H}/\Gamma(q)$, it is an elementary fact that $\lambda_0(\Gamma(q)) = \lambda_0(\Gamma)$ and the issue is a uniform gap $\lambda_1(\Gamma(q)) - \lambda_0$ when q varies. When $\Gamma = \mathrm{SL}_2(\mathbb{Z})$. Selberg proved that $\lambda_1(\Gamma(q)) \geq \frac{3}{16}$ and conjectured $\lambda_1(\Gamma(q)) \geq \frac{1}{4}$ (no exceptional eigenvalues). The current record seems to be

$$\lambda_1(\Gamma(q)) > \tfrac{1}{4} - \left(\tfrac{7}{64}\right)^2,$$

due to Kim and Sarnak [1995].

Returning to thin groups $\Gamma = \langle S \rangle$ with $\delta_\Gamma > \frac{1}{2}$, one has the following extension of Selberg's result.

Theorem 6. *There is $\varepsilon = \varepsilon(\Gamma) > 0$ such that $\lambda_1(\Gamma(q)) > \lambda_0 + \varepsilon$ for all $q \in \mathbb{Z}_+$.*

The proof relies essentially on the expander family $\mathcal{G}\bigl(\mathrm{SL}_2(q), \pi_q(S)\bigr)$ and the conversion of the combinatorial spectral gap to a geometric one follows arguments due to Burger and Brooks in the finite volume case.

Combining Theorem 5 and Theorem 6 leads to the following Archimedean-modular distributional property.

Corollary 7. *Let $\Gamma = \langle S\rangle \subset \mathrm{SL}_2(\mathbb{Z})$, and $\delta_\Gamma > \frac{1}{2}$. There is some $q_0 = q_0(\Gamma) \in \mathbb{Z}$ such that, if $(q, q_0) = 1$ and $g \in \mathrm{SL}_2(q)$,*

$$\big|\{\gamma \in \Gamma : \|\gamma\| \le N \text{ and } \pi_q(\gamma) = g\}\big| = \frac{cN^{2\delta}}{|\mathrm{SL}_2(q)|} + O(q^C N^{2\delta-\varepsilon}), \tag{3-5}$$

where ε, c and C only depend on Γ.

This type of result plays a key role in the diophantine applications discussed in the next sections of this exposé. A similar (though slightly weaker) statement may be obtained in the general case $\delta_\Gamma > 0$ following the thermodynamical approach which we review next.

Assuming $\Gamma = \langle T_1, \dots, T_k\rangle \subset \mathrm{SL}_2(\mathbb{Z})$ has no parabolic elements, one identifies Γ with the set Σ_* of finite sequences on the alphabet $\{\pm 1, \dots, \pm k\}$ compatible with a transition matrix and the limit set $L(\Gamma)$ with the corresponding set Σ of infinite sequences. The Nielsen map $F : L \to L$ corresponds to a finite type shift $\sigma : \Sigma \to \Sigma$ under the symbolic conversion and the distortion function $f = \log|F'|$ is expressed by $\tau(x) = d_{\mathbb{H}}(i, xi) - d_{\mathbb{H}}(i, x_2x_3\dots i)$ for $x = (x_1, x_2, \dots) \in \Sigma$. Taking $0 < \rho < 1$, we define a metric on Σ by

$$d(x, y) = \rho^m, \quad \text{with } m = \max\{j : x_j = y_j\}, \tag{3-6}$$

and denote by $\mathcal{F}$ the corresponding space of Hölder functions f, i.e., those satisfying

$$|f(x) - f(y)| \le Kd(x, y) \quad \text{for some } K > 0. \tag{3-7}$$

Given a function $f \in \mathcal{F}$, define the transfer operator $\mathcal{L}_f : \mathcal{F} \to \mathcal{F}$ by

$$(\mathcal{L}_f g)(x) = \sum_{\sigma(y)=x} e^{f(y)} g(y). \tag{3-8}$$

Recall Ruelle's theorem to the effect that for $f \in \mathcal{F}$ real, there is a simple largest eigenvalue $\lambda_f > 0$ of $\mathcal{L}_f$ with strictly positive eigenfunction h_f and a Borel probability measure ν_f on Σ satisfying

$$\mathcal{L}^*\nu = \lambda\nu \quad \text{and} \quad \int h\, d\nu = 1.$$

Denote by $P(f) = \log \lambda_f$ the pressure function. Taking $f = -s\tau$ with τ as above, $P(-s\tau)$ is strictly increasing in $s \in \mathbb{R}$ and, according to a result due to Patterson and Sullivan, vanishes at $s = \delta = \dim L(\Gamma)$.

The renewal approach to the counting problem consists in introducing a counting function

$$N_\phi(T, x) = \sum_{n=0}^{\infty} \sum_{\sigma^n y = x} \phi(y) 1_{\{S_n\tau(y) \le T\}}, \tag{3-9}$$

where

$$S_n f = f + (f \circ \sigma) + \cdots + (f \circ \sigma^{n-1}). \tag{3-10}$$

Thus in particular, for x the identity and $\phi = 1$, we obtain the number of elements of Γ for which $d_{\mathbb{H}}(\gamma i, i) \le T$.

The counting function satisfies the renewal equation

$$N_\phi(T, x) = \phi(x) 1_{\{T \ge 0\}} + \sum_{\sigma(y)=x} N_\phi(T - \tau(y), y). \tag{3-11}$$

Introducing its Laplace transform

$$F(s, x) = \int_{-\infty}^{\infty} e^{-sT} N(T, x)\, dT \quad (\operatorname{Re} s \gg 1), \tag{3-12}$$

(3-11) is converted to

$$F(s, x) = -\mathcal{R}_s \frac{\phi(x)}{s}, \tag{3-13}$$

where $\mathcal{R}_s = (I - \mathcal{L}_{-s\tau})^{-1}$ is the resolvent and is analytic for Re $s > \delta$. The following statement results from the work of Lalley and Naud (based on work of Dolgopyat).

Theorem 8. (i) *$\mathcal{R}_s$ has a meromorphic extension to a strip* Res $> \delta - \varepsilon$ *with a simple pole at $s = \delta$.*

(ii) *$\|\mathcal{R}_s\| < C(1 + |\operatorname{Im} s|^2)$ for $|s| \to \infty$.*

(iii) *$|\{\gamma \in \Gamma : d_{\mathbb{H}}(i, \gamma i) \le s\}| = Ce^{\delta s} + O(e^{(\delta - \varepsilon)s})$.*

Statement (iii) follows from (i) and (ii) by standard Tauberian arguments (see [Bourgain et al. 2011] for details) and a widening of the analyticity region leads to improved error terms in the counting.

The next step consists in extending this theory in order to capture congruence subgroups $\Gamma(q)$ of Γ with uniformity in q.

Replace Σ by $\Sigma \times \mathrm{SL}_2(q)$ and $\mathcal{F}$ by the space $\mathcal{F}_q = \mathcal{F}(\Sigma \times \mathrm{SL}_2(q))$ of vector-valued Hölder functions f on Σ with norm

$$\|f\|_\rho = \|f\|_\infty + |f|_\rho,$$

where

$$\|f\|_\infty = \max_x \Big(\sum_{g \in \mathrm{SL}_2(q)} |f(x, g)|^2 \Big)^{\frac{1}{2}} \tag{3-14}$$

and

$$|f|_\rho = \max_m \frac{\operatorname{Var}_m f}{\rho^m}, \tag{3-15}$$

with

$$\mathrm{Var}_m f = \sup\left\{\left(\sum_{g\in \mathrm{SL}_2(q)} |f(x,g)-f(y,g)|^2\right)^{\frac{1}{2}} : x_j = y_j \text{ for } j \le m\right\} \tag{3-16}$$

On the extended space $\mathcal{F}_q$, introduce the transfer operator as

$$(\mathcal{L}_{-s\tau} f)(x,g) = \sum_{\sigma(y)=x} e^{s\tau(y)} f(y, yx^{-1}g). \tag{3-17}$$

The resonance-free region obtained in [Bourgain et al. 2011] is of the form

$$\mathrm{Re}\, s > \delta - \frac{c}{\log(2+|\mathrm{Im}\, s|)}, \tag{3-18}$$

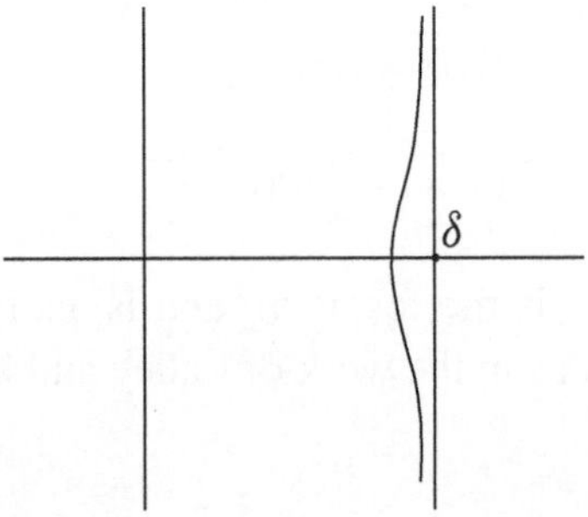

where (3-18) is uniform in q. The proof makes again essential use of the expansion in $\Gamma/\Gamma(q)$. Using (3-18), following analogue of Corollary 7 is deduced.

Theorem 9. *There is $q_0 = q_0(\Gamma) \in \mathbb{Z}$ such that for $(q, q_0) = 1$ and $g \in \mathrm{SL}_2(q)$*

$$\left|\{\gamma \in \Gamma : \|\gamma\| \le N \text{ and } \pi_q(\gamma) = g\}\right| = \frac{cN^{2\delta}}{|\mathrm{SL}_2(q)|}\left(1 + O(N^{-1/\log\log N})\right) + O(q^C N^{2\delta-\varepsilon}). \tag{3-19}$$

Compared with Corollary 7, the error term is a bit worse, but the term in $O\big(N^{-1/\log\log N}\big)$ may be further reduced (and even removed) if $1_{[\|\gamma\|\le N]}$ is replaced by a suitable weighted average.

In both Corollary 7 and Theorem 9, one may moreover specify the expanding direction v_+ of g to be in a sector I, with the effect of introducing a factor $\mu(I)$ in the main term, where μ is the Sullivan–Patterson measure.

4. Integral Apollonian circle packings

For background, see [Graham et al. 2003; Sarnak 2011]. Recall Apollonius' theorem: Given three mutually tangent circles in the plane, there are exactly two circles tangent to all three. Starting from four mutually tangent circles as depicted below and filling in repeatedly the lacunae with tangent circles leads to so-called Apollonian circle packings (ACPs):

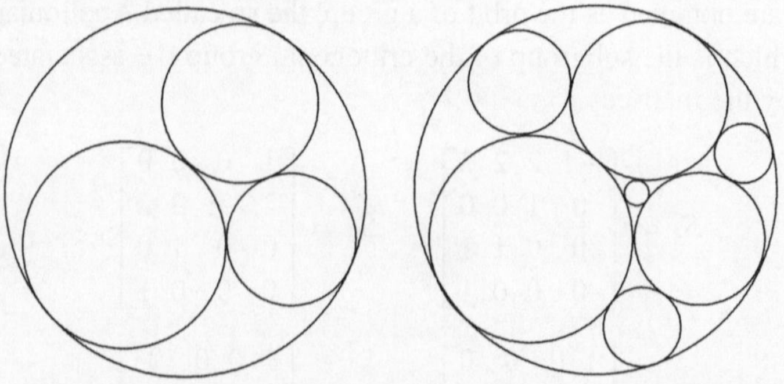

It was observed by F. Soddy that if the curvatures of the initial four circles — known as the *root quadruple* — are integers, all circles in the packing will have integral curvature, leading to integral ACPs. An example is shown in Figure 1.

This phenomenon may be explained by considering the Descartes quadratic form

$$F(x_1, x_2, x_3, x_4) = 2(x_1^2 + x_2^2 + x_3^2 + x_4^2) - (x_1 + x_2 + x_3 + x_4)^2, \quad (4\text{-}1)$$

since $F(x_1, x_2, x_3, x_4) = 0$ is tantamount to x_1, x_2, x_3, x_4 being curvatures of four mutually tangent circles.

Which integers are produced in a given integral ACP? That is the general question proposed in [Graham et al. 2003] and [Sarnak 2011], where the following conjectures were formulated.

(A) *The positive density conjecture*: the set of curvatures in any integral ACP is of positive density in $\mathbb{Z}$.

(B) *Local to global principle*: all integers are produced, up to a finite congruence condition.

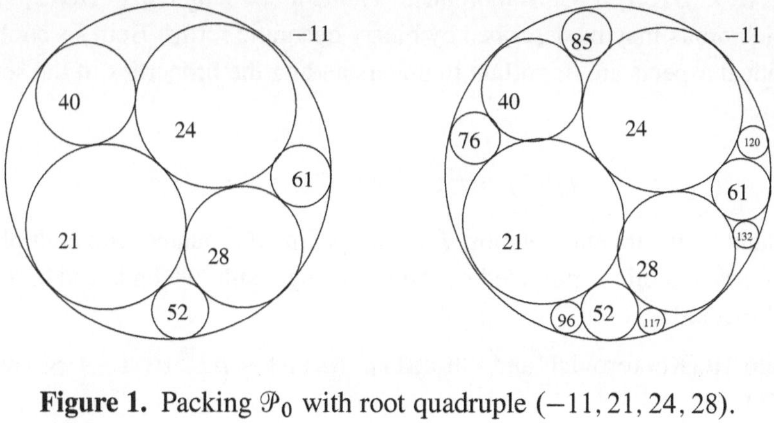

Figure 1. Packing $\mathcal{P}_0$ with root quadruple $(-11, 21, 24, 28)$.

Obviously (B) implies (A).

It turns out that the curvatures in a given integral ACP with root quadruple (a, b, c, d) are obtained as the orbit of a group, the so called Apollonian packing group A, which is the subgroup of the orthogonal group $\mathbb{O}_F$ associated to (4-1), generated by the matrices

$$S_1 = \begin{bmatrix} -1 & 2 & 2 & 2 \\ 0 & 1 & 0 & 0 \\ 0 & 0 & 1 & 1 \\ 0 & 0 & 0 & 1 \end{bmatrix}, \quad S_2 = \begin{bmatrix} 1 & 0 & 0 & 0 \\ 2 & -1 & 2 & 2 \\ 0 & 0 & 1 & 0 \\ 0 & 0 & 0 & 1 \end{bmatrix},$$

$$S_3 = \begin{bmatrix} 1 & 0 & 0 & 0 \\ 0 & 1 & 0 & 0 \\ 2 & 2 & -1 & 2 \\ 0 & 0 & 0 & 1 \end{bmatrix}, \quad S_4 = \begin{bmatrix} 1 & 0 & 0 & 0 \\ 0 & 1 & 0 & 0 \\ 0 & 0 & 1 & 0 \\ 2 & 2 & 2 & -1 \end{bmatrix}.$$

A first important feature of A is that, under the spin double cover, it identifies with a finitely generated subgroup Γ of $\mathrm{SL}_2(\mathbb{Z}+i\mathbb{Z})$ to which the spectral method discussed in the previous section apply. More specifically considering the action on $\mathbb{H}^3$, the limit set L has dimension

$$\delta = \delta_\Gamma = 1.30568\ldots, \tag{4-2}$$

and Lax–Phillips theory is applicable. Denoting $a(C)$ the curvature of the circle C, for any packing $\mathcal{P}$ the Poincaré series

$$\sum_{C \in \mathcal{P}} a(C)^{-s} \tag{4-3}$$

has exponent of convergence equal to δ, which is the dimension of the residual set of the packing $\mathcal{P}$ (and is independent of $\mathcal{P}$).

The other feature of A are its arithmetic subgroups $\langle S_1, S_2, S_3\rangle$, $\langle S_2, S_3, S_4\rangle$, $\langle S_3, S_4, S_1\rangle$, $\langle S_4, S_1, S_2\rangle$, isomorphic to a finite index subgroup of $SO(2,1)(\mathbb{Z})$, and which orbits may be described by binary quadratic forms. Both spectral and arithmetical aspects are important in understanding the properties of the sets of curvatures.

Let

$$N_{\mathcal{P}}(T) = \#\{C \in \mathcal{P} : a(C) \le T\}, \tag{4-4}$$

be the number of curvatures at most T in the packing $\mathcal{P}$, counted with multiplicity. The exact asymptotic is provided by the following result, which is and based on spectral methods.

Theorem 10 [Kontorovich and Oh 2011]. *$N_{\mathcal{P}}(T) \sim bT^\delta$ as $T \to \infty$, where $b = b(\mathcal{P})$.*

For instance, Fuchs and Sanden showed that $b(\mathcal{P}_0) = 0.0458\ldots$.

Assuming $\mathcal{P}$ primitive, that is, $(a, b, c, d) = 1$, denote further by $\pi^{\mathcal{P}}(T)$ the number of prime circles in $\mathcal{P}$ of curvature at most T and by $\pi_2^{\mathcal{P}}(T)$ the number of twin (= tangent) prime circles.

Theorem 11 [Kontorovich and Oh 2011].

$$\pi^{\mathcal{P}}(T) \ll \frac{T^\delta}{\log T} \quad \textit{and} \quad \pi_2^{\mathcal{P}}(T) \ll \frac{T^\delta}{(\log T)^2} \textit{ as } T \to \infty. \tag{4-5}$$

Lower bounds of the same order but for "pseudoprime circles" hold. This is a result in similar vein as [Bourgain et al. 2010a; Bourgain et al. 2011], based on sieving methods combined with a spectral input.

Let us now return to the diophantine aspects and conjectures (A) and (B).

Theorem 12 [Fuchs 2010; Fuchs and Sanders 2010]. *Determination of* $\pi_q(A) \subset \mathrm{GL}_4(\mathbb{Z}/q\mathbb{Z})$ *for every* q*: The ramified set consists only of the primes* 2 *and* 3.

Theorem 13 [Bourgain and Fuchs 2011]. *The positive density conjecture holds.*

Theorem 14 [Bourgain and Kontorovich 2012]. *The local/global principle holds up to an exceptional set of size at most* $T^{1-\varepsilon}$, *for some* $\varepsilon > 0$.

As an example, consider the packing $\mathcal{P}_0$ introduced above. The only congruence restrictions are

$$a(C) \in \{0, 4, 12, 13, 16, 21\} \pmod{24}, \tag{4-6}$$

and by Theorem 14

$$\#\{a < T : \pi_{24}(a) \in \{0, 4, 12, 13, 16, 21\} \text{ and } a \text{ is not a } \mathcal{P}_0\text{-curvature}\} < T^{1-\varepsilon}.$$

Initial results towards Theorem 13 were obtained by Sarnak [2011] and Fuchs [2010] and their work also introduced several of the methods needed to attack these questions.

Several ingredients are involved in the proof of Theorems 12, 13 and 14:

(i) representation of integers by binary quadratic forms;[1]

(ii) Lax–Phillips theory and spectral gap;

(iii) the Hardy–Littlewood circle method.

The use of binary quadratic forms may be briefly summarized as follows. Fix $(a, b, c, d) \in \mathcal{P}$ and set

$$A = a + b \quad B = \tfrac{1}{2}(a + b - c + d) \quad C = a + d. \tag{4-7}$$

[1] As originally pointed out by Sarnak in a letter to Lagarias.

Consider the quadratic form

$$f(x, y) = Ax^2 + 2Bxy + Cy^2 \text{ with discriminant } D_f = 4(B^2 - AC) = -4a^2.$$

Then all integers represented by $f - a$ appear as curvatures of circles C in $\mathcal{P}$ that are tangent to the circle C_a with curvature a.

This observation, going back to Sarnak and Fuchs, leads to the question of understanding how many integers can be represented by a given quadratic form of large discriminant.

Theorem 15 [Blomer and Granville 2006; Bourgain and Fuchs 2012]. *Let*

$$f(x, y) = ax^2 + bxy + cy^2 \in \mathbb{Z}[X, Y],$$

suppose $D = b^2 - 4ac < 0$ *and denote by* $U_f(M)$ *the number of integers not exceeding* M *that are representable by* f. *Assume further that*

$$\log |D| < O(\log \log M). \tag{4-8}$$

(i) *Assume*

$$|D| \ll (\log M)^{(\log 2) - \varepsilon}. \tag{4-9}$$

Then

$$U_f(M) \gg \frac{M}{(\log M)^{\frac{1}{2}+}}. \tag{4-10}$$

(ii) *Assuming* (4-9), f *represents all integers that are representable by the genus of* f, *up to an exceptional set of size at most* $M(\log M)^{-(1/2)-\varepsilon'}$.

(iii) *Assume*

$$|D| \gg (\log M)^{2(\log 2)+\varepsilon}. \tag{4-11}$$

Then

$$U_f(M) \asymp \frac{M}{\sqrt{|D|}}. \tag{4-12}$$

Statements (i) and (iii) appear in [Blomer and Granville 2006]. As explained there, the different behavior for small and large discriminant has to do with the interplay between the class group and the prime factorization of a typical integer $< M$ representable by the genus of f; the intermediate range being difficult to analyze. Statement (ii) is proven in [Bourgain and Fuchs 2012] and relies on a more general principle in combinatorial group theory.

It is worthwhile to point out that while the initial progress towards conjecture (A) exploits the range (4-9), the proof of Theorem 13 in [Bourgain and Fuchs 2012] uses the range (4-11). Combining part (ii) of Theorem 15 with Iwaniec's half-dimensional sieve [1972], [Friedlander and Iwaniec 2010] gives the following refinement of a result in [Sarnak 2011].

Theorem 16 [Bourgain and Fuchs 2012]. *Let C_a be a circle with curvature a in the packing $\mathcal{P}$ and assume*

$$a < (\log T)^{\frac{1}{2}\log 2-\varepsilon}. \tag{4-13}$$

There are at least $T(\log T)^{-(3/2)-\varepsilon}$ prime circles C in $\mathcal{P}$ with $a(C) < T$ that are tangent to C_a.

In order to discuss Theorem 14, which combines all three ingredients, we need to recall the basics around the circle method. This discussion is also relevant to the next section.

Let λ be a (probability) distribution on $\{1, 2, \dots, N\}$, that we expect to be roughly equidistributed and set

$$R = \operatorname{supp} \lambda = \{n : \lambda(n) > 0\}. \tag{4-14}$$

In order to estimate $N - |R|$, introduce the exponential sum

$$S(\theta) = \sum_{1}^{N} \lambda(n) e(n\theta), \quad \theta \in \mathbb{T} = \mathbb{R}/\mathbb{Z}, \tag{4-15}$$

so that

$$\lambda(n) = \hat{S}(n) = \int_0^1 S(\theta) e(-n\theta)\, d\theta.$$

Let $B = B(N)$ be a parameter and make a subdivision of $\mathbb{T}$ in major arcs

$$\mathcal{M} = \bigcup_{q<B} \bigcup_{(a,q)=1} \left[\frac{a}{q} - \frac{B}{N}, \frac{a}{q} + \frac{B}{N}\right] \tag{4-16}$$

and the minor arcs $\mathbb{T}\backslash\mathcal{M}$.

Define

$$\lambda_1(n) = \int_{\mathcal{M}} S(\theta) e(-n\theta)\, d\theta. \tag{4-17}$$

Assume that S is "small" on the minor arcs, in the sense that

$$\int_{\mathbb{T}\backslash\mathcal{M}} |S(\theta)|^2\, d\theta < \frac{\varepsilon(N)}{N}. \tag{4-18}$$

It follows then from Parseval that

$$\sum |\lambda(n) - \lambda_1(n)|^2 < \frac{\varepsilon(N)}{N}.$$

Assuming $\lambda_1(n) \sim 1/N$ for $n \in R_1 \subset \{1, \dots, N\}$, it follows that

$$N - |R| \le N - |R_1| + O(\varepsilon(N)N).$$

The goal is to have a precise description of $S(\theta)$ for $\theta \in \mathcal{M}$, leading to an arithmetical description of λ_1 of the form

$$\lambda_1(n) = \mathfrak{S}(n)\pi(n), \tag{4-19}$$

with

$$\mathfrak{S}(n) = \prod_p \mathfrak{S}_p(n) = \text{ the singular series}, \tag{4-20}$$

$$\pi(n) = \text{ singular integral} \sim \frac{1}{N}. \tag{4-21}$$

In order to achieve this, $B(N)$ will have to be taken sufficiently small. On the other hand, the larger $B(N)$, the smaller $\varepsilon(N)$ will be.

Since λ_1 satisfies the local to global principle, so will R, up to an exceptional set.

Returning to ACPs, let $\mathcal{P}$ be the primitive packing obtained from the root quadruple $v = (a, b, c, d)$. The goal is to introduce an appropriate generating set of curvatures in $\mathcal{P}$. We proceed as follows.

Let $T = T_0 X^2$ with T_0 a small power of T. Define

$$B_{T_0} = \{\gamma \in A : \|\gamma\| \sim T_0\} \tag{4-22}$$

and

$$\xi_{x,y} = \begin{pmatrix} 1 & 0 & 0 & 0 \\ * & * & * & * \\ * & * & * & * \\ 4x^2+2xy+y^2-1 & 4x^2+2xy & -2xy & 2xy+y^2 \end{pmatrix} \in \langle S_2, S_3, S_4 \rangle, \tag{4-23}$$

where $x, y \asymp X, (2x, y) = 1$. Hence

$$\langle \xi_{x,y}\gamma v, e_4 \rangle \in \mathcal{P}, \tag{4-24}$$

and produces the generating distribution. The corresponding exponential sum is given by

$$S_T(\theta) = \sum_{\gamma \in B_{T_0}} \sum_{\substack{x,y \asymp X \\ (2x,y)=1}} e\big(\langle \xi_{xy}\gamma v, e_4 \rangle \theta\big). \tag{4-25}$$

The estimates on the minor arcs exploits the arithmetical element ξ_{xy} and families of quadratic forms in x, y.

The major arcs analysis is based on precise Archimedean/modular distributional properties of B_{T_0} obtained from Lax–Phillips theory and congruence spectral gaps, as discussed in Section 3.

5. Partial quotients of rationals and Zaremba's conjecture

Use the standard notation $[a_1, \ldots, a_k]$ for the continued fraction expansion

$$\cfrac{1}{a_1 + \cfrac{1}{a_2 + \ddots + \cfrac{1}{a_k}}}$$

For given $A \in \mathbb{Z}_+$, define

$$\mathcal{R}_A = \left\{ \frac{b}{d} = [a_1, \ldots, a_k] : 1 \le a_j \le A \text{ for all } j = 1, \ldots, k \right\} \tag{5-1}$$

and let

$$\mathcal{D}_A = \left\{ d \in \mathbb{Z}_+ : \text{ there is } (b, d) = 1 \text{ with } \frac{b}{d} \in \mathcal{R}_A \right\} \tag{5-2}$$

the corresponding set of denominators.

Conjecture 1 [Zaremba 1966; 1972]. *For some sufficiently large A, we have*

$$\mathcal{D}_A = \mathbb{Z}_+.$$

In fact, $A = 5$ should suffice ($54 \notin \mathcal{D}_4$).

Conjecture 2 [Niederreiter 1978]. *$\mathcal{D}_3$ contains every sufficiently large integer.*

Conjecture 3 [Hensley 1996]. *Same statement for $\mathcal{D}_2$.*

Note that $\mathcal{R}_1 = \{f_n/f_{n+1}\}, \mathcal{D}_1 = \{f_n\}$ is the Fibonacci sequence.

Using an explicit construction, Niederreiter [1986] showed that $\{2^j\} \subset \mathcal{D}_3$ and similar results were established for other lacunary sequences.

Zaremba's interest in the problem comes from numerical integration, Monte Carlo methods and pseudo-random numbers.

Let us recall the notion of discrepancy. Let

$$X = \{x_j\}_{j \le N} \subset Q = [0, 1]^s,$$

and define

$$D(X) = \max_{\substack{I \subset Q \\ \text{box}}} \left| |I| - \frac{1}{N} \#\{j \le N : x_j \in I\} \right|. \tag{5-3}$$

The significance of this notion is illustrated by results such as the Hlawka–Koksma inequality. Let f be a function of bounded variation on Q; thus

$$V = \max_{\alpha \subset \{1, \ldots, s\}} \|\partial^{(\alpha)} f\|_{L^1(Q)} < \infty. \tag{5-4}$$

Then

$$\left| \int_Q f(x)\,dx - \frac{1}{N} \sum_{j=1}^{N} f(x_j) \right| \le C\,VD(X). \tag{5-5}$$

The problem is how to construct explicitly sequences with small discrepancy when $s \ge 2$. The following result addresses this question when $s = 2$.

Theorem 17 [Zaremba 1966; 1972]. *Take* $(b, d) = 1$ *with* $b/d \in \mathfrak{R}_A$. *Let*

$$X = \left\{ x_j = \left(\frac{j}{d}, \frac{bj}{d} \right) : 1 \le j \le d \right\}, \tag{5-6}$$

where bj/d *is reduced* (mod 1). *Then*

$$D(X) < \left(\frac{4A}{\log(A+1)} + \frac{4A+1}{\log d} \right) \frac{\log d}{d}. \tag{5-7}$$

Inequality (5-7) clarifies the role of the diophantine properties of b/d. One of the simplest pseudo-random number generators (PRNG) is the linear congruential

$$x \mapsto bx + c \,(\operatorname{mod} d), \tag{5-8}$$

where $(b, d) = 1$ and b primitive $(\operatorname{mod} d)$. Set $c = 0$ in (5-8). The quality of the corresponding PRNG depends in particular on the statistical properties of

$$X_1 = \left\{ \frac{b^j}{d} (\operatorname{mod} 1) : 1 \le j \le d \right\} \tag{5-9}$$

and especially

$$X_2 = \left\{ \left(\frac{b^j}{d}, \frac{b^{j+1}}{d} \right) (\operatorname{mod} 1) : 1 \le j \le d \right\} \tag{5-10}$$

(serial correlation of pairs). Note that $D(X_2)$ is essentially $D(X)$ with X defined in (5-6). Figure 2 on the next page shows two examples.

Returning to (5-7), apart from the prefactor depending on A, the discrepancy is essentially the optimal one, as implied by the following result due to W. Schmidt.

Theorem 18 [Schmidt 1972]. *Any sequence* $X = \{x_j\}_{1 \le j \le N}$ *in* $[0, 1]^2$ *satisfies*

$$D(X) > c \frac{\log N}{N}, \tag{5-11}$$

where c *is an absolute constant.*

Returning to Zaremba's conjecture, partial progress was achieved:

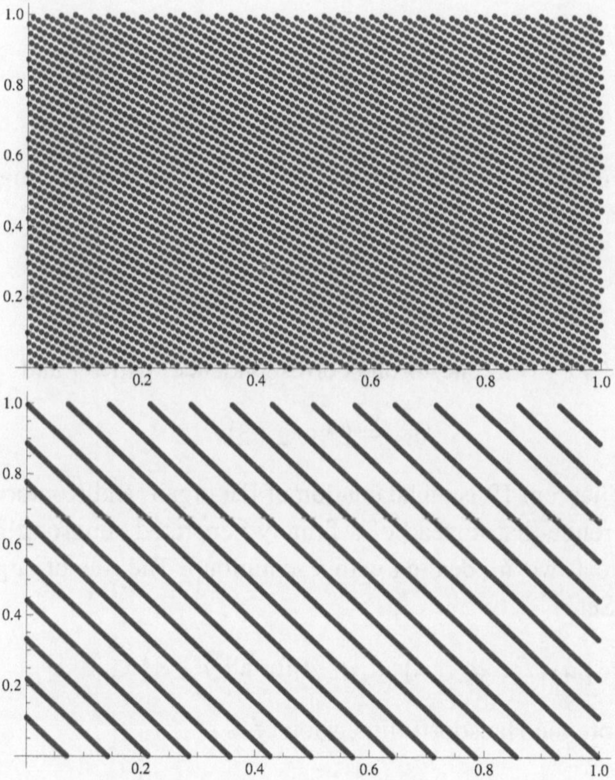

Figure 2. The set

$$\left\{\left(\frac{b^n}{d}, \frac{b^{n+1}}{d}\right) \bmod 1\right\}_{n=1}^{d},$$

where b/d equals $\frac{3523}{4547} = [1,3,2,3,1,2,3,2,1,3]$ (top) and $\frac{3535}{4547} = [1,3,2,35,1,1,1,4]$ (bottom).

Theorem 19 [Bourgain and Kontorovich 2011]. *For $A = 50$, $\mathfrak{D}_A$ is of density* 1. *Quantitatively, setting*

$$\mathfrak{D}_A(N) = \mathfrak{D}_A \cap [1, N],$$

we have

$$|\mathfrak{D}_A(N)| = N + 0\left(N^{1-\frac{c}{\log\log N}}\right). \tag{5-12}$$

Corollary 20 (linear congruential PRNG). *$\mathfrak{R}_{51}$ contains infinitely many fractions b/d with d prime, so that the multiplier b is a primitive root* (mod d).

How to produce elements from $\mathfrak{R}_A$? We note that

$$\frac{b}{d} = [a_1, \ldots, a_k]$$

is equivalent with

$$\begin{pmatrix} * & b \\ * & d \end{pmatrix} = \begin{pmatrix} 0 & 1 \\ 1 & a_1 \end{pmatrix} \begin{pmatrix} 0 & 1 \\ 1 & a_2 \end{pmatrix} \cdots \begin{pmatrix} 0 & 1 \\ 1 & a_k \end{pmatrix}. \tag{5-13}$$

Hence we are considering the orbit of $e_2 \equiv (0, 1)$ under the semigroup $\mathcal{G}_A \subset \mathrm{GL}_2(\mathbb{Z})$ generated by the matrices

$$\begin{pmatrix} 0 & 1 \\ 1 & a \end{pmatrix}, \quad \text{with } 1 \le a \le A. \tag{5-14}$$

Clearly $\mathcal{O}_A = \mathcal{G}_A.e_2$ is in one-to-one correspondence with $\mathcal{G}_A$ and

$$\mathcal{D}_A = \langle \mathcal{G}_A.e_2, e_2 \rangle. \tag{5-15}$$

The proof of Theorem 19 is an adaptation of [Bourgain and Kontorovich 2010] with this difference that instead of a finitely generated subgroup $\Gamma \subset \mathrm{SL}_2\, \mathbb{Z})$ with δ_Γ close to 1, we are dealing with a semigroup. The role of large A should be explained. Set

$$\mathcal{C}_A = \{[a_1, \dots, a_j, \dots] : a_j \le A \text{ for all } j \ge 1\} \subset [0, 1] \tag{5-16}$$

and let δ_A denote the Hausdorff dimension of $\mathcal{C}_A$.

Then

$$\mathcal{C}_1 = \left\{ \frac{1}{\varphi} \right\}, \quad \text{where } \varphi = \frac{1 + \sqrt{5}}{2},$$

and $\mathcal{C}_2$ is a Cantor set, which dimension $\delta_2 = 0,5312805\dots$ was calculated by Jenkinson and Pollicott [2001] up to high accuracy using the thermodynamical approach.

Theorem 21 [Hensley 1989; 1992]. *For large A,*

$$\delta_A = 1 - \frac{6}{\pi^2}\frac{1}{A} - \frac{72}{\pi^4}\frac{\log A}{A^2} + O\left(\frac{1}{A^2}\right). \tag{5-17}$$

Theorem 22 [Hensley 1989]. *Set*

$$\mathcal{R}_A(N) = \left\{ \frac{b}{d} \in \mathcal{R}_A : (b, d) = 1 \text{ and } 1 \le b < d < N \right\}. \tag{5-18}$$

Then

$$\#\mathcal{R}_A(N) \asymp N^{2\delta_A} \quad \text{for } N \to \infty. \tag{5-19}$$

This gives the complete analogy with the group setting with $\mathcal{C}_A$ and δ_A replacing $L(\Gamma)$ and δ_Γ.

Again, the diophantine analysis of the set $\mathfrak{D}_A$, based on (5-15), is carried out using the circle method. Rather than considering balls $B_N = \{\gamma \in \mathscr{G}_A : \|\gamma\| \le N\}$, we introduce sets of the form

$$\Omega_N = B_{N_1} B_{N_2} \dots B_{N_\ell}, \tag{5-20}$$

where $N \sim N_1 N_2 \dots N_\ell$ and N_j suitably chosen. Considering the exponential sums

$$S_N(\theta) = \sum_{\gamma \in \Omega_N} e(\langle \gamma e_2, e_2 \rangle \theta), \tag{5-21}$$

the interest of the multilinear structure (5-20) is to enable to carry out minor arcs estimates by means of general Vinogradov-type bilinear technology. This part of the analysis is completely similar to [Bourgain and Kontorovich 2010] and depends on having δ_A close enough to 1.

Again, the evaluation of $S_N(\theta)$ on major arcs depends on precise Archimedean/ modular distributional properties of the balls $B_N \subset \mathscr{G}_A$. However, since $\mathscr{G}_A$ is only a semigroup, the Lax–Phillips automorphic approach (as used in [Bourgain and Kontorovich 2010] in the setting of [Bourgain et al. 2010b] and in [Bourgain and Kontorovich 2012] using [Vinogradov 2012]; see also [Mohammadi and Oh 2012]), is not applicable. Instead, we rely on the thermodynamical method and in particular Theorem 9.

It turns out that the error in (5-12) may be further reduced to N^{1-c} for some $c > 0$. This stronger statement has some additional interest in the context of some questions proposed to the author by R. Kenyon (private communication) and related to extensions of Hall's theorem on sums of continued fractions. M. Hall [1947] showed that every number in the interval $[\sqrt{2}-1, 4\sqrt{2}-4[$ is sum of two continued fractions whose partial quotients do not exceed four.

A natural problem would be to obtain a result of the flavor of Hall's theorem for the rational numbers. There are several possible formulations.

One could ask for instance if there is an absolute constant C such that given $b/q \in \mathbb{Q} \cap I$, I a suitable interval, there is a representation of b as a sum of at most C positive integers b_i such that each of the fractions b_i/q has its partial quotients bounded by C. Of course, if we require moreover that $(b, q) = 1$, Zaremba's conjecture ought to be proven first. While we are not able to contribute directly to this question, the methods from [Bourgain and Kontorovich 2011] permit to show the following.

Theorem 23 [Bourgain 2012]. *There is an absolute constant C such that any rational $b/q \in]0, 1[$, $(b, q) = 1$, admits a representation as a finite sum,*

$$\frac{b}{q} = \sum_\alpha \pm \frac{b_\alpha}{q_\alpha}, \quad (b_\alpha, q_\alpha) = 1, \tag{5-22}$$

such that

$$\sum_{\alpha} \sum_{i} a_i \left(\frac{b_\alpha}{q_\alpha} \right) \leq C \log q, \tag{5-23}$$

where $\{a_i(x)\}$ *stands for the sequence of partial quotients of* x.

Note that since $\sum_i a_i(b_\alpha/q_\alpha) \geq \log q_\alpha$, (5-23) is essentially optimal. As pointed out by Kenyon, a statement of this kind may be viewed as a measure of complexity of rationals of given height.

References

[Blomer and Granville 2006] V. Blomer and A. Granville, "Estimates for representation numbers of quadratic forms", *Duke Math. J.* **135**:2 (2006), 261–302.

[Bourgain 2010] J. Bourgain, "New developments in combinatorial number theory and applications", pp. 233–251 in *European Congress of Mathematics*, edited by A. Ran et al., Eur. Math. Soc., Zürich, 2010.

[Bourgain 2012] J. Bourgain, "Partial quotients and representation of rational numbers", *C. R. Math. Acad. Sci. Paris* **350**:15–16 (2012), 727–792.

[Bourgain and Fuchs 2011] J. Bourgain and E. Fuchs, "A proof of the positive density conjecture for integer Apollonian circle packings", *J. Amer. Math. Soc.* **24**:4 (2011), 945–967.

[Bourgain and Fuchs 2012] J. Bourgain and E. Fuchs, "On representation of integers by binary quadratic forms", *Int. Math. Res. Not.* **2012**:24 (2012), 5505–5553.

[Bourgain and Gamburd 2008] J. Bourgain and A. Gamburd, "Uniform expansion bounds for Cayley graphs of $\mathrm{SL}_2(\mathbb{F}_p)$", *Ann. of Math.* (2) **167**:2 (2008), 625–642.

[Bourgain and Kontorovich 2010] J. Bourgain and A. Kontorovich, "On representations of integers in thin subgroups of $\mathrm{SL}_2(\mathbb{Z})$", *Geom. Funct. Anal.* **20**:5 (2010), 1144–1174.

[Bourgain and Kontorovich 2011] J. Bourgain and A. Kontorovich, "On a conjecture of Zaremba", preprint, 2011. arXiv 1107.3776

[Bourgain and Kontorovich 2012] J. Bourgain and A. Kontorovich, "On the strong density conjecture for integral Apollonian circle packings", preprint, 2012. arXiv 1205.4416

[Bourgain and Varjú 2012] J. Bourgain and P. P. Varjú, "Expansion in $\mathrm{SL}_d(\mathbf{Z}/q\mathbf{Z})$, q arbitrary", *Invent. Math.* **188**:1 (2012), 151–173.

[Bourgain et al. 2010a] J. Bourgain, A. Gamburd, and P. Sarnak, "Affine linear sieve, expanders, and sum-product", *Invent. Math.* **179**:3 (2010), 559–644.

[Bourgain et al. 2010b] J. Bourgain, A. Kontorovich, and P. Sarnak, "Sector estimates for hyperbolic isometries", *Geom. Funct. Anal.* **20**:5 (2010), 1175–1200.

[Bourgain et al. 2011] J. Bourgain, A. Gamburd, and P. Sarnak, "Generalization of Selberg's $3/16$ theorem and affine sieve", *Acta Math.* **207**:2 (2011), 255–290.

[Breuillard and Gamburd 2010] E. Breuillard and A. Gamburd, "Strong uniform expansion in $\mathrm{SL}(2, p)$", *Geom. Funct. Anal.* **20**:5 (2010), 1201–1209.

[Breuillard et al. 2011] E. Breuillard, B. Green, and T. Tao, "Approximate subgroups of linear groups", *Geom. Funct. Anal.* **21**:4 (2011), 774–819.

[Ellenberg et al. 2012] J. S. Ellenberg, C. Hall, and E. Kowalski, "Expander graphs, gonality, and variation of Galois representations", *Duke Math. J.* **161**:7 (2012), 1233–1275.

[Friedlander and Iwaniec 2010] J. Friedlander and H. Iwaniec, *Opera de cribro*, American Mathematical Society Colloquium Publications **57**, American Mathematical Society, Providence, RI, 2010.

[Fuchs 2010] E. Fuchs, *Arithmetic properties of Apollonian circle packings*, Ph.D. thesis, Princeton University, 2010, http://math.berkeley.edu/~efuchs/efuchsthesis.pdf.

[Fuchs and Sanders 2010] E. Fuchs and K. Sanders, "Some experiments with integral Apollonian circle packings", preprint, 2010. arXiv 1001.1406

[Gamburd 2002] A. Gamburd, "On the spectral gap for infinite index "congruence" subgroups of $SL_2(\mathbb{Z})$", *Israel J. Math.* **127** (2002), 157–200.

[Graham et al. 2003] R. L. Graham, J. C. Lagarias, C. L. Mallows, A. R. Wilks, and C. H. Yan, "Apollonian circle packings: Number theory", *J. Number Theory* **100**:1 (2003), 1–45.

[Green 2009] B. Green, "Approximate groups and their applications: Work of Bourgain, Gamburd, Helfgott and Sarnak", preprint, 2009. arXiv 0911.3354

[Hall 1947] M. Hall, Jr., "On the sum and product of continued fractions", *Ann. of Math.* (2) **48** (1947), 966–993.

[Helfgott 2008] H. A. Helfgott, "Growth and generation in $SL_2(\mathbb{Z}/p\mathbb{Z})$", *Ann. of Math.* (2) **167**:2 (2008), 601–623.

[Helfgott 2011] H. A. Helfgott, "Growth in $SL_3(\mathbb{Z}/p\mathbb{Z})$", *J. Eur. Math. Soc.* (*JEMS*) **13**:3 (2011), 761–851.

[Hensley 1989] D. Hensley, "The distribution of badly approximable numbers and continuants with bounded digits", pp. 371–385 in *Théorie des nombres* (Quebec, 1987), edited by J.-M. D. Koninck and C. Levesque, de Gruyter, Berlin, 1989.

[Hensley 1992] D. Hensley, "Continued fraction Cantor sets, Hausdorff dimension, and functional analysis", *J. Number Theory* **40**:3 (1992), 336–358.

[Hensley 1996] D. Hensley, "A polynomial time algorithm for the Hausdorff dimension of continued fraction Cantor sets", *J. Number Theory* **58**:1 (1996), 9–45.

[Hoory et al. 2006] S. Hoory, N. Linial, and A. Wigderson, "Expander graphs and their applications", *Bull. Amer. Math. Soc.* (*N.S.*) **43**:4 (2006), 439–561.

[Iwaniec 1972] H. Iwaniec, "Primes of the type $\phi(x, y) + A$ where ϕ is a quadratic form", *Acta Arith.* **21** (1972), 203–234.

[Jenkinson and Pollicott 2001] O. Jenkinson and M. Pollicott, "Computing the dimension of dynamically defined sets: E_2 and bounded continued fractions", *Ergodic Theory Dynam. Systems* **21**:5 (2001), 1429–1445.

[Kontorovich and Oh 2011] A. Kontorovich and H. Oh, "Apollonian circle packings and closed horospheres on hyperbolic 3-manifolds", *J. Amer. Math. Soc.* **24**:3 (2011), 603–648.

[Kowalski 2012] E. Kowalski, "Crible en expansion", exposé 1028 (pp. 17–64) in *Séminaire Bourbaki* 2010/2011, Astérisque **348**, Soc. Math. de France, Paris, 2012.

[Lax and Phillips 1982] P. D. Lax and R. S. Phillips, "The asymptotic distribution of lattice points in Euclidean and non-Euclidean spaces", *J. Funct. Anal.* **46**:3 (1982), 280–350.

[Lubotzky 1994] A. Lubotzky, *Discrete groups, expanding graphs and invariant measures*, Birkhäuser, Basel, 1994.

[Lubotzky 2012] A. Lubotzky, "Expander graphs in pure and applied mathematics", *Bull. Amer. Math. Soc.* (*N.S.*) **49**:1 (2012), 113–162.

[Margulis 1973] G. A. Margulis, "Explicit constructions of expanders", *Problemy Peredači Informacii* **9**:4 (1973), 71–80. In Russian; Translated in English in *Problems of Information Transmission* **9**:4 (1973), 325–332, 1975.

[Matthews et al. 1984] C. R. Matthews, L. N. Vaserstein, and B. Weisfeiler, "Congruence properties of Zariski-dense subgroups. I", *Proc. London Math. Soc.* (3) **48**:3 (1984), 514–532.

[Mohammadi and Oh 2012] A. Mohammadi and H. Oh, "Matrix coefficients, counting and primes for orbits of geometrically finite groups", preprint, 2012. To appear in *J. Eur. Math. Soc.* arXiv 1208.4139

[Niederreiter 1978] H. Niederreiter, "Quasi-Monte Carlo methods and pseudo-random numbers", *Bull. Amer. Math. Soc.* **84**:6 (1978), 957–1041.

[Niederreiter 1986] H. Niederreiter, "Dyadic fractions with small partial quotients", *Monatsh. Math.* **101**:4 (1986), 309–315.

[Pinsker 1973] M. S. Pinsker, "On the complexity of a concentrator", pp. 318/1–318/4 in *7th International Teletrafic Conference* (Stockholm, 1973), 1973.

[Pyber and Szabó 2010] L. Pyber and E. Szabó, "Growth in finite simple groups of Lie type of bounded rank", preprint, 2010. arXiv 1005.1858

[Salehi Golsefidy and Sarnak 2011] A. Salehi Golsefidy and P. Sarnak, "Affine sieve", preprint, 2011. arXiv 1109.6432

[Salehi Golsefidy and Varjú 2012] A. Salehi Golsefidy and P. P. Varjú, "Expansion in perfect groups", *Geom. Funct. Anal.* **22**:6 (2012), 1832–1891.

[Sarnak 1995] P. Sarnak, "Selberg's eigenvalue conjecture", *Notices Amer. Math. Soc.* **42**:11 (1995), 1272–1277.

[Sarnak 2008] P. Sarnak, "Equidistribution and primes", pp. 225–240 in *Géométrie différentielle, physique mathématique, mathématiques et société, II*, Astérisque **322**, 2008.

[Sarnak 2011] P. Sarnak, "Integral Apollonian packings", *Amer. Math. Monthly* **118**:4 (2011), 291–306.

[Sarnak 2014] P. Sarnak, "Notes on thin matrix groups", pp. 343–362 in *Thin groups and superstrong approximation*, edited by H. Oh and E. Breuillard, Math. Sci. Res. Inst. Publ. **61**, Cambridge Univ. Press, Cambridge, 2014.

[Schmidt 1972] W. M. Schmidt, "Irregularities of distribution, VII", *Acta Arith.* **21** (1972), 45–50.

[Varjú 2012] P. P. Varjú, "Expansion in $\mathrm{SL}_d(O_K/I)$, I square-free", *J. Eur. Math. Soc. (JEMS)* **14**:1 (2012), 273–305.

[Vinogradov 2012] I. Vinogradov, "Effective bisector estimate with application to Apollonian circle packings", preprint, 2012. arXiv 1204.5498

[Zaremba 1966] S. C. Zaremba, "Good lattice points, discrepancy, and numerical integration", *Ann. Mat. Pura Appl.* (4) **73** (1966), 293–317.

[Zaremba 1972] S. K. Zaremba, "La méthode des "bons treillis" pour le calcul des intégrales multiples", pp. 39–119 in *Applications of number theory to numerical analysis (Proc. Sympos., Univ. Montreal)* (Montreal, Que., 1971), edited by S. K. Zaremba, Academic Press, New York, 1972.

bourgain@ias.edu *Institute for Advanced Study, 1 Einstein Drive, Princeton, NJ 08540, United States*

Thin groups and superstrong approximation
MSRI Publications
Volume **61**, 2013

A brief introduction to approximate groups

EMMANUEL BREUILLARD

This introduction to approximate groups highlights their connection with superstrong approximation, the Freiman inverse problem, the Helfgott–Lindenstrauss conjecture, and the classification of approximate subgroups of simple algebraic groups over finite fields.

1. Approximate groups and their classification

A finite subgroup of a group G is a finite subset stable under product. It is tempting to investigate what happens if we consider finite subsets that are almost stable under products. Are they close to genuine subgroups in some meaningful sense? Before answering this question, we need to make precise what we mean by "almost stable". The subject of approximate groups first attempts to do just that, and then tackles the more general problem of classifying such objects.

The formal definition of an approximate group given in Definition 1.2 below was introduced by T. Tao [2008] and was in part motivated by its use in the groundbreaking work of Bourgain and Gamburd [2008b] on superstrong approximation for Zariski dense subgroups of $\mathrm{SL}(2,\mathbb{Z})$. However the origins of the concept can be traced much earlier and people have been studying approximate groups much before they were even defined. A significant part of additive number theory as it developed in the last fifty years, and in particular the study of sets of integers with small doubling (e.g., Freiman's theorem; see Theorem 1.9 below), was precisely about understanding abelian approximate groups. Similarly the sum-product phenomenon (see Theorem 2.5), which played a key role in superstrong approximation especially in the early developments of the subject (e.g., in [Helfgott 2008]), is equivalent to classifying approximate subgroups of the affine group $\{x \mapsto ax+b\}$.

This article intends to give a brief introduction to this subject and present some of its recent developments. Worthwhile expository readings on the same topics include [Green 2009; 2012; Breuillard et al. 2013b; Pyber and Szabó 2014].

1.1. ***Approximate groups: the definition.*** Given sets A, B in a group G, we write $AB = \{ab \mid a \in A,\, b \in B\}$ and $A^{n+1} = A^n A$. Also $|A|$ denotes the cardinality of A.

Definition 1.2 (approximate groups [Tao 2008]). Let $K \geqslant 1$ be a parameter, G a group and $A \subset G$ a finite subset. We say that A is a K-approximate subgroup of G if

(1) $1 \in A$,

(2) A is symmetric: $A = A^{-1}$, and

(3) there is a symmetric set X of size at most K such that $AA \subset XA$.

Remark. Observe that if $K = 1$, then we recover the definition of a finite subgroup of G. Namely 1-approximate subgroups are just genuine finite subgroups.

Tao's definition of a K-approximate subgroup is only one of several natural candidates. The following result says that all these notions are roughly equivalent.

Proposition 1.3 (Balog, Szemerédi, Gowers, Tao). *There is an absolute constant $C > 0$ such that, given a finite set A in an ambient group G and parameter $K \geqslant 1$, the following conditions are roughly equivalent in a sense about to be made precise*:

(1) $|AA| \leqslant K|A|$.

(2) $|AAA| \leqslant K|A|$.

(3) $|\{(a, b, c, d) \in A \times A \times A \times A \mid ab = cd\}| \geqslant |A|^3/K$.

(4) $|\{(a, b) \in A \times A : ab \in A\}| \geqslant |A|^2/K$.

(5) *A is a K-approximate subgroup of G*,

More precisely, if any of these conditions holds for A with a constant K, the other four conditions will hold, with a constant $K' \leqslant CK^C$, for a set $A' \subset G$ such that $|A \cap A'| \geqslant 1/(CK^C) \max\{|A|, |A'|\}$.

Conditions (1) and (2) are the small doubling and tripling conditions, respectively. Conditions (3) and (4) are statistical in nature. The rough equivalence above was proved in [Tao 2008], but it relies[1] on a tricky yet extremely useful graph-theoretical result of Balog and Szemerédi, which was later improved by Gowers, yielding the polynomial bound CK^C in the statement. See [Tao and Vu 2006, Section 6.4] for a proof of the Balog–Szemerédi–Gowers lemma.

This proposition also gives a hint at what kind of equivalence between sets one would like to impose when attempting to *classify* approximate groups. For example, passing to a large (say $\geqslant 1/CK^C$) proportion of a set A is allowed and does not significantly alter the structure of A (at least for our purposes).

[1] At least for the conditions involving (3) and (4); the rough equivalence between (1), (2) and (5) is easier.

Remark. It is often the case in various mathematical problems that one has to confront approximate versions of some well-known mathematical notion, namely objects that satisfy the usual axioms only most of the time, or perhaps just part of the time. When it happens most of the time, people say that we are in the 99% regime, while if the axioms are satisfied only part of the time, we are in the 1% regime. This distinction makes sense for approximate groups when using the statistical definitions (3) or (4). Approximate groups in the 99% regime are easily seen to be very close to genuine subgroups (see Proposition 1.7 below and the remark after it) and much of the subject of approximate groups concentrates on the 1% regime.

Tao's approximate groups are easier to handle than the other notions (1)–(4) defined in this proposition. In fact it is fair to say that Tao's definition is tailored so as to reduce to a maximum the appearance of purely additive combinatorial arguments in the proofs, so that classifying sets satisfying either of the conditions (1)–(4) above in a given group G is reduced to understanding approximate groups in Tao's sense. This usually boils down to a more familiar algebraic or geometric problem about the ambient group where combinatorics play a minor role.

1.4. ***Approximate groups and superstrong approximation.*** One of the main motivations for the subject of approximate groups and its recent fast development is its connection with superstrong approximation as was first made clear in [Bourgain and Gamburd 2008b], for the case of $\mathrm{SL}_2(\mathbb{Z}/p\mathbb{Z})$.

Superstrong approximation is the main topic of this volume and is discussed at length in [Sarnak 2014; Salehi 2014; Ellenberg 2014]. So I will not attempt here to give a detailed account of it, but let me only recall that if $\Gamma \leqslant \mathrm{SL}_d(\mathbb{Z})$ is a Zariski dense subgroup, then strong approximation for Γ is the statement that for every large enough prime number p, the subgroup Γ surjects onto $\mathrm{SL}_d(\mathbb{Z}/p\mathbb{Z})$ (see [Rapinchuk 2014; Matthews et al. 1984; Nori 1987]), while *superstrong approximation* asserts that given any fixed generating set S of Γ, the sequence of Cayley graphs $\mathrm{Cay}(\mathrm{SL}_d(\mathbb{Z}/p\mathbb{Z}), S \bmod p)$ forms a family of ε-expanders, for some $\varepsilon = \varepsilon(S) > 0$.

Up until 2005, the main tool for constructing such families of expanders was representation theory. This started with Margulis [1973] and his use of Kazhdan's property (T) to give the first construction of expander graphs, then was continued in the work of Lubotzky, Phillips and Sarnak on Ramanujan graphs [Lubotzky et al. 1988], and many others later on. This approach applied only to arithmetic lattices Γ and was essentially based on a transfer principle between the representation theory of $\mathbb{L}^2(G/\Gamma)$ and that of Γ.

A consequence of the expander property is that the simple random walk on Γ (i.e., the stochastic process one gets by multiplying on the left a generator from

S chosen at random uniformly among the generators), when projected onto the finite quotients, becomes equidistributed very fast, typically in logarithmic time (in the size of the quotient). And usually this fast equidistribution is the reason why one wants to prove a spectral gap (as in applications to sieving, for example). In 2005 Bourgain and Gamburd [2008b] reversed the idea: they proved the fast equidistribution of the random walk by combinatorial methods, then deduced the spectral gap. Indeed it is not difficult to prove (see Theorem 3.2 in [Hoory et al. 2006] or Appendix E in [Breuillard et al. 2013a]) that the spectral gap is in fact equivalent to the fast equidistribution of the random walk.

For this, the strategy is to control the random walk on the finite quotients $\mathrm{SL}_d(\mathbb{Z}/p\mathbb{Z})$ in three stages:

(1) For short times (typically $t < c \log p$, $c > 0$ small constant), one needs to show that the random walk *escapes proper subgroups*; that is, it does not give too much mass to any proper subgroup of $\mathrm{SL}_d(\mathbb{Z}/p\mathbb{Z})$.

(2) For medium times (with $c \log p < t < C \log p$, $c < C$), one shows that the walk *escapes from approximate subgroups*. This is sometimes phrased in terms of probability measures as the "ℓ^2-flattening" lemma.

(3) For long times one uses *quasirandomness* (i.e., the Frobenius–Landazuri–Seitz bounds on the dimension of complex linear representations of finite simple groups) to show that the walk covers the whole group very quickly.

The hard parts in this strategy are (1) and (2). In their original paper, Bourgain and Gamburd dealt only with $\mathrm{SL}_2(\mathbb{Z}/p\mathbb{Z})$, whose subgroup structure is very simple, so item (1) in this case was a simple consequence of Kesten's thesis [1959] on the decay of the probability of return to the identity of simple random walks on groups. Currently there are two (related) known methods to deal with (1) in higher rank: to use ping-pong and produce a free subgroup which has small intersection with every proper algebraic subgroup (see [Varjú 2012; Salehi and Varjú 2012]), or to use the theory of products of random matrices à la Furstenberg–Guivarc'h; see [Bourgain and Gamburd 2008a; 2009].

Item (2) is the subject of this article and amounts to understanding approximate subgroups of the finite quotients. This was first done in [Helfgott 2008] for $\mathrm{SL}_2(\mathbb{Z}/p\mathbb{Z})$ (see also [Helfgott 2011] for $\mathrm{SL}_3(\mathbb{Z}/p\mathbb{Z})$), which then allowed Bourgain and Gamburd to implement their strategy in the SL_2 case. Basically Helfgott's theorem says that there are no approximate subgroups of $\mathrm{SL}_2(\mathbb{Z}/p\mathbb{Z})$ except for bounded sets and cobounded sets (i.e., sets forming a significant proportion of the whole group). Hence the random walk at stage (2) cannot remain "stuck" in an approximate subgroup of intermediate size.

The higher rank case and the extension of Helfgott's theorem to arbitrary semisimple algebraic groups over arbitrary finite fields was done later in [Pyber and Szabó 2010b] and independently in [Breuillard et al. 2011]. See the second part of this article as well as [Pyber and Szabó 2014] to this volume for a detailed account of this story.

1.5. ***The Freiman inverse problem.*** While in part motivated by the superstrong approximation result of Bourgain and Gamburd, Tao's definition of an approximate group fitted well within a general line of thought coming from additive number theory and combinatorics, which was guided by the:

Freiman inverse problem. Given a group G and a parameter $K \geqslant 1$, describe the "structure" of finite subsets A of G such that $|AA| \leqslant K|A|$.

Sets with $|AA| \leqslant K|A|$ are said to have *doubling at most* K, and the ratio $|AA|/|A|$ is often called the *doubling constant* of A. Note that K-approximate groups are examples of sets with doubling at most K.

Later in this note, I will state a recent theorem of Green, Tao and myself [Breuillard et al. 2012], which provides an answer to Freiman's inverse problem for general groups. For the applications to superstrong approximation however (i.e., for step (2) of the Bourgain–Gamburd strategy outlined above) this general theorem is of no use, because it provides no explicit bounds in terms of the parameter K. Nevertheless it treats the general case while for these applications one only cares about approximate subgroups of linear groups (i.e., subgroups of GL_d for some fixed d). In the linear setting one has an entirely different set of tools and techniques (in particular algebraic geometry) that can be exploited. And they indeed yield explicit (even polynomial) bounds for the Freiman inverse problem as I will explain in Section 2.

Many people have contributed to the Freiman inverse problem in recent years in the noncommutative case. To name a few:

- Bourgain, Katz and Tao [Bourgain et al. 2004]: the *sum-product* theorem for finite fields (2003).
- Helfgott [2008]: the $\mathrm{SL}_2(\mathbb{F}_p)$ case using the sum-product (2005).
- Tao [2008] transposed to the noncommutative setting most of the apparatus of additive number theory previously used to tackle Freiman's problem in abelian groups and defined approximate groups (2005).
- Helfgott [2011]: $\mathrm{SL}_3(\mathbb{F}_p)$ then partial results for $\mathrm{SL}_d(\mathbb{F}_p)$ in [Gill and Helfgott 2011].
- Dinai [2010]: $\mathrm{SL}_2(\mathbb{F}_q)$.
- Tao [2010]: general solvable groups with bounded solvability length.
- Breuillard and Green [2011a; 2011b; 2012]: polynomial bounds for torsion-free nilpotent groups, solvable linear groups, compact Lie groups.

- Bourgain, Gamburd and Sarnak [2010]: $\mathrm{SL}_2(\mathbb{Z}/q\mathbb{Z})$ with q square-free integer.
- Hrushovski [2012b]: progress towards the Freiman inverse problem for arbitrary groups; also for general linear groups (no explicit bounds) using model theory.
- Pyber and Szabó [2010b]; Breuillard, Green and Tao [Breuillard et al. 2011]: $\mathbf{G}(\mathbb{F}_q)$, $\mathbf{G}$ a simple algebraic group over a finite field $\mathbb{F}_q$ (with polynomial bounds).
- Varjú [2012]: $\mathbf{G}(\mathbb{Z}/q\mathbb{Z})$, q square free, $\mathbf{G} = \mathrm{SL}_n$.
- Bourgain and Varjú [2012]: handled $\mathrm{SL}_d(\mathbb{Z}/n\mathbb{Z})$ for arbitrary modulus n.
- Salehi and Varjú [2012]: $\mathbf{G}(\mathbb{Z}/q\mathbb{Z})$, q square free, $\mathbf{G}$ perfect.
- Gill and Helfgott [2010]: solvable algebraic subgroups over $\mathbb{F}_p$.
- Breuillard, Green, and Tao [Breuillard et al. 2012]: arbitrary groups (but no explicit bounds).
- Tointon [2012]: arbitrary nilpotent groups (with polynomial bounds).
- Pyber and Szabó [2014]: general linear groups (with polynomial bounds).

1.6. ***Some examples of approximate groups.*** Having given the definition of approximate groups and stated Freiman's inverse problem, we now discuss some simple instances of this problem and give a number of examples of approximate groups. We begin with a simple observation.

Remark. Suppose A is a finite set of an ambient group G. It is a simple exercise to prove that the requirement that $|A| = |AA|$ is equivalent to saying that A is a *normalizing coset* of a finite subgroup, namely that $A = aH$ for some $a \in G$ and some finite subgroup H in G such that $aH = Ha$.

This remark answers completely Freiman's inverse problem when the doubling constant K equals 1. What if K is slightly bigger than 1? Here is an old result of Freiman, first published in [Freiman 1973a]:

Proposition 1.7 (Freiman inverse problem for $K < \frac{3}{2}$; see [Tao 2009], [Freiman 2012], or [Breuillard 2011b]). *Let A be a finite subset of an ambient group G such that $|AA| < \frac{3}{2}|A|$. Then there exists a finite subgroup H of G and $a \in G$ such that $aH = Ha$ and $A \subset aH$ with $|A| > \frac{2}{3}|H|$. The converse is clear.*

In other words if A has doubling $< \frac{3}{2}$, then A is contained in a coset of a genuine subgroup which is not much larger than A itself. This is certainly an instance of the Freiman problem, because starting only from a small doubling assumption, we have exhibited structure: there is a genuine subgroup that hangs around.

As an illustrative example, let us give a complete argument showing that if $|AA^{-1}A| \leqslant (1+\varepsilon)|A|$ for some $\varepsilon < 1$, then A is close to a (left coset of a) genuine subgroup in the sense that $H := A^{-1}A$ is a genuine subgroup of cardinality at most $(1+\varepsilon)|A|$. Indeed $H^{-1} = H$ and for every $x, y \in H$, Ax and Ay intersect nontrivially, because $\varepsilon < 1$. It follows that $xy^{-1} \in H$. Hence H is stable under multiplication and hence a genuine subgroup. Note finally that $|H| \leqslant (1+\varepsilon)|A|$ and $A \subset aH$ for every $a \in A$. The proof of Proposition 1.7 is more involved, but makes use of similar arguments.

If $K > \frac{3}{2}$, Freiman's problem is more tricky. However, as long as $K < 2$, it will remain the case that doubling at most K implies that A is contained in a bounded number of cosets of a genuine finite subgroup, which is itself not much bigger than A. This is a recent result of Y. Hamidoune [2013], which answered a question of Tao (see also [Tao 2011] for a proof).

It is clear that such a thing no longer holds if $K \geqslant 2$, because of the following other well-known example of a set with small doubling (besides finite subgroups), namely arithmetic progressions: the subset $A := [-N, N] \subset \mathbb{Z}$ has doubling at most 2.

This brings about the following family of approximate groups:

Example 1.8 (symmetric generalized arithmetic progressions). Let $N_1, \dots, N_d$ be positive integers and consider the box $B = \prod_1^d [-N_i, N_i] \subset \mathbb{Z}^d$. Let $\pi : \mathbb{Z}^d \to G$ be a group homomorphism. Then $A := \pi(B)$ is called a (symmetric) d-dimensional (generalized) arithmetic progression. Clearly B is a 2^d-approximate group. It follows that A too is a 2^d-approximate group and in particular $|AA| \leqslant 2^d|A|$.

For instance, $\{a, a+r, a+2r, \dots, a+2Nr\}$, an arithmetic progression of odd length in $\mathbb{Z}$, is nothing other than a translate of a symmetric generalized arithmetic progression of dimension $d = 1$, where $N_1 = N$ and the $\pi(x) = rx$.

Generalized arithmetic progressions can be generalized further to the setting of nilpotent groups. Basically any homomorphic image of a "box" in a finitely generated nilpotent group will have small doubling. This leads to the notion of *nilprogression* or nilpotent progression. It was investigated in [Breuillard and Green 2011a] as well as in Tao's paper on solvable groups [2010]. There are several natural definitions of nilprogressions which are all roughly equivalent. One can define them as the homomorphic image of a box in the free nilpotent group $N_{r,k}(\mathbb{Z})$ of step r and rank k. A natural definition for the box can be to take all elements that can be written as a word in the generators $e_1, \dots, e_k$ of $N_{r,k}(\mathbb{Z})$ with e_i appearing at most N_i times. Another more geometric way to proceed is to take as our box the integer points in the Lie group $N_{r,k}(\mathbb{R})$ that lie in the ball of radius 1 for the left-invariant Carnot–Caratheodory metric induced on $N_{r,k}(\mathbb{R})$ by the norm $\|(x_1, \dots, x_k)\| = \sum |x_i|/N_i$ on the abelianization $\mathbb{R}^k$ of $N_{r,k}(\mathbb{R})$.

The two definitions lead to two roughly equivalent notions of nilprogressions. See [Breuillard and Green 2011a], the appendix to [Breuillard et al. 2012], and Tointon's paper [2012] on approximate subgroups of nilpotent groups.

Let us leave the general case for a moment and say a word about approximate subgroups of $G = \mathbb{Z}$, the infinite cyclic group. In this case, the inverse Freiman problem was solved by Freiman himself in the late 1960s. There are no nontrivial finite subgroups of $\mathbb{Z}$, so finite groups will not appear. However there are generalized arithmetic progressions. Freiman's theorem [1973b] says that every approximate subgroup of $\mathbb{Z}$ is roughly equivalent to a generalized arithmetic progression.

Theorem 1.9 (Freiman's theorem). *Let A be a K-approximate subgroup of $\mathbb{Z}$. Then there is a d-dimensional generalized arithmetic progression P and a set X in $\mathbb{Z}$ such that*

(1) $A \subset X + P$,

(2) $|P| \leqslant C|A|$, *with* $C \leqslant O_K(1)$,

(3) $|X| \leqslant O_K(1)$,

(4) $d \leqslant O_K(1)$.

For a proof, see [Green 2002] or [Tao and Vu 2006].

Ruzsa [1994] gave a simplified proof of Freiman's theorem, which was improved by Chang [2002] and then pushed to all abelian groups by Green and Ruzsa [2007]. Ruzsa's proof gave bounds of the form $C \leqslant \exp(O(K^{O(1)}))$, $d \leqslant O(K^{O(1)})$ and $|X| \leqslant O(K^{O(1)})$. Note that, given the exponential bound on C, one could ignore the set X altogether by declaring it to be part of the progression P at the expense of increasing slightly the rank d of the progression. However the set X becomes important when one considers the following conjecture:

Conjecture 1.10 (polynomial Freiman–Ruzsa conjecture). *One can take*

$$C \leqslant O(K^{O(1)}),$$

while keeping $|X|$ and d of size $O(K^{O(1)})$.

Recently Tom Sanders [2012] gave almost polynomial bounds towards this conjecture: he has $d = O(\log^6 K)$, while $C \leqslant K^3$ and $|X| = O(K^{\log^6 K})$. See also his excellent recent survey [Sanders 2013], where these bounds are further improved.

1.11. ***The combinatorial toolbox of approximate groups.*** As Tao has observed, many combinatorial arguments from additive number theory actually work without modification in the noncommutative setting. This is the case for the celebrated

Ruzsa triangle inequality, which asserts that the *Ruzsa distance*

$$d(A, B) = \log \frac{|AB^{-1}|}{\sqrt{|A||B|}}$$

for any finite subsets A, B of an ambient group G satisfies the triangle inequality

$$d(A, C) \leqslant d(A, B) + d(B, C).$$

The proof of the Ruzsa triangle inequality is just a few lines, which I cannot resist including here. Note first that obviously

$$|AB^{-1}| \geqslant |A| \quad \text{and} \quad |AB^{-1}| \geqslant |B|,$$

so that $|AB^{-1}|^2 \geqslant |A||B|$, yielding $d(A, B) \geqslant 0$ always. Next consider the map $AC^{-1} \times B \to AB^{-1} \times BC^{-1}$, which sends (x, b) to $(a_x b^{-1}, bc_x^{-1})$, where we made a choice of $a_x \in A$ and $c_x \in C$ for each $x \in AC^{-1}$. Then quite obviously this map is injective. Hence $|B||AC^{-1}| \leqslant |AB^{-1}||BC^{-1}|$, which is another way to phrase the triangle inequality $d(A, C) \leqslant d(A, B) + d(B, C)$. QED.

Clearly this proof is much easier than that of the Balog–Szemerédi–Gowers lemma mentioned on page 24. Applying only the Ruzsa triangle inequality one can prove the following (see, e.g., [Tao 2008] or [Breuillard 2011b]):

Lemma 1.12. *Let A be a finite subset of a group G and $K \geqslant 1$ be a parameter.*

- *If $|A^3| \leqslant K|A|$, then $|A^n| \leqslant K^{2n}|A|$ for all $n \geqslant 1$.*
- *If $|A^3| \leqslant K|A|$, then $B := (A \cup A^{-1} \cup \{1\})^2$ is a $O(K^{O(1)})$-approximate group.*
- *If A is a K-approximate subgroup and B an L-approximate subgroup, then $A^2 \cap B^2$ is a $(KL)^2$-approximate subgroup.*

Note. Although the two notions are roughly equivalent in the sense of Proposition 1.3, small doubling is not enough to guarantee small tripling! If $A = H \cup \{x\}$ for some finite subgroup H such that $xHx^{-1} \cap H = \{1\}$ (this situation can arise), then $AA = H \cup xH \cup Hx \cup \{x^2\}$—a set of size at most $3|A|$—while AAA contains HxH, which has size $|H|^2$.

The polynomial bounds in Proposition 1.3 and Lemma 1.12 are crucial for the applications to superstrong approximation and to the classification of approximate subgroups of simple algebraic groups that we are about to describe. Also crucial is the following approximate version of the orbit-stabilizer lemma for group actions.

Lemma 1.13 (approximate orbit-stabilizer lemma). *Suppose a group G acts on a set X and let A be a K-approximate subgroup of G for some $K \geqslant 1$. For every integer $k \geqslant 2$, and $x \in X$, we have*

$$|A| \leqslant |A \cdot x|\,|\mathrm{Stab}(x) \cap A^k| \leqslant K^{k+1}|A|.$$

Observe that this lemma applies in particular to the action by left translations on the coset space G/H for any subgroup H. It follows from the lemma applied to this action that the size of $A^k \cap H$ is roughly (i.e., up to a factor K^k) independent of $k \geqslant 2$. See [Pyber and Szabó 2014], where this feature is exploited with great skill: growth in a subgroup implies growth of the set.

Proofs of the two lemmas above can be found in [Breuillard 2011b]; see also [Tao and Vu 2006; Tao 2008; Helfgott 2008]. They form the basic toolkit of approximate groups. Their use is widespread in the proofs of many classification results regarding approximate groups. They also allow to make many group-theoretical arguments (as used in classical group theory; at least those not involving divisibility properties of the order of the group) work in the approximate groups setting. And indeed the following principle will remain our slogan for the remainder of this article:

Philosophy. Group-theoretical arguments can often be successfully transferred to approximate groups.

1.14. ***Classification of approximate groups and the Helfgott–Lindenstrauss conjecture.*** We have seen two chief examples of approximate groups: finite subgroups, and generalized arithmetic progressions. We also mentioned that the latter is only a special case of the notion of nilprogression.

Furthermore one can build extensions of approximate groups: if A normalizes a finite subgroup H and A is an approximate subgroup, then AH is again an approximate subgroup. In particular any set of the form HL, where H is a finite subgroup normalized by L and L is a finite subset such that $H\backslash HL$ is a nilprogression is an approximate subgroup. Such HL sets are called *coset nilprogressions.*

The following conjecture and theorem say that every approximate group is roughly equivalent to an HL set as above. The conjecture was formulated by E. Lindenstrauss in a private communication. It is also implicit in Helfgott's $\mathrm{SL}_3(\mathbb{Z}/p\mathbb{Z})$ paper [2011], because it coincides with his description of an arbitrary approximate subgroup of $\mathrm{SL}_3(\mathbb{Z}/p\mathbb{Z})$.

Conjecture 1.15 (Helfgott–Lindenstrauss). *Let G be an arbitrary group. Let A be a K-approximate subgroup of G. Then there are finite subsets $P, X \subset G$ satisfying the following conditions:*

(1) $A \subset XP$.
(2) $|X| \leqslant O_K(1)$.
(3) $|P| \leqslant O_K(1)|A|$.
(4) *P is a coset nilprogression; that is, $P = HL$, where H is a finite subgroup of G and L a finite subset lying in the normalizer $N_G(H)$ of H in G such that $H\backslash HL$ generates a nilpotent subgroup of $H\backslash N_G(H)$ with complexity $O_K(1)$ (i.e., number of generators and nilpotency class are $O_K(1)$).*

This conjecture was proved by Green, Tao and myself:

Theorem 1.16 [Breuillard et al. 2012]. *The Helfgott–Lindenstrauss conjecture holds. In it, one can take $P \subset A^4$, a coset nilprogression of complexity $O_K(1)$.*

Note that the theorem not only proves the conjecture, but also generalizes Freiman's classification of approximate subgroups of $\mathbb{Z}$ (see Theorem 1.9), because nilprogressions in $\mathbb{Z}$ are just generalized arithmetic progressions. In fact our proof of Theorem 1.16 gives a new proof of Freiman's theorem.

The theorem also gives a strengthening of Gromov's polynomial growth theorem and has several applications to Riemannian geometry and nonnegative curvature. Gromov's theorem can be deduced in only a few lines from Theorem 1.16.

A proof of Theorem 1.16 can be found in [Breuillard et al. 2012]. The basic strategy was inspired by the preprint version of [Hrushovski 2012b], which outlines a way to tackle the Freiman inverse problem for general groups using model theory to construct limits of sequences of approximate groups. In particular Hrushovski showed that every infinite sequence of K-approximate groups (K fixed) yields a certain locally compact group in a certain model-theoretic limit. Studying this locally compact group, and in particular applying the Gleason–Montgomery–Zippin–Yamabe structure theorem (Hilbert fifth problem), already gets you a long way towards the above theorem and indeed Hrushovski was also able to improve on Gromov's polynomial growth theorem using these ideas. In [Breuillard et al. 2012], we delve into the proof of the Gleason–Montgomery–Zippin–Yamabe structure theorem and manage to transfer some of the group-theoretic arguments there to approximate groups in order to exhibit the coset nilprogression P. For readers with a taste for model theory, we also recommend Hrushovski's beautiful lecture notes [2012a], where the proof of Theorem 1.16 is presented in full from a model-theoretic viewpoint.

The proof of Theorem 1.16 does not give any explicit bounds on the complexity of the coset nilprogression[2] nor on the size of X. This is due to the inherently nonexplicit nature of the proof, which makes use of ultrafilters to take limits.

[2]If one does not require $P \subset A^4$, our proof does give a $O(\log K)$ bound on the dimension of P.

In view of the polynomial Freiman–Ruzsa conjecture (Conjecture 1.10) it is reasonable to expect that these bounds can be made polynomial in K.

We will see in Section 2 that this polynomiality of the bounds can be proven for approximate subgroups of GL_d with exponents depending only on the dimension d. But can they be made independent of d? This was asked by László Pyber at his workshop lecture (see also [Pyber and Szabó 2014]). For approximate subgroups of GL_d one can even hope to find a P which is normalized by A; this is what happens in Helfgott's $\mathrm{SL}_3(\mathbb{Z}/p\mathbb{Z})$ theorem. One should however bear in mind the following example due to Pyber (see also the end of [Pyber and Szabó 2010b]).

Example 1.17. Let $G = \mathcal{S}_{2n+1}$ be the symmetric group on $2n+1$ objects. Let H be the subgroup generated by all transpositions $(i, i+1)$ for $i = 1, \dots, n$. Let σ be the shift $i \mapsto i+2 \bmod 2n+1$. Let $A := H \cup \{\sigma^{\pm 1}\}$. Then A is a 10-approximate group which generates G. While it is contained in at most 10 cosets of H, it does not normalize any proper subgroup of G (except A_{2n+1}), because $\mathcal{S}_{2n+1}$ has no nontrivial normal subgroup (apart from A_{2n+1}).

In this example, the approximate group is roughly equivalent to a large finite subgroup which is *almost normalized* by A, but A does not normalize any subgroup (except trivial ones, which are either much smaller or much larger than A).

2. Approximate subgroups of linear groups: some proofs

In the remainder of the article, we give an introduction to the works [Breuillard et al. 2011] and [Pyber and Szabó 2010b], which classify approximate subgroups of simple algebraic groups over finite fields. We will give complete proofs save for the Larsen–Pink-type nonconcentration estimate (Theorem 2.15 below), whose proof we only briefly sketch as it is rather more involved.

A key tool is the use of quasirandomness through Gowers' trick, which shows that large subsets of a finite simple group of Lie type grow quickly. We include a proof of that via nonabelian Fourier analysis of finite groups.

We also include a presentation of the sum-product phenomenon and give a geometric proof of it, which is run in parallel with the proof of the classification of approximate subgroups of simple algebraic groups, emphasizing the parentage between the two problems.

2.1. ***Quasirandomness and Gowers' trick.*** A distinctive feature of finite simple groups (as opposed to abelian groups for instance) is that they have few complex linear representations of small dimension. The smallest dimension $m(G)$ of a nontrivial complex linear representation must tend to infinity with the size of the finite simple group G. This fact is a simple consequence of Jordan's theorem on finite subgroups of $\mathrm{GL}_d(\mathbb{C})$, which asserts that every such group must have an abelian normal subgroup of index at most some bound $c(d)$ which depends on d only. Indeed, since G is simple, any nontrivial linear representation of G must be faithful, hence the abelian normal subgroup must be trivial, and G must have cardinality at most $c(d)$ (see, e.g., [Breuillard 2011a] for some historical comments on Jordan's theorem).

This feature has played a very important role in the spectral theory of arithmetic surfaces; see [Sarnak and Xue 1991]. It also plays an important role in the Bourgain–Gamburd proof of the spectral gap for $\mathrm{SL}_2(\mathbb{Z}/p\mathbb{Z})$ in the last step of their proof, when one derives the spectral gap from the fast decay of the probability of return to the identity of the random walk at time $C \log p$.

For finite simple groups of Lie type (as opposed to the alternating groups) a very strong lower bound on the dimension of complex linear representations is known. This goes back to Frobenius, who showed that $m(\mathrm{PSL}_2(\mathbb{F}_p)) = \frac{1}{2}(p-1)$, and was established in full generality as follows:

Fact (Landazuri–Seitz bound [1974]). *There is a constant $c_d > 0$ such that $m(G) \geq c_d |G|^{r/d}$ for every finite simple group of Lie type $G = \mathbf{G}(q)$ over a finite field $\mathbb{F}_q$ with dimension $d = \dim \mathbf{G}$ and rank*[3] *r.*

In the early 2000s, Tim Gowers [2008] exploited this fact in order to answer a combinatorial question of Babai and Sós: Does every finite group G have a product free set of size $> c|G|$? A product free set is a subset $X \subset G$ such that $XX \subset G \setminus X$. Gowers shows that answer is no for $\mathrm{PSL}_2(\mathbb{F}_p)$ and for all finite simple groups of Lie type precisely thanks to the above fact about $m(G)$. And indeed this follows directly from the following formulation (first observed by Nikolov and Pyber in [2011]) of Gowers' result (take $A = B = X$ and $C = X^{-1}$ to answer the Babai–Sos question negatively):

Lemma 2.2 (Gowers' trick). *Suppose A, B, C are subsets of a finite group G such that $|A||B||C| > |G|^3/m(G)$. Then $ABC = G$.*

Gowers' proof (as well as the proof given later by Babai, Nikolov and Pyber [Babai et al. 2008]) is based on spectral analysis of bipartite graphs. We give a seemingly different argument based on the nonabelian Fourier transform.

Recall that the convolution of two functions f_1, f_2 on a finite group G is defined by

$$f_1 * f_2(x) = \sum_{\substack{a,b \in G \\ ab = x}} f_1(a) f_2(b).$$

---

[3]Recall that the rank of G is the dimension of a maximal torus, in particular it is $< d$.

The convolution product is associative: $(f_1 * f_2) * f_3 = f_1 * (f_2 * f_3)$. For a subset A in G we denote by 1_A the indicator function of A, namely $1_A(x) = 1$ if $x \in A$ and $1_A(x) = 0$ if $x \notin A$.

Proof. Let $f := 1_A * 1_B * 1_C$ be the convolution product of the indicator functions of the three subsets A, B and C. Note that the support of f is precisely the product set ABC. So in order to show that $ABC = G$ it is enough to prove that $f(g) > 0$ for every $g \in G$. To show that, the idea is very simple: expand f in Fourier.

Recall the nonabelian Fourier inversion and Parseval formulas (see [Serre 1977] on representation theory of finite groups, for example). Let $d_\pi = \dim(\mathcal{H}_\pi)$ be the dimension of the irreducible representation π of G. Then

$$\textit{Parseval}: \qquad \sum_{g \in G} |f(g)|^2 = \frac{1}{|G|} \sum_{\pi} d_\pi \|\pi(f)\|^2,$$

$$\textit{Fourier inversion}: \qquad f(g) = \frac{1}{|G|} \sum_{\pi} d_\pi \langle \pi(f), \pi(g) \rangle.$$

Here we have written $\pi(f) = \sum_{g \in G} f(g)\pi(g)$, where π is an irreducible complex linear representation of G, and the sums on the right extend over all such representations; moreover the scalar product is defined on $\mathrm{End}(\mathcal{H}_\pi)$ by $\langle X, Y \rangle = \mathrm{trace}(XY^*)$.

From the Parseval formula applied to 1_A and the bound $d_\pi \geqslant m(G)$ for every nontrivial π, we see that $|A| = (1/|G|) \sum_\pi d_\pi \|\pi(1_A)\|^2$, and thus

$$\|\pi(1_A)\| \leqslant \sqrt{\frac{|A||G|}{m(G)}}, \tag{2.2.1}$$

for every nontrivial π.

Now writing the Fourier inversion formula for f and splitting the sum into a main term (corresponding to the trivial representation) and a remainder term (corresponding to all other representations), we get

$$f(g) \geqslant \frac{|A||B||C|}{|G|} - \frac{1}{|G|} \sum_{\pi \neq 1} d_\pi \|\pi(1_A)\| \cdot \|\pi(1_B)\| \cdot \|\pi(1_C)\|.$$

Using (2.2.1) to control $\|\pi(1_A)\|$ and Cauchy–Schwarz inequality together with the Parseval identity to handle $\|\pi(1_B)\|$ and $\|\pi(1_C)\|$, we get

$$f(g) \geqslant \frac{|A||B||C|}{|G|} - \sqrt{\frac{|A||G|}{m(G)}} \frac{1}{|G|} \sqrt{|G||B|} \sqrt{|G||C|},$$

which is > 0 as soon as $|A||B||C| > |G|^3/m(G)$, as claimed. □

Combining Gowers' trick with the bound on $m(G)$ mentioned above, we obtain this result for approximate groups:

Corollary 2.3. *Let $G = \mathbf{G}(q)$ be a finite simple group of Lie type of dimension $d = \dim \mathbf{G}$ over a finite field $\mathbb{F}_q$. There is $\delta = \delta(d) > 0$ independent of q such that $AAA = G$ for every subset $A \subset G$ such that $|A| > |G|^{1-\delta}$.*

Gowers calls *quasirandom* a finite group G for which $m(G)$ is large. This terminology comes from the abelian case, where a quasirandom subset, say of $G = (\mathbb{F}_p, +)$, is by definition a subset $A \subset G$ such that $\chi(1_A) = \sum_{a \in A} \chi(a)$ is small compare to $|G|$ for every nontrivial character χ of G. Certainly random subsets of G (chosen by flipping independent coins for each element of G) are quasirandom. The bound (2.2.1) shows that if $m(G)$ is large, then every subset of G is quasirandom in the sense that $\|\pi(1_A)\|$ is small compared to $|G|$ for every nontrivial irreducible representation π.

2.4. ***The sum-product theorem.*** The story of approximate groups really began when Bourgain, Katz and Tao proved the following:

Theorem 2.5 (sum-product in $\mathbb{F}_p$; see [Bourgain et al. 2004]). *Let $\mathbb{F}_p$ be the finite field with p elements (p prime). Then for every $\delta > 0$, there is $\varepsilon > 0$ independent of p such that*

$$|SS| + |S+S| \geqslant |S|^{1+\varepsilon},$$

for every subset $S \subset \mathbb{F}_p$ such that $p^\delta < |S| < p^{1-\delta}$.

There are several proofs of this result (see, e.g., [Tao and Vu 2006]), most of them quite combinatorial. Konyagin [2003] gave a proof that does not require the $|S| > p^\delta$ assumption. We will give a more geometric proof (also not requiring the $|S| > p^\delta$ assumption) below. All proofs require some version of the following result:

Lemma 2.6 (Katz and Tao; see [Tao and Vu 2006] or [Breuillard 2011b]). *For every $n \geqslant 1$ there is a constant $C = C(n) > 0$ such that for every $K \geqslant 1$ and for any set $S \subset \mathbb{F}_p$ with $|S+S| + |SS| \leqslant K|S|$, there are $\lambda \in \mathbb{F}_p^*$ and a subset $S' \subset \lambda S$ with $|S'| \geqslant |S|/CK^C$ and $|F_n(S')| \leqslant CK^C|S|$, where $F_n(S')$ denotes the set of all elements of $\mathbb{F}_p$ one can obtain from 0 by applying at most n operations (i.e., additions, subtractions, multiplications, divisions) by elements from S' or from the previously constructed elements.*

It turns out that one can recast the sum-product theorem in terms of the Freiman inverse problem, which we discussed in the first lecture. This was first observed by Helfgott in his SL_3 paper [2011]. Consider the group of affine transformations of the $\mathbb{F}_p$-line, namely $\mathbb{F}_p \rtimes \mathbb{F}_p^\times$ viewed as a matrix group as

$$\left\{ \begin{pmatrix} \alpha & \beta \\ 0 & 1 \end{pmatrix} \,\middle|\, \alpha \in \mathbb{F}_p^\times, \beta \in \mathbb{F}_p \right\},$$

and inside this group consider the subset

$$B := \left\{ \begin{pmatrix} \alpha & \beta \\ 0 & 1 \end{pmatrix} \,\middle|\, \alpha \in S', \beta \in S' \right\},$$

where S' is the subset obtained from the Katz–Tao lemma (applied with $n = 4$, say). Then B satisfies

$$|BBB| \leqslant CK^C |B|$$

for some absolute constant C. So in view of Lemma 1.12 the subset

$$A := (B \cup B^{-1} \cup \{\mathrm{id}\})^2$$

is a $O(K^{O(1)})$-approximate subgroup of the affine group. So if we knew the solution to Freiman's inverse problem for the affine group, namely a complete description (with polynomial bounds) of its approximate subgroups, then we would derive the sum-product theorem as a corollary. We will pursue this strategy to the end, but before that I would like to describe the answer to Freiman's inverse problem inside simple algebraic groups.

2.7. *The product theorem.* In 2005, Helfgott established a seminal result:

Theorem 2.8 (product theorem [Helfgott 2008]). *For every $\delta > 0$ there is $\varepsilon > 0$ (independent of p) such that*

$$|SSS| \geqslant |S|^{1+\varepsilon}$$

for every finite generating subset of $\mathrm{SL}_2(\mathbb{Z}/p\mathbb{Z})$ such that $|S| < |\mathrm{SL}_2(\mathbb{Z}/p\mathbb{Z})|^{1-\delta}$.

Approximate groups are not mentioned in this statement. The bridge between product theorems and results about the classification of approximate groups is clear however: if one has $|SSS| \leqslant |S|^{1+\varepsilon}$, then S has tripling at most K, where $K = |S|^\varepsilon$, and thus by Lemma 1.12 $A := (S \cup S^{-1} \cup \{\mathrm{id}\})^2$ is a CK^C-approximate group. So Helfgott's theorem can be rephrased (somewhat abusively) as saying that: *there are no nontrivial approximate subgroups of* $\mathrm{SL}_2(\mathbb{F}_p)$.

Helfgott's proof was based on the Bourgain–Katz–Tao sum-product theorem and explicit 2×2 matrix calculations. It appeared clearly from the proof however that a key role was played by large subsets of simultaneously diagonalizable matrices in S. This idea was further exploited in Helfgott's SL_3 paper [2011].

After Helfgott's results (and also the partial results on $\mathrm{SL}_d(\mathbb{Z}/p\mathbb{Z})$ in [Gill and Helfgott 2011]) it became highly plausible that a product theorem should hold in full generality for subsets of arbitrary simple algebraic groups (such as SL_d) over an arbitrary field. Moreover a proof of such a theorem should be geometric and exploit the underlying algebraic geometry of simple algebraic groups and in particular the geometry of maximal tori.

The breakthrough came in 2009, with the preprint of [Hrushovski 2012b], where use was made of model-theoretic tools to give an essentially complete classification of approximate subgroups of simple algebraic groups, albeit with no explicit bounds. One statement is the following:

Theorem 2.9 [Hrushovski 2012b, Theorem 1.3]. *Let $\mathbf{G}$ be a simple algebraic group over an algebraically closed field k with* $\dim \mathbf{G} = d$. *Let $A \subset \mathbf{G}(k)$ be a K-approximate subgroup of $\mathbf{G}(k)$. Then there exists a subgroup $\mathbb{H}$ of $\mathbf{G}(k)$ and a number $f(d, K)$ depending only on d and K satisfying the following conditions*:

(i) *A intersects at most $f(d, K)$ cosets of $\mathbb{H}$.*

(ii) *Either $\mathbb{H}$ is a proper connected algebraic subgroup of $\mathbf{G}$, or $\mathbb{H}$ is finite and contained in A^4.*

Hrushovski's interest in approximate groups was triggered by his observation of the similarity between the Freiman inverse problem and some model-theoretic results, such as the Zilber stabilizer lemma, in stable group theory. His proof however (as often in model theory) gave no explicit bounds on the function $f(d, K)$ above in terms of d and K.

A few months after Hrushovski's paper appeared on the arXiv, however, Pyber and Szabó [2010a; 2010b] and independently Green, Tao and myself [Breuillard et al. 2010; 2011] managed to give a polynomial bound on $f(K, d)$ and to improve Hrushovski's conclusion slightly as follows:

Theorem 2.10 (classification of approximate subgroups of $\mathbf{G}(k)$). *There are constants $C(d)$ independent of K, k, and $f(d, K) \leqslant O_d(K^{O_d(1)})$, such that for every simple algebraic group $\mathbf{G}$ of dimension $d =$* $\dim \mathbf{G}$ *defined over an algebraically closed field k and every K-approximate subgroup $A \subset \mathbf{G}(k)$, we are in one of the following situations*:

(1) *There exists a proper closed algebraic subgroup $\mathbb{H}$ with at most $C(d)$ connected components and such that $A \subset \mathbb{H}(k)$.*

(2) $|A| \leqslant f(d, K)$.

(3) $|A| \geqslant |\langle A \rangle| / f(d, K)$.

In cases (1) or (2), the first alternative in Theorem 2.9(ii) holds. If we are in case (3) of Theorem 2.10, then $\langle A \rangle$ is finite and not much larger than A, a condition that is close to the condition in the second branch of Theorem 2.9(ii), yet not exactly the same (see however Corollary 2.12 below).

I will sketch a proof of Theorem 2.10 below. (For a beautiful exposition of this proof in the framework of model theory, we refer the reader to [Hrushovski 2013].) An explicit bound on the implied constants $f(d, K)$ and $C(d)$ is obtainable in principle from the proof (especially the version given by Pyber and Szabó), although it has not been worked out, mostly because tracking the constants throughout the proof would most likely not yield very sharp bounds.

It follows from the theorem that if conclusions (1) and (2) fail, then A generates a finite subgroup. In view of Jordan's lemma on finite linear groups in characteristic zero, which we already mentioned, this implies that k is of positive characteristic. Now a deep result of Larsen and Pink [2011] implies that $\langle A\rangle$ is then essentially (up to some bounded index issues) a finite simple group of Lie type. In particular the Landazuri–Seitz bound (page 35) on the dimension of complex linear representations holds and Gowers' trick kicks in.

One can then easily derive the following generalization of Helfgott's product theorem.

Corollary 2.11 (the product theorem). *Let* $\mathbf{G}$ *be a simple algebraic group over an algebraically closed field* k *with dimension* $d = \dim \mathbf{G}$. *There is a constant* $\varepsilon = \varepsilon(d) > 0$, *independent of the field* k, *such that for every finite subset* $S \subset \mathbf{G}(k)$,

(1) *either* S *is contained in a proper algebraic subgroup with at most* C *connected components, or*

(2) $|S^N| \geqslant \min\{|\langle S\rangle| : |S|^{1+\varepsilon}\}$, *where* $N = N(d)$ *and* $C = C(d)$ *are constants independent of the field* k.

One can take $N(d) \leqslant \max\{3, |Z|\}$, where $|Z|$ is the size of the center of the simply connected cover of $\mathbf{G}$.

Sketch of proof. Set $K = |S|^{\varepsilon}$ and apply Theorem 2.10 to $A := (S \cup S^{-1} \cup \{\mathrm{id}\})^2$, which is a $O(K^{O(1)})$-approximate group. Then, thanks to the polynomial bound on $f(d, K)$ obtained in Theorem 2.10, item (2) in that theorem cannot hold if ε is chosen small enough. So if (1) does not hold either, it must be that (3) holds. Then $\langle A\rangle$ is finite and k has positive characteristic (because the negation of (1) is incompatible with Jordan's lemma in characteristic zero). Larsen–Pink then tell us that $\langle A\rangle$ is (after taking the commutator subgroup and modding out by the center) a finite simple group of Lie type of bounded rank. If ε is small enough, we can apply Gowers' trick (see Corollary 2.3 above) to $[\langle A\rangle, \langle A\rangle]/\text{center}$ and the result follows. □

Finite simple groups of Lie type[4] are of form $G = \mathbf{G}(\mathbb{F}_q)/\text{center}$ for some absolutely almost simple (simply connected) algebraic group $\mathbf{G}$ defined over $\mathbb{F}_q$.

[4]Except the Suzuki and Ree families, which arise slightly differently and can also be handled similarly.

It can be shown that they (rather their lift to **G**) are not contained in a proper algebraic group of **G** with boundedly many connected components. Moreover finite simple groups are quasirandom in the sense of Gowers (compare the Landazuri–Seitz bound, page 35) and Gowers' trick applies to large subsets of G. The product theorem then takes the following simple form for generating sets of finite simple groups of Lie type.

Corollary 2.12 (product theorem for finite simple groups of Lie type). *Let $G = \mathbf{G}(q)$ be a finite simple group of Lie type over a finite field $\mathbb{F}_q$ with $d = \dim \mathbf{G}$. Let S be any generating set for G. Then*

$$|SSS| \geqslant \min\{|G|, |S|^{1+\varepsilon}\}$$

for some constant $\varepsilon = \varepsilon(d) > 0$ independent of the field $\mathbb{F}_q$.

I will now turn to the proof of Theorem 2.10. I will give an essentially complete proof, modulo the Larsen–Pink inequality, for which I will refer to [Breuillard et al. 2011]. I will start with a geometric proof of the sum-product theorem (Theorem 2.5), because this proof can easily be transformed into a proof of Theorem 2.10.

2.13. ***A geometric proof of the sum-product theorem.*** In this paragraph I give a proof of Theorem 2.5. I keep the notation of that theorem and of the discussion following it. In particular $G = \mathbb{F}_p \rtimes \mathbb{F}_p^\times$ is the group of affine transformations of the line over the finite field $\mathbb{F}_p$. In matrix notation

$$G = \left\{ \begin{pmatrix} \alpha & \beta \\ 0 & 1 \end{pmatrix} \,\middle|\, \alpha \in \mathbb{F}_p^\times, \beta \in \mathbb{F}_p \right\}.$$

This group admits two remarkable actions:

(a) its action on itself by conjugation $g^h := hgh^{-1}$, and

(b) its action on the affine line $\mathbb{F}_p$ by affine transformations $g \cdot x := \alpha x + \beta$.

In case (b) the stabilizers of a point $x \in \mathbb{F}_p$ are the *tori* T_x made of all homotheties fixing the point x. In case (a) the stabilizers are centralizer subgroups. Note that if g is a nontrivial homothety (i.e., fixes a point x and is not the identity), then its centralizer $C_G(g)$ is precisely the torus T_x. Finally note that $gT_xg^{-1} = T_{g \cdot x}$.

The sum-product theorem is a consequence of the tension between these two actions. The proof relies on the orbit-stabilizer lemma for approximate groups (i.e., Lemma 1.13) applied to both actions. The approximate group in consideration is obtained from the set S in the way that was described earlier, namely $A := (B \cup B^{-1} \cup \{\mathrm{id}\})^2$, where B is the set

$$B := \left\{ \begin{pmatrix} \alpha & \beta \\ 0 & 1 \end{pmatrix} \,\middle|\, \alpha \in S', \beta \in S' \right\},$$

and S' is the subset obtained from the Katz–Tao lemma applied with $n = 6$, say, and with $K = |S|^{\varepsilon}$ for some small ε to be determined later. Then B satisfies $|BBB| \leqslant CK^C|B|$ and by Lemma 1.12 A is a $O(K^{O(1)})$-approximate subgroup of G. Moreover $|S'| = |B|^{\frac{1}{2}} \geqslant |S|/O(K^{O(1)})$.

Let us apply the orbit-stabilizer lemma (Lemma 1.13) to both actions:

Action (a): suppose $g \in A^2 \cap T_x$ for some $x \in \mathbb{F}_p$ and $g \neq 1$, then computing the matrix g^h for $h \in A$, the Katz–Tao lemma implies that $|g^A| \leqslant O(K^{O(1)})|S|$ because the translation part of that matrix is an algebraic expression of small length involving only elements from S'. From the orbit-stabilizer lemma, we conclude that

$$|A^2 \cap T_x| \geqslant \frac{|S|}{O(K^{O(1)})}.$$

Action (b): we clearly have

$$|A \cdot x| \geqslant |S'| \geqslant \frac{|S|}{O(K^{O(1)})}$$

for every $x \in \mathbb{F}_p$, for example, because A contains many translations. From the orbit-stabilizer lemma applied to this action, we conclude that

$$|A^2 \cap T_x| \leqslant O(K^{O(1)})|S|.$$

Conclusion. *For every $x \in \mathbb{F}_p$, if $A^2 \cap T_x \neq \{1\}$, then $|A^2 \cap T_x| \asymp_K |S|$.*

Here the shorthand $|A_1| \asymp_K |A_2|$ signifies that $|A_1| \geqslant |A_2|/O(K^{O(1)})$ and vice versa.

This is where the miracle happens: if $A^2 \cap T_x$ has one nontrivial element, then it has many! Everything will follow easily from this. Let $\mathcal{T}$ be the set of tori T_x which intersect A^2 nontrivially (the so-called *involved tori* in the terminology of [Breuillard et al. 2010]).

Key claim. *If $|S'| \geqslant CK^C$ for some absolute constant C, then $\mathcal{T}$ is invariant under conjugation by A (and hence by the subgroup $\langle A \rangle$ generated by A).*

Proof. Recall again the orbit stabilizer lemma (Lemma 1.13) and its extra feature that $|A^k \cap \mathrm{Stab}(x)|$ is roughly of the same size as $|A^2 \cap \mathrm{Stab}(x)|$ for any given $k \geqslant 2$. So we may write:

$$|A^2 \cap aT_xa^{-1}| = |a^{-1}A^2a \cap T_x| \asymp_K |a^{-1}A^4a \cap T_x| \geqslant |A^2 \cap T_x|$$

If the $T_x \in \mathcal{T}$, then the right-hand side is large. Hence the left-hand side too is large, and is in particular > 1, so $aT_xa^{-1} \in \mathcal{T}$ as claimed. □

Note from the way we defined A that A contains a nontrivial translation (e.g., of the form $b_1^{-1}b_2$, where b_1 and b_2 have the same top left matrix entry). Since p is prime, $\mathbb{Z}/p\mathbb{Z}$ has no nontrivial proper subgroup and it follows that $\langle A \rangle$ contains all translations. Therefore, since $gT_xg^{-1} = T_{g\cdot x}$, every torus T_x belongs to $\mathcal{T}$, and $|\mathcal{T}| = p$.

To finish the proof it only remains to count A^2 by slicing it into different tori. Since tori are disjoint (except for the fact that they all contain the identity), we may write

$$\bigcup_{T_x \in \mathcal{T}} (A^2 \cap T_x \backslash \{1\}) \subset A^2;$$

thus

$$|\mathcal{T}| \frac{|S|}{O(K^{O(1)})} \leqslant \sum_{T_x \in \mathcal{T}} |A^2 \cap T_x \setminus \{1\}| \leqslant |A^2| \leqslant K|A| = O(K^{O(1)})|S|^2.$$

Hence

$$|S| \geqslant \frac{|\mathcal{T}|}{O(K^{O(1)})} = \frac{p}{O(K^{O(1)})}.$$

Thus (remember that $K = |S|^{\varepsilon}$) choosing ε small enough — $\varepsilon \leqslant \dfrac{\delta}{O(1)}$ will do — we obtain $|S| > p^{1-\delta}$ as claimed.

2.14. ***A proof of the product-theorem and the Larsen–Pink inequality.*** The proof of the product theorem (in the form of the classification theorem for approximate subgroups, that is, Theorem 2.10) follows exactly the same path as the above geometric proof of the sum product theorem. Here too there will be two different actions of the group and the tension between these two actions, via the orbit-stabilizer lemma for approximate groups (Lemma 1.13), will yield the proof.

It turns out that in order to implement this strategy, one needs one further ingredient, which was already present in a crucial way in Hrushovski's proof of Theorem 2.9 (although used differently). This is the celebrated dimension inequality of Larsen and Pink, devised in their 1995 preprint on finite subgroups of linear groups (which has now appeared as [Larsen and Pink 2011]) and then investigated in [Hrushovski and Wagner 2008] in the model-theoretic framework.

Theorem 2.15 (Larsen–Pink inequality). *Let* $\mathbf{G}$ *be a (connected) simple algebraic group over an algebraically closed field* k. *Given* $M \geqslant 1$, *there is* $C = C(M, \dim \mathbf{G}) > 0$ *independent of* k *such that the following holds. If* A *denotes a finite* K*-approximate subgroup of* $\mathbf{G}(k)$, *then either* A *is contained* *in some proper algebraic subgroup* $\mathbf{H}$ *of* $\mathbf{G}$ *such that* $[\mathbf{H}, \mathbf{H}^0] \leqslant C$, *or for every closed algebraic subvariety* $\mathcal{V}$ *of* $\mathbf{G}$ *with degree at most* M,

$$|A \cap \mathcal{V}| \leqslant O(K^{O(1)})|A|^{\frac{\dim T}{\dim \mathbf{G}}},$$

where the implied constants depend on M *and* $\dim \mathbf{G}$ *only.*

When $k = \overline{\mathbb{F}_p}$, which is the case of interest in the applications to approximate groups, if A generates a subgroup of the form $\mathbf{G}(\mathbb{F}_q)$, for some large enough $q = p^n$, then the first possibility that A may be contained in some proper algebraic subgroup with a bounded number of connected components is automatically ruled out.

Larsen and Pink proved this inequality in [2011, Theorem 4.2] in the case when A is a genuine finite subgroup of $\mathbf{G}(k)$ (namely when $K = 1$). Hrushovski and Wagner [2008] then gave a model theoretic proof as well as a vast generalization, which was subsequently used in [Hrushovski 2012b] to prove Theorem 2.9. It turns out that the proof in the approximate group case is no more difficult than in the group case and this is a very good example where the philosophy of transferring group-theoretical arguments to the approximate group setting is particularly successful.

A word on the proof. There are at least two cases where the inequality is obvious: when $\dim \mathcal{V} = 0$, because then $\mathcal{V}$ is finite and its degree is its number of elements; and when $\dim \mathcal{V} = \dim \mathbf{G}$, obviously. Now the proof proceeds by a double induction on the dimension of $\mathcal{V}$. Starting with two possible counterexamples, one of smallest possible dimension $\mathcal{V}_-$ and one of largest possible dimension $\mathcal{V}_+$ one uses the assumption on A (that A is *sufficiently Zariski-dense*, or *sufficiently general* in the Larsen–Pink terminology) and the simplicity of $\mathbf{G}$ to deduce that there is $a \in A^k$, where k depends only on the degree bound, such that $\mathcal{V}_- a \mathcal{V}_+$ has dimension $\dim \mathcal{V}_+ + 1$ at least. Indeed assuming as we may that $\mathcal{V}_-$ and $\mathcal{V}_+$ are irreducible, an equality between the dimensions $\dim \mathcal{V}_- a \mathcal{V}_+ = \dim \mathcal{V}_+$ for all $a \in A^k$ would imply that A^k is contained in the proper (because $\mathbf{G}$ is simple) subvariety of bounded degree $\{g \in \mathbf{G}(k) \mid g^{-1}\mathcal{V}_-^{-1}\mathcal{V}_- g \subset \mathrm{Stab}(\mathcal{V}_+)\}$, where $\mathrm{Stab}(\mathcal{V}_+)$ is the subgroup $\{g \in \mathbf{G}(k) | g\mathcal{V}_+ = \mathcal{V}_+\}$. Then one can use the induction hypothesis on $\mathcal{V}_- a \mathcal{V}_+$ to deduce a contradiction (it will have too many points in A^{k+2}).

In the proof of the product theorem, Theorem 2.15 will be applied to only three kinds of subvarieties $\mathcal{V}$, all of them of bounded degree (maximal tori and their normalizers, conjugacy classes of regular semisimple elements, and the set of nonregular semisimple elements).

We now move on to the proof of Theorem 2.10, which as we already said, is just a matter of adapting the geometric proof of the sum-product theorem that was given above. At this point I only have to point out the words that need to be changed in order to turn it into a proof of Theorem 2.10. At the blackboard this was very easy to do by simply erasing and replacing a couple of words here and there with a colored chalk. I cannot do this in this article, so let me briefly describe what remains to be done.

The group now is $G = \mathbf{G}(k)$ and as above we consider two actions of this group:

(a) the action of G on itself by conjugation, $g^h := hgh^{-1}$,

(b) the action of G on the variety of maximal tori $G/N_G(T)$, $g \cdot T := gTg^{-1}$.

Recall that maximal tori in G (i.e., maximal connected subgroups made of semisimple elements) are all conjugate to each other (k is algebraically closed), so the stabilizer of a maximal torus T in action (b) equals its normalizer $N_G(T)$. Moreover recall that $N_G(T)/T$ is finite (the Weyl group) and independent of k.

Now point stabilizers in action (a) are the centralizers of elements. Elements $g \in G$ such that the (connected component of the) centralizer of $g \in G$ is a maximal torus are called *regular semisimple* (e.g., the elements with distinct eigenvalues in case $\mathbf{G} = \mathrm{SL}_n$). They form a Zariski-open subset of G as well as of every maximal torus T (in a maximal torus nonregular semisimple elements are contained in a union of boundedly many proper subtori, the *root tori*). So here, in order to define a notion of involved torus T, we need to require A^2 to intersect T not only nontrivially, but in such a way that $A^2 \cap T$ contains a regular element. Denote by T_{reg} the regular semisimple elements of T. Then define

$$\mathcal{T} := \{T \text{ maximal torus } | A^2 \cap T_{\mathrm{reg}} \neq \varnothing\}.$$

We can now apply the orbit-stabilizer lemma (Lemma 1.13) in combination with the Larsen–Pink inequality to actions (a) and (b).

Action (a): Suppose $g \in A^2 \cap T_{\mathrm{reg}}$. The Larsen–Pink inequality applied to $\mathcal{V} = g^G$ the conjugacy class of g (it is a closed subvariety of bounded degree and of dimension $\dim \mathbf{G} - \dim T$ because g is regular semisimple) yields

$$|g^A| \leqslant |A^3 \cap \mathcal{V}| \leqslant O(K^{O(1)})|A|^{1-\frac{\dim T}{\dim \mathbf{G}}}.$$

So, by the orbit-stabilizer lemma, we conclude that[5]

$$|A^2 \cap T| \geqslant \frac{|A|^{\frac{\dim T}{\dim \mathbf{G}}}}{O(K^{O(1)})}.$$

[5]Note that $[C_G(g) : C_G(g)^\circ]$ is bounded above by a constant depending only on $\dim \mathbf{G}$ and not on g nor k.

Action (b): We can apply the Larsen–Pink inequality directly to the variety $\mathcal{V} = T$ and conclude that

$$|A^2 \cap T| \leqslant O(K^{O(1)})|A|^{\frac{\dim T}{\dim \mathbf{G}}}.$$

Since nonregular elements form a proper Zariski closed subset of bounded degree, another application of the Larsen–Pink inequality yields

$$|A^2 \cap (T \setminus T_{\mathrm{reg}})| \leqslant O(K^{O(1)})|A|^{\frac{\dim T}{\dim \mathbf{G}}-1}.$$

Conclusion. *For every maximal torus T, if $A^2 \cap T_{\mathrm{reg}} \neq \varnothing$, then*

$$|A^2 \cap T_{\mathrm{reg}}| \asymp_K |A|^{\frac{\dim T}{\dim \mathbf{G}}},$$

Precisely the same argument as in the proof of the sum-product theorem given above yields the analogous claim:

Key claim. *If $|A| \geqslant CK^C$, for some constant C depending on* $\dim \mathbf{G}$ *only, then $\mathcal{T}$ is invariant under conjugation by A (and hence by the subgroup $\langle A \rangle$ generated by A).*

The end of the proof is also the same: one can slice A^2 into different maximal tori and write, noting that $T_{\mathrm{reg}} \cap T'_{\mathrm{reg}} = \varnothing$ for two different tori T and T',

$$\bigcup_{T \in \mathcal{T}} (A^2 \cap T_{\mathrm{reg}}) \subset A^2,$$

thus

$$|\mathcal{T}| \frac{|A|^{\frac{\dim T}{\dim \mathbf{G}}}}{O(K^{O(1)})} \leqslant \sum_{T \in \mathcal{T}} |A^2 \cap T_{\mathrm{reg}}| \leqslant |A^2| \leqslant K|A|,$$

hence

$$|A|^{1-\frac{\dim T}{\dim \mathbf{G}}} \geqslant \frac{|\mathcal{T}|}{O(K^{O(1)})}. \tag{2.15.1}$$

However by the key claim and the orbit-stabilizer lemma (the original one for groups this time!) we have for any $T \in \mathcal{T}$:

$$|\mathcal{T}| \geqslant |T^{\langle A \rangle}| = \frac{|\langle A \rangle|}{|\langle A \rangle \cap N_G(T)|}.$$

Finally another application of the Larsen–Pink inequality (this time the original one for genuine subgroups) gives

$$|\langle A \rangle \cap N_G(T)| \leqslant O(K^{O(1)})|\langle A \rangle|^{\frac{\dim T}{\dim \mathbf{G}}}.$$

Combining this with (2.15.1) we obtain the desired conclusion:

$$|A| \geqslant \frac{|\langle A \rangle|}{O(K^{O(1)})}.$$

This ends the proof of the product theorem.

Acknowledgements

I thank Lior Silberman and Jitendra Bajpai for sharing with me their notes of my lectures. This helped me a lot in preparing this text. I am also grateful to Hee Oh and Nicolas de Saxcé for their careful reading of the manuscript.

References

[Babai et al. 2008] L. Babai, N. Nikolov, and L. Pyber, "Product growth and mixing in finite groups", pp. 248–257 in *Proceedings of the Nineteenth Annual ACM-SIAM Symposium on Discrete Algorithms* (San Francisco, CA, 2008), ACM, New York, 2008.

[Bourgain and Gamburd 2008a] J. Bourgain and A. Gamburd, "Expansion and random walks in $\mathrm{SL}_d(\mathbb{Z}/p^n\mathbb{Z})$. I", *J. Eur. Math. Soc.* (*JEMS*) **10**:4 (2008), 987–1011.

[Bourgain and Gamburd 2008b] J. Bourgain and A. Gamburd, "Uniform expansion bounds for Cayley graphs of $\mathrm{SL}_2(\mathbb{F}_p)$", *Ann. of Math.* (2) **167**:2 (2008), 625–642.

[Bourgain and Gamburd 2009] J. Bourgain and A. Gamburd, "Expansion and random walks in $\mathrm{SL}_d(\mathbb{Z}/p^n\mathbb{Z})$. II", *J. Eur. Math. Soc.* (*JEMS*) **11**:5 (2009), 1057–1103.

[Bourgain and Varjú 2012] J. Bourgain and P. P. Varjú, "Expansion in $\mathrm{SL}_d(\mathbf{Z}/q\mathbf{Z})$, q arbitrary", *Invent. Math.* **188**:1 (2012), 151–173.

[Bourgain et al. 2004] J. Bourgain, N. Katz, and T. Tao, "A sum-product estimate in finite fields, and applications", *Geom. Funct. Anal.* **14**:1 (2004), 27–57.

[Bourgain et al. 2010] J. Bourgain, A. Gamburd, and P. Sarnak, "Affine linear sieve, expanders, and sum-product", *Invent. Math.* **179**:3 (2010), 559–644.

[Breuillard 2011a] E. Breuillard, "An exposition of Jordan's original proof of his theorem on finite subgroups of $\mathrm{GL}_n(\mathbb{C})$", preprint, 2011, http://www.math.u-psud.fr/~breuilla/Jordan.pdf.

[Breuillard 2011b] E. Breuillard, "Lectures on approximate groups", lecture notes, 2011, http://www.math.u-psud.fr/~breuilla/ClermontLectures.pdf.

[Breuillard and Green 2011a] E. Breuillard and B. Green, "Approximate groups, I: The torsion-free nilpotent case", *J. Inst. Math. Jussieu* **10**:1 (2011), 37–57.

[Breuillard and Green 2011b] E. Breuillard and B. Green, "Approximate groups, II: The solvable linear case", *Q. J. Math.* **62**:3 (2011), 513–521.

[Breuillard and Green 2012] E. Breuillard and B. Green, "Approximate groups, III: The unitary case", *Turkish J. Math.* **36**:2 (2012), 199–215.

[Breuillard et al. 2010] E. Breuillard, B. Green, and T. Tao, "Linear approximate groups", *Electron. Res. Announc. Math. Sci.* **17** (2010), 57–67.

[Breuillard et al. 2011] E. Breuillard, B. Green, and T. Tao, "Approximate subgroups of linear groups", *Geom. Funct. Anal.* **21**:4 (2011), 774–819.

[Breuillard et al. 2012] E. Breuillard, B. Green, and T. Tao, "The structure of approximate groups", *Publ. Math. IHES* **116**:1 (2012), 115–221.

[Breuillard et al. 2013a] E. Breuillard, B. Green, R. Guralnick, and T. Tao, "Expansion in finite simple groups of Lie type", preprint, 2013. arXiv 1309.1975

[Breuillard et al. 2013b] E. Breuillard, B. J. Green, and T. Tao, "Small doubling in groups", preprint, 2013. To appear in and Erdös centennial volume. arXiv 1301.7718

[Chang 2002] M.-C. Chang, "A polynomial bound in Freiman's theorem", *Duke Math. J.* **113**:3 (2002), 399–419.

[Dinai 2010] O. Dinai, "Expansion properties of finite simple groups", preprint, 2010. MR 1001:5069

[Ellenberg 2014] J. S. Ellenberg, "Superstrong approximation for monodromy groups", pp. 51–71 in *Thin groups and superstrong approximation*, edited by H. Oh and E. Breuillard, Math. Sci. Res. Inst. Publ. **61**, Cambridge Univ. Press, Cambridge, 2014.

[Freiman 1973a] G. Freiman, "Number-theoretic studies in the Markov spectrum and in the structural theory of set addition", pp. 175–183 in *Groups and the inverse problems of additive number theory*, Kalinin. Gos. Univ., Moscow, 1973. in Russian.

[Freiman 1973b] G. A. Freiman, *Foundations of a structural theory of set addition*, Transl. Math. Monogr. **37**, Amer. Math. Soc., Providence, 1973.

[Freiman 2012] G. A. Freiman, "On finite subsets of nonabelian groups with small doubling", *Proc. Amer. Math. Soc.* **140**:9 (2012), 2997–3002.

[Gill and Helfgott 2010] N. Gill and H. A. Helfgott, "Growth in solvable subgroups of $\mathrm{GL}_r(\mathbb{Z}/p\mathbb{Z})$", preprint, 2010. arXiv 1008.5264

[Gill and Helfgott 2011] N. Gill and H. A. Helfgott, "Growth of small generating sets in $\mathrm{SL}_n(\mathbb{Z}/p\mathbb{Z})$", *Int. Math. Res. Not.* **2011**:18 (2011), 4226–4251.

[Gowers 2008] W. T. Gowers, "Quasirandom groups", *Combin. Probab. Comput.* **17**:3 (2008), 363–387.

[Green 2002] B. J. Green, "Structure theory of set addition", notes from the ICMS Instructional Conference in Combinatorial Aspects of Mathematical Analysis(Edinburgh), 2002, https://www.dpmms.cam.ac.uk/~bjg23/papers/icmsnotes.pdf.

[Green 2009] B. J. Green, "Approximate groups and their applications: work of Bourgain, Gamburd, Helfgott and Sarnak", preprint, 2009. arXiv 0911.5681

[Green 2012] B. Green, "What is... an approximate group?", *Notices Amer. Math. Soc.* **59**:5 (2012), 655–656.

[Green and Ruzsa 2007] B. Green and I. Z. Ruzsa, "Freiman's theorem in an arbitrary abelian group", *J. Lond. Math. Soc.* (2) **75**:1 (2007), 163–175.

[Hamidoune 2013] Y. O. Hamidoune, "Two inverse results", *Combinatorica* **33** (2013), 217–230.

[Helfgott 2008] H. A. Helfgott, "Growth and generation in $\mathrm{SL}_2(\mathbb{Z}/p\mathbb{Z})$", *Ann. of Math.* (2) **167**:2 (2008), 601–623.

[Helfgott 2011] H. A. Helfgott, "Growth in $\mathrm{SL}_3(\mathbb{Z}/p\mathbb{Z})$", *J. Eur. Math. Soc. (JEMS)* **13**:3 (2011), 761–851.

[Hoory et al. 2006] S. Hoory, N. Linial, and A. Wigderson, "Expander graphs and their applications", *Bull. Amer. Math. Soc. (N.S.)* **43**:4 (2006), 439–561.

[Hrushovski 2012a] E. Hrushovski, "Notes on Hilbert's 5th problem and approximate groups", 2012, http://www.ma.huji.ac.il/~ehud/notesH5-BGT.pdf.

[Hrushovski 2012b] E. Hrushovski, "Stable group theory and approximate subgroups", *J. Amer. Math. Soc.* **25**:1 (2012), 189–243.

[Hrushovski 2013] E. Hrushovski, "On pseudo-finite dimensions", in *Proceedings of a model theory conference* (Oléron, France, 2011), 2013. to appear in *Notre Dame J. Formal Logic*.

[Hrushovski and Wagner 2008] E. Hrushovski and F. Wagner, "Counting and dimensions", pp. 161–176 in *Model theory with applications to algebra and analysis, II*, edited by Z. Chatzidakis et al., London Math. Soc. Lecture Note Ser. **350**, Cambridge Univ. Press, 2008.

[Kesten 1959] H. Kesten, "Symmetric random walks on groups", *Trans. Amer. Math. Soc.* **92** (1959), 336–354.

[Konyagin 2003] S. Konyagin, "A sum-product estimate in fields of prime order", preprint, 2003. arXiv math.NT/0304217

[Landazuri and Seitz 1974] V. Landazuri and G. M. Seitz, "On the minimal degrees of projective representations of the finite Chevalley groups", *J. Algebra* **32** (1974), 418–443.

[Larsen and Pink 2011] M. J. Larsen and R. Pink, "Finite subgroups of algebraic groups", *J. Amer. Math. Soc.* **24**:4 (2011), 1105–1158.

[Lubotzky et al. 1988] A. Lubotzky, R. Phillips, and P. Sarnak, "Ramanujan graphs", *Combinatorica* **8**:3 (1988), 261–277.

[Margulis 1973] G. A. Margulis, "Explicit constructions of expanders", *Problemy Peredači Informacii* **9**:4 (1973), 71–80. In Russian; translated in *Probl. Inf. Transm.* **9**:4 (1973), 325–332 (1975).

[Matthews et al. 1984] C. R. Matthews, L. N. Vaserstein, and B. Weisfeiler, "Congruence properties of Zariski-dense subgroups. I", *Proc. London Math. Soc.* (3) **48**:3 (1984), 514–532.

[Nikolov and Pyber 2011] N. Nikolov and L. Pyber, "Product decompositions of quasirandom groups and a Jordan type theorem", *J. Eur. Math. Soc. (JEMS)* **13**:4 (2011), 1063–1077.

[Nori 1987] M. V. Nori, "On subgroups of $\mathrm{GL}_n(\mathbf{F}_p)$", *Invent. Math.* **88**:2 (1987), 257–275.

[Pyber and Szabó 2010a] L. Pyber and E. Szabó, "Growth in finite simple groups of Lie type", preprint, 2010. arXiv 1001.4556

[Pyber and Szabó 2010b] L. Pyber and E. Szabó, "Growth in finite simple groups of Lie type of bounded rank", preprint, 2010. arXiv 1005.1858

[Pyber and Szabó 2014] L. Pyber and E. Szabó, "Growth in linear groups", pp. 253–268 in *Thin groups and superstrong approximation*, edited by H. Oh and E. Breuillard, Math. Sci. Res. Inst. Publ. **61**, Cambridge Univ. Press, Cambridge, 2014.

[Rapinchuk 2014] A. Rapinchuk, "Strong approximation for algebraic groups", pp. 269–298 in *Thin groups and superstrong approximation*, edited by H. Oh and E. Breuillard, Math. Sci. Res. Inst. Publ. **61**, Cambridge Univ. Press, Cambridge, 2014.

[Ruzsa 1994] I. Z. Ruzsa, "Generalized arithmetical progressions and sumsets", *Acta Math. Hungar.* **65**:4 (1994), 379–388.

[Salehi 2014] A. Salehi Golsefidy, "Affine sieve and expanders", pp. 325–342 in *Thin groups and superstrong approximation*, edited by H. Oh and E. Breuillard, Math. Sci. Res. Inst. Publ. **61**, Cambridge Univ. Press, Cambridge, 2014.

[Salehi and Varjú 2012] A. Salehi Golsefidy and P. P. Varjú, "Expansion in perfect groups", *Geom. Funct. Anal.* **22**:6 (2012), 1832–1891.

[Sanders 2012] T. Sanders, "On the Bogolyubov–Ruzsa lemma", *Anal. PDE* **5**:3 (2012), 627–655.

[Sanders 2013] T. Sanders, "The structure theory of set addition revisited", *Bull. Amer. Math. Soc. (N.S.)* **50**:1 (2013), 93–127.

[Sarnak 2014] P. Sarnak, "Notes on thin matrix groups", pp. 343–362 in *Thin groups and superstrong approximation*, edited by H. Oh and E. Breuillard, Math. Sci. Res. Inst. Publ. **61**, Cambridge Univ. Press, Cambridge, 2014.

[Sarnak and Xue 1991] P. Sarnak and X. X. Xue, "Bounds for multiplicities of automorphic representations", *Duke Math. J.* **64**:1 (1991), 207–227.

[Serre 1977] J.-P. Serre, *Linear representations of finite groups*, Graduate Texts in Mathematics **42**, Springer, New York, 1977.

[Tao 2008] T. Tao, "Product set estimates for non-commutative groups", *Combinatorica* **28**:5 (2008), 547–594.

[Tao 2009] T. Tao, "An elementary non-commutative Freiman theorem", blog entry, 2009, http://tinyurl.com/tao-freiman.

[Tao 2010] T. Tao, "Freiman's theorem for solvable groups", *Contrib. Discrete Math.* **5**:2 (2010), 137–184.

[Tao 2011] T. Tao, "Hamidoune's Freiman–Kneser theorem for nonabelian groups", blog entry, 2011, http://tinyurl.com/tao-hamidoune.

[Tao and Vu 2006] T. Tao and V. Vu, *Additive combinatorics*, Cambridge Studies in Advanced Mathematics **105**, Cambridge University Press, 2006.

[Tointon 2012] M. Tointon, "Freiman's theorem in an arbitrary nilpotent group", preprint, 2012. arXiv 1211.3989

[Varjú 2012] P. P. Varjú, "Expansion in $\mathrm{SL}_d(O_K/I)$, I square-free", *J. Eur. Math. Soc.* (*JEMS*) **14**:1 (2012), 273–305.

emmanuel.breuillard@math.u-psud.fr *Laboratoire de Mathématiques, Université Paris Sud 11, Bâtiment 425, 91405 Orsay, France*

Thin groups and superstrong approximation
MSRI Publications
Volume **61**, 2013

Superstrong approximation for monodromy groups

JORDAN S. ELLENBERG

Finitely generated subgroups of arithmetic lattices are frequently encountered in geometry, where they appear as *monodromy groups* attached to the variation of cohomology in families of manifolds or algebraic varieties. Here we survey briefly the construction of monodromy groups, discuss our (limited) knowledge about whether such groups are arithmetic, and summarize the results of [Ellenberg et al. 2012], which derive an application to arithmetic geometry from recent advances in superstrong approximation [Helfgott 2008; Pyber and Szabó 2011; Breuillard et al. 2011; Salehi Golsefidy and Varjú 2012]. We conclude by indulging ourselves in some speculations about more general contexts, asking: what are the interesting questions about "nonabelian superstrong approximation" and "superstrong approximation for Galois groups"?

1. What is monodromy?

The simplest example of monodromy is provided by the topological notion of a *covering space*. Suppose given a path-connected base space B, endowed with a choice of basepoint b, and let $f : X \to B$ be a covering space of degree n; that is, each point $x \in X$ has a neighborhood U such that $f^{-1}(U)$ is homeomorphic to n disjoint copies of U. In particular, $f^{-1}(b)$ consists of n points. A paradigmatic example is provided by $B = X = \mathbb{C}^\times$, where the map f is given by $z \mapsto z^n$, and $b = 1$.

Now let γ be a loop in B beginning and ending at b, and let $\tilde{b}$ be a point in $f^{-1}(b)$. It follows from the definition of covering space that there is a unique path $\tilde{\gamma}$ which starts at $\tilde{b}$ and projects to γ. Intuitively, one imagines the point in X "following its shadow" in B, whose motion is specified by γ. In the paradigmatic example, we can take $\gamma : [0, 1] \to B$ to be a counterclockwise traversal of the unit circle

$$\gamma(t) = e^{2\pi i t}$$

and $\tilde{b} = 1$. Then $\tilde{b}$ starts at 1, and at time t is required to satisfy

$$\tilde{\gamma}(t)^n = e^{2\pi i t}.$$

It follows that $\tilde{\gamma}(t) = e^{2\pi i t/n}$. The path $\tilde{\gamma}$ starts at 1, but it ends at $e^{2\pi i/n}$. In fact, starting at any point $\tilde{b}$ and following your shadow around γ puts you at $\tilde{b}e^{2\pi/n}$. The permutation of the n points in $f^{-1}(b)$ induced by this operation is called the *monodromy* of the cover around γ; in this case, it is a cyclic permutation of the n points. It is easy to check that the map sending a path γ based at b to its monodromy is a homomorphism

$$\mu : \pi_1(B,b) \to \mathrm{Sym}(f^{-1}(b)),$$

which we call the *monodromy map of the cover*. The image of μ is called the *monodromy group*.

Any identification of $f^{-1}(b)$ with the set $[1,\dots,n]$ allows us to express the monodromy of the cover as a homomorphism from the fundamental group $\pi_1(B,b)$ to S_n; of course, changing the labeling on $f^{-1}(b)$ changes the monodromy by conjugation. Note that the monodromy map need not surject onto the symmetric group of $f^{-1}(b)$; for instance, in the paradigmatic example, the monodromy group is cyclic of order n. For that matter, the monodromy group can be trivial — this happens precisely when X is isomorphic to the trivial cover $B \times [1\dots,n]$. The monodromy group is transitive if and only if X is connected; in this case, there is a well-defined conjugacy class of point stabilizers in the monodromy group, and the preimage in $\pi_1(B,b)$ of such a stabilizer is an index-n subgroup of $\pi_1(B,b)$. In this setting, we are just rephrasing the usual Galois-theoretic identification between connected degree-n covers of B and index-n subgroups of $\pi_1(B,b)$.

We can generalize this story: suppose now that $f : X \to B$ is a *fibration* of manifolds. Denote the fiber over $b \in B$ by X_b. One can take hold of a homology class in $H_i(X_b,\mathbb{Z})$ and "move it along γ" just as we did in above with a single point of X, and this defines a monodromy map

$$\mu : \pi_1(B,b) \to \mathrm{GL}(H_i(X_b,\mathbb{Z})),$$

Again, the *monodromy group* of the fibration (or *the monodromy group on* H_i, if there is some ambiguity) is the image of $\pi_1(B,b)$ in the arithmetic group $\mathrm{GL}(H_i(X_b,\mathbb{Z}))$. Of course, we can define a monodromy group using the action on cohomology in just the same way.

Rather than define this process rigorously, we will sketch an illustrative case. Suppose that B is the moduli space $\mathcal{M}$ of lattices in $\mathbb{R}^2$ up to homothety, and let X be the universal torus over $\mathcal{M}$; in other words, for each point p of $\mathcal{M}$ parametrizing a lattice $\Lambda_p \subset \mathbb{R}^2$, the fiber X_p is the torus $\mathbb{R}^2/\Lambda_p$.

Now let the basepoint b be the point parametrizing the square lattice Λ_b

generated by $(1,0)$ and $(0,1)$, and define a loop $\gamma \subset \mathcal{M}$ by letting $\gamma(t)$ be the lattice generated by $(1,0)$ and $(t,1)$. Note that $\gamma(1) = \gamma(0) = b$.

The path from $(0,0)$ to $(0,1)$ represents a nontrivial homology class in $H_1(X_b, \mathbb{Z})$, which we call e_1. Write e_2 for the homology class of the path from $(0,0)$ to $(1,0)$; then e_1 and e_2 freely generate $H_1(X_b, \mathbb{Z})$.

What happens as we move along γ, carrying e_1 along with us as we go? At time t, our path moves along a straight line segment from $(0,0)$ to $(t,1)$. And when the loop is finished, our path goes from $(0,0)$ to $(1,1)$. The torus has returned to its starting point, but the homology class on X_b represented by our path is now $e_1 + e_2$.

What if we'd started with the path from $(0,0)$ to $(1,0)$? This path stays in place through our whole trip around γ.

In other words, the loop γ induces a *unipotent* monodromy transformation on $H_1(X_b, \mathbb{Z})$, sending e_1 to $e_1 + e_2$ and fixing e_2.

Below we present a few examples of monodromy maps attached to families of manifolds.

Example 1. If $X \to B$ is a finite covering space of degree n, then X_b is a set of n points, and the only nontrivial homology group of X_b is $H_0(X_b) \cong \mathbb{Z}^n$. In this case, after choosing an identification of X_b with $1, \dots, n$, the monodromy map factors as

$$\mu : \pi_1(B, b) \to S_n \to \mathrm{GL}_n(\mathbb{Z}),$$

where the second map is the standard permutation representation.

Example 2. Let B be the moduli space of genus g Riemann surfaces, and let X be the universal curve over B; that is, for each point p of B, the fiber X_p is the Riemann surface parametrized by p. In this case, $\pi_1(B, b)$ is the *mapping class group* Γ_g of genus g. (See [Farb and Margalit 2011] for a thorough introduction to the mapping class group and its basic properties.) We can identify $H_1(X_b, \mathbb{Z})$ with $\mathbb{Z}^{2g}$; having done so, the monodromy map takes the form

$$\mu : \Gamma_g \to \mathrm{GL}_{2g}(\mathbb{Z}).$$

This map is definitely not surjective once $d > 2$, because $H_1(X_b, \mathbb{Z})$ carries a symplectic intersection form which is preserved when moving around in moduli space. So the image lies in the subgroup of matrices preserving this form, and we can write:

$$\mu : \Gamma_g \to \mathrm{Sp}_{2g}(\mathbb{Z}).$$

In fact, the image of μ is known to be all of $\mathrm{Sp}_{2g}(\mathbb{Z})$; see [Farb and Margalit 2011, §6.3.2], for example.

Example 3. Let B be the configuration space parametrizing unordered n-tuples of distinct elements of $\mathbb{C}$, so that $\pi_1(B, b)$ is the Artin braid group B_n on n strands. Let X be the family whose fiber over a configuration $\{p_1, \ldots, p_n\}$ is the smooth proper algebraic curve with affine model

$$y^d = (x - p_1)(x - p_2)\ldots(x - p_n).$$

Then the monodromy map

$$\mu : B_n \to \mathrm{GL}(H_1(X_b))$$

is a representation of the braid group, which is studied extensively in [McMullen 2013]. Each fiber X_b is an algebraic curve with a $\mathbb{Z}/d\mathbb{Z}$-action: namely, 1 sends y to $e^{2\pi i/d}y$ and leaves x unchanged. Thanks to the intersection form, the image of μ is again contained in the group of symplectomorphisms $H_1(X_b)$. By contrast with the previous example, the monodromy group is not the full symplectic group, because it commutes with the canonical action of $\mathbb{Z}/d\mathbb{Z}$ on the fiber. So the monodromy group is contained in the centralizer of $\mathbb{Z}/d\mathbb{Z}$ in the symplectic group; we denote this group by $\mathrm{Sp}_T(H_1(X_b))$.

Remark. Another classical route to monodromy groups is via solutions of differential equations. We won't pursue examples of this kind here, but see [Sarnak 2014; Fuchs 2014] in this volume for a fuller account.

2. Big monodromy and superstrong approximation

Given a fibration $X \to B$, we might aim to describe, as best we can, the corresponding monodromy group $\Gamma = \mu(\pi_1(B, b))$ in $\mathrm{GL}(H_i(X_b, \mathbb{Z}))$. A crucial invariant is the Zariski closure G of Γ, which is an algebraic group over $\mathbb{Q}$ contained in a general linear group GL_n, where $n = \dim_{\mathbb{Q}} H_i(X_b, \mathbb{Q})$. We refer to G as the *algebraic monodromy group*. By $G(\mathbb{Z})$ we shall mean the intersection of $G(\mathbb{C})$ with $\mathrm{GL}(H_i(X_b, \mathbb{Z}))$.[1]

As a first step, we would like to be able to say whether it is *big*. Of course, there are several senses in which the word "big" could be meant. For instance, we can ask whether G is all of GL_n. As we've seen, the answer is often negative, sometimes for readily perceived reasons — for instance, in the case where $X \to B$ is a family of genus g curves, the intersection form on $H_1(X_b, \mathbb{Z})$ constrains G to be contained in Sp_{2g}. So one might also ask, more vaguely, "is G as big as it could be, given all the reasons I can think of for it to be small"? For a typical family of curves, this amounts to asking whether G is the full symplectic group. In the context of Example 3, it would mean to ask whether G is the whole of Sp_T.

[1] In this article, we will always be in a context where $H_i(X_b, \mathbb{Z})$ is torsion-free, and where any subtleties about the "right" model for G over Spec $\mathbb{Z}$ do not intervene.

It is not straightforward to compute the algebraic monodromy group G in general, but much progress has been made towards ensuring that G is big under various checkable conditions. See [Katz 2004] and [Hall 2008] for representative recent results.

2.1. *Thin monodromy.* Even if G is known, there are still important questions about the size of Γ itself. First of all, we can compare Γ with the full lattice $G(\mathbb{Z})$. The critical question is whether Γ has finite index in $G(\mathbb{Z})$.

Definition 4. A subgroup Γ of $G(\mathbb{Z})$ is *thin* if it is Zariski-dense and has infinite index.

There is, of course, a large literature on the geometry, group theory, and dynamics of arithmetic lattices — the main theme of this conference is to understand the extent to which some of these results extend to thin subgroups.

It can be difficult, in practice, to know whether the monodromy group of a family is thin, and both thin and nonthin cases do arise. For instance, the family of hyperelliptic curves of genus g has monodromy group which is finite-index in $\mathrm{Sp}_{2g}(\mathbb{Z})$, by a theorem of A'Campo [1979]. On the other hand, the family of curves described in Example 3 can have either thin or nonthin monodromy depending on the parameters; for instance, McMullen [2013, Section 11] shows (following Deligne and Mostow) that the monodromy is thin when $(n, d) = (4, 18)$. On the other hand, Venkataramana [2013] shows that the monodromy is an arithmetic lattice whenever $n \geq 2d$. Brav and Thomas [2012] have recently shown that a family of quintic three-folds constructed by Dwork has as its H^3-monodromy a thin subgroup of $\mathrm{Sp}_4(\mathbb{Z})$. Fuchs's article [2014] in this volume describes several senses in which thinness seems to be the "generic" situation, including many examples of differential equations whose monodromy groups appear on experimental grounds to be thin.

Example 5. Consider the family of complex algebraic curves given by

$$X_t : y^2 = f(x)(x-t),$$

where $f(x)$ is a fixed squarefree polynomial of degree $2g$, with $g > 1$, and t varies over the base B, which is the complement of the roots of f in $\mathbb{A}^1$. The monodromy group Γ of this family is Zariski dense in Sp_{2g}; see unpublished work of Yu [1997] and the more general theorem of Hall [2008, Theorem 4.1]. In fact, Yu proves that the monodromy group of this family is a specified finite-index subgroup of $\mathrm{Sp}_{2g}(\mathbb{Z})$, contained in the principal congruence subgroup of level 2 and containing the principal congruence subgroup of level 4. We take this opportunity to correct an error in Section 4.1 of [Ellenberg et al. 2012], where we make the incorrect assertion that the monodromy group is the whole of $\Gamma(2)$. (This error does not affect the argument in which it appears.)

Yu computes the monodromy group by direct computation with matrices. Ian Agol explained to us a very handsome pure-thought argument via the Margulis normal subgroup theorem; we include this argument here in the hope that the idea might be useful in other similar cases.

Let Conf_{2g+1} be the configuration space parametrizing squarefree complex polynomials of degree $2g+1$. The fundamental group of Conf_{2g+1} is the Artin braid group Br_{2g+1}, and there is a natural family of genus-g curves over Conf_{2g+1} whose fiber over a polynomial $g(x)$ is the hyperelliptic curve $y^2 = g(x)$. By the theorem of A'Campo already mentioned, the image of the corresponding monodromy map

$$\mathrm{Br}_{2g+1} \to \mathrm{Sp}_{2g}(\mathbb{Z})$$

is a finite-index subgroup of $\mathrm{Sp}_{2g}(\mathbb{Z})$ containing the full level-2 congruence subgroup $\Gamma(2)$.

The permutation action on strands induces a map $\mathrm{Br}_{2g+1} \to S_{2g+1}$ whose kernel P_{2g+1} is called the *pure braid group*. The image of P_{2g+1} under the monodromy image above is precisely $\Gamma(2)$.

Now the base space B considered in the present example maps to Conf_{2g+1} by sending t to $f(x)(x-t)$, and the monodromy map $\pi_1(B,b) \to \mathrm{Sp}_{2g}(\mathbb{Z})$ thus factors as a composition

$$\mu : \pi_1(B,b) \to \pi_1(\mathrm{Conf}_{2g+1}, b) \cong \mathrm{Br}_{2g+1} \to \mathrm{Sp}_{2g}(\mathbb{Z}).$$

Topologically speaking, the image of $\pi_1(B,b)$ in Br_{2g+1} is the *point-pushing group* H consisting of all braids in which the first $2g$ strands are fixed in place, and the last is free to wind around the others. The point-pushing group is the kernel in the *Birman exact sequence*

$$H \to P_{2g+1} \to P_{2g} \to 1,$$

and in particular is a *normal* subgroup of the pure braid group. Thus, $\mu(H)$ is a normal subgroup of the lattice $\Gamma(2)$. It then follows form the Margulis normal subgroup theorem that $\mu(H)$ is either finite or finite-index; but it is easy to check that a loop carrying t once around a root of f has unipotent (and thus infinite-order) monodromy, so $\mu(H)$ must be finite index, as claimed.

To sum up: the monodromy group Γ can be "big" in the sense that its Zariski closure G is as large as possible, but it might well be the case that Γ is very "small" by virtue of being thin, and in a given explicit geometric situation we may not know which is the case. The question is then whether thin groups are "big enough" for applications.

2.2. ***Strong approximation.*** One may ask how well Γ approximates $G(\mathbb{Z})$ locally — that is, with respect to the p-adic Lie groups $G(\mathbb{Z}_p)$. A natural question in this vein is whether the inclusion $\Gamma \to G(\mathbb{Z}_p)$ is dense for almost all p; that is, does Γ satisfy *strong approximation?* Of course, this requires that $G(\mathbb{Z})$ itself satisfies strong approximation; this is the case whenever G is semisimple and simply connected and $G(\mathbb{R})$ is noncompact. It is a remarkable fact, due in its strongest form to Weisfeiler, that (under mild conditions on G) *any* finitely generated Zariski-dense subgroup Γ of $G(\mathbb{Z})$ satisfies strong approximation. For details and history concerning these results, see [Rapinchuk 2014] in this volume.

A central theme of the lectures in this workshop has been that, despite the name "strong approximation", density of Γ in almost all $G(\mathbb{Z}_p)$ is not the strongest local approximation condition we can place on Γ. Indeed, strong approximation is *not* strong enough to make sieving arguments work. For this we need a stronger condition, called *superstrong approximation*, which we explain below. See [Salehi Golsefidy 2014; Kowalski 2014] in this volume for a survey of the *affine sieve* and other arguments which use superstrong approximation to derive quantitative statements on thin subgroups of arithmetic groups.

2.3. ***Superstrong approximation.*** Let Γ be a finitely generated subgroup of $G(\mathbb{Z})$, and let $\underline{\gamma} = \gamma_1, \dots, \gamma_r$ be a set of generators for Γ. If $\pi : G(\mathbb{Z}) \to Q$ is a finite quotient of $G(\mathbb{Z})$, then we can form a Cayley graph $X(Q, \underline{\gamma})$, whose vertices are labeled by the elements of Q, and which has q adjacent to q' precisely when $q' = q\pi(\gamma_i)^{\pm 1}$ for some generator γ_i.

In what follows, we will let Q range over the congruence quotients $G(\mathbb{Z}/p\mathbb{Z})$, and we will refer to the Cayley graph $X(G(\mathbb{Z}/p\mathbb{Z}), \underline{\gamma})$ simply as X_p.

Note first that Γ surjects on $G(Z/p\mathbb{Z})$ if and only if X_p is connected. The notion of connectedness, in turn, can be addressed spectrally: the (suitably normalized) Laplacian operator on X_p always has an eigenvalue of 0, attached to the space of constant eigenfunctions, and all other eigenvalues ≥ 0. In fact, the number of components of X_p is precisely the multiplicity of the eigenvalue 0. In other words, the second-largest eigenvalue λ_1 is 0 precisely when X_p is not connected. And strong approximation, from a spectral point of view, says that $\lambda_1(X_p) = 0$ for only finitely many p.

Replacing this condition with a quantitative bound yields the notion of *superstrong approximation.*

Definition 6. Let Γ be a finitely generated subgroup of $G(\mathbb{Z})$ and define a sequence of Cayley graphs X_p as above. We say Γ has *superstrong approximation* if there exists a constant $\epsilon > 0$ such that $\lambda_1 > \epsilon$ for all sufficiently large primes p.

In other words, superstrong approximation is the condition that the X_p form a family of *expander graphs*. By the usual theory of expander graphs, this condition does not depend on the choice of the generating set $\underline{\gamma}$; in other words, it is a property of $\Gamma \subset G(\mathbb{Z})$, as our phrasing suggests. We note that superstrong approximation follows immediately if Γ has Kazhdan's property T, as it must if it is a lattice in a higher-rank arithmetic group. But when Γ is thin, superstrong approximation doesn't follow from any abstract group-theoretic property of Γ. Indeed, in practice Γ is often free, so has many actions on finite sets whose corresponding Cayley graphs have small λ_1; superstrong approximation demands that the $G(\mathbb{Z}/p\mathbb{Z})$ are not among those finite sets.

The remarkable progress of the last few years towards uniform bounds for expansion in Cayley graphs of linear groups, as described elsewhere in this volume [Pyber and Szabó 2014], has brought us to the point where we now have a superstrong analogue of Weisfeiler's theorem; under mild conditions on the ambient algebraic group G, thin groups have superstrong approximation merely by virtue of their Zariski density in G.

3. Application: gonality growth in families of covers

We now describe an arithmetic application of superstrong approximation for monodromy groups. All the material here is explained in substantially more detail in our joint paper with Hall and Kowalski [Ellenberg et al. 2012].

Let U be an smooth (but not necessarily proper) algebraic curve over a number field k, and let $A \to U$ be an abelian scheme over U (i.e., a family of abelian varieties parametrized by U.) If t is a point of $U(\bar{k})$, we denote by A_t the fiber of the family over t; it is an abelian variety defined over the finite extension $k(t)/k$. If n is an integer, $A_t[n]$ denotes the group of n-torsion points of A_t.

For instance, we could fix a squarefree polynomial f, let U be the complement in $\mathbb{A}^1$ of the roots of f, and take A to be the family whose fiber A_t is the Jacobian of the hyperelliptic curve $y^2 = f(x)(x-t)$. (Though be warned that to prove Theorem 7 in this case doesn't use new results on thin groups, since in this case the monodromy of the family is finite-index in $\mathrm{Sp}_{2g}(\mathbb{Z})$, as explained in Example 5.)

The next result follows from Theorem 7 of [Ellenberg et al. 2012].

Theorem 7. *For every $d \geq 1$ there exists a constant $l(d)$ such that, for all primes $l > l(d)$, there are only finitely many $t \in U(\bar{k})$ such that $[k(t) : k] \leq d$ and $A_t[l](k(t))$ contains a nontrivial point.*

In words, "there are only finitely many l-torsion points on fibers of A over points of bounded degree".

We now sketch the proof. The first step is to construct finite covers of U as follows. For each integer n, let $f_n : U_n \to U$ be the map whose fiber is the finite set of nontrivial n-torsion points $A_t[n]$; this is a finite unramified covering of degree $n^{2\dim A} - 1$. One might think of these covers as analogues of the classical modular curves; for if U is the moduli space of elliptic curves $Y(1)^2$, the cover U_n is precisely the moduli space $Y_1(n)$ of elliptic curves with an n-torsion point. In fact, the argument below is very much in the spirit of those of [Abramovich 1996] and [Zograf 1984] which gave a lower bound for the gonality of the classical modular curves; where we use new developments in superstrong approximation, they use Selberg's theorem that (in modern language) $SL_2(\mathbb{Z})$ has property τ with respect to the family of congruence subgroups.

Theorem 7 is really a statement about algebraic points on the covers U_l: we are saying that, for all $l > l(d)$, the curve U_l has only finitely many points whose field of definition has degree at most d over k.

This statement might at first seem rather strong, since there are infinitely many degree-d field extensions of k. Even if a curve C has genus at least 2, so that $|C(k')| < \infty$ for each k'/k of degree d by Faltings' theorem, it is far from clear that the union of $C(k')$ over *all* such k' should be finite. Indeed, this is often not the case: for instance, the curve

$$y^2 = x^5 + x + 1$$

has infinitely many points over quadratic extensions of $\mathbb{Q}$, since one may produce such a point by specifying $x \in \mathbb{Q}$ arbitrarily. Of course, the story is the same for any hyperelliptic curve, which may have arbitrarily large genus. So the genus is not sufficient to distinguish curves with many degree-d algebraic points from curves with few such points. For that purpose, we need to study a different geometric invariant: the *gonality*.

Definition 8. The *gonality* of an algebraic curve C/k is the minimum integer γ such that there exists a morphism $C \to \mathbb{P}^1$ of degree γ.

(In this section, we take gonality to be a *geometric* invariant of the curve; that is, in the definition we allow the morphism $C \to \mathbb{P}^1$ to be defined over an extension of k.)

For example, a curve has gonality 1 precisely when it has genus 0, and gonality 2 precisely when it is hyperelliptic but not rational.

The gonality of a curve has a very strong effect on its points over fields of bounded degree over a fixed field, as we know from [Faltings 1983; Abramovich and Voloch 1996; Frey 1994]. Namely, if C/k has gonality greater than $2d$, then $C(\bar{k})$ has only finitely many points t such that $[k(t) : k] \leq d$.

[2] Warning: $Y(1)$ is not really an algebraic curve but a stack with nontrivial generic inertia. So it would be more precise to say that when $U = Y_1(p)$ for $p > 3$ and not dividing n, we have $U_n = Y_1(np)$.

To prove Theorem 7, it thus suffices to show that the gonality $\gamma(U_l)$ increases with l. This is accomplished using two main tools. First, we pass to the complex numbers and invoke an analytic theorem of Li and Yau [1982] which provides lower bounds for the gonality of a complex algebraic curve in terms of the spectrum of the Laplacian. In particular, it is enough to show that the family of Riemann surfaces U_l has a Laplacian spectral gap.

Why should this be so? This requires a combinatorial detour. The argument runs roughly as follows. Choose a basepoint $u \in U(\mathbb{C})$ and a basis $\gamma_1, \dots, \gamma_r$ for $\pi_1(U(\mathbb{C}), u)$. (For simplicity we assume here that U is a punctured sphere, so that $\pi_1(U)$ is free.) We can represent each γ_i by a loop in $U(\mathbb{C})$ based at u, and we can arrange our loops to be mutually disjoint. We denote the union of these loops by X; it is a one-vertex graph embedded in the Riemann surface $U(\mathbb{C})$.

Now the preimage of X under the cover $U_l \to U$ is a graph $X(U_l)$ embedded in the Riemann surface $U(\mathbb{C})$. The vertices of $X(U_l)$ are the points of U_l lying over u; we recall that these correspond to the $l^{2g} - 1$ nontrivial l-torsion points on A_u.

On the other hand, we have a monodromy representation

$$\mu : \pi_1(U(\mathbb{C}), u) \to \mathrm{Sp}(H_1(A, \mathbb{Z})) \cong \mathrm{Sp}_{2g}(\mathbb{Z}).$$

The l-torsion in A_u is canonically identified with the mod-l homology group $H_1(A_u, \mathbb{Z}/l\mathbb{Z})$. So what happens if we start at a point $\tilde{u}$ in $U_l(\mathbb{C})$ lying over u, and make $\tilde{u}$ "follow its shadow" around a loop γ_i in $U(\mathbb{C})$? The description of monodromy from the first section tells us that the path ends at $\mu(\gamma_i)\tilde{u}$. In other words, the permutation on the preimages of u induced by γ_i is the same as that induced by the reduction mod l of the symplectomorphism $\mu(\gamma_i)$. This is precisely the same as saying that two vertices $\tilde{u}$ and $\tilde{u}'$ of $X(U_l)$ are adjacent precisely when $\tilde{u}' = \mu(\gamma_i)\tilde{u}$ for some i. In other words, $X(U_l)$ is a quotient of the Cayley graph X_l we described in Section 2.

This is the crucial point. For a given family of abelian variety we may have no easy way of checking whether the monodromy group Γ is thin or arithmetic. But it doesn't matter! The Helfgott, Pyber–Szabó, Breuillard–Green–Tao, and Golsefidy–Varjú theorems work equally well in either case to tell us that Γ has superstrong approximation. It is precisely this robustness that makes Theorem 7 possible.

Superstrong approximation for Γ tells us that the X_l have a spectral gap; that is, the nontrivial eigenvalues of the *discrete* Laplacians on these graphs is bounded away from 0 as l grows. (Hidden here is a certain amount of work necessary to pin down the possibilities for G, the Zariski closure of monodromy; this is necessary in order to ensure that superstrong approximation applies. This is covered in Section 5 of [Ellenberg et al. 2012].)

What is the relationship between the Laplacian of the graph and the geometry of the surface in which the graph is embedded? It can be illustrative to visualize the case where all the U_l have genus 0; that is, they are homeomorphic to the sphere. Now a graph embedded in a sphere is a planar graph, and it is known that a sequence of planar graphs with increasingly many vertices cannot have a spectral gap; intuitively, graphs with a spectral gap have very quickly dispersing random walks, while the random walk on a planar graph is constrained to linger for a while in the neighborhood of the plane where it starts out.

More generally, we should expect X_l to serve as a kind of combinatorial approximation to the geometry of U_l. Speaking roughly: the graph is like a wire frame along which U_l is stretched, and the graph separates U_l into many small components, each one of which is a copy of a patch on U. So the geometry of U_l is, in a sense, a hybrid of the geometry of X_l and the geometry of U. As l grows, and along with it the number of vertices of X_l, the graph plays a proportionately greater and greater role in the geometry of U_l and the shape of U a smaller and smaller one.

This intuition is beautifully confirmed and extended by results of Kelner, Brooks, and Burger, which together provide us with a "comparison principle" between the Laplacian spectrum of the Riemann surface $U_l(\mathbb{C})$ and the (discrete) Laplacian spectrum of the embedded graph X_l. In particular, the combination of their theorems shows that the sequence of Riemann surfaces $U_l(\mathbb{C})$ has a spectral gap whenever the sequence of graphs X_l does. This completes the proof.

Remarks. (1) For the arithmetic applications in [Ellenberg et al. 2012], the full strength of superstrong approximation was not needed, and we used a weaker condition instead. We call a sequence of graphs $\{X_l\}$ *esperantist* when $\lambda_1(X_l)$ decays no more quickly than a power of the logarithm of $|X_l|$. At the time [Ellenberg et al. 2012] was written, results guaranteeing superstrong approximation in the required generality were not available, but the results of [Pyber and Szabó 2011] established the esperantist condition in the contexts we needed. Though we now have superstrong approximation for thin groups in much greater generality, it still seems useful to keep in mind that in applications it is often sufficient to have at hand the easier condition on Γ that the Cayley graphs attached to Γ are esperantist ("slightly superstrong approximation..."?) On the other hand, the full strength of expansion allows one to conclude, not only that the gonality of U_l grows without bound as $l \to \infty$, but that the gonality is bounded below by a constant multiple of $\deg(U_l/U)$; in this case we say the gonality has *linear growth*.

(2) Many more applications of the basic method appear in [Ellenberg et al. 2012]. For instance, if the monodromy group Γ is Zariski dense in $\mathrm{Sp}_{2g}(\mathbb{Z})$, it follows

from Theorem 4 of [Ellenberg et al. 2012] that there are only finitely many t of degree at most d over k such that A_t is isogenous to a product of abelian varieties of smaller dimension.

(3) The theorems in [Ellenberg et al. 2012] require that the families of abelian varieties be defined over fields of characteristic 0, since everything rests on the ability to compare the geometry of Riemann surfaces with that of graphs drawn thereon. The *statements* of the theorems, however, largely make sense in characteristic p; it seems very interesting to wonder whether the results continue to be true in this case, and in particular whether any analogue of the argument by superstrong approximation can be made to work.

(4) Work of Cadoret and Tamagawa, carried out independently around the same time as [Ellenberg et al. 2012], gives lower bounds for the gonality of some of the covers discussed here, by very different methods [Cadoret and Tamagawa 2009; Cadoret 2012]. In particular, their results are purely algebraic, and can be applied in characteristic p as well as characteristic 0.

4. Speculations: nonabelian superstrong approximation

We can phrase the approximation criteria of the previous sections in a different way. Think of $\mathrm{GL}_n(\mathbb{Z})$ as the automorphism group of the free abelian group $\mathbb{Z}^n$. Then, for any finite abelian group A which can be generated by n elements, $\mathrm{GL}_n(\mathbb{Z})$ acts by permutations on $\mathrm{Epi}(\mathbb{Z}^n, A)$, the set of surjections from $\mathbb{Z}^n$ to A: we thus have a homomorphism

$$\mathrm{Aut}\,\mathbb{Z}^n \to \mathrm{Sym}(\mathrm{Epi}(\mathbb{Z}^n, A)).$$

Strong approximation for a subgroup Γ of $\mathrm{GL}_n(\mathbb{Z})$ is essentially the statement that the restriction of this action to Γ is *transitive* for all A; equivalently, the Cayley–Schreier graphs attached to this action are connected. Superstrong approximation, on the other hand, implies the stronger statement that those graphs form an expander family.

It is natural to ask about nonabelian analogues of this story; for instance, one can replace $\mathbb{Z}^n$ with a free group F_n on n generators, and replace the finite abelian group A with a general finite group G. It turns out that the analogues of strong and superstrong approximation have already appeared in various interesting contexts.

4.1. ***The product replacement algorithm.*** (Main references: [Lubotzky and Pak 2001; Lubotzky 2011].) If G is a finite group, then $\mathrm{Epi}(F^n, G)$ is precisely the set of n-element generating sets $(g_1, \ldots, g_n)$ of G. The *product replacement algorithm* of Leedham-Green and Soicher is a random walk on this set, which proceeds as follows: choose distinct i, j in $1, \dots, n$ at random, and then multiply g_i either on the left or right (chosen by coin flip) by g_j or g_j^{-1} (also chosen by coin flip.)

This random walk appears in practice to converge extremely quickly to the uniform distribution on generating n-tuples, making it an effective way to generate random elements in a group without having to enumerate the entire group. In fact, the product replacement algorithm is known to converge to uniform distribution in polynomial time when n is allowed to grow (slowly) with $|G|$, by a result of Pak [2000]. In the present paper we will concentrate mostly on the case where n is fixed and G changes.

The product replacement graph is in fact a Cayley–Schreier graph for the permutation action

$$A^+ \to \mathrm{Sym}(\mathrm{Epi}(F^n, G)),$$

where A^+ is a subgroup of Aut F_n of index 2, generated by automorphisms corresponding to the $4n(n-1)$ operations of the product replacement algorithm. (We will casually fail to distinguish between Aut F_2 and its finite-index subgroup A^+ in what follows.) The rapid convergence of the random walk naturally suggests the question: are these graphs expanders? In other words, if $G_1, G_2, \dots$ is a sequence of finite groups, we may ask:

Question 9. Does Aut F^n have superstrong approximation with respect to its image in $\mathrm{Sym}(\mathrm{Epi}(F^n, G_i))$?

The above question would certainly have a positive answer if Aut F^n had property T (or even property τ for a suitable class of finite quotients.) Under such a hypothesis, quantitative bounds for the performance of the product replacement algorithm are given in [Lubotzky and Pak 2001]. Whether Aut F^n has property T in general is a well-known open question, about whose answer there is no clear consensus. We know that Aut F^n does *not* have property T when $n = 2, 3$ [Grunewald and Lubotzky 2009].

One very interesting partial result was proved in [Gamburd and Pak 2006]: Question 9 has a positive answer for the family of groups $\mathrm{PSL}_2(\mathbb{F}_p)$ under the hypothesis that $\mathrm{PSL}_2(\mathbb{F}_p)$ has *uniform expansion*; that is, there is a constant $\epsilon > 0$ such that, for all p, and all Cayley graphs of $\mathrm{PSL}_2(\mathbb{F}_p)$ with a specified number of generators, the spectral gap is at least ϵ. A very recent theorem of Breuillard and Gamburd shows that there is a family P of primes of density 1 such that $\mathrm{PSL}_2(\mathbb{F}_p)$ has uniform expansion as p ranges over P, but the general case remains mysterious.

It's important to keep in mind that, for all our talk about superstrong approximation, even the prior question of strong approximation is poorly understood in the nonabelian case. Recall that strong approximation has to do with the *transitivity* of the action on the set of epimorphisms. There are many (G, n) for which we know that Aut F_n acts transitively on $\mathrm{Epi}(F^n, G)$. But there are many examples of (G, n) such that the action is nontransitive; see [Pak 2001; Lubotzky 2011] for a full discussion of what is known and what is conjectured about this problem.

So far, no thin groups have appeared in this discussion. It is not even precisely clear how the notion of "thin subgroup" of Aut F_n should be defined. Still, we can imagine what kind of questions might be asked in this setting. For instance, one could ask about the following strengthening of Question 9.

Question 10. Let Γ be a finitely generated subgroup of Aut F_n whose image in $\mathrm{Sym}\big(\mathrm{Epi}(F^n, \mathrm{PSL}_2(\mathbb{F}_p))\big)$ is the same as that of Aut F_n for all p. Do the Cayley–Schreier graphs of the action of Γ on $\mathrm{Epi}(F^n, \mathrm{PSL}_2(\mathbb{F}_p))$ form an expander family?

It goes without saying that it would be of interest to ask all of the above questions with Aut F_n replaced by the braid group on n strands — or, for that matter, the mapping class group of any oriented surface whose fundamental group is isomorphic to F_n. This is more in the spirit of monodromy, since mapping class groups, unlike automorphism groups of free groups, are fundamental groups of natural moduli spaces — namely, moduli spaces of punctured curves.

4.2. ***Square-tiled surfaces and approximation for* Aut F_2.** The action of

$$\text{Aut } F_2 \quad \text{on } \mathrm{Epi}(F_2, G)$$

also appears in the theory of *square-tiled surfaces*. A square-tiled surface is just a connected degree-d cover of a torus branched at a single point. Such a cover can be thought of as a map from $\pi_{1,1}$ to S_d, where $\pi_{1,1}$ is the fundamental group of the once-punctured torus; it is a free group on 2 generators.

There is a natural obstruction to transitivity of the action of Aut F_2 on $\mathrm{Hom}(F_2, S_d)$; namely, the conjugacy class of the puncture on the torus (alternately: the commutator of a pair of free generators of F_2) is preserved up to conjugacy by Aut F_2. Since we have not demanded that our homomorphisms be surjective, we can break down $\mathrm{Hom}(F_2, S_d)$ further according to the image of the homomorphism.

Precisely: Let μ be a conjugacy class of S_d, that is, a partition of d, and let G be a transitive permutation group of rank d (i.e., a transitive subgroups of S_d defined up to conjugacy.) Then we denote by $Hom_\mu(F_2, S_d, G)$ the set of homomorphisms from F_2 to S_d which send the puncture class to μ and whose image is G. This set also carries an action of S_d, by conjugation on the right, and we denote the set of orbits of this action by $H^1_\mu(F_2, S_d, G)$.

Now Aut F_2 acts on $H^1_\mu(F_2, S_d, G)$, and in fact this action factors through its quotient Out F_2. We denote the associated Cayley–Schreier graph by $X(d, \mu, G)$, and we can now ask questions of the same genre we have considered before: for which (d, μ) is $X(d, \mu, G)$ connected? For which sequences of (d, μ, G) do the $X(d, \mu, G)$ form an expander family?

We remark that Out F_2 is naturally identified with $\mathrm{GL}_2(\mathbb{Z})$ via its action on $F_2^{\mathrm{ab}} = \mathbb{Z}^2$, and its index 2 subgroup is in turn naturally identified with the mapping class group $\Gamma_{1,1}$ of the once-punctured torus, which is the fundamental group of $\mathcal{M}_{1,1}$, the moduli space of genus-1 curves with a single marked point. Under these identifications, the action of $\Gamma_{1,1}$ on the finite set $H^1_\mu(F_2, S_d, G)$ corresponds to a finite cover of $\mathcal{M}_{1,1}$; this is precisely a moduli space $\mathcal{H}_{d,\mu,G}$ of square-tiled surfaces, where the map $\mathcal{H}_{d,\mu,G} \to \mathcal{M}_{1,1}$ sends a square-tiled surface to the punctured torus which it covers.

The moduli space $\mathcal{H}_{d,\mu,G}$ is a *Hurwitz space* and is also an example of a *Teichmüller curve*. Their geometry and their relationship with moduli spaces of higher genus curves has been much studied — see, for example, [McMullen 2005; Chen 2011]. Any superstrong approximation result for a family of $X(d, \mu, G)$ could in principle be used to give lower bounds for the gonality of $\mathcal{H}_{d,\mu,G}$ along the lines explained in the previous section. And of course one could go further and ask about approximation results for infinite-index subgroups of $\Gamma_{1,1}$ (though it is not clear at first glance whether such results have interesting implications in this context.)

We note that it is not possible that only finitely many $\mathcal{H}_{d,\mu,G}$ have a component with low gonality, for the simple reason that every algebraic curve over a number field is birational to a connected component of $\mathcal{H}_{d,\mu,G}$ for some G [Ellenberg and McReynolds 2012]. So any general statement of growing gonality would have to be restricted to a special family of (d, μ, G).

5. Speculations: superstrong approximation for Galois groups and the Bogomolov property

In Section 3 we explained how results on superstrong approximation could provide lower bounds for gonality in a family of covers of algebraic curves over characteristic 0 fields. The group to which the approximation results were applied was the image of the topological monodromy group of the base (a free group) in the arithmetic lattice $\mathrm{Sp}_{2g}(\mathbb{Z})$.

If one tries to do the same thing over a more general base — for instance, a base in characteristic p — one immediately encounters the problem that there is no longer a *discrete* fundamental group; all that remains is the étale fundamental group, a topologically finitely generated profinite group.

What might superstrong approximation mean for a profinite, rather than a discrete group? Let Π be a profinite group which is topologically finitely generated. If $\pi_1, \dots, \pi_k$ are topological generators, the subgroup Γ they generate is dense in Π. We say Γ has property τ with respect to Π if the Cayley graphs induced by Γ on the finite quotients of Π form an expander family. We might wish to say that Π satisfies superstrong approximation if the dense subgroup Γ has property τ with respect to Π. But there's a problem; it follows from a result of Alon, Lubotzky and Wigderson [Alon et al. 2001] show that there are profinite groups Π and dense finitely generated subgroups $\Gamma, \Gamma' \subset \Pi$ such that Γ has property τ with respect to Π, but Γ' does not. Kassabov (personal communication) has shown that this can hold even when the profinite completions of Γ and Γ' are actually *isomorphic* to Π.

One can instead place on Π the stronger condition that *every* dense finitely generated subgroup Γ has property τ with respect to Π.[3] In this case we say that Π has property $\hat{\tau}$. This is very close to asking that the Cayley graphs attached to generating sets of size k of finite quotients Q of Π satisfy a spectral gap which is independent of Q. In other words, Π needs to have *uniform expansion* as discussed in the previous section. Uniform expansion is for the moment out of reach, but it is widely believed to hold, for example, when $\Pi = \mathrm{Sp}_{2g}(\hat{\mathbb{Z}})$, the relevant case for the current discussion.

Suppose $\mathrm{Sp}_{2g}(\hat{\mathbb{Z}})$ has property $\hat{\tau}$, let U be a complex algebraic curve, and let $\rho : \pi_1(U(\mathbb{C}), u) \to \mathrm{Sp}_{2g}(\hat{\mathbb{Z}})$ be a representation of the fundamental group of U. For each l, the reduction $\bar{\rho}_l : \pi_1(U(\mathbb{C}), u) \to \mathrm{Sp}_{2g}(\mathbb{Z}/l\mathbb{Z})$ gives rise to a cover U_l of U, and the arguments of the previous section show that the U_l grow in gonality.

One might ask the same question for a curve U over a finite field $\mathbb{F}_q$. We slightly modify the question in order to avoid introducing the machinery of the étale fundamental group. Suppose given, for an infinite sequence of primes l, an étale cover $U_l \to U$ such that $(U_l)_{\bar{\mathbb{F}}_q} \to U_{\bar{\mathbb{F}}_q}$ is a Galois cover with Galois group $\mathrm{Sp}_{2g}(\mathbb{Z}/l\mathbb{Z})$. For instance, you might make such a cover by adjoining the l-torsion points of an abelian g-fold A over U.

Question 11. Is the gonality of U_l bounded below by a positive constant multiple of $[U_l : U]$? What if we impose the extra condition that the U_l are obtained by adjoining the l-torsion points on a fixed abelian g-fold A/U?

We have focused on the case of curves over finite fields because of an intriguing connection with a different topic in number theory, which we now explain. Write K_l for the function field of U_l. To say U_l has gonality greater than n is to say

[3]There is a natural temptation to call this criterion *superduperstrong approximation*, but this terminology is perhaps better kept quarantined in a footnote.

that there is no rational function f on U_l with at most n poles and n zeroes. This is equivalent to the assertion that there is no nonconstant element f of K_l^* whose (logarithmic) Weil height $h_{K_l}(f)$ is at most n. The *absolute height* of an element f of an extension of K_l is just $h_{K_l}(f)/[U_l : U]$. In other words, Question 11 asks whether the *absolute* height of an element of K_l^* is bounded below by a constant if it is nonzero.

This condition already has a name — it is called the *Bogomolov property*.

Definition 12. Let K be a global field. A set S of algebraic numbers over K has the *Bogomolov property* if the set of positive absolute heights of elements of S is bounded away from 0.

Note that this definition makes sense not only when K is a function field (as in the case described above) but also when K is a number field; in fact, it is the case $K = \mathbb{Q}$ that has been most energetically studied. Some sets known to have the Bogomolov property over $\mathbb{Q}$ are the field of totally real algebraic numbers (Schinzel), the maximal abelian extension of $\mathbb{Q}$ (Amoroso–Zannier) and the maximal extension of $\mathbb{Q}$ totally split at a finite place p (Bombieri–Zannier). The recent paper of Amoroso, David, and Zannier [Amoroso et al. 2011] contains a good summary of what is known along with several striking new results.

Speculation piled atop speculation: If we believe that $\mathrm{Sp}_{2g}(\mathbb{F}_l)$ has uniform expansion, and if we believe by analogy with the complex case that uniform expansion of Galois groups implies the Bogomolov property for the corresponding extensions of function fields, might this be the case for number fields as well? There are many ways one could analogize Question 11 to the number field case. Here is a minimalist version:

Question 13. Let $A/\mathbb{Q}$ be an abelian variety and let K_l be the number field $\mathbb{Q}(A[l])$. Does the compositum of all the K_l have the Bogomolov property?

And here is a maximalist version:

Question 14. Let F/Q be an infinite Galois extension such that $\mathrm{Gal}(F\mathbb{Q}^{\mathrm{ab}}/\mathbb{Q}^{\mathrm{ab}})$ has property $\hat{\tau}$. Does F have the Bogomolov property?

A striking result of Habegger [2013] shows that the answer to Question 13 is affirmative when A is an elliptic curve. The theorems from [Ellenberg et al. 2012] described in Section 3 can be used to show that the analogue of Question 13 with $\mathbb{Q}$ replaced by $\mathbb{C}(t)$ has an affirmative answer as well; this requires lower bounds for the gonality, not only of the covers U_l, but of fiber products of multiple U_{l_i} over U. In the case of $\mathbb{C}(t)$, the relevant expansion theorem (which was not available when [Ellenberg et al. 2012] was written) came with [Salehi Golsefidy and Varjú 2012]. It shows that expansion holds not only for the Cayley graphs associated to reduction modulo primes, but also modulo arbitrary squarefree integers.

As for the maximalist question, its interest is somewhat diminished by our lack of examples of groups known to have property $\hat{\tau}$. This need not keep us from speculating. For instance, S be a finite set of primes, and let K_S^t be the maximal extension of $\mathbb{Q}$ which is unramified outside S and *tamely* ramified at S, and let G_S^t be its Galois group. Then G_S^t has property *FAb*, which is to say that every finite-index subgroup of G_S^t has finite abelianization. This makes G_S^t a reasonable candidate for having some form of spectral expansion property. Caution is required, since G_S^t is only conjectured to be topologically finitely generated. One can circumvent this problem as follows. For each integer k, we can consider the family of all Cayley graphs on finite quotients G of G_S^t endowed with subsets $(g_1, \ldots, g_k)$ generating G. We say that G_S^t has $\hat{\tau}_k$ if there is a uniform spectral gap for all such Cayley graphs.

Question 15. Does G_S^t have property $\hat{\tau}_k$ for all k? Does K_S^t have the Bogomolov property?

Note that for G_S^t to have property $\hat{\tau}_2$ would require that there are only finitely many n such that G_S^t has a quotient isomorphic to S_n. It is not even known whether there are arbitrarily large S_n-extensions of $\mathbb{Q}$ unramified away from S; a recent preprint of Roberts and Venkatesh [$\geq$ 2012] argues that such extensions might indeed exist and proposes a construction, but at the same time suggests that these extensions are *not* likely to be tamely ramified at the primes in S. Another hint that the first question is not completely unreasonable comes from Ershov's work in [2011] on Golod–Shafarevich groups with infinite quotients with property T. But we should emphasize that we have no strong opinion as to the answer to either question.

Acknowledgements

This document is an expanded version of a lecture presented by the author at a conference on "Thin Groups and Superstrong Approximation" held at the Mathematical Sciences Research Institute in February 2012. The work described here was partially supported by NSF Grant DMS-1101267 and a Romnes Faculty Fellowship. The author is grateful to Ian Agol, Frank Calegari, and Peter Sarnak for useful conversations about the mathematics discussed here, and to MathOverflow for useful input concerning the image of the point-pushing subgroup in the symplectic group and the meaning of superstrong approximation for Galois groups.

References

[Abramovich 1996] D. Abramovich, "A linear lower bound on the gonality of modular curves", *Int. Math. Res. Not.* **1996**:20 (1996), 1005–1011.

[Abramovich and Voloch 1996] D. Abramovich and J. F. Voloch, "Lang's conjectures, fibered powers, and uniformity", *New York J. Math.* **2** (1996), 20–34.

[A'Campo 1979] N. A'Campo, "Tresses, monodromie et le groupe symplectique", *Comment. Math. Helv.* **54**:2 (1979), 318–327.

[Alon et al. 2001] N. Alon, A. Lubotzky, and A. Wigderson, "Semi-direct product in groups and zig-zag product in graphs: Connections and applications", pp. 630–637 in *42nd IEEE Symposium on Foundations of Computer Science* (Las Vegas, NV, 2001), IEEE Computer Soc., Los Alamitos, CA, 2001.

[Amoroso et al. 2011] F. Amoroso, S. David, and U. Zannier, "On fields with the property (b)", preprint, 2011, http://hal.archives-ouvertes.fr/docs/00/64/99/54/PDF/FieldsB.pdf. To appear in *Proc. Amer. Math. Soc.*

[Brav and Thomas 2012] C. Brav and H. Thomas, "Thin monodromy in Sp(4)", preprint, 2012. arXiv 1210.0523

[Breuillard et al. 2011] E. Breuillard, B. Green, and T. Tao, "Approximate subgroups of linear groups", *Geom. Funct. Anal.* **21**:4 (2011), 774–819.

[Cadoret 2012] A. Cadoret, "Note on the gonality of abstract modular curves", pp. 89–106 in *The arithmetic of fundamental groups*, edited by J. Stix, Contributions in Mathematical and Computational Sciences **2**, 2012.

[Cadoret and Tamagawa 2009] A. Cadoret and A. Tamagawa, "A uniform open image theorem for l-adic representations, II", preprint, 2009, http://www.math.polytechnique.fr/~cadoret/uoi2_Final.pdf.

[Chen 2011] D. Chen, "Square-tiled surfaces and rigid curves on moduli spaces", *Adv. Math.* **228**:2 (2011), 1135–1162.

[Ellenberg and McReynolds 2012] J. S. Ellenberg and D. B. McReynolds, "Arithmetic Veech sublattices of SL(2, **Z**)", *Duke Math. J.* **161**:3 (2012), 415–429.

[Ellenberg et al. 2012] J. S. Ellenberg, C. Hall, and E. Kowalski, "Expander graphs, gonality, and variation of Galois representations", *Duke Math. J.* **161**:7 (2012), 1233–1275.

[Ershov 2011] M. Ershov, "Kazhdan quotients of Golod–Shafarevich groups", *Proc. Lond. Math. Soc.* (3) **102**:4 (2011), 599–636.

[Faltings 1983] G. Faltings, "Endlichkeitssätze für abelsche Varietäten über Zahlkörpern", *Invent. Math.* **73**:3 (1983), 349–366.

[Farb and Margalit 2011] B. Farb and D. Margalit, *A primer on mapping class groups*, Princeton Mathematical Series **49**, Princeton University Press, 2011.

[Frey 1994] G. Frey, "Curves with infinitely many points of fixed degree", *Israel J. Math.* **85**:1-3 (1994), 79–83.

[Fuchs 2014] E. Fuchs, "The ubiquity of thin groups", pp. 73–92 in *Thin groups and superstrong approximation*, edited by H. Oh and E. Breuillard, Math. Sci. Res. Inst. Publ. **61**, Cambridge, New York, 2014.

[Gamburd and Pak 2006] A. Gamburd and I. Pak, "Expansion of product replacement graphs", *Combinatorica* **26**:4 (2006), 411–429.

[Grunewald and Lubotzky 2009] F. Grunewald and A. Lubotzky, "Linear representations of the automorphism group of a free group", *Geom. Funct. Anal.* **18**:5 (2009), 1564–1608.

[Habegger 2013] P. Habegger, "Small height and infinite non-abelian extensions", preprint, 2013. arXiv 1109.5859

[Hall 2008] C. Hall, "Big symplectic or orthogonal monodromy modulo l", *Duke Math. J.* **141**:1 (2008), 179–203.

[Helfgott 2008] H. A. Helfgott, "Growth and generation in $SL_2(\mathbf{Z}p\mathbf{Z})$", *Ann. of Math.* (2) **167**:2 (2008), 601–623.

[Katz 2004] N. M. Katz, "Larsen's alternative, moments, and the monodromy of Lefschetz pencils", pp. 521–560 in *Contributions to automorphic forms, geometry, and number theory* (Baltimore, MD, 2002), edited by H. Hida et al., Johns Hopkins Univ. Press, Baltimore, MD, 2004.

[Kowalski 2014] E. Kowalski, "Sieves in discrete groups, especially sparse", pp. 107–140 in *Thin groups and superstrong approximation*, edited by H. Oh and E. Breuillard, Math. Sci. Res. Inst. Publ. **61**, Cambridge, New York, 2014.

[Li and Yau 1982] P. Li and S. T. Yau, "A new conformal invariant and its applications to the Willmore conjecture and the first eigenvalue of compact surfaces", *Invent. Math.* **69**:2 (1982), 269–291.

[Lubotzky 2011] A. Lubotzky, "Dynamics of $\mathrm{Aut}(F_N)$ actions on group presentations and representations", pp. 609–643 in *Geometry, rigidity, and group actions*, edited by B. Farb and D. Fisher, Univ. Chicago Press, Chicago, IL, 2011.

[Lubotzky and Pak 2001] A. Lubotzky and I. Pak, "The product replacement algorithm and Kazhdan's property (T)", *J. Amer. Math. Soc.* **14**:2 (2001), 347–363.

[McMullen 2005] C. T. McMullen, "Teichmüller curves in genus two: Discriminant and spin", *Math. Ann.* **333**:1 (2005), 87–130.

[McMullen 2013] C. T. McMullen, "Braid groups and Hodge theory", *Math. Ann.* **355**:3 (2013), 893–946.

[Pak 2000] I. Pak, "The product replacement algorithm is polynomial", pp. 476–485 in *41st Annual Symposium on Foundations of Computer Science* (Redondo Beach, CA, 2000), IEEE Comput. Soc. Press, Los Alamitos, CA, 2000.

[Pak 2001] I. Pak, "What do we know about the product replacement algorithm?", pp. 301–347 in *Groups and computation, III* (Columbus, OH, 1999), edited by W. M. Kantor and Á. Seress, Ohio State Univ. Math. Res. Inst. Publ. **8**, de Gruyter, Berlin, 2001.

[Pyber and Szabó 2011] L. Pyber and E. Szabó, "Growth in finite simple groups of lie type of bounded rank", preprint, 2011. arXiv 1005.1858

[Pyber and Szabó 2014] L. Pyber and E. Szabó, "Growth in linear groups", pp. 253–268 in *Thin groups and superstrong approximation*, edited by H. Oh and E. Breuillard, Math. Sci. Res. Inst. Publ. **61**, Cambridge, New York, 2014.

[Rapinchuk 2014] A. Rapinchuk, "Strong approximation for algebraic groups", pp. 269–298 in *Thin groups and superstrong approximation*, edited by H. Oh and E. Breuillard, Math. Sci. Res. Inst. Publ. **61**, Cambridge, New York, 2014.

[Roberts and Venkatesh $\geq$ 2012] D. P. Roberts and A. Venkatesh, "Hurwitz number fields", in preparation; see http://cda.morris.umn.edu/~roberts/research/HNF.pdf for slides of a related lecture.

[Salehi Golsefidy 2014] A. Salehi Golsefidy, "Affine sieve and expanders", pp. 325–342 in *Thin groups and superstrong approximation*, edited by H. Oh and E. Breuillard, Math. Sci. Res. Inst. Publ. **61**, Cambridge, New York, 2014.

[Salehi Golsefidy and Varjú 2012] A. Salehi Golsefidy and P. P. Varjú, "Expansion in perfect groups", *Geom. Funct. Anal.* **22**:6 (2012), 1832–1891.

[Sarnak 2014] P. Sarnak, "Notes on thin matrix groups", pp. 343–362 in *Thin groups and superstrong approximation*, edited by H. Oh and E. Breuillard, Math. Sci. Res. Inst. Publ. **61**, Cambridge, New York, 2014.

[Venkataramana 2013] T. N. Venkataramana, "Monodromy of cyclic coverings of the projective line", preprint, 2013. To appear in *Invent. Math.* arXiv 1204.4778

[Yu 1997] J.-K. Yu, "Toward a proof of the Cohen–Lenstra conjecture in the function field case", preprint, 1997.

[Zograf 1984] P. G. Zograf, "Small eigenvalues of automorphic Laplacians in spaces of cusp forms", *Zap. Nauchn. Sem. Leningrad. Otdel. Mat. Inst. Steklov.* (*LOMI*) **134** (1984), 157–168. In Russian.

ellenber@math.wisc.edu *Department of Mathematics, University of Wisconsin, 480 Lincoln Drive, Madison, WI 53706, United States*

Thin groups and superstrong approximation
MSRI Publications
Volume **61**, 2013

The ubiquity of thin groups

ELENA FUCHS

This article deals with the question of whether typical subgroups of $\mathrm{GL}_n(\mathbb{Z})$ are thin. We discuss interpretations of what "typical" should mean, motivated by recently solved arithmetic problems involving thin groups. We give an overview of what is known about the ubiquity of thin groups that come up in these problems and describe the known tools to detect whether a given subgroup of $\mathrm{GL}_n(\mathbb{Z})$ is thin.

1. Introduction

While studying number theoretic properties of arithmetic groups is fairly classical, the number theory connected to thin groups (see Definition 1.1 below) is a relatively new and still developing area of mathematics, and our understanding of the theory is much more narrow than that of its classical counterpart. Various new arithmetic methods (see [Bourgain et al. 2010; Ellenberg et al. 2012; Salehi Golsefidy and Varjú 2012], for example) that apply to thin and arithmetic groups alike have already served to unify these two fields by showing that thin groups are often as rich as arithmetic groups in some sense. With this in mind, it is natural to try to further not only our understanding of the arithmetic of thin groups, but also of how thin groups fit into the general picture. For example, given a subgroup of $\mathrm{GL}_n(\mathbb{Z})$, we have a limited arsenal of tools to determine whether the group is thin or not. Similarly, given an infinite family of groups which come up in a specific kind of arithmetic problem, we know little about whether the generic group in such a family is thin. In this article, we discuss both of these issues and what is currently known towards resolving them.

Throughout what follows, we focus for simplicity on subgroups of $\mathrm{GL}_n(\mathbb{Z})$, and so we give the following definition of a thin group.

Definition 1.1. Let Γ be a subgroup of $\mathrm{GL}_n(\mathbb{Z})$, and let $G = \mathrm{Zcl}(\Gamma)$ be its Zariski closure. We say that Γ is *thin* if Γ is of infinite index in $G(\mathbb{Z})$.

We begin by reviewing briefly the tools used in various arithmetic problems associated to thin groups, as this will shed some light on what remains to be understood from the number theoretic point of view in this context. One of the

The author is supported by the Simons Foundation through the Postdoctoral Fellows Program.

main ingredients used in such problems is families of expander graphs. There are various equivalent definitions of such families of graphs (see [Lubotzky 2012]). One such definition can be given by considering the eigenvalues of the adjacency matrices of the graphs in the family. By the adjacency matrix M of a finite graph X on n vertices, we mean the $n \times n$ matrix whose rows and columns are indexed by the vertices of X, with

$$M_{ij} = \begin{cases} 1 & \text{if the vertices } v_i \text{ and } v_j \text{ are adjacent,} \\ 0 & \text{otherwise.} \end{cases}$$

Let $\lambda_0(M) \geq \cdots \geq \lambda_{n-1}(M)$ be the eigenvalues of M. If X is connected, as we will assume from now on, we have that $\lambda_0 > \lambda_1$. Furthermore, if X is a k-regular graph, we have that $\lambda_0 = k$. With this notation, we define a family of expander graphs as follows.

Definition 1.2. Let $\{X_i\}_{i \geq 1}$ be an infinite family of finite graphs (we assume here that each graph X_i is connected and k-regular for some fixed k) and let $|X_i| \to \infty$ as $i \to \infty$. Let M_i denote the adjacency matrix of X_i. We say that $\{X_i\}_{i \geq 1}$ is a family of expanders if

$$\limsup_{n \to \infty} \lambda_1(M_n) < k.$$

In other words, a family of graphs as above is an expander family if there is a spectral gap between the top two eigenvalues of the family's corresponding adjacency matrices. This notion of expander graphs is featured in the arithmetic of finitely generated subgroups of $\mathrm{GL}_n(\mathbb{Z})$ as follows. Given a group $\Gamma \subset \mathrm{GL}_n(\mathbb{Z})$ via a finite symmetric generating set S, we let Γ_d and S_d denote the image of Γ and, respectively, S in $\mathrm{GL}_n(\mathbb{Z}/d\mathbb{Z})$ where $d \in \mathbb{Z}$ is square free. To each such Γ_d we associate the Cayley graph $\mathrm{Cay}(\Gamma_d, S_d)$ and so we obtain an infinite family of connected finite graphs associated to Γ. Whether or not Γ is thin, if this infinite family of finite graphs is an expander family, we say that Γ satisfies the expander property, and this property plays a crucial role in various arithmetic problems, two of which we mention in the next section. We note that checking whether or not Γ satisfies the expander property is relatively straightforward due to the following theorem, which is the culminating result in a long line of work [Bourgain and Gamburd 2008; Bourgain et al. 2010; Breuillard et al. 2011; Gamburd 2002; Helfgott 2008; Pyber and Szabó 2010; Sarnak and Xue 1991; Varjú 2012].

Theorem 1.3 [Salehi Golsefidy and Varjú 2012]. *Let $\Gamma \leq \mathrm{GL}_n(\mathbb{Z})$ be a group with a finite symmetric generating set S and let G denote the Zariski closure of Γ. For $d \in \mathbb{Z}$ square free, let Γ_d and S_d denote the projection of Γ and S, respectively, in $\mathrm{GL}_n(\mathbb{Z}/d\mathbb{Z})$. Then the necessary and sufficient condition for*

$$\{\mathrm{Cay}(\Gamma_d, S_d) \mid d \in \mathbb{Z} \text{ square free}, (d, C) = 1\}$$

to be a family of expanders for some integer $C > 0$ is that the connected component of G is perfect.

Thus to determine whether Γ has the expander property one essentially needs information only about the Zariski closure of the group, which is completely independent of the thinness of the group and is in practice easier to determine than the index. So if this crucial expander property does not differentiate between thin and arithmetic groups in the above sense, what is the difference between the two in practice? One major difference is that there is currently no method to determine a good lower bound on the spectral gap associated to a thin group Γ, and such a lower bound can be important in applications. In fact, although for arithmetic groups there are more tools to determine the spectral gap, these tools are only useful if we can show are group Γ is arithmetic in the first place (which in general is quite difficult). So it is clear that there is a need to be able to detect thin versus arithmetic groups. In addition, it is interesting to obtain some measure of how ubiquitous thin groups are in number theoretic applications as opposed to arithmetic groups: if thin groups are generic in some sense, it is natural that we should try to develop further the methods we have to study them.

In this article we give a flavor of how to approach questions about the genericity of thin groups as well as how one might decide whether a given group is thin. Our discussion is motivated by two arithmetic applications of Theorem 1.3: one is the affine sieve as developed in [Bourgain et al. 2010; Salehi Golsefidy and Sarnak 2013], and the other is the application to monodromy groups of [Ellenberg et al. 2012]. We describe these applications via examples in the next section: the affine sieve will be introduced by considering Apollonian circle packings, and the application from [Ellenberg et al. 2012] will be introduced by considering a family of hyperelliptic curves. These examples are also meant to show the contrast between the study of thin and arithmetic groups. In Section 2 we then survey what has been done to answer the question of how generic thin groups are in various situations. Finally, in Section 3 we focus on the issue of determining when certain monodromy groups are thin.

Two examples: thin versus arithmetic groups. We now describe two examples of the applications of expanders discussed above: the first involves a thin group, and the second involves an arithmetic group.

Example 1. Our first example concerns Apollonian circle packings and is meant to illustrate how the affine sieve can be useful in diophantine problems connected to thin groups. Apollonian packings are constructed by starting with the Descartes configuration of four mutually tangent circles, one of which is on the outside of the other three, and repeatedly packing smaller circles into the resulting triangular interstices as in the picture in Figure 1. A theorem of Apollonius of Perga states that there is indeed a unique way to inscribe a circle into every interstice below, and so this construction is well defined.

These packings give rise to problems in number theory as follows: one can show that if the original four circles have integer curvature, all of the circles in the packing will have integer curvature: such packings are called integer Apollonian packings. One can ask many arithmetic questions in this context — see, for example, [Fuchs 2010; Graham et al. 2003; Kontorovich and Oh 2011] — but in this example we will focus on the study of quadruples of mutually tangent circles all of which have curvatures with few prime factors. The key to studying this and many other arithmetic properties of Apollonian packings is a theorem of Descartes that relates the curvatures of any four mutually tangent circles. Namely, if a, b, c, and d denote the curvatures of four mutually tangent circles (where a circle is taken to have negative curvature if it is internally tangent to the other three), then

$$Q(a,b,c,d) := 2(a^2+b^2+c^2+d^2)-(a+b+c+d)^2 = 0. \tag{1-1}$$

A fairly easy consequence of this fact (see [Fuchs 2010], for example) is that in a given packing P one can interpret the set of Descartes quadruples, or the curvatures of quadruples of mutually tangent circles, as the orbit of a group $A \subset \mathrm{O}_Q(\mathbb{Z})$ which is generated by reflections and known as the Apollonian group. Specifically, if $\boldsymbol{v}_P \in \mathbb{Z}^4$ is any Descartes quadruple in the packing P then there is a one to one correspondence between points of the orbit $A\boldsymbol{v}_P$ and Descartes quadruples in P, and every Descartes quadruple in P is a vector in $A\boldsymbol{v}_P$. So to study Apollonian packings one must study orbits of this group A. It can be shown (see [Fuchs 2010], for example) that A is a thin group: it is Zariski dense in $\mathrm{O}_Q(\mathbb{C})$ yet it is of infinite index in $\mathrm{O}_Q(\mathbb{Z})$.

Now, suppose given an Apollonian orbit $\mathbb{O}$ corresponding to a packing P and a polynomial $f(x_1,x_2,x_3,x_4) \in \mathbb{Z}[x_1,x_2,x_3,x_4]$ we are interested in the set

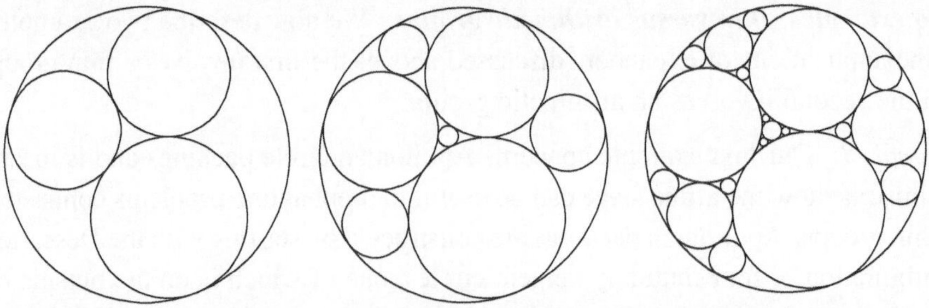

Figure 1. An Apollonian circle packing.

of points

$$\{x \in \mathbb{O} \mid f(x) \text{ has at most } r \text{ prime factors}\}, \tag{1-2}$$

where r is some positive integer. For example, we might wish to count the number of points in a ball in this set, or determine for which r this set is large in some sense. For a general thin group, questions of this kind have been only recently handled using the affine sieve as developed in [Bourgain et al. 2010; Salehi Golsefidy and Sarnak 2013]: for any integer orbit $\mathbb{O}$ of a finitely generated subgroup G of $\mathrm{GL}_n(\mathbb{Z})$ which satisfies the expander property it is known that there is some finite $r > 0$ such that the set in (1-2) is large in the sense that it is Zariski dense in $\mathrm{Zcl}(\mathbb{O})$. We call the smallest such r the saturation number of the pair $(\mathbb{O}, f)$. Furthermore, the affine sieve enables us to count the number of points in the set (1-2) in a ball, and the expander property described before controls the remainder term in this sieve. However, there is still work to be done if we wish to implement the affine sieve with any accuracy. For example, the saturation number of $(\mathbb{O}, f)$ depends on the spectral gap of the group G: the larger the spectral gap, the smaller the saturation number. As we mentioned before, there is currently no method to get a good lower bound on the spectral gap for thin groups — the Apollonian group A in particular. Thus the affine sieve does not give good upper bounds for saturation numbers connected to the orbits of A. Luckily, while A is thin it is a relatively nice thin group: for example, it contains many unipotent subgroups which are a main tool in the proof of the following theorem.

Theorem 1.4 [Fuchs 2010]. *Let $\boldsymbol{v_P}$ be a primitive integer Descartes quadruple, let A be the Apollonian group and let $\mathcal{P} = A\boldsymbol{v_P}$. For $\boldsymbol{x} = (x_1, x_2, x_3, x_4)^t \in \mathcal{P}$ let $f(\boldsymbol{x}) = x_1x_2x_3x_4/12$, and let $\mathcal{P}_{28}$ denote those points $\boldsymbol{x} \in \mathcal{P}$ for which $f(\boldsymbol{x})$ has at most 28 prime factors. Then $\mathcal{P}_{28}$ is Zariski dense in $\mathrm{Zcl}(\mathcal{P})$.*

We remark here that it is conjectured that the 28 above should be 4, and without a good grip on the spectral gap for the Apollonian group it is unclear how one could come much closer to this conjectured value.

Example 2. The second example comes from work of Ellenberg, Hall and Kowalski. Their results are quite general, but we will focus on the following very specific result for contrast with the previous example above.

Theorem 1.5 [Ellenberg et al. 2012, Corollary 5]. *Let k be a number field, and let $f \in k[X]$ be a squarefree polynomial of degree $2g$ with $g \geq 1$. Let U_f be the complement of the zeros of f in $\mathbb{A}^1$, and let $\mathcal{C}/U$ be the family of hyperelliptic curves given by*

$$\mathcal{C} : y^2 = f(x)(x - t),$$

with Jacobians $J_t = \mathrm{Jac}(\mathscr{C}_t)$. *Then for any* $d \geq 1$, *the set*

$$\bigcup_{[k_1:k]=d} \{t \in U(k_1) \mid \mathrm{End}_{\mathbb{C}}(J_t) \neq \mathbb{Z}\}$$

is finite.

In their proof of this theorem, Ellenberg et al. consider the monodromy group Γ associated to the family of hyperelliptic curves above. In this case $\Gamma \subset \mathrm{Sp}_{2g}(\mathbb{Z})$ and is Zariski dense in Sp_{2g}. A crucial ingredient in the proof is that Γ satisfies the expander property which is an immediate consequence of Theorem 1.3. However, because in this case it is known from [Yu 1995] that Γ is of finite index in $\mathrm{Sp}_{2g}(\mathbb{Z})$ — that is, it is arithmetic — one can show this expander property rather classically, without appealing to the intricate methods which go into the proof of Theorem 1.3. One way to do this for $g \geq 2$ is to note that $\mathrm{Sp}_{2g}(\mathbb{Z})$ and any finite index subgroup has Kazhdan property T (for a definition and discussion of property T; see [Lubotzky 1994]). Therefore Γ has property T as well, and the fact that Γ must then satisfy the expander property follows by a theorem of Margulis [1973], where expander graphs were explicitly constructed for the first time. In addition, in this case one can get a good bound on the spectral gap. However, this property T route is not available unless one knows that the group in question is arithmetic.

We should mention here that while Theorem 1.5 concerns just one monodromy group, [Ellenberg et al. 2012] deals with an infinite family of monodromy groups some of which are known to be thin.

2. Generic thin groups

We now turn to our question of how generic or ubiquitous thin groups are, pointing out two different variants of the question. The first variant is inspired by the affine sieve, which is a tool to count prime and almost prime points in integer orbits of subgroups of $\mathrm{GL}_n(\mathbb{Z})$: namely, given a suitable definition of "likely", which we specify shortly, we ask:

(1) How likely is it that a finitely generated subgroup of $\mathrm{GL}_n(\mathbb{Z})$ is thin?

On the other hand, we might wish to ask a more specific question about the genericity of thin groups by focusing on Ellenberg et al.'s work on monodromy groups in [2012]. Thus another variant of the question about the ubiquity of thin groups is this:

(2) Is the generic monodromy group appearing in [Ellenberg et al. 2012] thin?

Concerning Question 1, it is expected that the generic finitely generated subgroup of $\mathrm{GL}_n(\mathbb{Z})$ is indeed thin for many reasonable definitions of generic.

One can simplify the problem somewhat by considering this question for $G = \mathrm{SL}_n(\mathbb{Z})$. Specifically, for $n \geq 2$, one can consider pairs of elements $(\gamma_1, \gamma_2) \in G^2$ and ask for the probability that $\Gamma(\gamma_1, \gamma_2)$, the group generated by γ_1, γ_2, is an infinite index subgroup of G (one can also consider groups generated by k elements for a fixed $k \geq 2$). This is a little different, at first glance, than asking for the probability that $\Gamma(\gamma_1, \gamma_2)$ is thin — that is, that the group is of infinite index in its Zariski closure. However, for many reasonable definitions of "generic", it is known that the generic subgroup of $\mathrm{SL}_n(\mathbb{Z})$ is in fact Zariski dense in SL_n (see [Rivin 2010], for example). In particular, $\mathrm{Zcl}(\Gamma(\gamma_1, \gamma_2))$ is all of SL_n with high probability, and so considering the group's index in $\mathrm{SL}_n(\mathbb{Z})$ is synonymous with considering thinness in this case.

We now describe two ways to interpret "probability" above, and note that there are of course other ways. In both cases we define a sequence of measures μ_T on G which in turn defines the probability we want.

One way to interpret the probability is combinatorial. Namely, fix a finite generating set $\{g_1, \ldots, g_r\}$ of G and let μ_T denote the normalized counting measure on the set W_T^2 of elements of $(\gamma_1, \gamma_2) \in G^2$ such that γ_1, γ_2 can be written as words of length $\leq T$ in the generators g_i. We then consider the limit

$$\lim_{T\to\infty} \mu_T\bigl(\{(\gamma_1, \gamma_2) \in W_T^2 \mid [G : \Gamma(\gamma_1, \gamma_2)] = \infty\}\bigr).$$

Aoun [2011] showed that this limit is in fact 1: that is, if we obtain our γ_i by taking random walks on a fixed generating set of G then Γ will be infinite index with high probability.

Another natural way to interpret this probability is via an archimedean model (in fact, in some sense this is the most intuitive formulation of the problem, since in arithmetic problems we are often interested in counting in archimedean balls). Namely, we let μ_T be the normalized counting measure on the set B_T^2 of elements $(\gamma_1, \gamma_2) \in G^2$ such that, for $k = 1, 2$, we have $\|\gamma_k\| < T$ and $\|\gamma_k^{-1}\| < T$. Here we define $\|\gamma\| > 0$ as follows:

$$\|\gamma\| := \sqrt{\lambda_{\max}(\gamma^t \gamma)}, \tag{2-1}$$

where $\lambda_{\max}$ denotes the maximal eigenvalue. We then consider the limit

$$\lim_{T\to\infty} \mu_T\bigl(\{(\gamma_1, \gamma_2) \in B_T^2 \mid [G : \Gamma(\gamma_1, \gamma_2)] = \infty\}\bigr).$$

This limit is also shown to be 1 in [Fuchs and Rivin]. The general strategy in both of these interpretations of the question is to show that $\Gamma(\gamma_1, \gamma_2)$ is free with high probability. Since any free subgroup of $G = \mathrm{SL}_n(\mathbb{Z})$, where $n > 2$ is of infinite index, this is enough for nearly all of the cases (in the case that $n = 2$ one uses the fact that the generic subgroup Γ is free to show that its limit set will generically have small Hausdorff dimension, which then implies the infinite index we need).

So in these senses thin subgroups of $\mathrm{SL}_n(\mathbb{Z})$ are generic.

As for Question 2, a major obstacle in answering it is choosing an intuitive definition for a generic monodromy group which would make the question above approachable. As of yet, we have found no such definition, and nothing is known towards answering this very broad question. We discuss the very different issue of deciding whether a given monodromy group is thin in the next section.

3. Monodromy groups

Since we have no satisfactory definition of a generic monodromy group, we restrict here to a special class of the monodromy groups which come up in [Ellenberg et al. 2012] and study the issue of thinness within this class. Let $\theta = z(d/dz)$ and define the differential operator D on $\mathbb{P}^1(\mathbb{C})$ by

$$\begin{aligned} D := D(\alpha, \beta) &= D(\alpha_1, \dots, \alpha_n, \beta_1, \dots, \beta_n) \\ &= (\theta + \beta_1 - 1) \cdots (\theta + \beta_n - 1) - z(\theta + \alpha_1) \cdots (\theta + \alpha_n), \end{aligned}$$

where

$$\alpha, \beta \in \mathbb{Q}^n, \quad 0 \le \alpha_1 \le \cdots \le \alpha_n < 1, \quad \text{and} \quad 0 \le \beta_1 \le \cdots \le \beta_n < 1$$

(one can also consider the more general case of $\alpha_i, \beta_i \in \mathbb{C}$ but we do not discuss this here). The differential equation $D(\alpha, \beta)u = 0$ is hypergeometric and is regular outside of three singularities at $z = 0, 1, \infty$. One associates a monodromy group to such a hypergeometric equation as follows. For $x_0 \in \mathbb{P}^1(\mathbb{C}) \backslash \{0, 1, \infty\}$ let S_0 be the solution space to $D(\alpha, \beta)u = 0$ at x_0. We then have a representation of the fundamental group $\pi_1(\mathbb{P}^1(\mathbb{C}) \backslash \{0, 1, \infty\})$ in $\mathrm{GL}(S_0)$. We call the image of this representation $H(\alpha, \beta)$ and refer to it as the monodromy group associated to $D(\alpha, \beta)$. Beukers and Heckman have proved various results about these groups, which we mention below. The following theorem, describing $H(\alpha, \beta)$ via a generating set, is key both in their work and in our joint work with Meiri and Sarnak [Fuchs et al. 2013].

Theorem 3.1 [Levelt 1961]. *For $1 \le i \le n$ let α_i and β_i be as above. Define the complex numbers $A_1, \dots, A_n, B_1, \dots, B_n$ as the coefficients of the polynomials*

$$P(z) := \prod_{j=1}^{k} (z - e^{2\pi i \alpha_j}) = z^n + A_1 z^{n-1} + \cdots + A_n,$$

$$Q(z) := \prod_{j=1}^{k} (z - e^{2\pi i \beta_j}) = z^n + B_1 z^{n-1} + \cdots + B_n.$$

Then $H(\alpha,\beta)$ is the group generated by

$$A=\begin{pmatrix} 0 & 0 & \cdots & 0 & -A_n \\ 1 & 0 & \cdots & 0 & -A_{n-1} \\ 0 & 1 & \cdots & 0 & \vdots \\ 0 & 0 & \ddots & 0 & -A_2 \\ 0 & 0 & \cdots & 1 & -A_1 \end{pmatrix}, \quad B=\begin{pmatrix} 0 & 0 & \cdots & 0 & -B_n \\ 1 & 0 & \cdots & 0 & -B_{n-1} \\ 0 & 1 & \cdots & 0 & \vdots \\ 0 & 0 & \ddots & 0 & -B_2 \\ 0 & 0 & \cdots & 1 & -B_1 \end{pmatrix}.$$

Note that $H(\alpha,\beta)\subset \mathrm{GL}_n(\mathbb{Z})$ if the polynomials $P(z)$ and $Q(z)$ above factor as cyclotomic polynomials. Since we are interested in integral monodromy groups, we will assume this for the rest of the article.

With the notation above, it is shown in [Beukers and Heckman 1989] show that $H=H(\alpha,\beta)$ falls into one of the following categories:

(0) H is a finite group (a complete list is given in [loc. cit.]).

(1) If n is even, H is infinite, and $A_n/B_n=1$ then $H\subset \mathrm{Sp}_n(\mathbb{Z})$ and $\mathrm{Zcl}(H)=\mathrm{Sp}_n(\mathbb{C})$.

(2) If n is odd and H is infinite; or if n is even, H is infinite, and $A_n/B_n=-1$, then $H\subset \mathrm{O}_{f_{\alpha,\beta}}(\mathbb{Z})$ for some quadratic form $f=f_{\alpha,\beta}$ in n variables and $\mathrm{Zcl}(H)=\mathrm{O}_f(\mathbb{C})$.

In category (2) above we further consider two subcategories: one is that H fixes a quadratic form of signature $(n-1,1)$, and the other is that it fixes a quadratic form of signature (p,q), where $p,q>1$. We should mention that computing the corresponding signature in this case is easy given α,β: Beukers and Heckman show that if H is as in (2) above, H fixes a quadratic form of signature (p,q), where $p+q=n$ and

$$|p-q|=\left|\sum_{j=1}^{n}(-1)^{j+m_j}\right|,$$

where $m_j=|\{k\mid \beta_k<\alpha_j\}|$ and the α_i and β_i are ordered as described at the beginning of this section. Our goal in the rest of this paper is to give a flavor of what can be said about the thinness of H in each of these three infinite cases. In the symplectic case, the best known method to detect thinness is to apply ping pong to show that the group is free. Outside of this one can also narrow down the given group to two possibilities: a thin group or a specific congruence subgroup of $\mathrm{Sp}_n(\mathbb{Z})$. In the orthogonal signature $(n-1,1)$ case, one has more tools at one's disposal since these groups act on hyperbolic space, and this well developed geometry often allows us to detect whether the given group is thin or not. Finally, although we lose these geometric tools in the orthogonal signature (p,q) case where $p,q>1$, there are ways to show thinness besides applying ping pong: on page 85 we give an example of how one might do this in signature $(2,2)$.

Symplectic monodromy. The question of whether $H = H(\alpha, \beta)$ is thin in the case that H is symplectic is perhaps the most elusive of the three scenarios outlined above. One method to answer this is to show that H is free (although it need not be in order to be thin) by applying the ping pong lemma: this has recently been done for a concrete example in [Brav and Thomas 2012]. However, the method in [Brav and Thomas 2012] is specific to the example treated there and it is not clear how to find the appropriate regions for applying ping pong to a general group. Furthermore, in several cases we consider the group in question can be easily seen not to be free, in which case one must look elsewhere for a method to show thinness.

In ongoing work with Capdeboscq and Rivin, we have considered 14 examples of hypergeometric monodromy groups in $\mathrm{Sp}_4(\mathbb{Z})$ associated to Calabi Yau three-folds (these groups are in particular among the $H(\alpha, \beta)$'s we described above). Each of these groups is given via two generators, one of which is a transvection: these generators have been derived by Cheng, Yang, and Yui [Chen et al. 2008]. Notably, these authors remark that they do not know whether these subgroups are lattices in Sp_4, and that two experts they have consulted on the matter have given them opposite conjectures: one asserts that the groups are probably all thin, while the other guesses that they must be finite index. We are inclined to agree with the former assessment of the situation, although we have made almost no progress in actually proving that any of the groups in [Chen et al. 2008] are thin.

Our belief that these groups are thin is based on a series of experiments which seem to imply that the number of elements in each of these groups of norm $\leq T$ is much too small for a finite-index subgroup of $\mathrm{Sp}_4(\mathbb{Z})$. We present the data from one of these experiments in Figure 2: this particular graph corresponds to the group $H = \langle A, B \rangle$, where

$$A = \begin{pmatrix} 1 & 1 & 0 & 0 \\ 0 & 1 & 0 & 0 \\ 16 & 16 & 1 & 0 \\ 0 & -8 & -1 & 1 \end{pmatrix}, \quad B = \begin{pmatrix} 1 & 0 & 0 & 0 \\ 0 & 1 & 0 & 1 \\ 0 & 0 & 1 & 0 \\ 0 & 0 & 0 & 1 \end{pmatrix}. \tag{3-1}$$

Specifically, Figure 2 shows the graph of $\log N(T)$ versus $\log T$, where

$$N(T) = \#\{\gamma \in H \mid \|\gamma\| < T\}$$

with $\|\cdot\|$ as in (2-1). Because the graph of $\log(N(T))$ is only an approximation, we have denoted it by "$\log(N(T))$". The graph also depicts a linear fitting which has slope 1.396. The idea is that this slope is much too small for H to be finite index in $\mathrm{Sp}_4(\mathbb{Z})$: the same experiment for the full group $\mathrm{Sp}_4(\mathbb{Z})$ yields a graph of slope 6, and a finite index subgroup would have slope 6 as well.

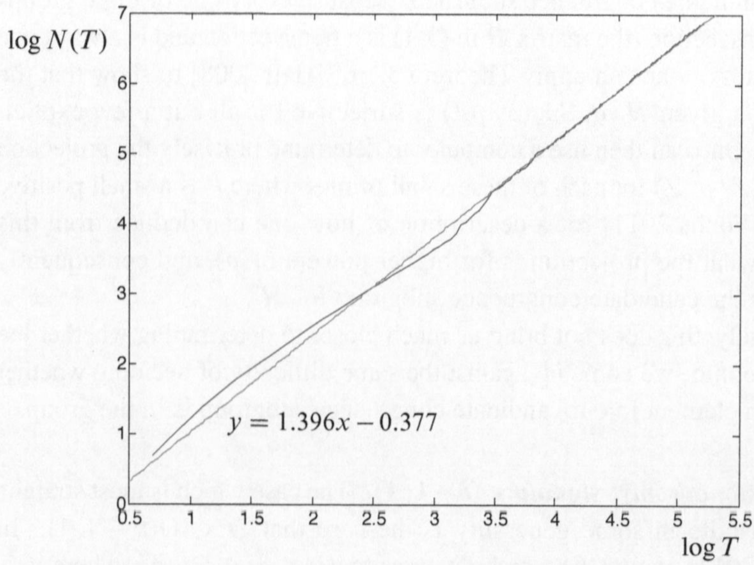

Figure 2. Counting elements in a symplectic monodromy group.

Be that as it may, we have no way of ascertaining that the values of $N(T)$ which we found are actually correct, and so the graph above, while suggestive, is merely a guess. Specifically, since the only information we have about H is its generators, our method of counting elements of norm $< T$ is as follows: we consider walks on the generators of H and count the number of elements we find this way that have norm $< T$; eventually, once the length of the walk is long enough, all elements we find have norm $> T$ and we assume that we have found all elements of norm $< T$. However, it may be that there are many words of length 1000, say, whose norm is small and these words, if they exist, are not captured by our graph, so at this point this data cannot prove the thinness of H (which, by the way, may well be free: we have found no relations in its generators). We have produced similar graphs with small slope for many of the other examples in [Chen et al. 2008], all of which suggest thinness but are inconclusive for the same reason: we do not have a method, given only the generators of a subgroup of Sp_4, to decide whether a given element is in the group or not (in fact, this might be undecidable in general). This issue has hindered various other attempts to show that the groups in [Chen et al. 2008] are thin.

One can, however, use the fact that $\mathrm{Sp}_4(\mathbb{Z})$ has the congruence subgroup property to prove a dichotomy statement for each of the 14 groups H_i given in [Chen et al. 2008]: namely, each of these groups is either thin or is a congruence subgroup of $\mathrm{Sp}_4(\mathbb{Z})$. Furthermore, it is not difficult to determine for each H_i

the precise candidate congruence subgroup. Because every one of these groups contains a transvection (the matrix B in (3-1) is a transvection and is always one of the generators), one can apply Theorem 3.1 of [Hall 2008] to show that the projection of a given H_i in $\mathrm{Sp}_4(\mathbb{Z}/p\mathbb{Z})$ is surjective for all but a few explicit small primes. One can then use a computer to determine precisely the projection of H_i in $\mathrm{Sp}_4(\mathbb{Z}/p^k\mathbb{Z})$ for each of these small primes where k is a small positive integer (see [Fuchs 2011] for a description of how one can deduce from this information what the projection is for higher powers of p), and consequently make explicit the candidate congruence subgroup for H_i.

Unfortunately, this does not bring us much closer to determining whether the groups H_i are thin: we come up against the same difficulty of deciding whether or not a given element in our candidate congruence subgroup is in the group.

Orthogonal monodromy: signature $(n-1,1)$. The case which is most straightforward to handle in some generality is the case that $H \subset \mathrm{O}(n-1,1)$. In [Fuchs et al. 2013], the (α,β) which give rise to $H(\alpha,\beta) \subset \mathrm{O}_f(\mathbb{Z})$, where f is a quadratic form in n variables with signature $(n-1,1)$ are described completely and are shown to belong to one of seven families outside of finitely many explicit cases in dimension $n \leq 7$. Using this description, the authors are able to write down the form f fixed by $H(\alpha,\beta)$ for in an infinite number of cases, as well as to produce infinite families of groups $H(\alpha,\beta)$ all of which are thin (we should note here that our group $H(\alpha,\beta) \subset \mathrm{O}(n-1,1)$ only when n is odd).

The strategy in doing this is to use the fact that in this case H is a group acting on hyperbolic space $\mathbb{H}^{n-1}$. As far as thinness goes, results of Nikulin, Vinberg and Prokhorov (see [Nikulin 1987] for a discussion of these results) say that if n is large enough certain such groups must be thin. Specifically, let f be a quadratic form in n variables of signature $(n-1,1)$ and let $\mathrm{R}_f(\mathbb{Z})$ denote the subgroup of $\mathrm{O}_f(\mathbb{Z})$ which is generated by all of the reflections in hyperplanes in $\mathrm{O}_f(\mathbb{Z})$. Then Nikulin shows that for $n > 300$ the group $\mathrm{R}_f(\mathbb{Z})$ must be thin (Vinberg had proven a similar result on cocompact groups generated by reflections in hyperplanes, and Prokhorov had proven that for $n > 900$ the group $\mathrm{R}_f(\mathbb{Z})$ is thin).

In particular, if we could show that our groups $H(\alpha,\beta)$ contain finite index subgroups generated by reflections, this would imply that H is automatically thin once the dimension is large. However, in general we are not so lucky, and one has to work harder.

Namely, for $H = \langle A, B\rangle$, where A is of finite order k (it is often the case that one of the generators is of finite order), we first obtain a finite index subgroup H' of H as follows. Let $R_0 = A^{-1}B$, and let $R_i = A^{-i}R_0A^i$ for $1 \leq i \leq k-1$. Note that each R_i is a Cartan involution of $\mathbb{H}^{n-1}$, meaning that it fixes a point and reverses all geodesics around that point. One can show that the group

$$H' := \langle R_i R_{i+1} \mid 0 \le i \le k-1 \rangle$$

is of finite index in H. Furthermore, in [Fuchs et al. 2013] we give an algorithm which in infinitely many cases expresses each generator of H' as a product of reflections in hyperplanes, so $H' \subset R_f(\mathbb{Z})$, where f is the quadratic form fixed by H. Using the result of Nikulin above we then have that H', and therefore H is thin.

While there are infinitely many groups $H(\alpha, \beta)$ in this signature $(n-1, 1)$ case which are still to be handled, it is conjectured that all but finitely many of these should be thin. The next step (after deciding thinness) is to clarify further the structure of these groups: for example, are they geometrically finite?

Orthogonal monodromy: the split* (2, 2) *case. In the case that $H \subset \mathrm{O}(p, q)$, where $p, q > 1$ one cannot appeal to the tools from hyperbolic geometry as we described in the previous section. In fact, it is not currently clear how to attack the question of whether H is thin in this situation, and in most cases the problem is as difficult as in the symplectic case. However, given a concrete example where the signature is $(2, 2)$ and the quadratic form has a perfect square determinant (we call this case split), one can rely on methods besides ping pong to decide thinness.

To demonstrate how one might tackle this problem, we consider the example $H = H(\alpha, \beta)$, where $\alpha = \{0, 0, 0, \frac{1}{2}\}$ and $\beta = \{\frac{1}{4}, \frac{1}{4}, \frac{3}{4}, \frac{3}{4}\}$ and show that it is in fact thin. We will do this in a fair amount of detail, since this computation, unlike the results discussed in the previous sections, is not contained in any other paper. By Theorem 3.1, the generators of H are

$$A = \begin{pmatrix} 0 & 0 & 0 & 1 \\ 1 & 0 & 0 & -2 \\ 0 & 1 & 0 & 0 \\ 0 & 0 & 1 & 2 \end{pmatrix}, \quad B = \begin{pmatrix} 0 & 0 & 0 & -1 \\ 1 & 0 & 0 & 0 \\ 0 & 1 & 0 & -2 \\ 0 & 0 & 1 & 0 \end{pmatrix}, \tag{3-2}$$

and $H \subset \mathrm{O}_Q(\mathbb{Z})$, where Q is

$$Q = \begin{pmatrix} 1 & 1 & -1 & -3 \\ 1 & 1 & 1 & -1 \\ -1 & 1 & 1 & 1 \\ -3 & -1 & 1 & 1 \end{pmatrix},$$

of determinant 16 and signature $(2, 2)$.

Theorem 3.2. *The group $H = H(\alpha, \beta)$ is thin.*

To prove this, we use the spin homomorphism mapping $\mathrm{SL}_2 \times \mathrm{SL}_2$ onto the connected component of the identity in $\mathrm{SO}(2,2)$ which we explain momentarily. Note that this is analogous to the strategy used in [Fuchs 2011] to study the congruence obstructions of the Apollonian group: in that case the spin homomorphism is a map from $\mathrm{SL}_2(\mathbb{C})$ to $\mathrm{SO}_{\mathbb{R}}(3,1)$. We wish to determine the preimage of $H \cap \mathrm{SO}_Q(\mathbb{Z})$ under this homomorphism (we will see that this intersection is indeed in the connected component of the identity). To this end, we have the following description of $H \cap \mathrm{SO}_Q(\mathbb{Z})$.

Claim. *The group* $H \cap SO_Q(\mathbb{Z})$ *equals* $\langle A^2, B\rangle$, *where* A, B *are as in* (3-2), *and this group is of index* 2 *in* H.

Proof. First we show that, for any $n \in \mathbb{Z}$, we have $AB^nA \in \langle A^2, B\rangle$. Namely, since $A^{-1}BA^{-1}B = AB^{-1}AB^{-1} = B^{-1}AB^{-1}A = BA^{-1}BA^{-1} = I$, we have

$$h = A^2A^{-1}BA^{-1}BB^{-1} = ABA^{-1} \in \langle A^2, B\rangle.$$

Since $AB^nA = h^nA^2$, we have $AB^nA \in \langle A^2, B\rangle$ as desired. Any element of $H \cap \mathrm{SO}_Q(\mathbb{Z})$ can be written as a word

$$A^{n_1}B^{m_1}A^{n_2}B^{m_2}\cdots A^{n_s}B^{m_s}, \tag{3-3}$$

for some $s \in \mathbb{N}$, $m_i, n_i \in \mathbb{Z}$ for $1 \le i \le s$, and $n_1 + n_2 + \cdots + n_s$ even since $\det(A) = -1$ and $\det(B) = 1$. We want to show that the word in (3-3) is in fact an element of $\langle A^2, B\rangle$.

If all of the exponents n_i for $1 \le i \le s$ are even, we are done. Suppose therefore that k is the smallest index for which n_k is odd. Then showing that the expression in (3-3) is in $\langle A^2, B\rangle$ is equivalent to showing

$$A^{n_k}B^{m_k}A^{n_{k+1}}B^{m_{k+1}}\cdots A^{n_s}B^{m_s} \in \langle A^2, B\rangle.$$

Suppose n_{k+1} is odd. Then for some $N, M \in \mathbb{Z}$, we have

$$A^{n_k}B^{m_k}A^{n_{k+1}} = A^{2M}AB^{m_k}AA^{2N} \in \langle A^2, B\rangle,$$

so it suffices to show that

$$A^{n_l}B^{m_l}A^{n_{l+1}}B^{m_{l+1}}\cdots A^{l_s}B^{m_s} \in \langle A^2, B\rangle,$$

where l is the smallest index such that n_l is odd and n_{l+1} is even. Noting that $A^{n_l}B^{m_l}A^{n_{l+1}} = A^{2M}AB^{m_l}AA^{2N}A$ for some $N, M \in \mathbb{Z}$, this reduces to showing

$$AB^{m_{l+1}}A^{n_{l+2}}\cdots A^{l_s}B^{m_s} \in \langle A^2, B\rangle.$$

Continuing the reduction process using the fact that $B^K, A^{2K} \in \langle A^2, B\rangle$ for all $K \in \mathbb{Z}$ and $AB^NA \in \langle A^2, B\rangle$ for all $N \in \mathbb{Z}$, we reduce to either the identity I, in which case we are done, or to the element $AB^{m_s}A^{n_s}$, where n_s is even. But the latter case contradicts the assumption that $n_1 + n_2 + \cdots + n_s$ is even, so it does not occur.

Using again the relation $AB^{-1}AB^{-1} = I$ and applying similar arguments we have that there are only two cosets, $A\langle A^2, B\rangle$ and $I\langle A^2, B\rangle$ of $\langle A^2, B\rangle$ in H, so $[H : \langle A^2, B\rangle] = 2$ as desired. □

The Claim implies in particular that if $\langle A^2, B\rangle$ is thin then H is thin as well as stated in Theorem 3.2. Now note that Q is equivalent over $\mathbb{Q}$ to

$$Q' = C^T QC = \begin{pmatrix} 1 & 0 & 0 & 0 \\ 0 & 0 & 2 & 0 \\ 0 & 2 & 0 & 0 \\ 0 & 0 & 0 & -4 \end{pmatrix}, \quad \text{where } C = \begin{pmatrix} 1 & -1 & 1 & 1 \\ 0 & 1 & 0 & 1 \\ 0 & 0 & 1 & -1 \\ 0 & 0 & 0 & 1 \end{pmatrix},$$

and that

$$Q'(x_1, x_2, x_3, x_4) = 2x_2 \cdot 2x_3 - (2x_4 - x_1)(2x_4 + x_1)$$

is the determinant of

$$M = \begin{pmatrix} 2x_2 & 2x_4 - x_1 \\ 2x_4 + x_1 & 2x_3 \end{pmatrix}.$$

The fact that one can bring Q to this form over $\mathbb{Q}$ makes this example particularly nice. Under this change of variables, $\langle A^2, B\rangle$ becomes $\langle A'^2, B'\rangle$, where

$$A'^2 = \begin{pmatrix} -1 & -2 & -8 & -8 \\ 0 & -1 & -4 & -4 \\ 1 & 0 & 3 & 2 \\ 0 & 1 & 2 & 3 \end{pmatrix}, \quad B' = \begin{pmatrix} 1 & -2 & -2 & 4 \\ 1 & -1 & 0 & 2 \\ 0 & 1 & 1 & -2 \\ 0 & 0 & 1 & -1 \end{pmatrix}.$$

We now define a map from $\rho : \mathrm{SL}_2 \times \mathrm{SL}_2 \to \mathrm{SO}_{Q'}$ as follows. Given a pair of elements g and h in SL_2, we have $\det(M) = \det(gMh^t)$, and in this sense a pair of elements in SL_2 fix Q'. As explained in Chapter 13.9 of [Cassels 1978], we map the pair $(g, h) \in \mathrm{SL}_2(\mathbb{R}) \times \mathrm{SL}_2(\mathbb{R})$ to an element $\gamma \in \mathrm{SO}_{Q'}(\mathbb{R})$ if

$$gMh^t = \begin{pmatrix} 2x'_2 & 2x'_4 - x'_1 \\ 2x'_4 + x'_1 & 2x'_3 \end{pmatrix},$$

where $(x'_1, x'_2, x'_3, x'_4)^t := \gamma(x_1, x_2, x_3, x_4)^t$. Given this, for each of our generators A'^2 and B above we solve for two matrices g, h in $\mathrm{SL}_2(\mathbb{R})$ which map to the generator. Note that there will not necessarily be a solution to this if one or both of the generators lie outside of the connected component of the identity. However, we are able to find solutions and we get that

$$\pm\left(\begin{pmatrix} 1 & 2 \\ 0 & 1 \end{pmatrix}, \begin{pmatrix} -1 & -2 \\ 2 & 3 \end{pmatrix}\right) \overset{\rho}{\longmapsto} A'^2 \tag{3-4}$$

and

$$\pm\left(\begin{pmatrix}-1 & -2\\ -1 & -1\end{pmatrix}, \begin{pmatrix}-1 & 0\\ 1 & -1\end{pmatrix}\right) \stackrel{\rho}{\longmapsto} B'. \tag{3-5}$$

If we consider the projections of this group in each factor $\mathrm{SL}_2(\mathbb{Z})$, we obtain the groups Γ_1 and Γ_2, where

$$\Gamma_1 = \left\langle \pm\begin{pmatrix}1 & 2\\ 0 & 1\end{pmatrix}, \pm\begin{pmatrix}-1 & -2\\ -1 & -1\end{pmatrix}\right\rangle \Gamma_2 = \left\langle \pm\begin{pmatrix}-1 & -2\\ 2 & 3\end{pmatrix}, \pm\begin{pmatrix}-1 & 0\\ 1 & -1\end{pmatrix}\right\rangle.$$

One might hope that either Γ_1 or Γ_2 are of infinite index in $\mathrm{SL}_2(\mathbb{Z})$, as this would immediately imply that the preimage of $H \cap \mathrm{SO}_Q(\mathbb{Z})$ in $\mathrm{SL}_2(\mathbb{Z}) \times \mathrm{SL}_2(\mathbb{Z})$ is also infinite index. Unfortunately, both Γ_1 and Γ_2 contain the finite index subgroup

$$\left\langle \begin{pmatrix}1 & 0\\ 2 & 1\end{pmatrix}, \begin{pmatrix}1 & 2\\ 0 & 1\end{pmatrix}\right\rangle$$

and so are both of finite index. So we must work with the preimage itself and show the following, which together with the Claim immediately implies Theorem 3.2.

Proposition 3.3. *Let*

$$S = \begin{pmatrix}0 & -1\\ 1 & 0\end{pmatrix}, \quad T = \begin{pmatrix}1 & 1\\ 0 & 1\end{pmatrix} \tag{3-6}$$

denote the generators of $SL_2(\mathbb{Z})$, with relations $S^2 = -I$ and $(ST)^3 = -I$. Let $G \subset \mathrm{SL}_2(\mathbb{Z}) \times \mathrm{SL}_2(\mathbb{Z})$ be the group generated by the elements in (3-4) *and* (3-5). *Then for any integer $P \neq 0$ we have $(\pm T^P, \pm T^P) \notin G$, meaning in particular that the left cosets of this group under multiplication by $(\pm T^k, \pm T^k)$, where $k \in \mathbb{Z}$ are all distinct and that G is of infinite index in* $\mathrm{SL}_2(\mathbb{Z}) \times \mathrm{SL}_2(\mathbb{Z})$.

Proof. We first note that the inverses of the generators of Γ in (3-4) and (3-5) are $\pm(T^{-2}, -ST^2S^{-1}T^2)$ and $\pm(TST^{-1}, ST^{-1}S)$, where S and T are as above, and we can rewrite this as

$$\Gamma = \langle \pm(-T^{-2}, T^2), \pm(TST^{-1}, ST^{-1}S)\rangle.$$

Ignoring for a moment the $\pm$, we denote by g_1 and g_2 the two generators above and suppose for contradiction that there is some integer $P \neq 0$ and some $\alpha, \beta \in \{\pm 1\}$ such that

$$(\alpha T^P, \beta T^P) = g_1^{m_1} g_2^{n_1} \cdots g_1^{m_s} g_2^{n_s}, \tag{3-7}$$

where $m_i, n_i \in \mathbb{Z}$ and $n_i, m_i \neq 0$ for $i > 1$. In particular,

$$T^P = \pm T^{-2m_1}(TST^{-1})^{n_1} \cdots T^{-2m_s}(TST^{-1})^{n_s}. \tag{3-8}$$

Note furthermore that $(TST^{-1})^2 = -I$.

Let $n_{i_1}, n_{i_2}, \ldots, n_{i_k}$ denote the odd exponents among $n_1, n_2, \ldots, n_s$ with $i_1 < i_2 < \cdots < i_k$. Suppose $k > 0$ (so there is at least one odd n_i). Then, given (3-8), there exist integers $M_1, M_2, \ldots, M_{k+1}$ with $|M_i| > 0$ for $1 < i < k+1$ such that

$$T^P = \pm T^{2M_1+1} S T^{2M_2} S T^{2M_3} \cdots S T^{2M_k} S T^{2M_{k+1}-1}. \tag{3-9}$$

We now consider two cases:

Case 1: $M_{k+1} \neq 0, 1$. In this case, (3-9) implies that

$$T^{P'} = \pm (ST^{2M_2})(ST^{2M_3}) \cdots (ST^{2M_k})(ST^{2M_{k+1}-1})$$

where $P' = P - 2M_1 - 1$ and all powers of T on the right are > 1 in absolute value. In other words,

$$\begin{pmatrix} 1 & P' \\ 0 & 1 \end{pmatrix} = \begin{pmatrix} 0 & -1 \\ 1 & N_1 \end{pmatrix} \begin{pmatrix} 0 & -1 \\ 1 & N_2 \end{pmatrix} \cdots \begin{pmatrix} 0 & -1 \\ 1 & N_k \end{pmatrix}, \tag{3-10}$$

where $|N_i| > 1$ for all $1 \leq i \leq k$. We now need the following lemma.

Lemma 3.4. *Let $\left(\begin{smallmatrix} a & b \\ c & d \end{smallmatrix}\right)$ be an integer matrix with $|b| > |a|$ and $|d| > |c|$. Let N be an integer such that $|N| > 1$. Then we have that*

$$\begin{pmatrix} a & b \\ c & d \end{pmatrix} \begin{pmatrix} 0 & -1 \\ 1 & N \end{pmatrix} = \begin{pmatrix} a' & b' \\ c' & d' \end{pmatrix},$$

where $|b'| > |a'| > |a|$ and $|d'| > |c'| > |c|$.

To prove this lemma, note that $a' = b$ and $b' = bN - a$. Then

$$|bN - a| \geq |bN| - |a| > 2|b| - |a| > |b|$$

(because $|b| > |a|$), so we have the desired statement about b'. A similar argument shows that $|d'| > |c'| > |c|$.

By the lemma, we have that the lower left entry of the expression on the right hand side of (3-10) cannot be 0, which contradicts (3-10).

Case 2: $M_{k+1} = 0$ or 1. First note that in this case the analog of (3-10) is that

$$\begin{pmatrix} 1 & P' \\ 0 & 1 \end{pmatrix} = \begin{pmatrix} 0 & -1 \\ 1 & N_1 \end{pmatrix} \begin{pmatrix} 0 & -1 \\ 1 & N_2 \end{pmatrix} \cdots \begin{pmatrix} 0 & -1 \\ 1 & N_k \end{pmatrix}, \tag{3-11}$$

where $|N_i| > 1$ for all $i < k$ and clearly we may assume $k > 1$. Again by the lemma, we have that

$$\begin{pmatrix} 0 & -1 \\ 1 & N_1 \end{pmatrix} \cdots \begin{pmatrix} 0 & -1 \\ 1 & N_{k-1} \end{pmatrix} = \begin{pmatrix} a & b \\ c & d \end{pmatrix},$$

where $c \neq 0$, $|b| > |a|$, and $|d| > |c|$. So in particular $d \neq 0$ and so the lower left entry of the expression on the right in (3-11) cannot be 0, contradicting (3-11).

Thus our assumption that there exist odd exponents among the n_i in (3-7) was false, and so we have $n_i = 2n_i'$ for some $n_i' \in \mathbb{Z}$ for all i, and the sum of the exponents m_i must be $-P/2$ for (3-7) to hold. With this restriction on the n_i we now go to the first factor, where (3-7) implies that

$$T^P = \pm\gamma_2^{m_1}\gamma_1^{n_1'} \cdots \gamma_2^{m_s}\gamma_1^{n_s'},$$

where P is even and $\gamma_1 = \left(\begin{smallmatrix}1&0\\2&1\end{smallmatrix}\right)$, $\gamma_2 = \left(\begin{smallmatrix}1&1\\0&1\end{smallmatrix}\right)$. Since $T^P = \gamma_2^{P/2}$ and $-I \notin \langle\gamma_1, \gamma_2\rangle$, we have that in fact

$$T^P = \gamma_2^{m_1}\gamma_1^{n_1'} \cdots \gamma_2^{m_s}\gamma_1^{n_s'}.$$

This gives a nontrivial relation on γ_1 and γ_2 as long as one of the $n_i > 0$. But γ_1 and γ_2 generate a free subgroup of $\mathrm{SL}_2(\mathbb{Z})$, and so we must have that $n_i = 0$ for all i, and that the sum of the exponents m_i is $P/2$ in order for (3-7) to hold. However, if $P \neq 0$ this contradicts the fact that the sum of the exponents m_i must be $-P/2$ from above and so our assumption that $(\pm T^P, \pm T^P) \in G$ for some $P \neq 0$ was false, and we are done. □

Therefore the group is of infinite index in $\mathrm{SL}_2(\mathbb{Z}) \times \mathrm{SL}_2(\mathbb{Z})$, and so $H \cap \mathrm{SO}_Q(\mathbb{Z})$ has infinite index in $\mathrm{SO}_Q(\mathbb{Z})$ and in $\mathrm{O}_Q(\mathbb{Z})$. Since $(H : H \cap \mathrm{SO}_Q(\mathbb{Z})) = 2$ (see above), we have that H is thin as stated in Theorem 3.2.

Acknowledgements

We thank MSRI as well as the organizers of the "Thin Groups and Super Strong Approximation" workshop for a fruitful and stimulating workshop. We also thank I. Capdeboscq, C. Meiri, I. Rivin, and P. Sarnak for the collaborations that led to much of the work summarized in this article.

References

[Aoun 2011] R. Aoun, "Random subgroups of linear groups are free", *Duke Math. J.* **160**:1 (2011), 117–173.

[Beukers and Heckman 1989] F. Beukers and G. Heckman, "Monodromy for the hypergeometric function ${}_nF_{n-1}$", *Invent. Math.* **95**:2 (1989), 325–354.

[Bourgain and Gamburd 2008] J. Bourgain and A. Gamburd, "Uniform expansion bounds for Cayley graphs of $\mathrm{SL}_2(\mathbb{F}_p)$", *Ann. of Math.* (2) **167**:2 (2008), 625–642.

[Bourgain et al. 2010] J. Bourgain, A. Gamburd, and P. Sarnak, "Affine linear sieve, expanders, and sum-product", *Invent. Math.* **179**:3 (2010), 559–644.

[Brav and Thomas 2012] C. Brav and H. Thomas, "Classical and derived monodromy of the quintic threefold", preprint, 2012, http://goo.gl/hVlWUA.

[Breuillard et al. 2011] E. Breuillard, B. Green, and T. Tao, "Approximate subgroups of linear groups", *Geom. Funct. Anal.* **21**:4 (2011), 774–819.

[Cassels 1978] J. W. S. Cassels, *Rational quadratic forms*, London Mathematical Society Monographs **13**, Academic Press, London, 1978.

[Chen et al. 2008] Y.-H. Chen, Y. Yang, and N. Yui, "Monodromy of Picard–Fuchs differential equations for Calabi–Yau threefolds", *J. Reine Angew. Math.* **616** (2008), 167–203.

[Ellenberg et al. 2012] J. S. Ellenberg, C. Hall, and E. Kowalski, "Expander graphs, gonality, and variation of Galois representations", *Duke Math. J.* **161**:7 (2012), 1233–1275.

[Fuchs 2010] E. Fuchs, *Arithmetic properties of Apollonian circle packings*, Ph.D. thesis, Princeton University, 2010, http://search.proquest.com/docview/759597603.

[Fuchs 2011] E. Fuchs, "Strong approximation in the Apollonian group", *J. Number Theory* **131**:12 (2011), 2282–2302.

[Fuchs and Rivin] E. Fuchs and I. Rivin, "Finitely generated subgroups of $SL_n(\mathbb{Z})$ are generically thin".

[Fuchs et al. 2013] E. Fuchs, C. Meiri, and P. Sarnak, "Hyperbolic monodromy groups for the hypergeometric equation and Cartan involutions", preprint, 2013. arXiv 1305.0729

[Gamburd 2002] A. Gamburd, "On the spectral gap for infinite index "congruence" subgroups of $SL_2(\mathbb{Z})$", *Israel J. Math.* **127** (2002), 157–200.

[Graham et al. 2003] R. L. Graham, J. C. Lagarias, C. L. Mallows, A. R. Wilks, and C. H. Yan, "Apollonian circle packings: Number theory", *J. Number Theory* **100**:1 (2003), 1–45.

[Hall 2008] C. Hall, "Big symplectic or orthogonal monodromy modulo l", *Duke Math. J.* **141**:1 (2008), 179–203.

[Helfgott 2008] H. A. Helfgott, "Growth and generation in $SL_2(\mathbb{Z}/p\mathbb{Z})$", *Ann. of Math.* (2) **167**:2 (2008), 601–623.

[Kontorovich and Oh 2011] A. Kontorovich and H. Oh, "Apollonian circle packings and closed horospheres on hyperbolic 3–manifolds", *J. Amer. Math. Soc.* **24**:3 (2011), 603–648.

[Levelt 1961] A. H. M. Levelt, *Hypergeometric functions*, Ph.D. thesis, University of Amsterdam, 1961.

[Lubotzky 1994] A. Lubotzky, *Discrete groups, expanding graphs and invariant measures*, Progress in Mathematics **125**, Birkhäuser, Basel, 1994.

[Lubotzky 2012] A. Lubotzky, "Expander graphs in pure and applied mathematics", *Bull. Amer. Math. Soc.* (*N.S.*) **49**:1 (2012), 113–162.

[Margulis 1973] G. A. Margulis, "Explicit constructions of expanders", *Probl. Peredachi Inf.* **9**:4 (1973), 71–80. In Russian; translated in *Problems of Inform. Transmission* **9**:4 (1973), 325–332.

[Nikulin 1987] V. V. Nikulin, "Discrete reflection groups in Lobachevsky spaces and algebraic surfaces", pp. 654–671 in *Proceedings of the International Congress of Mathematicians, I* (Berkeley, CA, 1986), edited by A. M. Gleason, Amer. Math. Soc., Providence, RI, 1987.

[Pyber and Szabó 2010] L. Pyber and E. Szabó, "Growth in finite simple groups of Lie type", preprint, 2010. arXiv 1001.4556

[Rivin 2010] I. Rivin, "Zariski density and genericity", *Int. Math. Res. Not.* **2010**:19 (2010), 3649–3657.

[Salehi Golsefidy and Sarnak 2013] A. Salehi Golsefidy and P. Sarnak, "The affine sieve", *J. Amer. Math. Soc.* **26**:4 (2013), 1085–1105.

[Salehi Golsefidy and Varjú 2012] A. Salehi Golsefidy and P. P. Varjú, "Expansion in perfect groups", *Geom. Funct. Anal.* **22**:6 (2012), 1832–1891.

[Sarnak and Xue 1991] P. Sarnak and X. X. Xue, "Bounds for multiplicities of automorphic representations", *Duke Math. J.* **64**:1 (1991), 207–227.

[Varjú 2012] P. P. Varjú, "Expansion in $\mathrm{SL}_d(\mathbb{O}_K/I)$, I square-free", *J. Eur. Math. Soc. (JEMS)* **14**:1 (2012), 273–305.

[Yu 1995] J.-K. Yu, "Toward a proof of the Cohen–Lenstra conjecture in the function field case", preprint, 1995.

efuchs@math.berkeley.edu *Department of Mathematics, University of California, Berkeley, CA 94720, United States*

Thin groups and superstrong approximation
MSRI Publications
Volume **61**, 2013

The orbital circle method

ALEX KONTOROVICH

We discuss several recent applications of the circle method to the study of integers represented by thin orbits.

1. "Almost every" local-global theorems in thin orbits

A number of articles in these proceedings will discuss the *affine sieve* method which in general produces some (usually unspecified) number of almost-primes in an orbit of affine linear maps [Bourgain et al. 2010a; Salehi Golsefidy and Sarnak 2011]. Our goal in these notes is to highlight several occasions in which actual primes can be produced in the thin setting.[1] In the results surveyed below, these come from the stronger statement of an "almost every" local-to-global phenomenon, which has recently been established in several a priori unrelated settings by Bourgain and the author. The key idea is to replace sieving with the circle method, and develop the latter in the setting of thin orbits. We now state these results.

1A. ***Representing numbers by thin subgroups of*** $\mathbf{SL_2(\mathbb{Z})}$**.** In [Bourgain and Kontorovich 2010] we studied integers represented by certain "thin" subgroups $\Gamma < \mathrm{SL}(2,\mathbb{Z})$, where thin means that Γ is of infinite index in $\mathrm{SL}(2,\mathbb{Z})$ (equivalently, the hyperbolic surface $\Gamma\backslash\mathbb{H}$ has infinite volume). Fix vectors $\boldsymbol{v}_0, \boldsymbol{w}_0 \in \mathbb{Z}^2$, assumed to be primitive (their coordinates are coprime). We shall say that $n \in \mathbb{Z}$ is *represented* by the triple $(\Gamma, \boldsymbol{v}_0, \boldsymbol{w}_0)$ if there is a $\gamma \in \Gamma$ such that $n = \langle \boldsymbol{v}_0\gamma, \boldsymbol{w}_0\rangle$. Here the inner product is the standard one on $\mathbb{R}^2$. Let S be the set of represented integers,

$$S = \langle \boldsymbol{v}_0 \cdot \Gamma, \boldsymbol{w}_0\rangle. \tag{1.1}$$

Partially supported by NSF grants DMS-1209373, DMS-1064214 and DMS-1001252.

[1] For our purposes, "thin" means the following. Let $\Gamma \subset \mathrm{GL}_n(\mathbb{Z})$ be a finitely generated semigroup and for $\boldsymbol{v}_0 \in \mathbb{Z}^n$, let $\mathcal{O} := \boldsymbol{v}_0 \cdot \Gamma$ be a Γ-orbit. Fix a norm $\|\cdot\|$ on $\mathbb{R}^n$, and let $\bar{\mathcal{O}}$ be the Zariski closure of $\mathcal{O}$. We call the orbit $\mathcal{O}$ *thin* if there is an $\eta > 0$ such that, for all T large,

$$|\{\boldsymbol{v} \in \mathcal{O} : \|\boldsymbol{v}\| < T\}| < |\{\boldsymbol{v} \in \bar{\mathcal{O}}(\mathbb{Z}) : \|\boldsymbol{v}\| < T\}|^{1-\eta}.$$

Other authors can simply define a Γ-orbit to be thin if Γ has infinite covolume in its Zariski closure, but this definition only makes sense when Γ is a group. Since our applications require us to consider semigroups Γ, we reformulate thinness as above.

For example, if $v_0 = w_0 = (0, 1)$, then $n \in S$ exactly when it is the bottom right entry of some matrix in Γ. We will see in Section 1B–1C that this setting is, roughly speaking, the starting point, and our other examples will boil down to similar problems. It was shown in [Bourgain and Kontorovich 2010], under some conditions on Γ, that S satisfies an "almost every" local-global principle. First we explain the conditions on Γ, before stating the theorem precisely.

Assume that Γ is free, finitely generated with no parabolic elements, and assume that Γ, while thin, is "not too thin." To explain what this last condition means, recall that one can attach a Poincare series $\mathcal{P}_\Gamma(s)$ to Γ, given by

$$\mathcal{P}_\Gamma(s) := \sum_{\gamma \in \Gamma} \|\gamma\|^{-2s}, \tag{1.2}$$

with $s \in \mathbb{C}$ and $\|\gamma\|$ the Euclidean norm of the matrix entries of γ. This series converges in some half-plane $\operatorname{Re} s \geq C$, and its abscissa of convergence is called the *critical exponent* $\delta = \delta(\Gamma)$. For example, if $\Gamma = \mathrm{SL}_2(\mathbb{Z})$ or any lattice, then $\delta(\Gamma) = 1$, while for Γ thin as above, it is known that $\delta < 1$. The condition that Γ is not too thin is then encoded in the requirement that

$$\delta > 1 - \varepsilon_0, \tag{1.3}$$

for some explicit $\varepsilon_0 > 0$; in particular $\varepsilon_0 = 5 \times 10^{-5}$ suffices.

We now explain the "almost every" local-global principle. Call $n \in \mathbb{Z}$ *admissible* if it is everywhere locally represented, meaning $n \in S \bmod q$ for all q. Let $\mathcal{A}$ denote the set of all admissible numbers. Though it is not *a priori* obvious, it follows from strong approximation that only finitely many q need be checked for admission to $\mathcal{A}$.

Theorem 1.4 [Bourgain and Kontorovich 2010]. *Let Γ be as above. For almost every n (in the sense of natural density),*

$$n \text{ is represented if and only if } n \text{ is admissible.} \tag{1.5}$$

Quantitatively, the number of exceptions to (1.5) *up to N, is at most $O(N^{1-\eta_0})$, for some $\eta_0 > 0$, that is,*

$$\frac{\#(S \cap [1, N])}{\#(\mathcal{A} \cap [1, N])} = 1 + O(N^{-\eta_0}) \quad \text{as } N \to \infty.$$

In particular, almost every admissible prime is produced, since up to N, the exceptional set of order $O(N^{1-\eta_0})$ is too small to contain even a positive proportion of primes, of cardinality $N/\log N$.

On probabilistic grounds, one may expect something like (1.5) to hold with a more relaxed condition on δ in (1.3). Indeed, if $\|\gamma\|$ is about T, then so is $n = \langle v_0 \cdot \gamma, w_0 \rangle$, roughly speaking. The number of such γ is on the order of $T^{2\delta}$, so conspiracies notwithstanding, n is expected to occur with multiplicity $T^{2\delta-1}$. As long as $\delta > \frac{1}{2}$, this multiplicity should grow, giving heuristic evidence that (1.3) can be relaxed to any $\varepsilon_0 < \frac{1}{2}$. If δ falls below $\frac{1}{2}$, then certainly (1.5) is false; the set S, even counted with multiplicity, is already too thin to be even a positive proportion of the integers.

1B. *Zaremba's conjecture.* If $x \in (0, 1)$ has the continued fraction expansion

$$x = \cfrac{1}{a_1 + \cfrac{1}{a_2 + \ddots}},$$

we write $x = [a_1, a_2, \ldots]$. The numbers $a_j \in \mathbb{N}$ are called the *partial quotients* of x. For $A \geq 1$, let $\mathfrak{R}_A$ denote by $\mathfrak{R}_A$ the set of all rational numbers b/d, $(b, d) = 1$, whose partial quotients are bounded by A:

$$\mathfrak{R}_A := \left\{ \frac{b}{d} = [a_1, \ldots, a_k] : (b, d) = 1 \text{ and } a_j \leq A \text{ for all } j \right\}.$$

Let $\mathfrak{D}_A$ be the set of denominators appearing in $\mathfrak{R}_A$:

$$\mathfrak{D}_A := \left\{ d \geq 1 : \text{there exists } b \text{ such that } (b, d) = 1 \text{ and } \frac{b}{d} \in \mathfrak{R}_A \right\}.$$

Conjecture 1.6 [Zaremba 1972, p. 76]. Every number is the denominator of a reduced fraction whose partial quotients are absolutely bounded. That is, there is some absolute $A > 1$ (possibly $A = 5$ suffices) such that

$$\mathfrak{D}_A = \mathbb{N}.$$

This seemingly innocuous conjecture has important applications to the theory of good lattice points in multidimensional quasi-Monte Carlo numerical integration, as well as to the linear congruential method for generating pseudorandom numbers; see, for example, [Niederreiter 1978].

The conjecture is known to be true for a set of density one:

Theorem 1.7 [Bourgain and Kontorovich 2011]. *Almost every number is the denominator of a reduced fraction whose partial quotients are bounded by* 50. *That is, for* $A = 50$,

$$\frac{\#(\mathfrak{D}_A \cap [1, N])}{N} \to 1 \quad \text{as } N \to \infty. \tag{1.8}$$

The best previously known bound, due to Hensley [2006, Theorem 3.2], states that, for any fixed $\varepsilon > 0$, there is some large $A = A(\varepsilon)$ such that

$$\#(\mathfrak{D}_A \cap [1, N]) > C_\varepsilon \cdot N^{1-\varepsilon}.$$

This problem turns out to be nearly identical to that in Section 1A. Observe that $b/d = [a_1, a_2, \ldots, a_k]$ if and only if

$$\begin{pmatrix} * & b \\ * & d \end{pmatrix} = \begin{pmatrix} 0 & 1 \\ 1 & a_1 \end{pmatrix} \begin{pmatrix} 0 & 1 \\ 1 & a_2 \end{pmatrix} \cdots \begin{pmatrix} 0 & 1 \\ 1 & a_k \end{pmatrix}. \tag{1.9}$$

Hence we set Γ_A to be the semigroup generated by the matrices of the above type:

$$\Gamma_A := \left\langle \begin{pmatrix} 0 & 1 \\ 1 & a \end{pmatrix} : a \le A \right\rangle^{+}. \tag{1.10}$$

(The superscript plus sign denotes generation as a semigroup.) Then from (1.9), the set $\mathfrak{D}_A$ of denominators is nothing more than the set of bottom right entries,

$$\mathfrak{D}_A = \langle \boldsymbol{v}_0 \cdot \Gamma_A, \boldsymbol{w}_0 \rangle,$$

with $\boldsymbol{v}_0 = \boldsymbol{w}_0 = (0, 1)$. That is, this set is precisely of the same form as (1.1). One can again attach a Poincaré series (1.2) to Γ_A; it has some critical exponent δ_A, which approaches 1 as $A \to \infty$ [Hensley 1992]. In fact, the condition $A = 50$ in (1.8) is derived from the condition

$$\delta_A > 1 - \tfrac{5}{312}, \tag{1.11}$$

as in (1.3).

1C. ***The local-global conjecture for Apollonian gaskets.*** In Figure 1 we see a bounded integral Apollonian gasket $\mathcal{G}$. This is constructed by starting with three tangent circles in the plane, and iteratively packing new circles into the

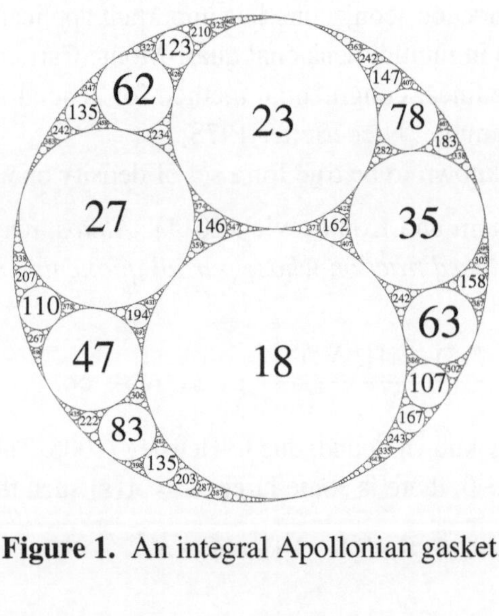

Figure 1. An integral Apollonian gasket.

complementary interstices. The closure of this circle packing in the plane is a fractal-like gasket of Hausdorff dimension

$$\delta \approx 1.3. \tag{1.12}$$

In the special configuration shown, all circles in the gasket have integral curvatures (one over their radii); these are the numbers illustrated. The author and Oh [2011] proved an asymptotic formula for the number of circles C in such a gasket $\mathcal{G}$ whose curvatures $b(C)$ are bounded by a growing parameter T: There is a constant $c > 0$ such that

$$\#\{C \in \mathcal{G} : b(C) < T\} = c\, T^{\delta}(1 + o(1)) \quad \text{as } T \to \infty. \tag{1.13}$$

This result was made effective in Vinogradov's thesis [2012], and independently by Lee and Oh [2013], replacing the error $o(1)$ in (1.13) by an effective rate $O(T^{-\varepsilon_0})$ with $\varepsilon_0 > 0$; see also [Sarnak 2011; Oh and Shah 2013].

In a lovely series of papers on Apollonian packings and generalizations, Graham et al. [2003] posed a different question: what numbers appear in Figure 1? Let $\mathcal{B} = \mathcal{B}(\mathcal{G})$ be the set of curvatures in $\mathcal{G}$,

$$\mathcal{B} := \{n \in \mathbb{Z} : \text{the exists } C \in \mathcal{G} \text{ with } b(C) = n\}.$$

As before, we say n is *represented* if $n \in \mathcal{B}$. A moment's reflection reveals that there are local obstructions, for example in the gasket $\mathcal{G}$ of Figure 1, every number is

$$\equiv 2, 3, 6, 11, 14, 15, 18, \text{ or } 23 \bmod 24. \tag{1.14}$$

As before, call such a number *admissible*, and let $\mathcal{A} = \mathcal{A}(\mathcal{G})$ be the set of admissible numbers. These obstructions, observed empirically in [Graham et al. 2003], are determined rigorously by Fuchs [2011], who proved that for any gasket, admissibility is a condition mod 24.

Conjecture 1.15 [Graham et al. 2003; Fuchs and Sanden 2011]. Every sufficiently large admissible number is represented.

This is again an instance of a local-global problem in thin orbits. (The orbit and thinness will be revealed below.)

Theorem 1.16 [Bourgain and Kontorovich 2013]. *Almost every admissible number is represented. That is,*

$$\frac{\#(\mathcal{B} \cap [1, N])}{\#(\mathcal{A} \cap [1, N])} \to 1 \quad \textit{as } N \to \infty.$$

The best previously known result, due to Bourgain and Fuchs [2011], is the so-called positive density conjecture, which states that

$$\#(\mathcal{B} \cap [1, N]) \gg N.$$

To relate this problem to the previous two, we assume that the reader is familiar with the Apollonian group and the root quadruple of a packing; see for example [Fuchs 2014] in this volume. Specifically, the Apollonian group Γ is the integer matrix group

$$\Gamma := \langle S_1, S_2, S_3, S_4 \rangle \tag{1.17}$$

generated by the following four explicit matrices:

$$S_1 = \begin{pmatrix} -1 & & & \\ 2 & 1 & & \\ 2 & & 1 & \\ 2 & & & 1 \end{pmatrix}, \; S_2 = \begin{pmatrix} 1 & 2 & & \\ & -1 & & \\ & 2 & 1 & \\ & 2 & & 1 \end{pmatrix}, \; S_3 = \begin{pmatrix} 1 & & 2 & \\ & 1 & 2 & \\ & & -1 & \\ & & 2 & 1 \end{pmatrix}, \; S_4 = \begin{pmatrix} 1 & & & 2 \\ & 1 & & 2 \\ & & 1 & 2 \\ & & & -1 \end{pmatrix}.$$

The Apollonian group sits inside the orthogonal group of integer matrices preserving a certain quadratic form Q of signature $(3, 1)$:

$$\Gamma < O_Q(\mathbb{Z}). \tag{1.18}$$

The index of Γ in $O_Q(\mathbb{Z})$ is infinite, but nevertheless Γ is Zariski dense in O_Q.

The root quadruple $\boldsymbol{v}_0 \in \mathbb{Z}^4$ is the set of curvatures corresponding to a configuration of four mutually tangent circles in $\mathcal{G}$ of largest radius (smallest curvature). For the packing $\mathcal{G}$ in Figure 1, the root quadruple is $\boldsymbol{v}_0 = (-10, 18, 23, 27)$. Here the negative sign attached to the outermost circle's curvature accounts for its opposite orientation, being internally tangent to the other circles. It is the only circle with negative curvature.

It is a consequence of Descartes' kissing circles theorem that the orbit

$$\mathcal{O} := \boldsymbol{v}_0 \cdot \Gamma$$

of the root quadruple under the Apollonian group consists of all quadruples of curvatures corresponding to four mutually tangent circles in $\mathcal{G}$. (Note that this orbit is thin, since Γ has infinite index in the ambient group $O_Q(\mathbb{Z})$.) In particular, the entries of $\mathcal{O}$ contain all curvatures in $\mathcal{G}$, which one can recover taking the union of sets of the form

$$\langle \boldsymbol{v}_0 \cdot \Gamma, \boldsymbol{w}_0 \rangle,$$

as $\boldsymbol{w}_0$ ranges through the standard basis vectors

$$\boldsymbol{w}_0 \in \{e_1, \ldots, e_4\}.$$

Here $e_1 = (1, 0, 0, 0), \dots, e_4 = (0, 0, 0, 1)$. So this problem again turns into the form (1.1).

2. Sketches of proofs

2A. ***The circle method.*** We now describe some of the ingredients going into the proofs of Theorems 1.4, 1.7, and 1.16. Specifically, we will highlight the following tools:

- the circle method; decomposition into major and minor arcs;
- infinite volume spectral and representation theory; Lax–Phillips and Patterson–Sullivan theory;
- the thermodynamic formalism of Ruelle transfer operators and their "congruence" versions;
- expansion and uniform spectral gaps;
- strong approximation, and explicit versions thereof;
- thin bisector counting with effective rates;
- Vinogradov's method of estimating bilinear forms;
- estimates of Gauss sums and a Kloosterman refinement.

Recall that in each of our applications, the set of represented numbers is of the form (1.1). We begin with the Hardy–Littlewood circle method, looking at the exponential sum

$$\mathcal{S}_N(\theta) := \sum_{\gamma \in \Omega_N} e(\theta \langle \boldsymbol{v}_0 \gamma, \boldsymbol{w}_0 \rangle), \tag{2.1}$$

where $\theta \in \mathbb{R}/\mathbb{Z}$, $e(x) := e^{2\pi i x}$, and

$$\Omega_N \subset \{\gamma \in \Gamma : \|\gamma\| < N\}$$

is a certain carefully chosen ensemble. For now, one can think of it as the whole Γ-ball of radius N. Observe that n is certainly represented if the n-th Fourier coefficient of $\mathcal{S}_N$,

$$\widehat{\mathcal{S}_N}(n) = \int_0^1 \mathcal{S}_N(\theta) e(-n\theta)\, d\theta = \sum_{\gamma \in \Omega_N} \mathbf{1}_{\{n = \langle \boldsymbol{v}_0 \gamma, \boldsymbol{w}_0 \rangle\}}$$

is nonzero. Here $\mathbf{1}_{\{\cdot\}}$ is the indicator function. According to the circle method, one breaks the range $[0, 1]$ of integration into the major arcs $\mathfrak{M}$ (where $\theta \approx a/q$ with q small) and the complementary minor arcs $\mathfrak{m} = [0, 1] \setminus \mathfrak{M}$, writing

$$\widehat{\mathcal{S}_N}(n) = \mathcal{M}_N(n) + \mathcal{E}_N(n). \tag{2.2}$$

Here

$$\mathcal{M}_N(n) := \int_{\mathfrak{M}} \mathcal{S}_N(\theta) e(-n\theta)\, d\theta$$

should be the *main term* and

$$\mathcal{E}_N(n) := \int_{\mathfrak{m}} \mathcal{S}_N(\theta) e(-n\theta)\, d\theta \tag{2.3}$$

is the *error*. For the main term $\mathcal{M}_N$, one attempts to prove an asymptotic formula, or at least a good lower bound. In particular, since every admissible number is expected to be represented, one would like to show, say, for $N/2 < n < N$, that

$$\mathcal{M}_N(n) \gg \mathfrak{S}(n) \frac{\#\Omega_N}{N}. \tag{2.4}$$

Here $\mathfrak{S}(n) \geq 0$ is a certain product of local densities called the *singular series*. In particular, $\mathfrak{S}(n)$ vanishes if n is not admissible, and otherwise is $\gg N^{-\varepsilon}$ for any $\varepsilon > 0$. The singular series is well-understood, and for ease of exposition, we pretend henceforth that every n is admissible and

$$\mathfrak{S}(n) = 1. \tag{2.5}$$

For the minor arcs integral $\mathcal{E}_N$, one wishes to give an upper bound, asymptotically smaller than the lower bound given for $\mathcal{M}_N$. If one is able to do this at the level of individual n, then one can show that every sufficiently large admissible n is represented. At present, we do not know how to give such strong upper bounds, instead settling for a sharp upper bound on average, in particular in L^2-norm, as follows. From the definition (2.3), Parseval's theorem states that

$$\sum_n |\mathcal{E}_N(n)|^2 = \int_{\mathfrak{m}} |\mathcal{S}_N(\theta)|^2\, d\theta. \tag{2.6}$$

A trivial bound for $\mathcal{S}_N(\theta)$ in (2.1) is just $\leq |\Omega_N|$, giving a trivial bound of $\leq |\Omega_N|^2$ for (2.6). We claim that it suffices for our applications to save a little more than $\sqrt{N}$ on average over $\mathfrak{m}$ off of each term $\mathcal{S}_N$ in (2.6), that is, we seek the bound

$$\int_{\mathfrak{m}} |\mathcal{S}_N(\theta)|^2\, d\theta = o\left(\frac{|\Omega_N|^2}{N}\right). \tag{2.7}$$

First we claim this suffices. Let $\mathfrak{E}(N)$ be the set of exceptional n up to N, that is, those that are admissible but not represented. Certainly the n-th Fourier coefficient (2.2) is nonzero if $\mathcal{M}_N(n) > |\mathcal{E}_N(n)|$, so the number of exceptions is bounded by

$$\#\mathfrak{E}(N) \leq \sum_{\substack{|n|<N \\ n \text{ is admissible}}} \mathbf{1}_{\{|\mathcal{E}_N(n)| \geq \mathcal{M}_N(n)\}}.$$

For admissible n, we have (2.4) and (2.5), thus

$$\#\mathfrak{E}(N) \le \sum_n \mathbf{1}_{\{|\mathcal{E}_N(n)| \gg |\Omega_N|/N\}} \ll \frac{N^2}{|\Omega_N|^2} \sum_n |\mathcal{E}_N(n)|^2.$$

Now applying (2.6) and (2.7) gives

$$\#\mathfrak{E}(N) = o\left(\frac{N^2}{|\Omega_N|^2} \frac{|\Omega_N|^2}{N} \right) = o(N),$$

so almost every admissible number is represented.

In the next two subsections, we focus individually on the tools needed to establish the major arcs bound (2.4) and the error bound (2.7).

2B. ***The major arcs.*** We now sketch an argument leading to (2.4). Recall that $\mathcal{M}_N$ is an integral over the major arcs $\theta \in \mathfrak{M}$, and we may pretend that θ is just equal to a fraction a/q. Then the analysis reduces to evaluating (2.1) at $\theta = a/q$. Pretend now that Ω_N is just the whole Γ-ball:

$$\Omega_N = \{\gamma \in \Gamma : \|\gamma\| < N\}.$$

Then

$$\mathcal{S}_N\left(\frac{a}{q}\right) = \sum_{\substack{\gamma \in \Gamma \\ \|\gamma\| < N}} e\left(\frac{a}{q} \langle \boldsymbol{v}_0 \cdot \gamma, \boldsymbol{w}_0 \rangle\right).$$

The exponent only depends on the residue class of γ mod q. Let $\Gamma_q = \Gamma \bmod q$ be the set of such residue classes. We split the sum as

$$\mathcal{S}_N\left(\frac{a}{q}\right) = \sum_{\gamma_0 \in \Gamma_q} e\left(\frac{a}{q} \langle \boldsymbol{v}_0 \gamma_0, \boldsymbol{w}_0 \rangle\right) \Bigg[\sum_{\substack{\gamma \in \Gamma \\ \|\gamma\| < N}} \mathbf{1}_{\{\gamma \equiv \gamma_0 \bmod q\}} \Bigg]. \tag{2.8}$$

Expansion (that certain corresponding Cayley graphs have a uniform spectral gap) is used critically here to estimate the bracketed term. Roughly speaking, expansion ensures that, for q sufficiently small relative to N, the random walk in Γ mod q is rapidly uniformly mixing, implying that the bracketed term in (2.8) is, up to acceptable error, just

$$\frac{1}{|\Gamma_q|} \sum_{\substack{\gamma \in \Gamma \\ \|\gamma\| < N}} 1.$$

In the group cases (Section 1A and the Apollonian case, Section 1C), one uses uniform spectral gaps [Gamburd 2002; Bourgain et al. 2010a; Bourgain and Varjú 2012; Varjú 2012], while in the semigroup case for Zaremba (Section 1B) one invokes the thermodynamic formalism and "congruence" Ruelle transfer operators; see [Bourgain et al. 2011].

The rest of the analysis of $\mathcal{S}_N$ reduces to a certain local analysis via explicit versions of strong approximation, Goursat's lemma, and some other ingredients; see [Fuchs 2011; Bourgain and Kontorovich 2010, Section 4.3]. Inserting this estimation of $\mathcal{S}_N$ into $\mathcal{M}_N$, one needs a final ingredient of independent interest to establish (2.4), namely an effective infinite-volume bisector estimate, as follows.

Considering first the setting of Section 1A, let $G = \mathrm{SL}(2, \mathbb{R})$, and write the Cartan decomposition $G = KA^+K$,

$$g = k_1(g)a(g)k_2(g)^{-1}.$$

Here $k_1, k_2 \in K = \mathrm{SO}(2)$ and

$$a \in A^+ = \left\{ \begin{pmatrix} e^{-t} & 0 \\ 0 & e^t \end{pmatrix} : t \geq 0 \right\}.$$

Given a nonelementary (not virtually abelian), finitely generated, discrete group $\Gamma < G$ of infinite covolume, and two intervals $\mathcal{I}_1, \mathcal{I}_2 \subset K$, a bisector count is an estimate for

$$\mathcal{N}_{\Gamma,\mathcal{I}_1,\mathcal{I}_2}(T) := \#\{\gamma \in \Gamma \ : \ \|\gamma\| < T,\ k_1(\gamma) \in \mathcal{I}_1,\ k_2(\gamma) \in \mathcal{I}_2\},$$

as $T \to \infty$. In the finite covolume case, such an estimate is given with best known error terms by Good [1983]. To explain the answer in infinite volume, we need to recall a few more notions.

Since Γ acts discontinuously on the hyperbolic plane $\mathbb{H}$, any Γ-orbit has a limit set $\Lambda = \Lambda(\Gamma)$ in the boundary $\partial\mathbb{H} \cong S^1$, and this limit set has some Hausdorff dimension. By Patterson–Sullivan theory [Patterson 1976; Sullivan 1984], this dimension is equal to the critical exponent δ of Γ. It is also connected to the spectral resolution of the hyperbolic Laplacian Δ acting on the Hilbert space of square-integrable functions on $\Gamma\backslash\mathbb{H}$. Namely, it was shown in [Lax and Phillips 1982], under the standard normalization of Δ, that the spectrum below $\frac{1}{4}$ consists of a finite number of discrete eigenvalues, and the spectrum above $\frac{1}{4}$ is purely continuous. Moreover, the discrete spectrum is empty unless $\delta > \frac{1}{2}$, in which case the base eigenvalue is related to the critical exponent by

$$\lambda_0 = \delta(1 - \delta).$$

Corresponding to λ_0 is the base eigenfunction, which can be realized explicitly as the integral of a Poisson kernel against the so-called Patterson–Sullivan measure μ on the boundary. Roughly speaking, μ is the weak* limit as $s \to \delta^+$ of the measures

$$\mu_s(x) := \frac{\sum_{\gamma\in\Gamma} \exp(-s\, d(\mathfrak{o}, \gamma \cdot \mathfrak{o}))\mathbf{1}_{x=\gamma\mathfrak{o}}}{\sum_{\gamma\in\Gamma} \exp(-s\, d(\mathfrak{o}, \gamma \cdot \mathfrak{o}))},$$

where $d(\cdot,\cdot)$ is the hyperbolic distance, and $\mathfrak{o}$ is the fixed point of K in $\mathbb{H}$. Then the theorem proved by Bourgain, Kontorovich and Sarnak [2010b] can be stated as follows. Assuming $\delta > \frac{1}{2}$ and $\partial\mathcal{I}_1 \cap \Lambda = \partial\mathcal{I}_2 \cap \Lambda = \varnothing$, there are some $C > 0$ and $\varepsilon_0 > 0$ such that

$$\mathcal{N}_{\Gamma,\mathcal{I}_1,\mathcal{I}_2}(T) = C\,\mu(\mathcal{I}_1)\,\mu(\mathcal{I}_2)\,T^{2\delta} + O(T^{2\delta-\varepsilon_0}), \tag{2.9}$$

as $T \to \infty$. Here $\varepsilon_0 > 0$ depends only on the spectral gap for Γ. Without effective rates, such an asymptotic statement can be proved by other means — see [Sharp 2001; Oh and Shah 2013], for example — but the error terms are used crucially in our applications. The method of proof is a combination of the above-mentioned spectral theory, as well as ergodic and representation theory of $L^2(\Gamma\backslash G)$.

A similar counting statement is established in [Vinogradov 2012] for the Apollonian case (Section 1C) of $G = \mathrm{SL}(2,\mathbb{C})$ (which is the spin double cover of the ambient orthogonal group; see (1.18)). For Zaremba's problem (Section 1B), we have only a *semigroup* Γ_A, so none of the spectral, ergodic, and representation-theoretic tools are available. Instead, one uses the counting technique pioneered by Lalley [1989] (see also [Dolgopyat 1998; Naud 2005]), analytically continuing the resolvent of the transfer operator, and its congruence version studied in [Bourgain et al. 2011].

These techniques and some more standard circle method analysis lead eventually to (2.4).

2C. *The minor arcs.* We use different strategies to prove (2.7) for the setting of Sections 1A–1B versus the Apollonian setting of Section 1C, so we present them separately. The details are quite involved and will quickly fall outside the scope of this survey. We will only give a few key ideas, encouraging the interested reader to consult the original references.

The setting of Sections 1A–1B. To handle the minor arcs here, we make the observation that the ensemble Ω_N in the definition of $\mathcal{S}_N$ from (2.1) need not be a full Γ-ball. In fact, the definition of $\mathcal{S}_N$ can be changed to, say,

$$\mathcal{S}_N(\theta) := \sum_{\substack{\gamma_1\in\Gamma \\ \|\gamma_1\|<\sqrt{N}}} \sum_{\substack{\gamma_2\in\Gamma \\ \|\gamma_2\|<\sqrt{N}}} e(\theta\langle \boldsymbol{v}_0\gamma_1\gamma_2, \boldsymbol{w}_0\rangle), \tag{2.10}$$

without damaging the major arcs analysis. This new sum encodes much more of the (semi)group structure of Γ, while preserving the property that a nonvanishing n-th Fourier coefficient implies that n is represented. (In reality, we use an even more complicated exponential sum.) The advantage of (2.10) is that, fixing one of the variables, we can apply the Cauchy–Schwarz inequality in the other, and begin to exploit the structure of (2.10) à la Vinogradov's method [Vinogradov 1937] for estimating bilinear forms. In so doing, we pass at some point from the thin and mysterious group Γ (or semigroup Γ_A) to the full ambient group $\mathrm{SL}(2,\mathbb{Z})$. This type of perturbation argument only succeeds when δ is near 1, explaining the restrictions (1.3) and (1.11).

The Apollonian case, Section 1C. The above strategy fails for the Apollonian problem, because the Hausdorff dimension (1.12) is a fixed invariant which cannot be adjusted. Instead, we use the observation from [Sarnak 2007] that the Apollonian group, while of infinite index in $O_Q(\mathbb{Z})$ (see (1.18)), contains arithmetic subgroups of $\mathrm{SO}(2,1)$. In particular, consider the group consisting of all even-length words in three of the generators of page 98 — say, S_1, S_2, S_3:

$$\Xi := \langle S_1, S_2, S_3 \rangle \cap \mathrm{SO}_Q < \Gamma.$$

Sarnak observed that it is isomorphic to the principal congruence subgroup of level 2:

$$\Lambda(2) := \{\gamma \in \mathrm{SL}(2,\mathbb{Z}) : \gamma \equiv I \bmod 2\}.$$

Then, like (2.10), we change the definition of the exponential sum to

$$\mathcal{S}_N(\theta) := \sum_{\substack{\xi\in\Xi \\ \|\xi\|<X}} \sum_{\substack{\gamma\in\Gamma \\ \|\gamma\|<T}} e(\theta\langle v_0 \cdot \xi\,\gamma, w_0\rangle), \tag{2.11}$$

for certain parameters X and T chosen optimally in relation to N. One uses the full sum over the group Γ to capture the major arcs. For the minor arcs bound, one keeps γ essentially fixed (though at some point even the γ sum is used), and tries to get cancellation from the sum on $\xi \in \Xi$. Since Ξ is an arithmetic group, this sum can be converted into a more classical exponential sum, allowing use of more standard tools (bounds for certain Gauss-like sums, and a Kloosterman-like refinement, among other ingredients). In this way, one gets the requisite cancellation in (2.11) and (2.6) to prove the desired bound (2.7).

Acknowledgements

It is a pleasure to thank MSRI and the organizers for the warm hospitality during the "Hot Topics" workshop. I am grateful to Jean Bourgain for the collaboration, and the referees for comments and corrections.

References

[Bourgain and Fuchs 2011] J. Bourgain and E. Fuchs, "A proof of the positive density conjecture for integer Apollonian circle packings", *J. Amer. Math. Soc.* **24**:4 (2011), 945–967.

[Bourgain and Kontorovich 2010] J. Bourgain and A. Kontorovich, "On representations of integers in thin subgroups of $SL_2(\mathbb{Z})$", *Geom. Funct. Anal.* **20**:5 (2010), 1144–1174.

[Bourgain and Kontorovich 2011] J. Bourgain and A. Kontorovich, "On Zaremba's conjecture", preprint, 2011. To appear in *Annals Math.* arXiv 1107.3776

[Bourgain and Kontorovich 2013] J. Bourgain and A. Kontorovich, "On the local-global conjecture for integral Apollonian gaskets", *Inventiones mathematicae* (2013).

[Bourgain and Varjú 2012] J. Bourgain and P. P. Varjú, "Expansion in $SL_d(\mathbf{Z}/q\mathbf{Z})$, q arbitrary", *Invent. Math.* **188**:1 (2012), 151–173.

[Bourgain et al. 2010a] J. Bourgain, A. Gamburd, and P. Sarnak, "Affine linear sieve, expanders, and sum-product", *Invent. Math.* **179**:3 (2010), 559–644.

[Bourgain et al. 2010b] J. Bourgain, A. Kontorovich, and P. Sarnak, "Sector estimates for hyperbolic isometries", *Geom. Funct. Anal.* **20**:5 (2010), 1175–1200.

[Bourgain et al. 2011] J. Bourgain, A. Gamburd, and P. Sarnak, "Generalization of Selberg's $\frac{3}{16}$ theorem and affine sieve", *Acta Math.* **207**:2 (2011), 255–290.

[Dolgopyat 1998] D. Dolgopyat, "On decay of correlations in Anosov flows", *Ann. of Math.* (2) **147**:2 (1998), 357–390.

[Fuchs 2011] E. Fuchs, "Strong approximation in the Apollonian group", *J. Number Theory* **131**:12 (2011), 2282–2302.

[Fuchs 2014] E. Fuchs, "The ubiquity of thin groups", pp. 73–92 in *Thin groups and superstrong approximation*, edited by H. Oh and E. Breuillard, Math. Sci. Res. Inst. Publ. **61**, Cambridge, New York, 2014.

[Fuchs and Sanden 2011] E. Fuchs and K. Sanden, "Some experiments with integral Apollonian circle packings", *Exp. Math.* **20**:4 (2011), 380–399.

[Gamburd 2002] A. Gamburd, "On the spectral gap for infinite index "congruence" subgroups of $SL_2(\mathbb{Z})$", *Israel J. Math.* **127** (2002), 157–200.

[Good 1983] A. Good, *Local analysis of Selberg's trace formula*, Lecture Notes in Mathematics **1040**, Springer, Berlin, 1983.

[Graham et al. 2003] R. L. Graham, J. C. Lagarias, C. L. Mallows, A. R. Wilks, and C. H. Yan, "Apollonian circle packings: number theory", *J. Number Theory* **100**:1 (2003), 1–45.

[Hensley 1992] D. Hensley, "Continued fraction Cantor sets, Hausdorff dimension, and functional analysis", *J. Number Theory* **40**:3 (1992), 336–358.

[Hensley 2006] D. Hensley, *Continued fractions*, World Scientific, Hackensack, NJ, 2006.

[Kontorovich and Oh 2011] A. Kontorovich and H. Oh, "Apollonian circle packings and closed horospheres on hyperbolic 3–manifolds", *J. Amer. Math. Soc.* **24**:3 (2011), 603–648.

[Lalley 1989] S. P. Lalley, "Renewal theorems in symbolic dynamics, with applications to geodesic flows, non-Euclidean tessellations and their fractal limits", *Acta Math.* **163**:1-2 (1989), 1–55.

[Lax and Phillips 1982] P. D. Lax and R. S. Phillips, "The asymptotic distribution of lattice points in Euclidean and non-Euclidean spaces", *J. Funct. Anal.* **46**:3 (1982), 280–350.

[Lee and Oh 2013] M. Lee and H. Oh, "Effective circle count for Apollonian packings and closed horospheres", *Geom. Funct. Anal.* **23**:2 (2013), 580–621.

[Naud 2005] F. Naud, "Expanding maps on Cantor sets and analytic continuation of zeta functions", *Ann. Sci. École Norm. Sup.* (4) **38**:1 (2005), 116–153.

[Niederreiter 1978] H. Niederreiter, "Quasi-Monte Carlo methods and pseudo-random numbers", *Bull. Amer. Math. Soc.* **84**:6 (1978), 957–1041.

[Oh and Shah 2013] H. Oh and N. A. Shah, "Equidistribution and counting for orbits of geometrically finite hyperbolic groups", *J. Amer. Math. Soc.* **26**:2 (2013), 511–562.

[Patterson 1976] S. J. Patterson, "The limit set of a Fuchsian group", *Acta Math.* **136**:3-4 (1976), 241–273.

[Salehi Golsefidy and Sarnak 2011] A. Salehi Golsefidy and P. Sarnak, "Affine sieve", preprint, 2011. arXiv 1109.6432

[Sarnak 2007] P. Sarnak, Letter to J. Lagarias, 2007, http://tinyurl.com/SarnakToLagarias.

[Sarnak 2011] P. Sarnak, "Integral Apollonian packings", *Amer. Math. Monthly* **118**:4 (2011), 291–306.

[Sharp 2001] R. Sharp, "Sector estimates for Kleinian groups", *Port. Math. (N.S.)* **58**:4 (2001), 461–471.

[Sullivan 1984] D. Sullivan, "Entropy, Hausdorff measures old and new, and limit sets of geometrically finite Kleinian groups", *Acta Math.* **153**:3-4 (1984), 259–277.

[Varjú 2012] P. P. Varjú, "Expansion in $\mathrm{SL}_d(\mathcal{O}_K/I)$, I square-free", *J. Eur. Math. Soc. (JEMS)* **14**:1 (2012), 273–305.

[Vinogradov 1937] I. M. Vinogradov, "Representation of an odd number as a sum of three primes", *Dokl. Akad. Nauk SSSR* **15** (1937), 291–294. In Russian; translated in *C. R. (Dokl.) Acad. Sci. URSS* **15** (1937), 169–172.

[Vinogradov 2012] I. Vinogradov, *Effective bisector estimate with application to Apollonian circle packings*, Ph.D. thesis, Princeton University, 2012. arXiv 1204.5498v1

[Zaremba 1972] S. K. Zaremba, "La méthode des "bons treillis" pour le calcul des intégrales multiples", pp. 39–119 in *Applications of number theory to numerical analysis* (Montreal, 1971), edited by S. K. Zaremba, Academic Press, New York, 1972.

alex.kontorovich@yale.edu *Department of Mathematics, Yale University, P.O. Box 208283, New Haven, CT 06520-8283, United States*

Thin groups and superstrong approximation
MSRI Publications
Volume 61, 2013

Sieves in discrete groups, especially sparse

EMMANUEL KOWALSKI

We survey the recent applications and developments of sieve methods related to discrete groups, especially in the case of infinite index subgroups of arithmetic groups.

1. Introduction

Sieve methods appeared in number theory as a tool to try to understand the additive properties of prime numbers, and then evolved over the twentieth century into very sophisticated tools. Not only did they provide extremely strong results concerning the problems most directly relevant to their origin (such as Goldbach's conjecture, the twin primes conjecture, or the problem of the existence of infinitely many primes of the form $n^2 + 1$), but they also became tools of crucial important in the solution of many problems which were not so obviously related (examples are the first proof of the Erdős–Kac theorem, and more recently sieve appeared in the progress, and solution, of the quantum unique ergodicity conjecture of Rudnick and Sarnak).

It is only quite recently that sieve methods have been applied to new problems, often obviously related to the historical roots of sieve, which involve complicated infinite discrete groups (of exponential growth) as basic substrate instead of the usual integers. Moreover, both "small" and "large" sieves turn out to be applicable in this context to a wide variety of very appealing questions, some of which are rather surprising. We will attempt to present this story in this survey.

The basic outline is the following: in Section 2, we present a sieve framework that is general enough to describe both the classical examples and those involving discrete groups; in Section 3, we show how to implement a sieve, with emphasis on small sieves. In Section 5, we take up the large sieve, which we discuss in a fair amount of details since it is only briefly mentioned in [Kowalski 2010] and has the potential to be a very useful general tool even outside of number-theoretic

MSC2010: primary 11N35; secondary 11N36, 22E40.
Keywords: expander graphs, Cayley graphs, sieve methods, prime numbers, thin sets, random walks on groups, large sieve.

contexts. Finally, we conclude with a sampling of problems and further questions in Section 6.

We include a general version of the Erdős–Kac theorem in the context of affine sieve (Theorem 4.12), which follows easily from the method of [Granville and Soundararajan 2007] (it generalizes a result from [Djanković 2011] for Apollonian circle packings).

Apart from this, the writing will follow fairly closely the notes for the course at MSRI, and in particular there will be relatively few details and no attempts at the greatest known generality. The final section had no parallel in the actual lectures, for reasons of time. More information can be gathered from the author's Bourbaki lecture [Kowalski 2010], or from [Salehi Golsefidy 2014] in these proceedings, and of course from the original papers. Overall, we have tried to emphasize general principles and some specific applications, rather than to repeat the more comprehensive survey of known results found in [Kowalski 2010].

Notation. The letter p will always refer to a prime number; for a prime p, we write $\mathbb{F}_p$ for the finite field $\mathbb{Z}/p\mathbb{Z}$. For a set X, $|X|$ is its cardinality, a nonnegative integer or $+\infty$.

The Landau and Vinogradov notation $f = O(g)$ and $f \ll g$ are synonymous, and $f(x) = O(g(x))$ for all $x \in D$ means that there exists an "implied" constant $C \geqslant 0$ (which may be a function of other parameters) such that $|f(x)| \leqslant Cg(x)$ for all $x \in D$. This definition *differs* from that of [Bourbaki 1976, Chapter V] since the latter is of topological nature. We write $f \asymp g$ if $f \ll g$ and $g \ll f$. On the other hand, the notation $f(x) \sim g(x)$ and $f = o(g)$ are used with the asymptotic meaning of loc. cit.

References. As a general reference on sieves in general, the best book available today is the masterful [Friedlander and Iwaniec 2010]. Concerning the large sieve, [Kowalski 2008] contains very general results. We also recommend [Sarnak 2010] on the affine sieve. Another survey of sieves in discrete groups, with a particular emphasis on small sieves, is the Bourbaki seminar [Kowalski 2010], and Salehi Golsefidy's paper [2014] in this volume gives an account of the most general version of the affine sieve, due to Sarnak and him.

2. The setting for sieve in discrete groups

Sieve methods attempt to obtain estimates on the size of sets constructed using local-global and inclusion-exclusion principles. We start by describing a fairly general framework for this type of questions, tailored to applications to discrete groups (there are also other settings of great interest, for example, concerning the distribution of Frobenius conjugacy classes related to families of algebraic varieties over finite fields; see [Kowalski 2008, Chapter 8]).

We will consider a group Γ, viewed as a discrete group, which will usually be finitely generated, and which is given either as a subgroup $\Gamma \subset \mathrm{GL}_r(\mathbb{Z})$ for some $r \geqslant 1$, or more generally is given with a homomorphism

$$\phi : \Gamma \to \mathrm{GL}_r(\mathbb{Z}),$$

which may not be injective (and of course is typically not surjective). Here are three examples.

Examples 2.1. (1) We can take $\Gamma = \mathbb{Z}$, embedded in $\mathrm{GL}_2(\mathbb{Z})$ for instance, using the map

$$\phi(n) = \begin{pmatrix} 1 & n \\ 0 & 1 \end{pmatrix}.$$

This case is of course the most classical.

(2) Consider a finite symmetric set $S \subset \mathrm{SL}_r(\mathbb{Z})$, and let $\Gamma = \langle S \rangle \subset \mathrm{GL}_r(\mathbb{Z})$. Of particular interest for us is the case when Γ is large, in the sense that it is Zariski-dense in SL_r. Recall that this means that there exist no polynomial relations among all elements $g \in \Gamma$ except for those which are consequence of the equation $\det g = 1$. A concrete example is as follows: for $k \geqslant 1$, let

$$S_k = \left\{ \begin{pmatrix} 1 & \pm k \\ 0 & 1 \end{pmatrix}, \begin{pmatrix} 1 & 0 \\ \pm k & 1 \end{pmatrix} \right\},$$

and let $\Gamma^{(k)}$ be the subgroup of $\mathrm{SL}_2(\mathbb{Z})$ generated by S_k. It is well-known that for $k \geqslant 1$, this is a Zariski-dense subgroup of SL_2.

We are especially interested in situations where Γ is nevertheless "small", in the sense that the index of Γ in the arithmetic group $\mathrm{SL}_r(\mathbb{Z})$ is *infinite*. We will call this the *sparse* case (though the terminology *thin* is also commonly used, we will wish to speak later of thin subsets of SL_r, as defined by Serre, and Γ is not thin in this sense).

In the example above, the groups $\Gamma^{(1)} = \mathrm{SL}_2(\mathbb{Z})$ and $\Gamma^{(2)}$ are of finite index in $\mathrm{SL}_2(\mathbb{Z})$ (the latter is the kernel of the reduction map modulo 2), but $\Gamma^{(k)}$ is sparse for all $k \geqslant 3$. In particular, the subgroup $\Gamma^{(3)}$ is sometimes known as the Lubotzky group.

(3) Here is an example where the group Γ is not given as a subgroup of a linear group: for an integer $g \geqslant 1$, let Γ be the mapping class group of a closed surface Σ_g of genus g, and let

$$\phi : \Gamma \to \mathrm{Sp}_{2g}(\mathbb{Z}) \subset \mathrm{GL}_{2g}(\mathbb{Z})$$

be the map giving the action of Γ on the first homology group $H_1(\Sigma_g, \mathbb{Z}) \simeq \mathbb{Z}^{2g}$, which is symplectic with respect to the intersection pairing on $H_1(\Sigma_g, \mathbb{Z})$. Here it is known (for instance, through the use of specific generators of Γ mapping to elementary matrices in $\mathrm{Sp}_{2g}(\mathbb{Z})$) that ϕ is surjective. (All facts on mapping class groups that we will use are fairly elementary and are contained in [Farb and Margalit 2011].)

The next piece of data are surjective maps

$$\pi_p : \Gamma \to \Gamma_p,$$

where p runs over prime numbers (or possibly over a subset of them) and Γ_p are finite groups. We view each such map as giving "local" information at the prime p, typically by reduction modulo p. Indeed, in all cases in this text, the homomorphism π_p is the composition

$$\Gamma \xrightarrow{\phi} \mathrm{GL}_r(\mathbb{Z}) \longrightarrow \mathrm{GL}_r(\mathbb{F}_p)$$

of ϕ with the reduction map of matrices modulo p, and Γ_p is defined as the image of this map.

Examples 2.2. (1) For $\Gamma = \mathbb{Z}$, reduction modulo p is surjective onto $\Gamma_p = \mathbb{Z}/p\mathbb{Z}$ for all primes.

(2) If Γ is Zariski-dense in SL_r, and we use reduction modulo p to define π_p, it is a consequence of general *strong approximation* statements that there exists a finite set of primes $T(\Gamma)$ such that π_p has image *equal* to $\mathrm{SL}_r(\mathbb{F}_p)$ for all $p \notin T(\Gamma)$, and in particular for all primes large enough.[1] For instance, in the case of the subgroups $\Gamma^{(k)} \subset \mathrm{SL}_2(\mathbb{Z})$, this property is visibly valid with

$$T(\Gamma^{(k)}) = \{\text{primes } p \text{ dividing } k\}.$$

We refer to the survey [Rapinchuk 2014] in these proceedings for a general account of strong approximation.

(3) For the mapping class group Γ of Σ_g, and ϕ given by the action on homology, the image of reduction modulo p is equal to $\mathrm{Sp}_{2g}(\mathbb{F}_p)$ for all primes p (simply because ϕ is onto, and $\mathrm{Sp}_{2g}(\mathbb{Z})$ surjects to $\mathrm{Sp}_{2g}(\mathbb{F}_p)$ for all p).

We want to combine the maps π_p, corresponding to local information, modulo many primes in order to get "global" results. This clearly only makes sense if using more than a single prime leads to an increase of information. Intuitively, this is the case when the reduction maps π_p, π_q, associated to distinct primes p and q are *independent*: knowing the reduction modulo p of an element of Γ should give no information concerning the reduction modulo q. We therefore make the following assumption on the data:

[1] This is directly related to the fact that SL_r is, as a linear algebraic group, connected and simply connected.

Assumption 2.3 (independence). There exists a finite set of primes $T_1(\Gamma)$, sometimes called the Γ-*exceptional primes*, such that for any finite set I of primes $p \notin T_1(\Gamma)$, the simultaneous reduction map

$$\pi_I : \Gamma \to \prod_{p \in I} \Gamma_p$$

modulo primes in I is onto.

We will write

$$\Gamma_I = \prod_{p \in I} \Gamma_p, \quad q_I = \prod_{p \in I} p.$$

Note that q_I is a squarefree integer, coprime with $T_1(\Gamma)$.

Examples 2.4. (1) For $\Gamma = \mathbb{Z}$, the Chinese remainder theorem shows that for any finite set of primes I, we have

$$\prod_{p \in I} \mathbb{Z}/p_i\mathbb{Z} \simeq \mathbb{Z}/q_I\mathbb{Z},$$

and hence the map π_I above can be identified with reduction modulo q_I. In particular, it is surjective, so that the assumption holds with an empty set of exceptional primes.

(2) If $\Gamma \subset \mathrm{GL}_r(\mathbb{Z})$ has Zariski closure SL_r, then the independence assumption holds for the same set of primes $T_1(\Gamma) = T(\Gamma)$ such that π_p is surjective onto $\mathrm{SL}_r(\mathbb{F}_p)$ for $p \notin T(\Gamma)$, simply for group-theoretic reasons: any subgroup of a finite product

$$\prod_{p \in I} \mathrm{SL}_r(\mathbb{F}_p),$$

which surjects to each factor $\mathrm{SL}_r(\mathbb{F}_p)$ is equal to the whole product. (This type of result is known as Goursat's lemma, as in [Chavdarov 1997, Proposition 5.1], or Hall's lemma, as in [Dunfield and Thurston 2006, Lemma 3.7].) Again a similar property holds if the Zariski closure of Γ is an almost simple, connected, simply connected algebraic group.

(3) In particular, the independence assumption holds with $T_1(\Gamma) = \varnothing$ for the mapping class group of Σ_g acting on the homology of the surface, because Goursat's lemma applies to the finite groups $\mathrm{Sp}_{2g}(\mathbb{F}_p)$.

(4) The independence assumption may fail, for instance in the context of orthogonal groups, when there is a global invariant which can be read off any reduction. The simplest example of such an invariant is the determinant: if $\Gamma \subset \mathrm{GL}_r(\mathbb{Z})$ is not contained in $\mathrm{SL}_r(\mathbb{Z})$, the compatibility condition

$$\det \pi_p(g) = \det g \in \{\pm 1\} \subset \mathbb{F}_p^\times,$$

valid for all p and $g \in \mathrm{GL}_r(\mathbb{Z})$, shows that the image of π_I is always contained in the proper subgroup

$$\{(g_p) \in \Gamma_I \mid \det g_p = \det g_q \text{ for all } p, q \in I\}$$

(identifying all copies of $\{\pm 1\}$). This issue appears, concretely, in the example of the Apollonian group and Apollonian circle packings, since the latter is a subgroup of an indefinite orthogonal group intersecting both cosets of the special orthogonal group, see [Fuchs 2011; Fuchs and Sanden 2011] for a precise analysis of this case, and [Fuchs 2013] for a survey.

It should be emphasized that this failure of the independence assumption is not dramatic: one can replace Γ by $\Gamma \cap \mathrm{SL}_r$ for instance, or by the other coset of the determinant (with some adaptation since this is not a group).

We can now define the sifted sets $\mathcal{S} \subset \Gamma$ constructed by inclusion-exclusion using local information: given a set $\mathcal{P}$ of primes (usually finite), and subsets

$$\Omega_p \subset \Gamma_p,$$

for $p \in \mathcal{P}$, we let

$$\mathcal{S} = \mathcal{S}(\mathcal{P}; \Omega) = \{g \in \Gamma \mid \pi_p(g) \notin \Omega_p \text{ for all } p \in \mathcal{P}\} = \bigcap_{p \in \mathcal{P}} (\Gamma - \pi_p^{-1}(\Omega_p)).$$

We want to know something about the size, or maybe more ambitiously the structure, of such sifted sets. In fact, quite often, we wish to study sets which are not exactly of this shape, but are closely related.

Frequently, we have an integer parameter $Q \geqslant 1$, and we take $\mathcal{P} = \{p \leqslant Q\}$, the set of primes up to Q. In that case, we will often denote $\mathcal{S}(Q; \Omega) = \mathcal{S}(\mathcal{P}; \Omega)$, and we may even sometimes simplify this to $\mathcal{S}(Q)$ if it is clear that Ω is fixed.

Examples 2.5. (1) Let $\Gamma = \mathbb{Z}$, and let $\Omega_p = \{0, -2\} \subset \mathbb{F}_p$ for all primes $p \leqslant Q$, where $Q \geqslant 2$ is some parameter. Taking $\mathcal{P} = \{p \leqslant Q\}$, we have by definition

$$\mathcal{S}(Q) = \mathcal{S}(\mathcal{P}; \Omega) = \{n \in \mathbb{Z} \mid \text{neither } n \text{ nor } n+2 \text{ has a prime factor } \leqslant Q\}.$$

In particular, for $N \geqslant 1$, the initial segment $\mathcal{S}(Q) \cap \{1, \ldots, N\}$ contains all twin primes n between Q and N, that is, all primes p with $Q < p \leqslant N$ such that $p+2$ is also prime. Hence an upper bound on the size of this initial segment will be an upper bound for the number of twin primes in this range. This is valid independently of the value of Q. Furthermore, if $Q \geqslant \sqrt{N+2}$, we have in fact equality: an integer $n \in \mathcal{S}(\sqrt{N+2}) \cap \{1, \ldots, N\}$ must be prime, as well as $n+2$, since both integers only have prime factors larger than their square-root. More generally, if $Q = N^{\beta}$ for some $\beta > 0$, we see that $\mathcal{S}(Q) \cap \{1, \ldots, N\}$ contains only integers n such that both n and $n+2$ have less than $1/\beta$ prime factors.

(2) The first example is the prototypical example showing how sieve methods are used to study prime patterns of various type. Bourgain, Gamburd and Sarnak [2010] extended this type of questions to discrete subgroups of $\mathrm{GL}_r(\mathbb{Z})$. We present here a special case of what is called the *affine linear sieve* or the *sieve in orbits*. There will be a few other examples below, and we refer to the original paper or to [Kowalski 2010] for a more general approach.

We assume for simplicity, as before, that Γ is Zariski-dense in SL_r. Let

$$f : \mathrm{SL}_r(\mathbb{Z}) \to \mathbb{Z}$$

be a nonconstant polynomial function, for instance the product of the coordinates. We want to study the multiplicative properties of the integers $f(g)$ when g runs over Γ. Consider

$$\Omega_p = \{g \in \Gamma_p \mid f(g) \equiv 0 \pmod p)\} \subset \Gamma_p \subset \mathrm{SL}_r(\mathbb{F}_p), \tag{2-1}$$

for $p \leqslant Q$. Then $\mathcal{S}(Q; \Omega)$ (recall that this is the sifted set for $\mathcal{P} = \{p \leqslant Q\}$) is the set of $g \in \Gamma$ such that $f(g)$ has no prime factor $\leqslant Q$. In particular, for any $\Delta > 0$, the intersection

$$\mathcal{S}(Q; \Omega) \cap \{g \in \Gamma \mid |f(g)| \leqslant Q^{\Delta}\}$$

consists of elements where $f(g)$ has $< \Delta$ prime factors. For instance, when f is the product of coordinates, this set contains elements $g \in \Gamma$ where all coordinates have less than Δ prime factors.

(3) For our last example, consider the mapping class group Γ of Σ_g. Let $\mathcal{H}_g$ be a handlebody with boundary Σ_g. For a mapping class $\phi \in \Gamma$, we denote by $\mathcal{M}_\phi$ the compact 3-manifold obtained by Heegaard splitting using $\mathcal{H}_g$ and ϕ, that is, it is the union of two copies of $\mathcal{H}_g$ where the boundaries are identified using (a representative of) ϕ (see [Dunfield and Thurston 2006] for more about this construction).

The image J of $H_1(\mathcal{H}_g, \mathbb{Z}) \simeq \mathbb{Z}^g$ in $H_1(\Sigma_g, \mathbb{Z}) \simeq \mathbb{Z}^{2g}$ is a lagrangian subspace (i.e., a subgroup of rank g such that the intersection pairing is identically zero on J). We denote by $J_p \subset \mathbb{F}_p^{2g}$ its reduction modulo p. It follows from algebraic topology that

$$H_1(\mathcal{M}_\phi, \mathbb{Z}) \simeq H_1(\Sigma_g, \mathbb{Z})/\langle J, \phi \cdot J\rangle,$$
$$H_1(\mathcal{M}_\phi, \mathbb{F}_p) \simeq H_1(\mathcal{M}_\phi, \mathbb{Z}) \otimes \mathbb{F}_p \simeq H_1(\Sigma_g, \mathbb{F}_p)/\langle J_p, \phi \cdot J_p\rangle.$$

Thus if we let

$$\begin{aligned}\Omega_p &= \{\gamma \in \mathrm{Sp}_{2g}(\mathbb{F}_p) \mid \gamma \cdot J_p \cap J_p = \varnothing\} \\ &= \{\gamma \in \mathrm{Sp}_{2g}(\mathbb{F}_p) \mid \langle J_p, \gamma \cdot J_p\rangle = \mathbb{F}_p^{2g}\},\end{aligned} \tag{2-2}$$

we see that any sifted set $\mathcal{S}(\mathcal{P};\Omega)$ contains all mapping classes such that $\mathcal{M}_\phi$ has first rational Betti number positive.

We will discuss this example further in Section 5. The reader who is not familiar with sieve is however encouraged to try to find the answer to the following question: What is the great difference that exists between this example and the previous ones?

3. Conditions for sieving

Having defined sifted sets and seen that they contain information of great potential interest, we want to say something about them. The basic question is "How large is a sifted set $\mathcal{S}$?" In order to make this precise, some truncation of $\mathcal{S}$ is needed, since in general this is (or is expected to be) an infinite set. In fact, we saw in the simplest examples (e.g., twin primes) that this truncation (in that case, the consideration of an initial segment of a sifted set) is crucially linked to deriving interesting information from $\mathcal{S}$, as one needs usually to handle a truncation which is correlated with the size of the primes in the set $\mathcal{P}$ defining the sieve conditions.

When sieving in the generality we consider, it is a striking fact that there are different ways to truncate the sifted sets, or indeed to measure subsets of Γ in general (although those we describe below seem, ultimately, to be closely related.) We will speak of *counting methods* below to refer to these various truncation techniques.

Method 1 (Archimedean balls). Fix a norm $\|\cdot\|$ (or some other metric) on the ambient Lie group $\mathrm{GL}_r(\mathbb{R})$ (for instance the operator norm as linear maps on euclidean space, but other choices are possible) and consider

$$\mathcal{S}\cap\{g\in\Gamma \mid \|g\|\leqslant T\},$$

for some parameter $T\geqslant 1$. This is a finite set, and one can try to estimate (from above or below, or both) its cardinality.

Example 3.1. Let Γ be a Zariski-dense subgroup of $\mathrm{SL}_r(\mathbb{Z})$ and f a nonconstant polynomial function on $\mathrm{SL}_r(\mathbb{Z})$. For some $d\geqslant 1$, we have

$$|f(g)|\ll \|g\|^d,$$

for all $g\in\Gamma$. Hence if we consider the sifted set (2-1) for $Q=T^\beta$, the elements in

$$\mathcal{S}(Q)\cap\{g\in\Gamma \mid \|g\|\leqslant T\}$$

are such that $f(g)$ has at most d/β prime factors.

Counting in archimedean balls in subgroups of arithmetic groups, even without involving sieve, is a delicate matter, especially in the sparse case, which involves deep ideas from spectral theory, harmonic analysis and ergodic theory. We refer to [Gorodnik and Nevo 2009] for the case of arithmetic groups, and to [Oh 2011; 2014] for the sparse case, as well as to [Mohammadi and Oh 2013], concerning geometrically finite subgroups of isometries of hyperbolic spaces.

Method 2 (combinatorial balls). Since the groups Γ of interest are most often finitely generated, and indeed sometimes given with a set of generators, one can replace the archimedean metric of the first method with a combinatorial one. Thus if $S = S^{-1}$ is a generating set of Γ, we denote by $l_S(g)$ the word-length metric on Γ defined using S. The sets

$$\mathcal{S} \cap \{g \in \Gamma \mid l_S(g) \leqslant T\} \quad \text{or} \quad \mathcal{S} \cap \{g \in \Gamma \mid l_S(g) = T\}$$

are again finite, and one can attempt to estimate their size.

This method is particularly interesting when S is a set of free generators of Γ (and their inverses), because one knows precisely the size of the balls for the combinatorial metric in that case. And even if this is not the case, one can often find a subgroup of Γ which is free of rank $\geqslant 2$, and use this subgroup instead of the original Γ. (This technique is used in [BGS 2010]; in that case, the necessary free subgroup is found using the Tits alternative, a very specific case of which says that if Γ is Zariski-dense in SL_r, then it contains a free subgroup of rank 2.)

Method 3 (random walks). Instead of trying to reduce to free groups using a subgroup, one can replace Γ by the free group $F(S)$ generated by S and use the obvious homomorphisms

$$\phi : F(S) \to \Gamma \to \mathrm{GL}_r(\mathbb{Z}) \quad \text{and} \quad F(S) \to \Gamma \to \Gamma_p$$

to define sieve problems and sifted sets. An alternative to this description is to use the generating set S and count elements in balls for the word-length metric l_S *with multiplicity*, the multiplicity being the number of representations of $g \in \Gamma$ by a word of given (or bounded) length. This means one measures the size of a set $X \subset \Gamma$ truncated to the sphere of radius $N \geqslant 1$ around the origin by its density

$$\mu_N(X) = \frac{1}{|S|^N} \big|\{(s_1, \ldots, s_N) \in S^N \mid s_1 \cdots s_N \in X\}\big|,$$

and therefore one tries to measure the density of the sifted set $\mu_N(\mathcal{S})$, as a way of measuring its size within a given ball. If one wishes to measure balls instead of spheres, a simple expedient is to replace S by $S_1 = S \cup \{1\}$ (since the sphere of radius N for l_{S_1} is the ball of radius N for l_S).

It is often convenient to think of this in terms of a random walk: one assumes given, on a probability space Ω, a sequence of independent S-valued random variables ξ_n, and one defines a random walk (γ_n) on Γ by

$$\gamma_0 = 1,$$
$$\gamma_{n+1} = \gamma_n \xi_{n+1} \quad \text{for } n \geqslant 0.$$

If all steps ξ_n are uniformly distributed on S, it follows that

$$\mu_N(X) = \mathbf{P}(\gamma_N \in X),$$

or in other words, the density μ_N is the probability distribution of the N-th step of this random walk.

Example 3.2. The analogue (for methods 2 and 3) of the argument in Example 3.1 is the following: given a function f as in that example, there exists $C \geqslant 1$ such that, for all $g \in \Gamma$, we have

$$|f(g)| \leqslant C^{l_S(g)}$$

(simply because the operator norm of g is submultiplicative and hence grows at most exponentially with the word-length metric). Thus elements which have word-length at most N and belong to a sifted set $\mathcal{S}(Q;\Omega)$ with Q of size A^N, for some $A > 1$, have at most $(\log A)/(\log C)$ prime factors.

Example 3.3 (Dunfield–Thurston random manifolds). This third counting method is the least familiar to classical analytic number theory. This random walk approach was however already considered by Dunfield and Thurston [2006] as a way of studying random 3-manifolds, using the Heegaard-splitting construction based on mapping class groups as in Example 2.5(3): given an integer $g \geqslant 1$, they consider a finite generating set S of the mapping class group Γ of Σ_g and the associated random walk (ϕ_n). The 3-manifolds $\mathcal{M}_{\phi_n}$ are then "random 3-manifolds" and some of their properties can be studied using sieve methods.

It is of course useful to have a way of considering these three methods in parallel. This can be done by assuming that one has a sequence (μ_N) of finite measures on Γ, and by considering the problem of estimating $\mu_N(\mathcal{S})$, the measure of the sifted set. In Method 1, these measures would be the uniform counting measure on the intersection of Γ with the balls of radius N in $\mathrm{GL}_r(\mathbb{R})$, in Method 2, the uniform counting measure on the combinatorial ball of radius N, and in Method 3, the probability law of the N-th step of the random walk.

4. Implementing sieve with expanders

We will now explain how all this relates to expanders. The one-line summary is that the expander condition will allow us to apply classical results of sieve theory to settings of discrete groups with exponential growth (one might prefer to say "in nonamenable settings"). We can motivate this convincingly as follows.

The simplest possible sieve problem occurs when the set $\mathcal{P}$ of conditions is restricted to a single prime, and one is asking for

$$\mu_N(\{g \in \Gamma \mid \pi_p(g) = g_0\}),$$

for a fixed prime p and a fixed $g_0 \in \Gamma_p$. One sees that, assuming p is fixed, this elementary-looking question concerns the distribution of the image of the sequence $\pi_{p,*}\mu_N$ of measures on the finite group Γ_p. This may well be expected to have a good answer.

Example 4.1. Consider (one last time) the classical case $\Gamma = \mathbb{Z}$. If we truncate by considering initial segments $\{1, \dots, N\}$, we are asking here about the number of positive integers $\leqslant N$ congruent to a given a modulo p. The proportion of these converges of course to $1/p$, and this is usually so self-evident that one never mentions it specifically. (But, still in classical cases, note that if one starts the sieve from the set of primes instead of $\mathbb{Z}$, then this basic question is resolved by Dirichlet's theorem on primes in arithmetic progressions, and the uniformity in this question is basically the issue of the generalized Riemann hypothesis.)

On intuitive grounds as well as theoretically, one can expect that the probability that g reduces modulo p to g_0 should be about $1/|\Gamma_p|$. This amounts to expecting that the probability measures $\pi_{p,*}(\mu)/\mu_N(\Gamma)$ converge weakly to the uniform (Haar) probability measure on this finite group. It is when considering uniformity of such convergence that expander graphs enter the picture.

We can already deduce from this intuition the following heuristic concerning the size of a sifted set $\mathcal{S}(\mathcal{P}; \Omega)$: each condition $\pi_p(g) \notin \Omega_p$ should hold with probability approximately

$$1 - \frac{|\Omega_p|}{|\Gamma_p|},$$

and these sieving conditions, for distinct primes, should be independent. Hence one may expect that

$$\mu_N(\mathcal{S}(\mathcal{P}; \Omega)) \approx \mu_N(\Gamma) \prod_{p \in \mathcal{P}} \left(1 - \frac{|\Omega_p|}{|\Gamma_p|}\right). \tag{4-1}$$

(The symbol $\approx$ here only means that the right side is a first guess for the left...)

The simplest counting method to explain this is Method 3, where the argument is very transparent. We therefore assume in the remainder of this section that μ_N is the probability law of the N-th step of a random walk on Γ as above.

It is then an immediate corollary of the theory of finite Markov chains (applied to the random walk on the Cayley graph of Γ_p induced by that on Γ) that, if

$1 \in S$ (or more generally if this Cayley graph is not bipartite, that is, if there exists no surjective homomorphism $\Gamma_p \to \{\pm 1\}$ such that each generator $s \in S$ maps to -1), we have exponentially fast convergence to the probability Haar measure. Precisely, let M_p be the Markov operator acting on functions on Γ_p by

$$(M_p\varphi)(x) = \frac{1}{|S|}\sum_{s\in S}\varphi(xs).$$

This operator also acts on functions of mean 0, that is, on the space $L_0^2(\Gamma_p)$ of functions such that

$$\sum_{g\in\Gamma_p}\varphi(g) = 0,$$

and has real eigenvalues. Let $\rho_p < 1$ be its spectral radius (it is *less* than 1 because the eigenvalue 1 is removed by restricting to L_0^2, while -1 is not an eigenvalue because the graph is not bipartite). We then have

$$\left|\mu_N(\pi_p(g) = g_0) - \frac{1}{|\Gamma_p|}\right| \leqslant \rho_p^N,$$

for all $N \geqslant 1$.

More generally, under the independence assumption (Assumption 2.3), if I is a finite set of primes not in $T_1(\Gamma)$, the same argument applied to the quotient

$$\Gamma \to \Gamma_I = \prod_{p\in I}\Gamma_p$$

shows that for any $(g_p) \in \Gamma_I$, we have

$$\left|\mu_N\big(\pi_p(g) = g_p \text{ for } p \in I\big) - \prod_{p\in I}\frac{1}{|\Gamma_p|}\right| \leqslant \rho_I^N, \tag{4-2}$$

where $\rho_I < 1$ is the corresponding spectral radius for Γ_I. It follows by summing over $x = (g_p) \in \Gamma_I$ that we have a quantitative equidistribution

$$\int \varphi\big((\pi_p(g))_{p\in I}\big)d\mu_N(g) = \frac{1}{|\Gamma_I|}\sum_{x\in\Gamma_I}\varphi(x) + O(|\Gamma_I|\|\varphi\|_\infty\rho_I^N) \tag{4-3}$$

(with an absolute implied constant) for any function φ on Γ_I.

In particular, we see that if $\mathcal{P}$ is a fixed set of primes (not in $T_1(\Gamma)$), then as $N \to +\infty$, the basic heuristic (4-1) is valid asymptotically:

$$\lim_{N\to+\infty}\mu_N(\mathcal{S}(\mathcal{P};\Omega)) = \lim_{N\to+\infty}\mathbf{P}(\gamma_N \in \mathcal{S}(\mathcal{P};\Omega)) = \prod_{p\in\mathcal{P}}\left(1 - \frac{|\Omega_p|}{|\Gamma_p|}\right). \tag{4-4}$$

We will call this a *bounded sieve statement.*

The difficulty (and fun!) of sieve methods is that the sifted sets of most interest are such that the primes involved in $\mathcal{P}$ are *not* fixed as $N \to +\infty$: they are in ranges increasing with the size of the elements being considered (as shown already by the example of the twin primes). It is clear that in order to handle such sifted sets, we need a uniform control of the equidistribution properties modulo primes, and modulo finite sets of primes simultaneously. The best we can hope for is that (4-2) hold with the spectral radius bounded away from one *independently of* I. This is, of course, exactly the conditions under which the family of Cayley graphs of Γ_I with respect to the generators S is a family of (absolute) expander graphs.

Remark 4.2. We have discussed the example of the random walk counting method. It is a fact that analogues of (4-2) hold in all cases where sieve methods have been successfully applied. Moreover, these analogues hold uniformly with respect to I, and ultimately, the source is always equivalent to the expansion property of the Cayley graphs, although the proofs and the equivalence might be much more involved than the transparent argument that exists in the case of random walks.

Example 4.3. The first case beyond the classical examples (or the case of arithmetic groups, where property (T) or (τ) can be used,[2] although this also had not been done before) where sieve in discrete groups was implemented is due to Bourgain, Gamburd and Sarnak [BGS 2010], who (based on [Helfgott 2008; Bourgain and Gamburd 2008]) proved that if Γ is a finitely generated Zariski-dense subgroup of $\mathrm{SL}_2(\mathbb{Z})$ (or even of $\mathrm{SL}_2(\mathbb{O})$, where $\mathbb{O}$ is the ring of integers in a number field), the Cayley graphs of Γ_I, where I runs over finite subsets of $T_1(\Gamma)$, form a family of (absolute) expanders. The problem of generalizing this to SL_r, or to Zariski-dense subgroups of other algebraic groups, was one of the motivations for the recent developments of this result, and of the basic growth theorem of Helfgott, to more general groups. We now know an essentially best possible result; see [Salehi Golsefidy and Varjú 2012; Varjú 2012] and the surveys [Salehi Golsefidy 2014; Breuillard 2014; Pyber and Szabó 2014] in this volume for introductions to this area).

Theorem 4.4 (Salehi Golsefidy and Varjú). *Let $\Gamma \subset \mathrm{GL}_r(\mathbb{Z})$ be finitely generated by $S = S^{-1}$, with Zariski-closure $\boldsymbol{G}$. For p prime, let Γ_p be the image of Γ under reduction modulo p, and for a finite set of primes I, let Γ_I be the image of Γ in $\prod_{p\in I} \Gamma_p$, under the simultaneous reduction homomorphism.*

If the connected component of the identity in $\boldsymbol{G}$ is a perfect group, then there exists a finite set of primes $T_1(\Gamma)$ such that the family of Cayley graphs of Γ_I, for $I \cap T_1(\Gamma) = \varnothing$, is an expander family.

[2]See [Gorodnik and Nevo 2012] for the best known in this direction.

We can now describe the implication of some classical sieve results in the context of discrete groups. We assume formally the following:

Assumption 4.5 (expansion). There exists a finite set of primes $T_2(\Gamma)$ such that Γ satisfies the independence assumption 2.3 for primes not in $T_2(\Gamma)$, and furthermore the family of Cayley graphs of Γ_I, for $I \cap T_2(\Gamma) = \varnothing$, is an expander family; that is, there exists $\rho < 1$, such that for any finite set I of primes $p \notin T_2(\Gamma)$, the spectral radius for the Markov operator on Γ_I satisfies

$$\rho_I \leqslant \rho.$$

By (4-2), this assumption implies that the asymptotic formula

$$\mathbf{P}\big(\pi_p(\gamma_n) = g_p \text{ for } p \in I\big) \sim \prod_{p \in I} \frac{1}{|\Gamma_p|} \tag{4-5}$$

holds uniformly for $n \geqslant 1$ and sets I such that $|\Gamma_I| \leqslant \tilde{\rho}^{-n}$, for any $\tilde{\rho} > \rho$. If we assume that

$$|\Gamma_p| \leqslant p^B, \tag{4-6}$$

for some fixed $B \geqslant 1$, this means that we can control simultaneously and uniformly all reductions of the N-th step as long as $q_I \leqslant \tilde{\rho}^{-n/B}$. Note that (4-6) is not very restrictive: it holds (with $B = r^2$) if π_p is just the reduction modulo p on $\mathrm{GL}_r(\mathbb{Z})$, which is the case in all our applications.

The most classical types of sieve are those when the sieving conditions determined by Ω_p hold with probability approximately κ/p, at least on average, were κ is a fixed real number traditionally called the *dimension* of the sieve. Precisely, we say that (Ω_p) is of dimension κ if we have[3]

$$\sum_{p \leqslant X} \frac{|\Omega_p|}{|\Gamma_p|} \log p = \kappa \log X + O(1), \tag{4-7}$$

for $X \geqslant 2$. This is certainly true, by the prime number theorem, if

$$\frac{|\Omega_p|}{|\Gamma_p|} = \frac{\kappa}{p} + O\Big(\frac{1}{p^{1+\delta}}\Big),$$

for some $\delta > 0$ and all p prime.

We then have the following basic result:

Theorem 4.6 (small sieve in discrete groups). *Let Γ be a discrete group finitely generated by $S = S^{-1}$, given with $\phi : \Gamma \to \mathrm{GL}_r(\mathbb{Z})$ and surjective homomorphisms π_p to finite groups Γ_p as above, in particular with* (4-6) *for some*

[3]There are other weaker conditions that are enough to allow an efficient sieve, but we just refer to [Friedlander and Iwaniec 2010, Section 5.5] for a discussion of these aspects.

fixed $B \geqslant 1$. Assume that Γ satisfies the independence assumption 2.3 and the expansion assumption 4.5. Let (γ_n) be a random walk on Γ using steps from S, and let μ_n denote the probability law of the n-th step. Let $\Omega_p \subset \Gamma_p$ be finite sets such that (4-7) *holds for some $\kappa > 0$.*

There exists $A > 0$ such that, for all $n \geqslant 1$, if we let $Q = A^n$ and take $\mathcal{P}$ to be the set of primes $p \leqslant Q$ with $p \notin T_2(\Gamma)$, then we have

$$\mathbf{P}(\gamma_n \in \mathcal{S}(\mathcal{P}; \Omega)) \asymp \frac{1}{n^\kappa},$$

for all N large enough.

This is essentially a direct consequence of the standard Brun-type sieve, building on the independence and expansion assumptions. The mechanism is explained in [Kowalski 2010], and to avoid repetition, we will not give further details here. We simply add a few remarks. First, this result confirms the heuristic (4-1) as far as the order of magnitude is concerned, i.e., up to multiplicative constants. Indeed, the right side of (4-1) is, in this case, given by

$$\prod_{\substack{p \leqslant A^n \\ p \notin T_2(\Gamma)}} \left(1 - \frac{|\Omega_p|}{|\Gamma_p|}\right) \sim \prod_{p \leqslant A^n} \left(1 - \frac{\kappa}{p}\right) \asymp n^{-\kappa},$$

as $n \to +\infty$, by (4-7) and the Mertens formula (or the prime number theorem.) Secondly, the result is best possible in the sense that one cannot replace the inequalities up to multiplicative constants by an asymptotic formula in this generality (this is also seen from the Mertens formula and the prime number theorem). Finally, the result is by no means an easy consequence of (4-2) and the uniformity afforded by expansion.

Example 4.7 (sieve in orbits). We illustrate the above result by deriving, as a corollary, a special case of the sieve in orbits (or affine linear sieve) of [BGS 2010].

Let Γ be Zariski-dense in $\mathrm{SL}_r(\mathbb{Z})$ with $r \geqslant 2$, and generated by the finite set $S = S^{-1}$. We take for π_p the reduction maps. Let

$$f : \mathrm{SL}_r(\mathbb{Z}) \to \mathbb{Z}$$

be a nonconstant polynomial map and let $\Omega_p \subset \mathrm{SL}_r(\mathbb{F}_p)$ be the set of zeros of f. Since f is nonconstant, the algebraic subvariety Z_f of SL_r defined by the equation $f = 0$ is a hypersurface in SL_r. Relatively elementary considerations of algebraic geometry, together with the Lang–Weil estimates for the number of points on algebraic varieties over finite fields, show that we have

$$\frac{|\Omega_p|}{|\Gamma_p|} = \frac{\kappa_p}{p} + O(p^{-3/2}), \tag{4-8}$$

for some $\kappa_p \geqslant 0$ which depends on the splitting of p in the field of definition of the geometrically irreducible components of Z_f (if all geometrically irreducible components of Z_f are defined over $\mathbb{Q}$, then κ_p is the number of these irreducible components, as is well-known; the general case is carefully explained in Proposition 15 and Corollary 17 of [Salehi Golsefidy and Sarnak 2013]). A further application of the Chebotarev density theorem (see Lemma 21 of the same reference) shows that

$$\sum_{p \leqslant X} \kappa_p = \kappa\pi(X) + O(X/(\log X)^2),$$

where κ is the number of $\mathbb{Q}$-irreducible components of Z_f (if all geometrically irreducible components are defined over $\mathbb{Q}$, we have $\kappa_p = \kappa$ for all but finitely many p).

Examples 4.8. (1) Consider the function

$$f\left(\begin{smallmatrix} a & b \\ c & d \end{smallmatrix}\right) = a^2 + d^2$$

on SL_2. For $p \equiv 3 \pmod 4$, we have $\kappa_p = 0$ (since $a^2 + d^2 = 0 \in \mathbb{F}_p$ implies $a = d = 0$ in this case), while for $p \equiv 1 \pmod 4$, we have $\kappa_p = 2$, reflecting the fact that Z_f has then, over $\mathbb{F}_p$, the two geometrically irreducible components defined by $a + \varepsilon d = 0$ and $a - \varepsilon d = 0$, where $\varepsilon^2 = -1$. The average of κ_p over p is then equal to $\kappa = 1$.

(2) Consider the function

$$f(g_{i,j}) = \prod_{i,j} g_{i,j}$$

on SL_r. Then the irreducible components are defined by $g_{i,j} = 0$ for a fixed (i, j), and are all defined over $\mathbb{Q}$. Thus we have $\kappa = \kappa_p = r^2$ for all primes p.

Thus all assumptions of Theorem 4.6 hold (the expansion assumption coming from Theorem 4.4), and we deduce that there exists a finite set of primes T and $A > 1$ such that if $\mathcal{P}$ is the set of primes not in T and $\leqslant A^n$, we have

$$\mathbf{P}(\gamma_n \in \mathcal{S}(\mathcal{P}; \Omega)) \asymp n^{-\kappa}.$$

Using Example 3.2, we therefore deduce this result of Bourgain, Gamburd and Sarnak:

Theorem 4.9 (sieve in orbits [BGS 2010]). *Let Γ and f be as above. There exists $\omega \geqslant 1$ such that the set $\mathcal{O}_f(\omega)$ of all $g \in \Gamma$ such that $f(g)$ has at most ω prime factors satisfies*

$$\mathbf{P}(\gamma_n \in \mathcal{O}_f(\omega)) \asymp n^{-\kappa}, \tag{4-9}$$

for n large enough.

One of the insights of [BGS 2010] was that such a statement has a more qualitative corollary which is already very interesting and doesn't require any consideration of a special counting method:

Corollary 4.10. *Let* Γ *and* f *be as above. There exists* $\omega \geqslant 1$ *such that the set* $\mathcal{O}_f(\omega)$ *is Zariski-dense in* SL_r.

Proof. It is enough to check that if a subset $X \subset \Gamma$ is not Zariski-dense, then a lower bound

$$\mathbf{P}(\gamma_n \in X) \gg n^{-\kappa}$$

does not hold for any $\kappa > 0$, since $\mathcal{O}_f(\omega) \subset Z$ would then contradict the sieve lower bound (4-9) (note that here (γ_n) is just an auxiliary tool).

Given X, there exists a nontrivial polynomial f such that $X \subset Z_f$ (recall that this is the zero set of f). Then, for any prime p (large enough so that reduction of f modulo p makes sense) the image of X modulo p is contained in the zeros of f modulo p. But using (4-8) and summing (4-5) over the zeros of f modulo p, we have

$$\mathbf{P}(\pi_p(\gamma_n) \in Z_f \ (\mathrm{mod}\ p)) \sim \kappa_p p^{-1}$$

uniformly for $p \leqslant A^n$ for some $A > 1$. Taking p of size A^n, we deduce

$$\mathbf{P}(\gamma_n \in X) \leqslant \mathbf{P}(\pi_p(\gamma_n) \in Z_f \ (\mathrm{mod}\ p)) \ll A^{-n},$$

for n large enough. Thus the probability to be a zero of a given function f is in fact exponentially small for a long walk, and this contradicts the lower bounds for $\mathcal{O}_f(\omega)$. □

In fact, as noted in [Kowalski 2010] and as we will see in the next section, this has a natural refinement where "Zariski-dense" is replaced by "not thin" in the sense of Serre.

Salehi Golsefidy and Sarnak [2013] have extended the basic small sieve statement to much more general groups, not necessarily reductive, using the full power of Theorem 4.4 together with special considerations to handle unipotent groups.

Example 4.11. Theorem 4.6 also applies in the context of Dunfield–Thurston manifolds, as in Example 3.3. Indeed, the expansion assumption 4.5 is here a consequence of property (T) for $\mathrm{Sp}_{2g}(\mathbb{Z})$ for $g \geqslant 2$. As observed in [Kowalski 2010], a consequence of Theorem 4.6 which is similar in spirit to the affine linear sieve is that there exists $\omega \geqslant 1$ such that

$$\mathbf{P}\big(H_1(\mathcal{M}_{\phi_n}, \mathbb{Z}) \text{ is finite and has order divisible by} \leqslant \omega \text{ primes}\big) \asymp n^{-1},$$

for n large enough. (We recall that the genus g defining the Heegaard splitting is fixed).

One can certainly use the sieve setting for many other purposes. As one further example, we show how the method of [Granville and Soundararajan 2007, Proposition 3] gives a version of the Erdős–Kac theorem for discrete groups. For simplicity, we only state the result for the affine sieve, and give one further example afterwards. (A version of this for curvatures of Apollonian circle packings was proved in [Djanković 2011].)

Theorem 4.12 (Erdős–Kac central limit theorem for affine sieve). *Assume that $\Gamma \subset \mathrm{SL}_r(\mathbb{Z})$ is Zariski-dense in SL_r and f is a nonconstant polynomial function satisfying the assumptions of Theorem 4.9. For a random walk (γ_n) on Γ, let $\omega_f(\gamma_n) = \omega(f(\gamma_n))$ if $f(\gamma_n) \neq 0$, and $\omega_f(\gamma_n) = 0$ otherwise. Then the random variables*

$$\frac{\omega_f(\gamma_n) - \kappa \log n}{\sqrt{\kappa \log n}}$$

converge in law to the standard normal random variable as $n \to +\infty$.

Proof. We proceed exactly as in [Granville and Soundararajan 2007], leaving some details to the reader. This uses the method of moments to prove convergence in law to the normal distribution: classical probability results imply that it is enough to prove that for all integers $k \geqslant 0$, we have

$$\mathbf{E}\left(\left(\frac{\omega_f(\gamma_n) - \kappa \log n}{\sqrt{\kappa \log n}}\right)^k\right) \to c_k,$$

as $n \to +\infty$, where $c_k = \mathbf{E}(\mathcal{N}(0,1)^k)$ is the k-th moment of a standard normal random variable.

We first deal with the possibility that $f(\gamma_n) \neq 0$. By bounding

$$\mathbf{P}(f(\gamma_n) = 0) \leqslant \mathbf{P}\big(f(\pi_p(\gamma)) = 0b\big),$$

for any prime p large enough, and arguing as in the proof of Corollary 4.10, we get

$$\mathbf{P}(f(\gamma_n) = 0) \ll c^{-n},$$

for some $c > 1$. Thus the expectation above, restricted to the set $f(\gamma_n) = 0$, is

$$\ll (\kappa \log n)^{k/2} c^{-n} \to 0,$$

as $n \to +\infty$.

Below we use the notation $\tilde{\mathbf{E}}$ to denote expectation restricted to $f(\gamma_n) \neq 0$. We fix some integer $k \geqslant 0$, and fix some auxiliary $A > 1$. We will compare

$$M_k = \tilde{\mathbf{E}}\big((\omega_f(\gamma_n) - \kappa \log n)^k\big)$$

with the moment of "truncated" count of primes dividing $f(\gamma_n)$ defined by

$$N_k = \tilde{\mathbf{E}}\Big(\Big(\sum_{\substack{p \leqslant A^n \\ \pi_p(\gamma_n) \in \Omega_p}} 1 - \sum_{p \leqslant A^n} \delta_p\Big)^k\Big),$$

where

$$\delta_p = \frac{|\Omega_p|}{|\Gamma_p|},$$

and then estimate asymptotically N_k when $A > 1$ is small enough with respect to k.

For the first step, we note that when $f(\gamma_n) \neq 0$, we have

$$\omega_f(\gamma_n) - \kappa \log n = A_1 + A_2 + A_3,$$

where

$$A_1 = \sum_{\substack{p \leqslant A^n \\ \pi_p(\gamma_n) \in \Omega_p}} 1 - \sum_{p \leqslant A^n} \delta_p, \qquad A_2 = \omega(f(\gamma_n)) - \sum_{\substack{p \leqslant A^n \\ \pi_p(\gamma_n) \in \Omega_p}} 1,$$

$$A_3 = \sum_{p \leqslant A^n} \delta_p - \kappa \log n.$$

If $C > 1$ is such that $|f(g)| \leqslant C^{l_S(g)}$, then we get

$$0 \leqslant A_2 \leqslant \frac{\log C}{\log A},$$

while, by (4-7), we have

$$A_3 = \sum_{p \leqslant A^n} \delta_p - \kappa \log n = \sum_{p \leqslant A^n} \frac{|\Omega_p|}{|\Gamma_p|} - \kappa \log n = O(1),$$

so that $A_2 + A_3$ is uniformly bounded for a fixed choice of A. Using the multinomial theorem, it follows that

$$M_k = N_k + O(\max_{j \leqslant k-1} \tilde{N}_j),$$

where

$$\tilde{N}_j = \tilde{\mathbf{E}}\Big(\Big|\sum_{\substack{p \leqslant A^n \\ \pi_p(\gamma_n) \in \Omega_p}} 1 - \sum_{p \leqslant A^n} \delta_p\Big|^k\Big).$$

We have $\tilde{N}_j = N_j$ if j is even and if j is odd, we get

$$\tilde{N}_j \leqslant \sqrt{\tilde{N}_{j-1}\tilde{N}_{j+1}} = \sqrt{N_{j-1}N_{j+1}}$$

by the Cauchy–Schwarz inequality, showing that a good understanding of N_j for $j \leqslant k$ will suffice to estimate M_k.

For the second step, we write

$$X_p = \mathbf{1}_{\pi_p(\cdot)\in\Omega_p} - \delta_p,$$

for $p \leqslant A^n$, sum over p, and open the k-th power defining N_k. Note that $|X_p| \leqslant 1$. Exchanging the multiple sum over primes and the expectation, we get

$$N_k = \sum_{p_1,\dots p_k \leqslant A^n} \cdots \sum \tilde{\mathbf{E}}\Big(\prod_{j=1}^{k} X_{p_j}\Big).$$

For any fixed $(p_1, \dots, p_k)$, we note that

$$\tilde{\mathbf{E}}\Big(\prod_{j=1}^{k} X_{p_j}\Big) = \mathbf{E}\Big(\prod_{j=1}^{k} X_{p_j}\Big) - \mathbf{E}\Big(\prod_{j=1}^{k} X_{p_j}\mathbf{1}_{f(\gamma_n)=0}\Big),$$

and the second term is bounded by $\mathbf{P}(f(\gamma_n)=0) \ll c^{-n}$ since $0 \leqslant X_{p_j} \leqslant 1$. Thus the total change in replacing $\tilde{\mathbf{E}}$ by $\mathbf{E}$ in the formula above for N_k is $\ll A^{nk}c^{-n}$, which is negligible if A is chosen small enough.

Having written

$$N_k = \sum_{p_1,\dots p_k \leqslant A^n} \cdots \sum \tilde{\mathbf{E}}\Big(\prod_{j=1}^{k} X_{p_j}\Big) = \sum_{p_1,\dots p_k \leqslant A^n} \cdots \sum \mathbf{E}\Big(\prod_{j=1}^{k} X_{p_j}\Big) + O(A^{nk}c^{-n}),$$

we can apply the equidistribution (4-3) to each expectation term, obtaining a main term which we will discuss in a moment and a total error term E which is bounded by

$$E \ll A^{nk(1+B)}\rho^n + A^{nk}c^{-n},$$

where B is as in (4-6). Therefore E tends to 0 as $n \to +\infty$ if A is chosen small enough (in terms of k), which we assume to be done.

There remains the main term. However, the latter is, by the independence assumption 2.3 and by retracing our steps, almost tautologically the same as

$$\mathbf{E}\Big(\Big(\sum_{p \leqslant A^n} Y_p - \delta_p\Big)^k\Big)$$

where the (Y_p) are independent Bernoulli random variables with expectation

$$\mathbf{E}(Y_p) = \delta_p = \frac{|\Omega_p|}{|\Gamma_p|}.$$

It is a basic probabilistic fact that the sum

$$\sum_{p \leqslant A^n} Y_p$$

satisfies the central limit theorem, with mean $\kappa \log n$ and variance $\kappa \log n$ (because of (4-7) again). Therefore this sum has the right k-th moment for all $k \geqslant 0$, and this easily concludes the proof (or see [Granville and Soundararajan 2007] for a direct analysis of this type of main terms to see the combinatorics from which the normal moments explicitly appear). □

Example 4.13 (Erdős–Kac theorem for random 3-manifolds). It is clear that the argument can be applied in greater generality (including for other counting methods, provided the analogue of quantitative and suitably uniform equidistribution is known). For instance, one sees that, for Dunfield–Thurston random 3-manifolds, the number $\omega(\mathcal{M}_{\phi_n})$ of primes p such that $H_1(\mathcal{M}_{\phi_n}, \mathbb{F}_p) \neq 0$ is such that

$$\frac{\omega(\mathcal{M}_{\phi_n}) - \log n}{\sqrt{\log n}}$$

converges to a standard normal random variable, with the convention $\omega(\mathcal{M}_{\phi_n}) = 0$ if $H_1(\mathcal{M}_{\phi_n}, \mathbb{Q}) \neq 0$.

5. The large sieve

We begin with a motivating example.

Example 5.1. Consider Corollary 4.10. Although the Zariski topology contains a fair amount of information (see [BGS 2010] for examples of distinction it makes concerning the sieve in orbits), it is not very arithmetic. By itself, the fact that $\mathcal{O}_f(\omega)$ is Zariski-dense in SL_r does not exclude the possibility that this set is contained, for instance, in the subset X of $\mathrm{SL}_r(\mathbb{Z})$ of matrices where the top-left coefficient is a perfect square (since X is Zariski-dense in SL_r.) It is natural to try to study this and similar possibilities. The following definition is relevant (see [Serre 2008, Chapter 3]):

Definition 5.2 (thin set). A subset $X \subset \mathrm{SL}_r(\mathbb{Q})$ is *thin* if there exists an algebraic variety $W/\mathbb{Q}$ with $\dim(W) \leqslant r^2 - 1$ and a morphism $\pi : W \to \mathrm{SL}_r$ such that $X \subset \pi(W(\mathbb{Q}))$ and π has no rational section.

Examples 5.3. (1) The set $X = \{g \in \mathrm{SL}_r(\mathbb{Q}) \mid g_{1,1} \text{ is a square}\}$ is thin. Indeed, we have a $\mathbb{Q}$-morphism

$$\pi : \mathbb{A}^{r^2} \to \mathbb{A}^{r^2}$$

mapping $(g_{i,j})$ to the matrix $(h_{i,j})$ with $h_{1,1} = g_{1,1}^2$ and all other coordinates unchanged. The pullback of this morphism to $\mathrm{SL}_r \subset \mathbb{A}^{r^2}$ gives a morphism

$$\pi : W \to \mathrm{SL}_r,$$

where

$$W = \{g \in \mathbb{A}^{r^2} \mid \det \pi(g) = 1\},$$

for which we have $X \subset \pi(W(\mathbb{Q}))$ by construction (and $\dim W \leqslant \dim \mathrm{SL}_r$ is clear since π has finite fibers.)

(2) A subset X which is not Zariski-dense is thin.

We wish to prove:

Proposition 5.4. *Let Γ and f be as in Corollary 4.10. Then there exists $\omega \geqslant 1$ such that $\mathcal{O}_f(\omega)$ is not thin in SL_r.*

The natural idea to prove this is to prove that if X is a thin set, then for a random walk on Γ, the probability

$$\mathbf{P}(\gamma_n \in X)$$

is too small to be compatible with (4-9). For this, we observe, as in the proof of the Zariski-density, that if $X \subset \pi(W(\mathbb{Q}))$ for some

$$\pi : W \to \mathrm{SL}_r,$$

as in the definition, we have

$$\pi_p(X) \subset \pi(W(\mathbb{F}_p)),$$

for all primes p large enough (such that W and π make sense modulo p). Hence if $g \in X$, we have

$$\pi_p(g) \notin \Omega_p = \mathrm{SL}_r(\mathbb{F}_p) - \pi(W(\mathbb{F}_p)),$$

for all p large enough. This implies a sieve upper bound

$$X \subset \mathcal{S}(\mathcal{P}; \Omega),$$

where $\mathcal{P}$ contains all but finitely many primes.

However, the size of Ω_p is typically much larger than the number of points of an algebraic variety, as one can guess by just looking at the example of squares in $\mathbb{Q}$, where the image modulo p contains roughly half of all residue classes. Indeed, in general we have:

Lemma 5.5. *Let $\pi : W \to \mathrm{SL}_r$ be a $\mathbb{Q}$-rational morphism with* $\dim W \leqslant r^2 - 1$ *and with no $\mathbb{Q}$-rational section. There exists $\delta < 1$ such that, for p large enough, we have*

$$\frac{|\pi(W(\mathbb{F}_p))|}{|\mathrm{SL}_r(\mathbb{F}_p)|} \leqslant \delta.$$

For the proof, see [Serre 2008, Theorem 3.6.2], for example.

Example 5.6 (homology of Dunfield–Thurston random manifolds). We consider the situation of Example 2.5(3). Here we found sifting conditions Ω_p defined in (2-2) such that, if $\mathcal{M}_{\phi_n}$ denotes the manifold obtained from the n-th step of a random walk on the mapping class group Γ (as in Example 3.3), we have

$$\mathbf{P}(H_1(\mathcal{M}_{\phi_n}, \mathbb{Q}) \neq 0) \leqslant \mathbf{P}(\phi_n \in \mathcal{S}(Q, \Omega_p)),$$

for any $Q \geqslant 1$, where Q refers to using all primes $p \leqslant Q$ as sieve conditions. It is an interesting computation to show that

$$\frac{|\Omega_p|}{|\Gamma_p|} = 1 - \prod_{j=1}^{g} \frac{1}{1 + p^{-j}}$$

(see [Dunfield and Thurston 2006, Theorem 8.4]) so that, for fixed g, there exists $\delta_g > 0$ for which

$$\frac{|\Omega_p|}{|\Gamma_p|} \geqslant \delta_g,$$

for all p.

We now revert to the general setting of a discrete group Γ with local information $\pi_p : \Gamma \to \Gamma_p$. We have found above two natural instances of *large sieves*, a terminology which refers to sieving problems where the sets Ω_p are "large", something which most commonly means that they contain a positive proportion of Γ_p: for some $\delta > 0$, we have

$$\frac{|\Omega_p|}{|\Gamma_p|} \geqslant \delta > 0, \tag{5-1}$$

for all $p \in \mathcal{P}$. This is to be compared with the "small" sieve assumption (4-7), and this leads to an interesting remark (answering the question to the reader at the end of Example 2.5(3)): the primes occur explicitly on both sides of (4-7), but as far as the left side is concerned, they are just indices that could be replaced with any other countable set. However, on the right side, the actual size of primes (and hence their distribution) is involved. This feature disappears in (5-1). This suggests that the large sieve could be of interest in much wider contexts outside of number theory. This is indeed the case, as was shown already partly in [Kowalski 2008], and even more convincingly in the recent works of Lubotzky, Meiri and Rosenzweig that we will discuss, some of which prove general algebraic statements about linear groups using some forms of sieve methods.

To present the large sieve in the context of discrete groups, we will use here the very simple version from [Lubotzky and Meiri 2012a], adapted to our setting.

Theorem 5.7 (large sieve). *Let Γ be a group generated by a finite symmetric set S with $1 \in S$. Let $\Gamma \to \Gamma_p$ be surjective homomorphisms onto finite groups for $p \geqslant p_0$. Assume that:*

(1) *For any $p \neq q$ primes $\geqslant p_0$, the induced homomorphisms*

$$\Gamma \to \Gamma_p \times \Gamma_q = \Gamma_{p,q} \tag{5-2}$$

are onto.

(2) *The family of Cayley graphs of $\Gamma_{p,q}$ and Γ_p with respect to S is an expander family, for $p, q \geqslant p_0$.*

(3) *For some $B \geqslant 1$ we have*

$$|\Gamma_p| \leqslant p^B.$$

Let $\Omega_p \subset \Gamma_p$ be such that

$$\frac{|\Omega_p|}{|\Gamma_p|} \geqslant \delta, \tag{5-3}$$

for some $\delta > 0$ independent of p.

Then there exists $A > 1$ and $c > 1$ such that for $Q = A^n$, we have

$$\mathbf{P}(\gamma_n \in \mathcal{S}(Q; \Omega)) \ll c^{-n},$$

for n large enough, where the sieving is done using primes $p_0 \leqslant p \leqslant Q$.

Note how the assumptions concerning the group and the Γ_p are slightly weaker versions of those used for the small sieve in Theorem 4.6, since expansion and independence is only required for pairs of primes instead of all squarefree integers. Thus this version of the large sieve applies whenever Theorem 4.6 is applicable.

In particular, in view of the example at the beginning of this section, we see that this theorem proves Proposition 5.4. Similarly, for the Dunfield–Thurston random manifolds of Example 5.6, this implies the following:

Proposition 5.8. *Let $g \geqslant 1$ be an integer, and let (ϕ_n) be a random walk on the mapping class group Γ of Σ_g associated to a finite generating set S. Then there exists $c > 1$ such that*

$$\mathbf{P}(H_1(\mathcal{M}_{\phi_n}, \mathbb{Q}) \neq 0) \ll c^{-n} \quad \text{for } n \geqslant 1.$$

That the probability tends to zero is Corollary 8.5 of [Dunfield and Thurston 2006], and the exponential decay is Proposition 7.19 of [Kowalski 2008].

Remark 5.9. It would be unreasonable to expect lower bounds for the size of sifted sets in the large sieve situation, unless the set $\mathcal{P}$ determining the sieving conditions is extremely small (so the situation essentially reverts to a bounded sieve (4-4)). Indeed, if we consider integers and sieve by removing the nonsquare residue classes modulo p for all $p \in \mathcal{P}$, which is certainly a large sieve, the right side of the heuristic size of the remaining set is $\left(\frac{1}{2}\right)^{|\mathcal{P}|}$. If $\mathcal{P}$ is the set of primes $\leqslant Q$, then this is much smaller than the number of squares in $\{1, \ldots, N\}$, which certainly remain after the sieve, if $Q = N^{\varepsilon}$ for any fixed $\varepsilon > 0$. (See [Green 2008] for a discussion of the fascinating question of the possibility of an inverse large sieve statement for integers.)

We adapt the simple proof of Theorem 5.7 given in [Lubotzky and Meiri 2012a] (due to R. Peled; it is reminiscent of some classical arguments going back to Rényi and Turán; see [Kowalski 2008, Proposition 2.15].)

Proof. For a fixed n, let X_p denote the Bernoulli random variable equal to 1 when $\pi_p(\gamma_n) \in \Omega_p$ and 0 otherwise, and let

$$X = \sum_{p_0 \leqslant p \leqslant Q} X_p.$$

We see that $\gamma_n \in \mathcal{S}(Q; \Omega)$ is tautologically equivalent to the condition $X = 0$. We can compute easily the expectation of X, namely

$$\mathbf{E}(X) = \sum_p \mathbf{P}(\pi_p(\gamma_n) \in \Omega_p),$$

which, by the expansion assumption 4.5 for (Γ_p), satisfies

$$\mathbf{E}(X) = \sum_p \frac{|\Omega_p|}{|\Gamma_p|} + O(Q^{1+B} \rho^n),$$

where $\rho < 1$ is an upper bound for the spectral radius of the expansion of the Cayley graphs. If $Q^{1+B} \ll \rho^n$, this gives

$$\mathbf{E}(X) \gg \pi(Q) \gg \frac{Q}{\log Q},$$

using the large sieve assumption on the size of Ω_p.

We will now use the Chebyshev inequality

$$\mathbf{P}(\gamma_n \in \mathcal{S}(Q; \Omega)) = \mathbf{P}(X = 0) \leqslant \frac{\mathbf{V}(X)}{\mathbf{E}(X)^2},$$

where $\mathbf{V}(X)$ is the variance of X. We compute

$$\mathbf{V}(X) = \mathbf{E}((X - \mathbf{E}(X))^2) = \sum_{p,q} W(p,q)$$

by expanding the square, where

$$\begin{aligned} W(p,q) &= \mathbf{E}(X_p X_q) - \mathbf{E}(X_p)\,\mathbf{E}(X_q) \\ &= \mathbf{P}(\pi_p(\gamma_n) \in \Omega_p \text{ and } \pi_q(\gamma_n) \in \Omega_q) - \mathbf{P}(\pi_p(\gamma_n) \in \Omega_p)\,\mathbf{P}(\pi_q(\gamma_n) \in \Omega_q) \end{aligned}$$

(a measure of the correlation between two primes). We isolate the diagonal terms where $p = q$, for which we use the trivial bound $|W(p, p)| \leqslant 1$, and obtain

$$\mathbf{V}(X) \leqslant Q + Q^2 \max_{p \neq q} |W(p,q)|.$$

Finally, to estimate $W(p,q)$ when $p \neq q$, we can apply the assumption (5-2) and the expansion of the Cayley graphs of $\Gamma_{p,q}$ and Γ_p: we have

$$\mathbf{P}\big(\pi_p(\gamma_n) \in \Omega_p \text{ and } \pi_q(\gamma_n) \in \Omega_q\big) = \frac{|\Omega_p||\Omega_q|}{|\Gamma_p||\Gamma_q|} + O(Q^{2B} \rho^n),$$

while, by the same argument used for computing the expectation, we have

$$\mathbf{P}(\pi_p(\gamma_n) \in \Omega_p)\,\mathbf{P}(\pi_q(\gamma_n) \in \Omega_q) = \frac{|\Omega_p||\Omega_q|}{|\Gamma_p||\Gamma_q|} + O(Q^B \rho^n).$$

The main terms cancel, and therefore

$$Q^2 \max_{p \neq q} |W(p,q)| \ll Q^{2+2B} \rho^n.$$

Take Q as large as possible so that $Q^{2+2B}\rho^n < 1$, so that $Q \geqslant A^n$ for some $A > 1$. Then the Chebyshev inequality gives

$$\mathbf{P}(\gamma_n \in \mathcal{S}(Q;\Omega)) \ll \frac{(Q + Q^{2+2B}\rho^n)(\log Q)^2}{Q^2} \ll \frac{(\log Q)^2}{Q},$$

which is of exponential decay in terms of n. □

Remarks 5.10. (1) Clearly, one can restrict the large sieve assumption (5-3) to a subset of primes with positive natural density (e.g., some arithmetic progression) without changing the conclusion, and this is often useful.

(2) This very simple proof is well-suited to situations where precise information on the expansion constant of the relevant Cayley graphs is missing (as is most often the case). When such information is available, one gets from this argument an explicit constant $c > 1$, and one may wish to get it as large as possible. For this, one can use rather more precise inequalities, as discussed in [Kowalski 2008].

The point of the large sieve is really the exponential decay it provides. If one is interested in a statement of qualitative decay

$$\lim_{n\to+\infty} \mathbf{P}(\gamma_n \in X) = 0, \tag{5-4}$$

for a subset $X \subset \Gamma$ such that $X \subset \mathcal{S}(Q;\Omega)$, for *all* Q large enough, where the Ω_p satisfy (5-3), then one can more easily apply the bounded sieve (4-4) to a finite set I, getting

$$\begin{aligned}\limsup_{n\to+\infty} \mathbf{P}(\gamma_n \in X) &\leqslant \lim_{n\to+\infty} \mathbf{P}\big(\pi_p(\gamma_n) \notin \Omega_p \text{ for } p \in I\big)\\ &= \prod_{p\in I}\Big(1 - \frac{|\Omega_p|}{|\Gamma_p|}\Big) \leqslant (1-\delta)^{|I|}.\end{aligned}$$

Then, letting $|I| \to +\infty$, we obtain (5-4). As an example, note that this qualitative decay is *not* sufficient to prove Proposition 5.4.

Lubotzky and Meiri introduce the following convenient definition:

Definition 5.11 (exponentially small sets). Let Γ be a finitely generated group. A subset $X \subset \Gamma$ is *exponentially small* if, for any finite symmetric generating set S containing 1, and with (γ_n) the corresponding random walk on Γ, there exists a constant $c_S > 1$ such that

$$\mathbf{P}(\gamma_n \in X) \ll c_S^{-n} \quad \text{for } n \geqslant 1.$$

Remark 5.12. We can summarize part of our previous discussion by stating that if X is a thin subset of $\mathrm{SL}_r(\mathbb{Q})$, and Γ is a finitely generated Zariski-dense subgroup of $\mathrm{SL}_r(\mathbb{Z})$, then $X \cap \Gamma$ is exponentially small in Γ, and by saying that the set of mapping classes (in a fixed mapping class group Γ of genus $g \geqslant 1$) for which the corresponding manifold obtained by Heegaard splitting has positive first rational Betti number is exponentially small.

The first inkling of the large sieve in nonabelian discrete groups is found in applications of the qualitative argument above in [Dunfield and Thurston 2006; Rivin 2008; 2009] in geometric contexts.[4] We illustrate further the large sieve with an example, and then discuss briefly two other applications.

Example 5.13 (pseudo-Anosov elements of the mapping class group [Rivin 2008]). Let $g \geqslant 1$ be given and let Γ be the mapping class group of Σ_g. Thurston's celebrated theory classifies the elements $\gamma \in \Gamma$ as reducible, finite-order, or pseudo-Anosov. To quantify the feeling that most elements are of the third type, Rivin used a criterion based on the action of γ on $H_1(\Sigma_g, \mathbb{Z})$, which says that *if* (but not only if) the characteristic polynomial P_γ of this action is

[4]Exponential decay was first obtained in [Rivin 2009], though the paper's publication was delayed by a long backlog. We thank I. Rivin for clarifying the priority in publication here.

P_γ is irreducible, and satisfies further easy conditions, then γ is pseudo-Anosov. One then notes that if P_γ is reducible, then so is its reduction modulo any prime, so $\pi_p(\gamma)$ is not in the subset Ω_p of elements of $\mathrm{Sp}_{2g}(\mathbb{F}_p)$ for which the characteristic polynomial is irreducible. A computation that goes back to [Chavdarov 1997, Section 3] shows that, for some $\delta > 0$, we have

$$\frac{|\Omega_p|}{|\mathrm{Sp}_{2g}(\mathbb{F}_p)|} \geqslant \delta > 0,$$

for all p, and hence the large sieve applies. A simple further argument deals with the other necessary conditions in the pseudo-Anosov criterion, and one concludes that the set of non-pseudo-Anosov elements is exponentially small in Γ.

It should be said, however, that this proof is to some extent unsatisfactory, because it doesn't use the deeper structural and dynamical properties of pseudo-Anosov elements. For instance, using the action on homology means that one cannot argue similarly for subgroups $\tilde{\Gamma} \subset \Gamma$ for which the action on homology is small, especially subgroups of the Torelli group, which is defined precisely as the kernel of the homomorphism

$$\Gamma \to \mathrm{Sp}_{2g}(\mathbb{Z}),$$

giving this action.

However, Maher [2011; 2012] has shown, using more geometric methods, that non-pseudo-Anosov elements are exponentially small in any subgroup of Γ, except those for which this property is false for obvious reasons, and his work applies in particular to the Torelli subgroup.

On the other hand, Lubotzky and Meiri [2012b] and Malestein and Souto [2013] (independently) have recently found proofs that non-pseudo-Anosov elements are exponentially small in the Torelli group using ideas similar to those above.

Example 5.14 (powers in linear groups). Lubotzky and Meiri prove the following statement using the large sieve. The reader should note that this is, on the face of it, a purely algebraic property of finitely generated linear groups.

Theorem 5.15 [Lubotzky and Meiri 2012a]. *Let Γ be a finitely generated subgroup of $\mathrm{GL}_r(\mathbb{C})$ for some $r \geqslant 2$. If Γ is not virtually solvable,[5] then the set X of proper powers, that is, the set of those $g \in \Gamma$ such that there exists $k \geqslant 2$ and $h \in \Gamma$ with $g = h^k$, is exponentially small in Γ.*

This strengthens considerably some earlier work of a Hrushovski, Kropholler, Lubotzky and Shalev [1996]. The proof is also very instructive, in particular by showing how sieve should be considered as a *tool* among others: here, one

[5] That is, there is no finite-index solvable subgroup of Γ.

can use the large sieve to control elements which are k-th powers for a *fixed* $k \geqslant 2$, but taking the union over all $k \geqslant 2$ cannot be done with sieve alone. So Lubotzky and Meiri use other tools to deal with large values of k, in that case based on ideas related to the work of Lubotzky, Mozes and Raghunathan comparing archimedean and word-length metrics [Lubotzky et al. 2000].

Example 5.16 (typical Galois groups of characteristic polynomials). Our last example has been studied by Rivin [2008], Jouve, Kowalski and Zywina [JKZ 2013], Gorodnik and Nevo [2011] and most recently Lubotzky and Rosenzweig [2012], who were the first to explicitly consider the case of sparse subgroups. However, the underlying idea of probabilistic Galois theory goes back to versions of Hilbert's irreducibility theorem, and especially to Gallagher's introduction of the large sieve in this context [Gallagher 1973]. (There are also relations with works of Prasad and Rapinchuk [Prasad and Rapinchuk 2009; Prasad and Rapinchuk 2003].)

In the (most general) version of [Lubotzky and Rosenzweig 2012], one considers a finitely generated field $K \subset \mathbb{C}$ and a finitely generated subgroup $\Gamma \subset \mathrm{GL}_r(K)$ for some $r \geqslant 2$. The basic question is: what is the "typical" behavior of the splitting field of the characteristic polynomial $\det(T - g) \in K[T]$ for some element $g \in \Gamma$?

This can be studied using the large sieve, as we explain in the simplest case when $\Gamma \subset \mathrm{SL}_r(\mathbb{Z})$. Let $\boldsymbol{G}$ be the Zariski-closure of Γ, and assume that $\boldsymbol{G}$ is connected and split over $\mathbb{Q}$, for instance $\boldsymbol{G} = \mathrm{SL}_r$. Let W be the Weyl group of $\boldsymbol{G}$: this will turn out to be the typical Galois group in this case.

To see this, the first ingredient is the existence, for any prime p large enough (such that $\boldsymbol{G}$ can be reduced modulo p), of a certain map

$$\varphi : \boldsymbol{G}(\mathbb{F}_p)_r^\sharp \to W^\sharp,$$

going back to Carter and Steinberg, where $G^\sharp$ denotes the set of conjugacy classes of a finite group and the subscript r restricts to regular semisimple elements in the finite group $\boldsymbol{G}(\mathbb{F}_p)$.

This map is used to detect elements in the Galois groups of elements in Γ in the following way. First, for $g \in \mathrm{SL}_r(\mathbb{Z})$, let P_g be the characteristic polynomial and let K_g be its splitting field, Gal_g its Galois group over $\mathbb{Q}$. The point is that, if g is a regular semisimple element of Γ, it is shown in [JKZ 2013] that there exists an injective homomorphism

$$j_g : \mathrm{Gal}_g \hookrightarrow W,$$

canonical up to conjugation, such that if p is any prime unramified in K_g, the Frobenius conjugacy class at p maps under j_g to the conjugacy class $\varphi(\pi_p(g)) \in$

$W^\sharp$. Thus one can detect whether the image of Gal_g in W intersects various conjugacy classes by seeing where the reduction modulo p of g lies with respect to φ. As it turns out, the image of φ becomes equidistributed among the conjugacy classes in W as p becomes large. Using this, it is not too hard to show that if $\alpha \in W^\sharp$ is a given conjugacy class and if Ω_p denotes the set of $g \in \boldsymbol{G}(\mathbb{F}_p)$ such that $\varphi(g) \notin \alpha$, then these sets satisfy a large sieve density assumption

$$\frac{|\Omega_g|}{|\boldsymbol{G}(\mathbb{F}_p)|} \geqslant \delta_\alpha > 0,$$

for some $\delta_\alpha > 0$ and all p large enough. It follows by the large sieve that the probability that the element γ_n at the n-th step of a random walk on Γ has Galois group such that $j_g(\mathrm{Gal}_g) \cap \alpha = \varnothing$ is exponentially small. This holds for all the finitely many classes in W, and a well-known lemma of Jordan[6] allows us to conclude that the set of $g \in \Gamma$ where j_g is not onto is exponentially small.

The general case treated by Lubotzky–Rosenzweig is quite a bit more involved. In particular, new phenomena appear when $\boldsymbol{G}$ is *not* connected, and the different cosets of the connected component of the identity then usually have different typical Galois groups. We refer to their paper for details.

6. Problems and questions

We discuss here a few questions and problems, selected to a large extent according to the author's own interests and bias.

(1) *Effective results.* A striking aspect of the results we have described is how little they use the many refinements and developments of sieve theory, as described in [Friedlander and Iwaniec 2010] for small sieves, and in [Kowalski 2008] for the large sieve. This is due to the almost complete absence of explicit forms of the expansion assumption for sparse groups, from which it follows that one can not, for instance, give a numerical value of the integer ω guaranteed to exist in Corollary 4.10 (recall that in classical sieves, the current state of the art is very refined indeed: one knows, for instance, that the number of primes $p \leqslant x$ such that $p+2$ has at most two prime factors is of the right order of magnitude). In fact, when implementing the combinatorial counting methods (either word-length or random walks), there is *no* known explicit sieve statement, as far as the author knows[7] (whereas a few explicit bounds do exist for archimedean balls, based on spectral or ergodic methods; see, for example, [Kontorovich 2009; Kontorovich

[6]In a finite group G, there is no proper subgroup H such that $H \cap \alpha \neq \varnothing$ for all conjugacy classes α in G.

[7]The remarkable results of Bourgain and Kontorovich [2010] are explicit, but not directly related to the sieve as we have considered here; see [Kontorovich 2014] for a survey in these proceedings.

and Oh 2011; 2012; Nevo and Sarnak 2010; Liu and Sarnak 2010; Gorodnik and Nevo 2012], or for random walks in a few arithmetic groups [Kowalski 2008]). It seems clear that the current proofs of expansion for sparse groups, although they are effective, would lead to dreadful bounds on a suitable ω (see [Kowalski 2012] for a numerical upper bound on the spectral radius for Cayley graphs of Zariski-dense subgroups of $\mathrm{SL}_2(\mathbb{Z})$ modulo primes, which suggests, for example, that one could not get better than ω of size at least $2^{2^{40}}$ or so for the product of coordinates function on the Lubotzky group...).

(2) *Average expansion*? One possibility suggested by the classical Bombieri–Vinogradov theorem is to attempt a proof of expansion "on average" for the relevant Cayley graphs: for many applications, it would be sufficient to prove estimates for quantities like

$$\sum_{q_I \leqslant Q} \max_{(g_p)\in\Gamma_I} \left| \mu_N(\pi_p(g) = g_p \text{ for } p \in I) - \frac{1}{|\Gamma_I|} \right|,$$

and such estimates could conceivably be provable without resorting to individual estimates for each q_I. They could also, optimistically, be of better quality than what is true for individual I. (Such a property is known for classical sieve, by work of Fouvry, Bombieri, Friedlander and Iwaniec).

(3) *Combinatorial balls*. It would be very interesting to have equidistribution and sieve results using truncations based on word-length balls, without resorting to random walks. Here, the hope is that one might not need to compute the asymptotics of the size of the combinatorial balls, since one is only interested in relative proportions of elements in a ball mapping to a given $g \in \Gamma_p$.

(4) *Reverse power*. This question is related to (1): at least in some cases, one has very convincing conjectures for the counting function of primes arising from small sieve in orbits (see, for example, [Fuchs 2011; Fuchs and Sanden 2011]). Suppose one assumes such conjectures. What does this imply for prime numbers? In other words, can one exploit information on primes represented using the sieve in orbits to derive other properties of prime numbers? Here the reference to keep in mind is the result of Gallagher (see [Gallagher 1976] and the generalization in [Kowalski 2011]) that shows that uniform versions of the Hardy–Littlewood k-tuples conjecture imply that the number of primes $p \leqslant x$ in intervals of length $\lambda \log x$, for fixed $\lambda > 0$, is asymptotically Poisson-distributed.

Acknowledgements

Thanks to A. Kontorovich and A. Salehi Golsefidy for pointing out imprecisions and minor mistakes in the first drafts of this paper, and to H. Oh for suggestions that helped clarify the text.

References

[BGS 2010] J. Bourgain, A. Gamburd, and P. Sarnak, "Affine linear sieve, expanders, and sum-product", *Invent. Math.* **179**:3 (2010), 559–644.

[Bourbaki 1976] N. Bourbaki, *Éléments de mathématique, IV: Fonctions d'une variable réelle*, Hermann, Paris, 1976.

[Bourgain and Gamburd 2008] J. Bourgain and A. Gamburd, "Uniform expansion bounds for Cayley graphs of $\mathrm{SL}_2(\mathbb{F}_p)$", *Ann. of Math.* (2) **167**:2 (2008), 625–642.

[Bourgain and Kontorovich 2010] J. Bourgain and A. Kontorovich, "On representations of integers in thin subgroups of $\mathrm{SL}_2(\mathbb{Z})$", *Geom. Funct. Anal.* **20**:5 (2010), 1144–1174. Erratum in *Geom. Funct. Anal.* **20** (2010), 1548–1549.

[Breuillard 2014] E. Breuillard, "A brief introduction to approximate groups", pp. 23–50 in *Thin groups and superstrong approximation*, edited by H. Oh and E. Breuillard, Math. Sci. Res. Inst. Publ. **61**, Cambridge, New York, 2014.

[Chavdarov 1997] N. Chavdarov, "The generic irreducibility of the numerator of the zeta function in a family of curves with large monodromy", *Duke Math. J.* **87**:1 (1997), 151–180.

[Djanković 2011] G. Djanković, "The Erdős–Kac theorem for curvatures in integral Apollonian circle packings", *Publ. Inst. Math. (Beograd) (N.S.)* **89**:103 (2011), 11–17.

[Dunfield and Thurston 2006] N. M. Dunfield and W. P. Thurston, "Finite covers of random 3–manifolds", *Invent. Math.* **166**:3 (2006), 457–521.

[Farb and Margalit 2011] B. Farb and D. Margalit, *A primer on mapping class groups*, Princeton Mathematical Series **49**, Princeton University Press, 2011.

[Friedlander and Iwaniec 2010] J. Friedlander and H. Iwaniec, *Opera de cribro*, American Mathematical Society Colloquium Publications **57**, Amer. Math. Soc., Providence, RI, 2010.

[Fuchs 2011] E. Fuchs, "Strong approximation in the Apollonian group", *J. Number Theory* **131**:12 (2011), 2282–2302.

[Fuchs 2013] E. Fuchs, "Counting problems in Apollonian packings", *Bull. Amer. Math. Soc. (N.S.)* **50**:2 (2013), 229–266.

[Fuchs and Sanden 2011] E. Fuchs and K. Sanden, "Some experiments with integral Apollonian circle packings", *Exp. Math.* **20**:4 (2011), 380–399.

[Gallagher 1973] P. X. Gallagher, "The large sieve and probabilistic Galois theory", pp. 91–101 in *Analytic number theory* (St. Louis, MO, 1972), edited by H. G. Diamond, Amer. Math. Soc., Providence, R.I., 1973.

[Gallagher 1976] P. X. Gallagher, "On the distribution of primes in short intervals", *Mathematika* **23**:1 (1976), 4–9.

[Gorodnik and Nevo 2009] A. Gorodnik and A. Nevo, *The ergodic theory of lattice subgroups*, Annals of Mathematics Studies **172**, Princeton University Press, 2009.

[Gorodnik and Nevo 2011] A. Gorodnik and A. Nevo, "Splitting fields of elements in arithmetic groups", *Math. Res. Lett.* **18**:6 (2011), 1281–1288.

[Gorodnik and Nevo 2012] A. Gorodnik and A. Nevo, "Lifting, restricting and sifting integral points on affine homogeneous varieties", *Compos. Math.* **148**:6 (2012), 1695–1716.

[Granville and Soundararajan 2007] A. Granville and K. Soundararajan, "Sieving and the Erdős–Kac theorem", pp. 15–27 in *Equidistribution in number theory, an introduction*, edited by A. Granville and Z. Rudnick, NATO Sci. Ser. II Math. Phys. Chem. **237**, Springer, Dordrecht, 2007.

[Green 2008] B. Green, "On a variant of the large sieve", preprint, 2008. arXiv 0807.5037

[Helfgott 2008] H. A. Helfgott, "Growth and generation in $\mathrm{SL}_2(\mathbb{Z}/p\mathbb{Z})$", *Ann. of Math.* (2) **167**:2 (2008), 601–623.

[Hrushovski et al. 1996] E. Hrushovski, P. H. Kropholler, A. Lubotzky, and A. Shalev, "Powers in finitely generated groups", *Trans. Amer. Math. Soc.* **348**:1 (1996), 291–304.

[JKZ 2013] F. Jouve, E. Kowalski, and D. Zywina, "Splitting fields of characteristic polynomials of random elements in arithmetic groups", *Israel J. Math.* **193**:1 (2013), 263–307.

[Kontorovich 2009] A. V. Kontorovich, "The hyperbolic lattice point count in infinite volume with applications to sieves", *Duke Math. J.* **149**:1 (2009), 1–36.

[Kontorovich 2014] A. V. Kontorovich, "The orbital circle method", pp. 93–106 in *Thin groups and superstrong approximation*, edited by H. Oh and E. Breuillard, Math. Sci. Res. Inst. Publ. **61**, Cambridge, New York, 2014.

[Kontorovich and Oh 2011] A. Kontorovich and H. Oh, "Apollonian circle packings and closed horospheres on hyperbolic 3–manifolds", *J. Amer. Math. Soc.* **24**:3 (2011), 603–648.

[Kontorovich and Oh 2012] A. Kontorovich and H. Oh, "Almost prime Pythagorean triples in thin orbits", *J. Reine Angew. Math.* **667** (2012), 89–131.

[Kowalski 2008] E. Kowalski, *The large sieve and its applications: Arithmetic geometry, random walks and discrete groups*, Cambridge Tracts in Mathematics **175**, Cambridge University Press, 2008.

[Kowalski 2010] E. Kowalski, "Sieve in expansion", preprint, 2010. arXiv 1012.2793

[Kowalski 2011] E. Kowalski, "Averages of Euler products, distribution of singular series and the ubiquity of Poisson distribution", *Acta Arith.* **148**:2 (2011), 153–187.

[Kowalski 2012] E. Kowalski, "Explicit growth and expansion for SL_2", preprint, 2012, http://www.math.ethz.ch/~kowalski/explicit-expansion.pdf.

[Liu and Sarnak 2010] J. Liu and P. Sarnak, "Integral points on quadrics in three variables whose coordinates have few prime factors", *Israel J. Math.* **178** (2010), 393–426.

[Lubotzky and Meiri 2012a] A. Lubotzky and C. Meiri, "Sieve methods in group theory, I: Powers in linear groups", *J. Amer. Math. Soc.* **25**:4 (2012), 1119–1148.

[Lubotzky and Meiri 2012b] A. Lubotzky and C. Meiri, "Sieve methods in group theory, II: The mapping class group", *Geom. Dedicata* **159** (2012), 327–336.

[Lubotzky and Rosenzweig 2012] A. Lubotzky and L. Rosenzweig, "The Galois group of random elements of linear groups", preprint, 2012. arXiv 1205.5290

[Lubotzky et al. 2000] A. Lubotzky, S. Mozes, and M. S. Raghunathan, "The word and Riemannian metrics on lattices of semisimple groups", *Inst. Hautes Études Sci. Publ. Math.* **91** (2000), 5–53.

[Maher 2011] J. Maher, "Random walks on the mapping class group", *Duke Math. J.* **156**:3 (2011), 429–468.

[Maher 2012] J. Maher, "Exponential decay in the mapping class group", *J. Lond. Math. Soc.* (2) **86**:2 (2012), 366–386.

[Malestein and Souto 2013] J. Malestein and J. Souto, "On genericity of pseudo-Anosovs in the Torelli group", *Int. Math. Res. Not.* **2013**:6 (2013), 1434–1449.

[Mohammadi and Oh 2013] A. Mohammadi and H. Oh, "Matrix coefficients, counting and primes for orbits of geometrically finite groups", preprint, 2013. arXiv 1208.4139

[Nevo and Sarnak 2010] A. Nevo and P. Sarnak, "Prime and almost prime integral points on principal homogeneous spaces", *Acta Math.* **205**:2 (2010), 361–402.

[Oh 2011] H. Oh, "Dynamics on geometrically finite hyperbolic manifolds with applications to Apollonian circle packings and beyond", pp. 1308–1331 in *Proceedings of the International Congress of Mathematicians, III* (Hyperabad, 2010), edited by R. Bhatia et al., Hindustan Book Agency, New Delhi, 2011.

[Oh 2014] H. Oh, "Harmonic analysis, ergodic theory and counting for thin groups", pp. 179–210 in *Thin groups and superstrong approximation*, edited by H. Oh and E. Breuillard, Math. Sci. Res. Inst. Publ. **61**, Cambridge, New York, 2014.

[Prasad and Rapinchuk 2003] G. Prasad and A. S. Rapinchuk, "Existence of irreducible $\mathbb{R}$-regular elements in Zariski-dense subgroups", *Math. Res. Lett.* **10**:1 (2003), 21–32.

[Prasad and Rapinchuk 2009] G. Prasad and A. S. Rapinchuk, "Weakly commensurable arithmetic groups and isospectral locally symmetric spaces", *Publ. Math. Inst. Hautes Études Sci.* **109** (2009), 113–184.

[Pyber and Szabó 2014] L. Pyber and E. Szabó, "Growth in linear groups", pp. 253–268 in *Thin groups and superstrong approximation*, edited by H. Oh and E. Breuillard, Math. Sci. Res. Inst. Publ. **61**, Cambridge, New York, 2014.

[Rapinchuk 2014] A. Rapinchuk, "Strong approximation for algebraic groups", pp. 269–298 in *Thin groups and superstrong approximation*, edited by H. Oh and E. Breuillard, Math. Sci. Res. Inst. Publ. **61**, Cambridge, New York, 2014.

[Rivin 2008] I. Rivin, "Walks on groups, counting reducible matrices, polynomials, and surface and free group automorphisms", *Duke Math. J.* **142**:2 (2008), 353–379.

[Rivin 2009] I. Rivin, "Walks on graphs and lattices—effective bounds and applications", *Forum Math.* **21**:4 (2009), 673–685.

[Salehi Golsefidy 2014] A. Salehi Golsefidy, "Affine sieve and expanders", pp. 325–342 in *Thin groups and superstrong approximation*, edited by H. Oh and E. Breuillard, Math. Sci. Res. Inst. Publ. **61**, Cambridge, New York, 2014.

[Salehi Golsefidy and Sarnak 2013] A. Salehi Golsefidy and P. Sarnak, "The affine sieve", *J. Amer. Math. Soc.* **26**:4 (2013), 1085–1105.

[Salehi Golsefidy and Varjú 2012] A. Salehi Golsefidy and P. Varjú, "Expansion in perfect groups", *Geom. Funct. Anal.* **22**:6 (2012), 1832–1891.

[Sarnak 2010] P. Sarnak, "Affine sieve", Slides from lectures given in June 2010, 2010, http://web.math.princeton.edu/sarnak/Affine sieve summer 2010.pdf.

[Serre 2008] J.-P. Serre, *Topics in Galois theory*, 2nd ed., Research Notes in Mathematics **1**, A K Peters, Wellesley, MA, 2008.

[Varjú 2012] P. P. Varjú, "Expansion in $\mathrm{SL}_d(\mathbb{O}_K/I)$, I square-free", *J. Eur. Math. Soc.* **14**:1 (2012), 273–305.

kowalski@math.ethz.ch

Department of Mathematics,
Eidgenössische Technische Hochschule Zürich,
Rämistrasse 101, 8092 Zürich, Switzerland

Thin groups and superstrong approximation
MSRI Publications
Volume 61, 2013

How random are word maps?

MICHAEL LARSEN

Any word w in the free group on d generators determines a function $G^d \to G$ for every group G. If w is a fixed nontrivial word and G ranges over the finite simple groups, the resulting sequence of functions can be expected to enjoy some properties of a random sequence of functions. In this paper, we review the current state of the art, emphasizing open problems.

1. Introduction

Let F_d denote the free group on d generators $x_1, \dots, x_d$ and $w \in F_d$ a nontrivial element. For every group G, w defines a *word map* $f_{w,G}\colon G^d \to G$ obtained by substituting for $x_1, \dots, x_d$ respectively the coordinates $g_1, \dots, g_d$ of a given element of G^d. A number of authors have examined the behavior of $f_{w,G}$ when w is fixed and G ranges over some set of groups, especially the set of all (nonabelian) finite simple groups. A unifying theme behind a good deal of recent work is this question: for a given word w, do the maps $f_{w,G}$ behave like random functions $G^d \to G$? The answer depends partly on the choice of w but also on what properties of random functions are desired. This paper examines some recent progress in understanding basic randomness properties of word maps, with an emphasis on open questions.

Given a positive integer d and an infinite sequence $X_1, X_2, X_3, \dots$ of finite sets, we consider sequences $f_1, f_2, f_3, \dots$ of functions $f_i\colon X_i^d \to X_i$ chosen uniformly and independently. We are interested in properties of such sequences of functions which hold with probability 1. We will limit ourselves to the case that

$$\sum_{n=1}^{\infty} \frac{1}{|X_n|} < \infty. \tag{1-1}$$

It is an easy consequence of classification that the finite simple groups satisfy this property. The following theorem gives some typical examples of properties which hold for almost all sequences of random functions for X_i satisfying (1-1).

Partially supported by the National Science Foundation.

Theorem 1.1. *Let d be a positive integer. Let $X_1, X_2, X_3, \dots$ denote an infinite sequence of finite sets satisfying* (1-1). *Then for almost every sequence of functions $f_i \colon X_i^d \to X_i$, the following conditions hold*:

(1) (Image size.)

(a) *If $d = 1$,*

$$\lim_{n\to\infty} \frac{|f_n(X_n)|}{|X_n|} = 1 - \frac{1}{e}.$$

(b) *If $d > 1$, then $f_n(X_n) = X_n$ for all but finitely many n.*

(2) (Measure preservation in the limit.) *If $d > 1$,*

$$\lim_{n\to\infty} \sup_{S\subset X_n} \left| \frac{|S|}{|X_n|} - \frac{|f_n^{-1}(S)|}{|X_n^d|} \right| = 0.$$

(3) (Fiber size.) *If $d > 1$ and $a > (d-1)/2$, then for all $n \gg 0$ and all $x \in X_n$,*

$$\bigl| |f_n^{-1}(x)| - |X_n|^{d-1} \bigr| < |X_n|^a.$$

Note that item (3) implies (1b) and (2). The proof of this theorem is given in the Appendix. In the body of the paper, we will examine whether these properties of random sequences of maps in fact hold for the $f_{w,G}$ as G ranges over the finite simple groups. Because they are all limiting properties, we can omit the sporadic groups, restricting our attention to alternating groups and groups of Lie type.

If $\underline{G}$ is an algebraic group, the word map $f_{w,\underline{G}}$ is a morphism of algebraic varieties. A theorem of Borel [1983] asserts that if $\underline{G}$ is a semisimple algebraic group and w is nontrivial, then $f_{w,\underline{G}}$ is a dominant morphism, so that its image contains a nonempty open subset of $\underline{G}$. (Note that in general $f_{w,\underline{G}}$ need not be surjective; for example, the matrix

$$\begin{pmatrix} -1 & 1 \\ 0 & -1 \end{pmatrix}$$

does not lie in the image of $f_{x_1^2,\mathrm{SL}_2}$.) Information about $f_{w,\underline{G}}$ can translate into information about $f_{w,G}$ when G is of Lie type, especially when G is a large group of small rank. This point of view is especially useful in dealing with groups of exceptional Lie type, such as $E_8(q)$ but of course breaks down completely in dealing with alternating groups.

2. Image size

In this section, we consider the images $w(G) := f_{w,G}(G^d)$ of word maps. We consider first the case $d = 1$. In this case $w = x_1^m$ for some nonzero integer m.

Of course the cases $m = \pm 1$ are special, with $f_{w,G}$ surjective for every G. For $|m| \geq 2$, we have [Larsen 1997]

$$|w(S_n)| \sim c_{|m|} n^{-\phi(m)/m} n!,$$

where ϕ is Euler's function, so

$$\lim_{n\to\infty} \frac{|w(A_n)|}{|A_n|} = 0.$$

The factor $n^{\phi(m)/m}$ grows very slowly compared to $n!$, so we might hope for some result that says that in general $|w(G)|$ is not much smaller than $|G|$. Borel's theorem implies that if G ranges over finite simple groups of bounded rank, $|w(G)|/|G|$ is bounded below by a positive constant. Starting from this point and systematically exploiting the fact that $w(G)$ contains $w(H)$ for every subgroup $H \subset G$, one can prove the following theorem:

Theorem 2.1 [Larsen 2004]. *For w fixed, we have*

$$\lim_{|G|\to\infty} \frac{\log |w(G)|}{\log |G|} = 1.$$

One might ask for something stronger:

Question 2.2. *For w fixed, do we have*

$$|w(G)| \geq \frac{|G|}{\log |G|},$$

for all G sufficiently large?

Here is a result in this direction, by Aner Shalev and the author:

Theorem 2.3 [Larsen and Shalev 2009]. *For w and $\epsilon > 0$ fixed, we have*

$$|w(G)| \geq \frac{|G|}{\log^{29/9+\epsilon} |G|},$$

for all sufficiently large finite simple groups, except possibly those of the form $\mathrm{PSL}_n(\mathbb{F}_q)$ *and* $\mathrm{PSU}_n(\mathbb{F}_q)$.

For $d \geq 2$, randomness would imply that $w(G) = G$ for all sufficiently large G. In special cases, this is known to be true. We mention two such results, the first due to Martin Liebeck, Eamonn O'Brien, Shalev and Pham Tiep, and the second by Shalev, Tiep, and the author:

Theorem 2.4 (Ore's conjecture [Liebeck et al. 2010]). *The commutator map is surjective for all finite simple groups.*

Theorem 2.5 [Larsen et al. 2011]. *If $w = w_1 w_2$, where w_1 and w_2 are nontrivial words such that no generator x_i of F_d appears in both w_1 and w_2, then $w(G) = G$ for all G sufficiently large.*

Words of the type described in Theorem 2.5 are particularly easy to handle because $w(G) = w_1(G)w_2(G)$. By analogy with the classical Waring problem in analytic number theory, we say that words of the form $w = x_1^n x_2^n$, and more generally, all words of the form $w = w_1 w_2$, where the generators appearing in w_1 and w_2 are disjoint, are of *Waring type*.

These special examples might raise the hope that if $d \geq 2$, (1b) holds for all G sufficiently large, but this cannot be true if w is of the form w_0^m for $m \geq 2$. Words of this type will be called *power words*; among all words, they seem in some sense to be the ones whose word maps are the farthest from random.

Recent examples due to Sebastian Jambor, Liebeck, and O'Brien [Jambor et al. 2012] show that even if w is not a power word, there may be arbitrarily large G for which $w(G) \neq G$. There seems to be a general expectation that if w is not a power word and G has sufficiently high rank, then $w(G) = G$. In a related direction, one might ask the following question:

Question 2.6. *If w is a nonpower word, is it true that*

$$\lim_{|G|\to\infty} \frac{|w(G)|}{|G|} = 1? \tag{2-1}$$

3. Measure preservation in the limit

For any finite set X, let μ_X denote the uniform probability measure on X. A function $f\colon X \to Y$ is *measure preserving* if $f_*\mu_X = \mu_Y$, or equivalently, if $|f^{-1}(y)|$ is constant on Y. The L^1 norm of $f_*\mu_X - \mu_Y$ therefore quantifies the failure of f to preserve measure. (To date, nobody seems to have considered the behavior of $\| f_*\mu_X - \mu_Y\|_p$ except in the cases $p = 1$ and $p = \infty$, the latter to be discussed in §4.) It is easy to see that

$$\| f_*\mu_X - \mu_Y\|_1 = 2 \sup_{S\subset Y} |\mu_X(f^{-1}(S)) - \mu_Y(S)|,$$

so that (2) can be reformulated as

$$\lim_{n\to\infty} \| f_{n*}\mu_{X_n^d} - \mu_{X_n}\|_1 = 0.$$

Question 3.1. *If w is a nonpower word, is it true that*

$$\lim_{|G|\to\infty} \| f_{w,G*}\mu_{G^d} - \mu_G\|_1 = 0? \tag{3-1}$$

Note that for any word w, (3-1) implies (2-1), so an affirmative answer to Question 3.1 implies an affirmative answer to Question 2.6.

Here is a geometric analogue of Question 3.1:

Question 3.2. *If w is a nonpower word, is it true that for all simply connected almost simple algebraic groups $\underline{G}$, the generic fiber of the morphism $f_{w,\underline{G}}$ is geometrically irreducible?*

One can ask equivalently whether $\underline{G}$ has a nonempty open set of points whose inverse images are geometrically irreducible of dimension $(d-1)\dim \underline{G}$. If the answer is affirmative, then by the Lang–Weil bound, almost all the fibers of $f_{w,\underline{G}(\mathbb{F}_q)}$ have approximate size $q^{(d-1)\dim \underline{G}}$ when q is large, and this implies that almost all fibers of $f_{w,G}$ have approximate size $|G|^{d-1}$ where G is the (simple) quotient of $\underline{G}(\mathbb{F}_q)$ by its center. It follows that an affirmative answer to Question 3.2 leads to an affirmative answer to Question 3.1 for bounded rank.

At present, little is known about these questions. We mention two results; the first is due to Shelly Garion and Shalev.

Theorem 3.3 [Garion and Shalev 2009]. *If $w = x_1x_2x_1^{-1}x_2^{-1}$ then* (3-1) *holds.*

Theorem 3.4 [Larsen and Shalev 2013]. *If $w = x_1^{n_1}x_2^{n_2}$ for integers n_1, n_2, then* (3-1) *holds.*

It seems likely that Theorem 3.4 should generalize at least to all words of Waring type. In this setting, the answer to Question 3.2 is known to be yes [Larsen and Shalev 2009].

4. Fiber size

Any nontrivial lower bound on fiber size immediately implies the surjectivity of $f_{w,G}$ for large G, so at present, we can only hope to prove such bounds for very special words. For example, by a theorem of Liebeck and Shalev [2005], for the "surface" words

$$w = \prod_{i=1}^{g} x_{2i-1}x_{2i}x_{2i-1}^{-1}x_{2i}^{-1}$$

of genus $g \geq 2$ we have

$$|f_{w,G}^{-1}(e)| = |G|^{2g-1} + o(|G|^{2g-1}),$$

and the proof gives the same estimate uniformly for all fibers of $f_{w,G}$. This, of course, is much weaker than (3), but it does give both lower and upper bounds for fiber size.

The best behaved words, in some sense, are the *primitive* words. Recall that an element $w_1 \in F_d$ is primitive if and only if there exist $w_2, \ldots, w_d$ such that $w_1, \ldots, w_d$ generate F_d, in which case they do so freely. There is a one-to-one correspondence between d-tuples $(g_1, \ldots, g_d) \in G^d$ and homomorphisms

$\phi\colon F_d \to G$ given by substituting $x_1 = g_1, \dots x_d = g_d$ and likewise a one-to-one correspondence between homomorphisms ϕ and d-tuples $(h_1, \dots, h_d)$ obtained by setting $h_1 = \phi(w_1), \dots, h_d = \phi(w_d)$. Thus $|f_{w,G}^{-1}(h_1)| = |G|^{d-1}$ for all $h_1 \in G$. Thus (3) holds for primitive words.

The situation for upper bounds is only a little better. Some words which behave fairly well with respect to other randomness properties can have quite large fibers. For instance, if $w = x_1 x_2 x_1^{-1} x_2^{-1}$ and $e \in G$ is the identity, then $|f_{w,G}^{-1}(e)| = |G| \cdot |G^\natural|$, where $G^\natural$ denotes the set of conjugacy classes of G. For G of Lie type of fixed rank r, $|G^\natural|$ grows like q^r. So, for instance, if $G = \mathrm{PSL}_2(\mathbb{F}_q)$, then $|f_{w,G}^{-1}(e)|$ grows like $|G|^{4/3}$.

For general words, we have:

Theorem 4.1 [Larsen and Shalev 2012]. *For every nontrivial word $w \in F_d$, there exists $\epsilon > 0$ such that for all sufficiently large finite simple groups G and all $g \in G$,*

$$|f_{w,G}^{-1}(g)| < |G|^{d-\epsilon}.$$

When $w = x_1^m$ for $m > 1$ a fixed odd integer, and $G = \mathrm{PSL}_m(\mathbb{F}_q)$ for $q \equiv 1 \pmod{m}$, the identity fiber $f_{w,G}^{-1}(e)$ contains a regular semisimple element, namely the image $\bar{\delta}$ in G of any diagonal element $\delta \in \mathrm{SL}_m(\mathbb{F}_q)$ whose diagonal entries are all the m-th roots of unity in $\mathbb{F}_q$. Thus, the identity fiber contains the image in G of the conjugacy class of δ in $\mathrm{SL}_m(\mathbb{F}_q)$, so

$$|F_{w,G}^{-1}(e)| \geq \frac{1}{m} \frac{|\mathrm{SL}_m(\mathbb{F}_q)|}{(q-1)^{m-1}} \geq \frac{q^{m^2-m}}{m}.$$

Taking the limit $q \to \infty$, we see that ϵ must be taken no greater than $1/(m+1)$ in Theorem 4.1. Taking the limit $m \to \infty$, we see that in some sense Theorem 4.1 is as strong as possible, though it remains an interesting question how ϵ depends on w, for example on the length of w as a word. One might also ask whether a stronger bound holds for nonpower words.

It turns out that, except when w is primitive, condition (3) *never* holds. Alexandru Nica [1994] proved that given a word w, there exists a rational function $R(x)$ such that for n sufficiently large, the average number of fixed points of $f_{w,S_n}(g_1, \dots, g_d)$ is $R(1/n)$. The argument holds for every sufficiently transitive permutation group and therefore for alternating groups A_n if n is sufficiently large. Doron Puder and Ori Parzanchevski [2012; 2013] proved that $R(x)$ is identically 1 if and only if w is primitive. Otherwise, it admits a power series expansion in x. On the other hand, if every fiber of f_{w,A_n} has order $|A_n|^{d-1} + |A_n|^a$ and $a < d-1$, then the expected number E_n of fixed points of $f_{w,A_n}(g_1, \dots, g_d)$ is $1 + O(n!^{a-d})$, so for every k, $|E_n - 1| < n^{-k}$ for large n. It follows that $R(x) = 1$, and w is primitive.

In conclusion, the only word maps which satisfy (1), (2), and (3) are those associated to primitive elements. However, even the primitive elements cannot boast true randomness. Indeed, if $f_n\colon X_n^d \to X_n$ is a random sequence, with probability 1, there exists an arbitrarily large n and a pair of elements $x_n, y_n \in X_n$ such that $|f_n^{-1}(x_n)| \neq |f_n^{-1}(y_n)|$.

Appendix: Proof of Theorem 1.1

Let X be a finite set and $n = |X|$. Let Y denote the set of functions $X \to X$. Define $f\colon Y \to \mathbb{Z}$ by $f(g) := |g(X)|$. Let $F\colon X \times Y \to \{0,1\}$ be the function such that $F(x,g) = 1$ if and only if $g^{-1}(x) = \varnothing$. Thus

$$n - f(g) = \sum_{x \in X} F(x,g),$$

and

$$(n-f(g))^2 = \sum_{x_1,x_2 \in X} F(x_1,g)F(x_2,g) = n - f(g) + \sum_{x_1 \neq x_2} F(x_1,g)F(x_2,g).$$

Thus

$$\begin{aligned}\sum_{g\in Y}\left(\frac{1}{e}-\left(1-\frac{f(g)}{n}\right)\right)^2 &= e^{-2}|Y| - 2e^{-1}n^{-1}\sum_{x\in X}\sum_{g\in Y} F(x,g)\\ &\quad + n^{-2}\sum_{x\in X}\sum_{g\in Y} F(x,g)\\ &\quad + n^{-2}\sum_{x_1\neq x_2}\sum_{g\in Y} F(x_1,g)F(x_2,g)\\ &= e^{-2}n^n + \left(\frac{1}{n}-\frac{2}{e}\right)(n-1)^n + \frac{n-1}{n}(n-2)^n. \qquad (4\text{-}1)\end{aligned}$$

As

$$\frac{(n-k)^n}{n^n} = e^{-k} + O(1/n),$$

we conclude that the average value of $(e^{-1} - (1 - f(g)/n))^2$ on Y is $O(1/n)$. Thus, with respect to the uniform probability distribution on Y, the measure of the set of functions $g\colon X \to X$ such that

$$|g(X)| \notin [(1-e^{-1}-\epsilon)n, [(1-e^{-1}+\epsilon)n]$$

is bounded above by a constant multiple of $1/\epsilon^2 n$. By (1-1), the Borel–Cantelli lemma implies (1a).

For (3) (and therefore also (1b) and (2)), we observe that the probability that

$$|f^{-1}(x) - n^{d-1}| \geq n^a$$

for any particular $x \in X$ is bounded above by

$$n^d \sup_{\{k:\,|k-n^{d-1}|\geq n^a\}} \binom{n^d}{k}\left(\frac{1}{n}\right)^k\left(\frac{n-1}{n}\right)^{n^d-k}.$$

Now,

$$\binom{n^d}{k}\left(\frac{1}{n}\right)^k\left(\frac{n-1}{n}\right)^{n^d-k} \leq \frac{\binom{n^d}{k}\left(\frac{1}{n}\right)^k\left(\frac{n-1}{n}\right)^{n^d-k}}{\binom{n^d}{n^{d-1}}\left(\frac{1}{n}\right)^{n^{d-1}}\left(\frac{n-1}{n}\right)^{n^d-n^{d-1}}}$$

$$= \frac{n^{d-1}!(n^d-n^{d-1})!}{(n-1)^{k-n^{d-1}}k!(n^d-k)!}. \tag{4-2}$$

Writing $k = n^{d-1} + j$, if j is positive (and therefore $j \geq n^a$), the right hand side of (4-2) equals

$$\prod_{i=1}^{j} \frac{(n-1)n^{d-1} - i}{(n-1)(n^{d-1}+i)} \leq \left(\frac{(n-1)n^{d-1} - \lfloor\frac{n^a}{2}\rfloor}{(n-1)(n^{d-1}+\lfloor\frac{n^a}{2}\rfloor)}\right)^{\lfloor\frac{n^a}{2}\rfloor} \ll e^{-\frac{n^{2a-(d-1)}}{2+\epsilon}}.$$

for every $\epsilon > 0$. If j is negative (and therefore $j \leq -n^a$), the right hand side of (4-2) equals

$$\prod_{i=1}^{-j} \frac{(n-1)(n^{d-1} - i)}{(n-1)n^{d-1}+i} \leq \left(\frac{(n-1)(n^{d-1} - \lfloor\frac{n^a}{2}\rfloor)}{(n-1)n^{d-1}+\lfloor\frac{n^a}{2}\rfloor}\right)^{\lfloor\frac{n^a}{2}\rfloor} \ll e^{-\frac{n^{2a-(d-1)}}{2+\epsilon}}.$$

Since $2a > d - 1$, we can apply Borel–Cantelli to obtain (3). □

References

[Borel 1983] A. Borel, "On free subgroups of semisimple groups", *Enseign. Math.* (2) **29**:1-2 (1983), 151–164.

[Garion and Shalev 2009] S. Garion and A. Shalev, "Commutator maps, measure preservation, and T-systems", *Trans. Amer. Math. Soc.* **361**:9 (2009), 4631–4651.

[Jambor et al. 2012] S. Jambor, M. W. Liebeck, and E. A. O'Brien, "Some word maps that are non-surjective on infinitely many finite simple groups", preprint, 2012. arXiv 1205.1952

[Larsen 1997] M. Larsen, "How often is a permutation an nth power?", preprint, 1997. arXiv math/9712223

[Larsen 2004] M. Larsen, "Word maps have large image", *Israel J. Math.* **139** (2004), 149–156.

[Larsen and Shalev 2009] M. Larsen and A. Shalev, "Word maps and Waring type problems", *J. Amer. Math. Soc.* **22**:2 (2009), 437–466.

[Larsen and Shalev 2012] M. Larsen and A. Shalev, "Fibers of word maps and some applications", *J. Algebra* **354** (2012), 36–48.

[Larsen and Shalev 2013] M. Larsen and A. Shalev, "On the distribution of values of certain word maps", preprint, 2013. arXiv 1308.1286

[Larsen et al. 2011] M. Larsen, A. Shalev, and P. H. Tiep, "The Waring problem for finite simple groups", *Ann. of Math.* (2) **174**:3 (2011), 1885–1950.

[Liebeck and Shalev 2005] M. W. Liebeck and A. Shalev, "Fuchsian groups, finite simple groups and representation varieties", *Invent. Math.* **159**:2 (2005), 317–367.

[Liebeck et al. 2010] M. W. Liebeck, E. A. O'Brien, A. Shalev, and P. H. Tiep, "The Ore conjecture", *J. Eur. Math. Soc.* (*JEMS*) **12**:4 (2010), 939–1008.

[Nica 1994] A. Nica, "On the number of cycles of given length of a free word in several random permutations", *Random Structures Algorithms* **5**:5 (1994), 703–730.

[Puder 2012] D. Puder, "Primitive words, free factors and measure preservation", preprint, 2012. arXiv 1104.3991

[Puder and Parzanchevski 2013] D. Puder and O. Parzanchevski, "Measure preserving words are primitive", preprint, 2013. arXiv 1202.3269

mjlarsen@indiana.edu

Department of Mathematics, Indiana University, Bloomington, IN 47405 United States

Thin groups and superstrong approximation
MSRI Publications
Volume **61**, 2013

Constructing thin groups

D. D. LONG AND A. W. REID

1. Introduction

Let G be a semisimple Lie group and $\Gamma < G$ a lattice. Following Sarnak [25], a subgroup Δ of Γ is called *thin* if Δ has infinite index in Γ, but is Zariski dense. Since it is straightforward to exhibit free subgroups of lattices that are Zariski dense, we shall, in addition insist that a thin group Δ is finitely generated and does not decompose as a free product.

There has been a good deal of interest recently in thin groups. This has been motivated in part by work on expanders, and in particular the so-called *affine sieve* of Bourgain, Gamburd and Sarnak [5]. Other recent work that study thin groups can be found in [13; 14; 15]. We also refer the reader to other papers in this volume.

The aim of this article is to summarize work in [22; 20; 21], as well as discussing some other results regarding (sometimes conjectural) constructions of thin subgroups in lattices.

We begin by outlining our strategy for generating some thin subgroups. We do this first in broad terms, and then specialize to some cases that we describe in more detail. In the remainder of this section, Γ will denote a Fuchsian group (i.e., a discrete subgroup of $\mathrm{PSL}(2, \mathbb{R})$) or a Kleinian group (i.e., a discrete subgroup of $\mathrm{PSL}(2, \mathbb{C})$) of finite covolume, and G a semisimple Lie group not locally isomorphic to $\mathrm{PSL}(2, \mathbb{R})$ or $\mathrm{PSL}(2, \mathbb{C})$.

Let

$$\mathrm{Hom}_0(\Gamma, G) = \{\rho : \Gamma \to G : \rho \text{ is an irreducible representation}\}$$

and $X(\Gamma, G)$ the quotient of $\mathrm{Hom}_0(\Gamma, G)$ under G-conjugation. Then, depending on G, $X(\Gamma, G)$ is a real or complex algebraic set.

Basic Idea. *If $X(\Gamma, G)$ contains a component of positive dimension then look for specializations that result in* ***integral*** *representations.*

In broad terms, integrality is necessary for the image to live in a lattice; however this does not suffice and subsequent work is usually necessary to arrange the

Both authors partially supported by the NSF.

image to lie in a lattice of a prescribed Lie group G. Zariski density is typically fairly simple to arrange. The main issue is exhibiting a certificate to demonstrate that the image has infinite index; once this is done, the nature of the constructions here will ensure the free indecomposability of the image.

In general the question of whether the image has infinite index in the lattice touches on decidability issues. For example, the following question is raised in [7]: *Is the finite presentation problem unsolvable in* $\mathrm{SL}(n, \mathbb{Z})$ *for some* n? (where one says that the finite presentation problem is solvable if there is an algorithm that, given a finite set of matrices of $\mathrm{SL}(n, \mathbb{Z})$ generating a finitely presentable subgroup Γ of $\mathrm{SL}(n, \mathbb{Z})$ outputs a finite presentation for Γ.) However the geometrical constructions offered below give extra information that one can exploit.

Although this article is for the most part survey, one new result that exploits the strategy outlined above is the following (we refer to Section 5 for an explanation of the notation).

Theorem 1.1. *Let d be a square-free positive integer and let $L = \mathbb{Q}(\sqrt{d})$ be a real quadratic number field.*

Then the lattice $\mathrm{SU}(J, \mathcal{O}_L, \tau) < \mathrm{SL}(3, \mathbb{R})$ *contains a thin subgroup isomorphic to the fundamental group of some closed orientable surface of genus* ≥ 2.

This result, taken together with [22] and the classification of nonuniform lattices in $\mathrm{SL}(3, \mathbb{R})$ (see Section 5 for more on this) shows that every nonuniform lattice inside $\mathrm{SL}(3, \mathbb{R})$ contains a thin hyperbolic surface group.

All the constructions that we outline here produce thin subgroups in nonuniform lattices — that is, those lattices Γ for which G/Γ has finite volume but is not compact. Finding thin subgroups in uniform lattices is apparently a good deal harder. One can certainly attempt to follow the strategy given by the Basic Idea, but the additional conditions for arithmeticity of uniform lattices thus far seem prohibitive.

Indeed, it is worth remarking that only recently has the existence of thin surface subgroups in all lattices in $\mathrm{PSL}(2, \mathbb{C})$ been completed (by the recent work of Kahn and Markovic [17] in the uniform case and previously in [11] in the nonuniform case).

2. $G = \mathrm{SL}(3, \mathbb{R})$ and Γ a cocompact Fuchsian group

We first consider the case where $G = \mathrm{SL}(3, \mathbb{R})$ with Γ a cocompact Fuchsian group so that $\Sigma = \mathrm{H}^2/\Gamma$ is a closed orientable hyperbolic 2-orbifold. Let $|\Sigma|$ denote the underlying space, k_c the number of cone points on Σ and b_c the number of cone points with cone angle π.

For each such Γ, there is a natural embedding of $\Gamma \hookrightarrow \mathrm{SL}(3,\mathbb{R})$ given by

$$\Gamma < \mathrm{PSL}(2,\mathbb{R}) \cong \mathrm{SO}_0(2,1) \hookrightarrow \mathrm{SL}(3,\mathbb{R}).$$

Thus the Teichmüller component of $X(\Gamma, \mathrm{SO}_0(2,1))$ is naturally part of $X(\Gamma, \mathrm{SL}(3,\mathbb{R}))$, and determines the *Hitchin component* $X^{\mathrm{Hit}}(\Gamma, \mathrm{SL}(3,\mathbb{R}))$. It was proven in [16] that *all* the characters on this component correspond to discrete and faithful representations of Γ. In fact, Theorem A of [8] shows that $X^{\mathrm{Hit}}(\Gamma, \mathrm{SL}(3,\mathbb{R}))$ is homeomorphic to a cell of dimension $-8\chi(|\Sigma|) + (6k_c - 2b_c)$, and so unless Γ is a triangle group of type (p,q,r) with one of p, q and r equal 2, then $X(\Gamma, \mathrm{SL}(3,\mathbb{R}))$ will contain a component as in the Basic Idea.

Moreover we also have (see for example Theorem 2.1 of [22]):

Theorem 2.1. *Suppose $\chi_\rho \in X^{\mathrm{Hit}}(\Gamma, \mathrm{SL}(3,\mathbb{R}))$ is not the character of the hyperbolic structure.*

Then the image $\rho(\Gamma)$ is Zariski dense in $\mathrm{SL}(3,\mathbb{R})$.

One needs to address the issue of specializations that have images that lie in a lattice in $\mathrm{SL}(3,\mathbb{R})$, but if this can be arranged then the image is necessarily thin, since a faithful representation of a Fuchsian group must have infinite index in a rank two lattice.

2.1. The most obvious lattice in $\mathrm{SL}(3,\mathbb{R})$ is $\mathrm{SL}(3,\mathbb{Z})$ and in [22] we analyzed the above discussion in much more detail for the case of the triangle group of type $(3,3,4)$. We proved:

Theorem 2.2. *The family of representations of the triangle group*

$$\Delta(3,3,4) = \langle a, b \mid a^3 = b^3 = (a.b)^4 = 1\rangle$$

given by

$$a \mapsto \begin{pmatrix} 0 & 0 & 1 \\ 1 & 0 & 0 \\ 0 & 1 & 0 \end{pmatrix} \quad \text{and} \quad b \mapsto \begin{pmatrix} 1 & 2-t+t^2 & 3+t^2 \\ 0 & -2+2t-t^2 & -1+t-t^2 \\ 0 & 3-3t+t^2 & (-1+t)^2 \end{pmatrix}$$

are discrete and faithful for every $t \in \mathbb{R}$.

It follows that for all integral values of t the image groups are subgroups of $\mathrm{SL}(3,\mathbb{Z})$, and using Theorem 2.1 can be shown to be thin. In fact, this line of representations determines a line of characters which embeds into $X^{\mathrm{Hit}}(\Gamma, \mathrm{SL}(3,\mathbb{R}))$, so that for distinct values of t, the images are nonconjugate.

The proof of this result relies on exploiting the method of [10], which allows for computation of the representation variety and identifies this component explicitly. This, together with some Diophantine analysis, yields the curve of representations defined above.

2.2. By Margulis' arithmeticity theorem [23], all lattices in $\mathrm{SL}(3,\mathbb{R})$ are arithmetic, and in the case of $\mathrm{SL}(3,\mathbb{R})$ the totality of the commensurability classes of nonuniform arithmetic lattices in $\mathrm{SL}(3,\mathbb{R})$ is described by Witte [28] (this is discussed in Section 5 below). We now discuss the proof of the result stated in Section 1. As mentioned, this will provide thin surface subgroups in all nonuniform lattices in $\mathrm{SL}(3,\mathbb{R})$ (again we refer the reader to Section 5 for notation).

Theorem 2.3. *Let d be a square-free positive integer and let $L = \mathbb{Q}(\sqrt{d})$ be a real quadratic number field. Then the lattice $\mathrm{SU}(J,\mathcal{O}_L,\tau)$ contains a thin subgroup isomorphic to the fundamental group of some closed orientable surface of genus ≥ 2.*

Proof. The proof is very much in the spirit of [22], applied in this case to the $(3,4,4)$ triangle group.

We follow the Basic Idea, and so first note that from the discussion above, $X^{\mathrm{Hit}}(\Gamma,\mathrm{SL}(3,\mathbb{R}))$ is again 2-dimensional. Following the analysis in [10; 22], one gets a description of $X^{\mathrm{Hit}}(\Gamma,\mathrm{SL}(3,\mathbb{R}))$:

$$a \mapsto \begin{pmatrix} 1 & 1 & \dfrac{a_1}{\eta} \\ 0 & \dfrac{(1-v)}{(-1+u)} & 1 \\ 0 & \dfrac{(-1+u-u^2+v+uv-v^2)}{(-1+u)^2} & \dfrac{(-u+v)}{(-1+u)} \end{pmatrix},$$

$$b \mapsto \begin{pmatrix} \dfrac{(-1+v)}{(-1+u)} & 0 & \dfrac{b_1}{\eta} \\ \dfrac{1-3u+(2+u+u^2)v-2v^2-(-1+u)\sqrt{D}}{(2(-1+u)^2)} & 1 & -1 \\ \dfrac{-b_2}{2(-1+u)^3} & 0 & \dfrac{(1-v)}{(-1+u)} \end{pmatrix},$$

where

$$D = -7-4u^3+4v-4v^2-4v^3+2u(2+7v)+u^2(-4+v^2),$$

$$\eta = 2\big(2+u^4-4v+5v^2-3v^3+v^4-u^3(3+v) + u^2(5-v+2v^2)-u(4-2v+v^2+v^3)\big),$$

$$a_1 = -(-1+u)\big(-1+u^3(-2+v)+2v-2v^2+2v^3+\sqrt{D} + u(v-4v^2-2\sqrt{D})+u^2(1+2v+\sqrt{D})\big),$$

$$b_1 = 3-8u+11u^2-8u^3+2u^4+(-5+5u-u^2+u^3)v - (-6+7u-2u^2+u^3)v^2+2(-1+u)v^3-(-1+u)^2(-1+v)\sqrt{D},$$

$$b_2 = -3+5u-6u^2+2u^3+(5+u^2)v-(6-u+u^2)v^2+2v^3 + (-1+u)(-1+v)\sqrt{D}.$$

The hyperbolic representation occurs at $u=7$, $v=7$.

The reader may easily verify that upon setting $u=v$, the representation above may be conjugated to the representation

$$\rho_v(a) = \begin{pmatrix} \frac{1}{4}(-v+\sqrt{(v-7)(v+1)}-1) & \frac{1}{4}(v-\sqrt{(v-7)(v+1)}+5) & 0 \\ \frac{1}{4}(-v+\sqrt{(v-7)(v+1)}-1) & \frac{1}{4}(v-\sqrt{(v-7)(v+1)}+1) & 1 \\ -\dfrac{4}{v+\sqrt{(v-7)(v+1)}-3} & \frac{1}{4}(v-\sqrt{(v-7)(v+1)}+1) & 0 \end{pmatrix},$$

$$\rho_v(b) = \begin{pmatrix} 1 & 0 & \frac{1}{2}(-v-\sqrt{(v-7)(v+1)}-1) \\ 0 & 0 & -1 \\ 0 & 1 & 0 \end{pmatrix}.$$

We will require that these matrices have entries in the ring of integers of a real quadratic number field. Hence, assuming that a square-free positive integer d is given, we require that $(v-7)(v+1) = Y^2 d$ for a positive integer Y. Notice that $(v-7)(v+1)$ can be rewritten as $(v-3)^2-16$, and so we need to solve $(v-3)^2-16 = Y^2 d$; that is, the equation $X^2 - Y^2 d = 16$. In particular, given any square-free positive integer d, choosing a solution $(X_0, Y_0) \in \mathbb{N}^2$ of the Pell's equation $X^2 - Y^2 d = 1$, then $v = 4X_0 + 3$ is the desired solution (which is clearly odd).

To conclude the proof we need to establish that these constraints can be arranged so that these matrices also preserve a Hermitian form of the required type. Further computation shows that the image of ρ_v (where we assume that $v = 4X_0 + 3$ is an odd positive integer) preserves the Hermitian form H_v defined by the matrix

$$J_v = \begin{pmatrix} 2(v+1) & -\frac{1}{2}(v+1)(v+\omega+1) & -\frac{1}{2}(v+1)(v+\omega+1) \\ -\frac{1}{2}(v+1)(v-\omega+1) & 2(v+1) & (v+1)^2 \\ -\frac{1}{2}(v+1)(v-\omega+1) & (v+1)^2 & 2(v+1) \end{pmatrix},$$

where we have set

$$\omega = \sqrt{(v-7)(v+1)}.$$

As it stands the form J_v is not of the type described in Section 5, however, we can argue as follows to show that J_v is equivalent to a matrix as in Section 5. To see this, first note that H_v determines a 6-dimensional quadratic form over $\mathbb{Q}$ (i.e., the form q_v defined by $H_v(x,x)$ for $x \in \mathbb{Q}(\omega)$). Checking the characteristic polynomial of J_v shows that H_v has signature $(2,1)$ and so q_v is a 6-dimensional indefinite quadratic form over $\mathbb{Q}$, and as such is isotropic. Hence H_v is isotropic as a Hermitian form over $\mathbb{Q}(\omega)$. It follows from the classification of Hermitian forms (see, for example, [28, Section 15J]) that H_v is equivalent to a form as described in Section 5.

Note that by Theorem 2.1, away from the solution $u = v = 7$, the surface subgroups will be thin. □

Example. For concreteness we describe an example in the case when $d = 30$.

In this case an appropriate unit is given by $11 + 2\sqrt{30}$ which from above implies that we take $v = 47$. With this we have

$$\rho_{47}(a) = \begin{pmatrix} -12+2\sqrt{30} & 13-2\sqrt{30} & 0 \\ -12+2\sqrt{30} & 12-2\sqrt{30} & 1 \\ -11+2\sqrt{30} & 12-2\sqrt{30} & 0 \end{pmatrix}, \quad \rho_{47}(b) = \begin{pmatrix} 1 & 0 & -24-4\sqrt{30} \\ 0 & 0 & -1 \\ 0 & 1 & 0 \end{pmatrix},$$

and the Hermitian form is determined by the matrix

$$J_{47} = \begin{pmatrix} 1 & -(12+2\sqrt{30}) & -(12+2\sqrt{30}) \\ -(12-2\sqrt{30}) & 1 & 24 \\ -(12-2\sqrt{30}) & 24 & 1 \end{pmatrix}.$$

3. $G = \mathrm{SL}(4, \mathbb{R})$ and Γ a cocompact Kleinian group

The starting point for this section is to consider certain closed hyperbolic 3-manifolds $M = \mathrm{H}^3/\Gamma$ described in [10] as *flexible*, by which one means that the hyperbolic structure may be deformed when it is regarded as a strictly convex real projective structure. (See [10] for more details.)

The certificate for thinness in this case comes from the application of deep results of Koszul [18], Benoist [2; 3], which taken together imply that all of these deformations are the holonomy of convex real projective structures and in particular, they are all discrete, faithful representations of the fundamental group in question. (This part of the construction plays the role of the work of Choi and Goldman [8] described in [22].) It is also argued in [21] that the resulting image groups are Zariski dense in $\mathrm{SL}(4, \mathbb{R})$.

This part of the argument is quite general and applies to any flexible hyperbolic 3-manifold (even if these are perhaps quite rare, see [10]), but as in Section 3 some specialization is now necessary to ensure that the deformed group lies inside a lattice. To this end, we fix attention upon one particular closed hyperbolic 3-manifold, traditionally known as vol 3. Again, the lattices in question are described in Section 5.

Theorem 3.1. *For infinitely many real quadratic number fields L, there exist lattices $\mathrm{SU}(J, \mathcal{O}_L, \tau)$ that contain a thin subgroup isomorphic to a subgroup of finite index in $\pi_1(\mathrm{vol}\,3)$.*

We will not discuss anything about the proof here, nor describe the matrices other than to say that this proof is similar to that described for Theorem 2.3 (although integrality requires some more delicate computation in this case). We refer the reader to [21] for more details.

However, for convenience we record one example of the representation.

Example. The fundamental group of vol 3 has a presentation

$$\langle a, b \mid aabbABAbb, aBaBabaaab \rangle,$$

where $A = a^{-1}$ and $B = b^{-1}$. As in [21] (following [10]), we work with an orbifold $Q = \text{vol}\,3/\langle u \rangle$ which is fourfold covered by vol 3. We denote by Γ_Q the orbifold fundamental group of Q. Notice that a representation of Γ_Q is discrete and faithful only if it is discrete and faithful when restricted to $\pi_1(\text{vol}\,3)$, so that it suffices to work with Γ_Q.

From [10] one sees that Γ_Q is generated by two elements of finite order u and c. The group $\pi_1(\text{vol}\,3)$ is recovered as $a = u^2c$ and $b = (aua)^{-1}u$. One such representation ρ of Γ_Q given by Theorem 3.1 is

$$\rho(u) = \begin{pmatrix} 0 & -1 & -4+3\sqrt{3} & -1+2\sqrt{3} \\ 1 & 0 & -2+\sqrt{3} & -1 \\ 0 & 0 & 1 & 0 \\ 0 & 0 & 0 & 1 \end{pmatrix},$$

$$\rho(c) = \begin{pmatrix} 0 & 0 & -1+\sqrt{3} & 0 \\ 0 & 0 & 0 & -1+\sqrt{3} \\ (1+\sqrt{3})/2 & 0 & 0 & 0 \\ 0 & (1+\sqrt{3})/2 & 0 & 0 \end{pmatrix}.$$

This does not have integral entries, but there is a surjection from $\pi_1(\text{vol}\,3)$ to the dihedral group with ten elements where one sends a to a reflection and b to a rotation. A direct calculation shows that the kernel of this map consists of elements whose entries lie in $\mathbb{Z}[\sqrt{3}]$. One can also check that a nondegenerate τ-Hermitian form for the image of ρ is

$$J = \begin{pmatrix} 2 & 0 & 2-2\sqrt{3} & -2\sqrt{3} \\ 0 & 2 & 6-4\sqrt{3} & 2-2\sqrt{3} \\ 2+2\sqrt{3} & 6+4\sqrt{3} & -4 & 0 \\ 2\sqrt{3} & 2+2\sqrt{3} & 0 & -4 \end{pmatrix},$$

which in turn can be checked as being equivalent to the form $\text{diag}(1, 1, 1, -5)$.

Remarks. (i) The manifold vol 3 is rather well understood and it is shown in [24] that vol 3 has a finite sheeted cover that fibers over S^1 with fiber a closed surface F. The image of this fiber group will have the same Zariski closure as $\pi_1(\mathrm{vol}\,3)$; that is, this exhibits a thin surface subgroup H in an $\mathrm{SL}(4,\mathbb{R})$-lattice, which has the additional property that its normalizer N in $\mathrm{SL}(4,\mathbb{R})$ contains an element δ for which $\delta^n \notin \mathrm{H}$ for all $n \in \mathbb{Z} \setminus \{0\}$. Motivated by the situation in $\mathrm{PSL}(2,\mathbb{C})$ we refer to such a surface group as *geometrically infinite*. Summarizing, we have as a corollary of Theorem 3.1:

Corollary 3.2. *For infinitely many real quadratic number fields L, there exist nonuniform lattices* $\mathrm{SU}(J,\mathbb{O}_L,\tau)$ *in* $\mathrm{SL}(4,\mathbb{R})$ *which contain a geometrically infinite, thin surface subgroup.*

As in the case of $n = 3$, for a surface group $\pi_1(\Sigma_g)$ $(g \geq 2)$, there is a component of $X^{\mathrm{Hit}}(\pi_1(\Sigma_g), \mathrm{SL}(4,\mathbb{R}))$ called the Hitchin component. Now Labourie [19] showed that the mapping class group of Σ_g acts properly discontinuously on this component. Hence for such surface group representations, there can never be an element δ as above (i.e., they are never geometrically infinite). Hence we are led to ask:

Question 3.3. Does there exist a (nonuniform) lattice Γ in $\mathrm{SL}(3,\mathbb{R})$ that contains a thin geometrically infinite surface subgroup?

(ii) Compared to $\mathrm{SL}(3,\mathbb{Z})$, the structure of the nonuniform lattices of Theorem 3.1 are far less well understood. For example, unlike $\mathrm{SL}(3,\mathbb{Z})$, these lattices are not yet known to be boundedly generated by unipotent elements, nor are they known to be not left-orderable (see [29] for more on this). For general countable groups, being left-orderable is equivalent to having an orientation-preserving, faithful action on $\mathbb{R}$. In the context of lattices in $\mathrm{SL}(n,\mathbb{R})$, $n \geq 3$, this is also equivalent to having no faithful action on S^1. More precisely, if Γ is a lattice in $\mathrm{SL}(n,\mathbb{R})$, $n \geq 3$, then Proposition 2.8 of [29] shows that the following two conjectures are equivalent:

Γ *is not left-orderable.*

Γ *has no faithful action on* S^1.

We note that the proof of equivalence uses that these statements hold for all lattices, and not just an individual one.

There has also been a great deal of interest recently in left-orderability (and lack thereof) of 3-manifold groups (see, for example, [6; 9; 12]). In light of the aforementioned conjectures, we note the following. We begin with a lemma.

Lemma 3.4. $\pi_1(\mathrm{vol}\,3)$ *is not left-orderable.*

Proof. Since subgroups of left-orderable groups are left-orderable, the lemma will be deduced from the following observation.

It is shown in [12, Theorem 1] that if $M^{(n)}$ denotes the n-fold cyclic branched cover of the figure-eight knot, then $\pi_1(M^{(n)})$ is not left orderable. In the case when $n = 4$, it is shown in [24] that vol 3 is double covered by $M^{(4)}$. Hence $\pi_1(\text{vol } 3)$ is not left-orderable as required. □

Now from the discussion above $\pi_1(\text{vol } 3)$ contains a subgroup of finite index contained in a lattice $\text{SU}(J, \mathcal{O}_L, \tau)$. It is now a standard argument (which we sketch below) that $\pi_1(\text{vol } 3)$ is a subgroup of a lattice Γ commensurable with $\text{SU}(J, \mathcal{O}_L, \tau)$. Hence Γ is not left-orderable.

To make this observation more interesting, requires ensuring that Γ is torsion-free, which for the moment seems hard to arrange.

We now sketch the proof that $\pi_1(\text{vol } 3)$ is contained in a lattice. Throughout this discussion, let Δ denote the image of vol 3 constructed implicitly in Theorem 3.1, $\Lambda = \text{SU}(J, \mathcal{O}_L, \tau)$ and Δ_0 a normal subgroup of Δ of finite index in $\Delta \cap \Lambda$. Consider

$$\mathfrak{D} = \mathcal{O}\Delta_0 = \{\Sigma a_i \gamma_i \mid a_i \in \mathcal{O}_L, \gamma_i \in \Delta_0\},$$

where the sums are finite. It is shown in [1, Proposition 2.2 and Corollary 2.3], that $\mathfrak{D}$ is an order of a central simple algebra B defined over L and contained in $M(4, L)$. Indeed, since Δ_0 is Zariski dense, $B = M(4, L)$. Then $\mathfrak{D}^1$ is an arithmetic group commensurable with $\text{SL}(4, \mathcal{O}_L)$. Furthermore, since Δ normalizes Δ_0, Δ is contained in the normalizer N of $\mathfrak{D}$ in $\text{SL}(4, \mathbb{R})$. Then N is the required arithmetic lattice commensurable with $\text{SL}(4, \mathcal{O}_L)$.

4. Further constructions

In the previous sections, we were able to exhibit thin subgroups, using the certificate of faithfulness. In this section, we describe some other constructions where the representation might not be faithful, but the image group could be thin coming from structural considerations. Nonetheless, the material here is a good deal more speculative.

The basis of this construction is the 4-strand braid group, B_4. It is a standard property of B_4 (see [4]) that there is a surjective homomorphism $B_4 \to B_3$ obtained by setting $\sigma_1 = \sigma_3$ and that the kernel K is a finitely generated free group of rank two generated by $\{\sigma_1\sigma_3^{-1}, \sigma_2\sigma_1\sigma_3^{-1}\sigma_2^{-1}\}$.

The connection with matrix representations comes by consideration of the classical reduced Burau representation [4]. This representation does not have determinants equal to one, but one can easily arrange this by scaling and one obtains $\beta : B_4 \to \mathrm{SL}(3, \mathbb{Z}[x, 1/x])$ as

$$\beta(\sigma_1) = \begin{pmatrix} x^2 & 1/x & 0 \\ 0 & 1/x & 0 \\ 0 & 0 & 1/x \end{pmatrix}, \quad \beta(\sigma_2) = \begin{pmatrix} 1/x & 0 & 0 \\ -x^2 & x^2 & 1/x \\ 0 & 0 & 1/x \end{pmatrix},$$

$$\beta(\sigma_3) = \begin{pmatrix} 1/x & 0 & 0 \\ 0 & 1/x & 0 \\ 0 & -x^2 & x^2 \end{pmatrix}.$$

It is an old result of Squier [26] that the Burau representation can be regarded as Hermitian for a certain form. In the case of the representation β, it can be checked that the matrices are Hermitian for the form

$$J = \begin{pmatrix} -2+1/x^3+x^3 & 1-1/x^3 & 0 \\ 1-x^3 & -2+1/x^3+x^3 & 1-1/x^3 \\ 0 & 1-x^3 & -2+1/x^3+x^3 \end{pmatrix},$$

by which we mean that they satisfy $g^* J g = J$ where g^* means apply the ring automorphism $x \to 1/x$ and then transpose.

Now let k be a real quadratic number field with ring of integers $\mathcal{O}_k$, and $u \in \mathcal{O}_k$ a nontrivial unit for which $\sigma(u) = 1/u$. Setting $x = u$ gives a representation

$$\beta_u : B_4/Z_4 \to \mathrm{SL}(3, \mathcal{O}_k),$$

and we will denote the image $\beta_u(B_4/Z_4)$ by $B(u)$ and the form J specialized at u by $J(u)$.

We prove the following (referring to Section 5 for arithmetic lattices considered).

Theorem 4.1. (i) *$B(u)$ is a subgroup of a nonuniform lattice in* $\mathrm{SL}(3, \mathbb{R})$;

(ii) *$B(u)$ is not virtually free.*

(iii) *$B(u)$ is Zariski dense in* $\mathrm{SL}(3, \mathbb{R})$.

Proof. It follows from our choice of u and Squier's result that $B(u)$ is unitary for the nondegenerate form $J(u)$. However as it stands, this is not a form defined over $\mathbb{Z}$: the diagonal entries are rational integers since they are σ-invariant integers of $\mathcal{O}_k$, but the off diagonal entries need not be.

This can be rectified by doing the Gram–Schmidt process which is a $\mathrm{GL}(3, k)$ change of basis. Denoting the integer $1 - 1/u^3 = \tau$, one checks easily that the matrix given by mapping the standard basis $\{e_1, e_2, e_3\}$ to $\{e_1, \sigma(\tau)e_2, \sigma(\tau^2)e_3\}$ gives a similarity of $J(u)$ with a nondegenerate symmetric $\mathbb{Z}$-matrix, $\Psi(u)$. Identifying $B(u)$ with its conjugate preserving the form $\Psi(u)$, this discussion shows that $B(u)$ contains a subgroup of finite index contained in the lattice

$\mathrm{SU}(\Psi(u), \mathbb{O}_k, \tau)$. Arguing as in Remark (ii) of Section 3, we see that $B(u)$ is also contained in a lattice. That the lattices are nonuniform can be checked by an argument similar to that used in the proof of Theorem 2.3; that is, checking that the associated 6-dimensional form over $\mathbb{Q}$ is isotropic. This proves part (i).

To prove (ii), we note that in a free group, commuting elements lie in a cyclic subgroup. However, the matrices $\beta_u(\sigma_1)$ and $\beta_u(\sigma_3)$ commute but

$$\langle \beta_u(\sigma_1), \beta_u(\sigma_3) \rangle \cong \mathbb{Z} \times \mathbb{Z}.$$

This persists in a finite index subgroup.

Finally, using the proof of (ii), we deduce that the Zariski closure of $B(u)$ is a noncompact Lie subgroup of $\mathrm{SL}(3, \mathbb{R})$ of $\mathbb{R}$-rank two. Hence it must be $\mathrm{SL}(3, \mathbb{R})$. □

We ask:

Question 4.2. Are the groups $B(u)$ thin?

Potential certificates come from the following observation:

Lemma 4.3. *In the notation established above,*

$$\beta(B_4)/\beta(K) \cong B_3.$$

Proof. As remarked above, $B_4/K \cong B_3$. Hence, the required statement follows from the first isomorphism theorem and the fact that $\ker(\beta) < K$; see [4]. □

In particular, this bypasses the long-standing open problem as to whether the Burau representation of B_4 is faithful, since, independently of whether β is faithful, if there is any algebraic specialization for which the map $\beta(B_4) \to B(u)$ is faithful, the image group $B(u)$ has an infinite quotient and therefore cannot be a rank 2 lattice. It seems rather likely that such specializations exist, but identifying one seems formidably difficult.

4.1. ***The figure eight knot group.*** In some real sense the simplest finite volume hyperbolic 3-manifold group is the fundamental group of the figure eight knot complement. This is a hyperbolic 1-punctured torus bundle so that one can present the fundamental group as

$$\pi_1(M) = \langle x, y, z \mid zxz^{-1} = xy, zyz^{-1} = yxy \rangle.$$

It's an intriguing question whether this group admits a discrete, faithful representation in $\mathrm{SL}(3, \mathbb{R})$ and this is a special case of a question due to Labourie.

Question 4.4. Does there exist a finite volume hyperbolic 3-manifold M for which $\pi_1(M)$ admits a discrete, faithful representation in $\mathrm{SL}(3, \mathbb{R})$?

Unlike the cases considered in Sections 3 and 4, there is apparently no obvious "natural geometric" representation of a hyperbolic 3-manifold group into $\mathrm{SL}(3,\mathbb{R})$ to attempt to deform. Nevertheless, one can attempt to look for representations as we now describe.

Given the presentation above, the following two propositions can be checked directly by matrix multiplication.

Proposition 4.5. *Define a map* $\rho_k : \Gamma \to \mathrm{SL}(3,\mathbb{Z}[k])$ *by*

$$\rho_k(x) = X_k = \begin{pmatrix} 1 & -2 & 3 \\ 0 & k & -1-2k \\ 0 & 1 & -2 \end{pmatrix}, \quad \rho_k(y) = Y_k = \begin{pmatrix} -2-k & -1 & 1 \\ -2-k & -2 & 3 \\ -1 & -1 & 2 \end{pmatrix},$$

$$\rho_k(z) = Z_k = \begin{pmatrix} 0 & 0 & 1 \\ 1 & 0 & -k \\ 0 & 1 & -1-k \end{pmatrix}.$$

Then ρ_k *is a homomorphism.*

Proposition 4.6. *Define a map* $\beta_T : \Gamma \to \mathrm{SL}(3,\mathbb{Z}[T])$ *by*

$$\beta_T(x) = X_T = \begin{pmatrix} -1+T^3 & -T & T^2 \\ 0 & -1 & 2T \\ -T & 0 & 1 \end{pmatrix}, \quad \beta_T(y) = Y_T = \begin{pmatrix} -1 & 0 & 0 \\ -T^2 & 1 & -T \\ T & 0 & -1 \end{pmatrix},$$

$$\beta_T(z) = Z_T = \begin{pmatrix} 0 & 0 & 1 \\ 1 & 0 & T^2 \\ 0 & 1 & 0 \end{pmatrix}.$$

Then β_T *is a homomorphism.*

Note that, in either case an integral specialization gives:

Corollary 4.7. *For integral* k *or* T, $\rho_k(\Gamma), \beta_T(\Gamma) \le \mathrm{SL}(3,\mathbb{Z})$.

Now it is shown in [20] that the representations ρ_k and β_T are each irreducible except possibly for four exceptional values of their parameter, and in particular, ρ_k is irreducible for all $k \in \mathbb{Z}$, and β_T is irreducible for all nonzero $T \in \mathbb{Z}$. Moreover, it is also shown in [20] that for $k \in \mathbb{Z}$ (resp. nonzero $T \in \mathbb{Z}$), the image groups $\rho_k(\Gamma)$ (resp. $\beta_T(\Gamma)$) are Zariski dense subgroups of $\mathrm{SL}(3,\mathbb{Z})$ (and similarly for the images of the fiber group F). The issue is in deciding whether the image is thin. One of the main results of [20] is the following.

Theorem 4.8. *Fix a nonzero integer value of* T.

Then the group $\beta_T(F)$ *(and therefore* $\beta_T(\Gamma)$*) has finite index in* $\mathrm{SL}(3,\mathbb{Z})$. *Furthermore,* $\bigcap_{T>0} \beta_T(F) = 1$.

From the perspective of thin subgroups, this is undeniably disappointing, however, it is important to note that simply being able to decide that the images of the β_T-representations are finite index in $\mathrm{SL}(3, \mathbb{Z})$ is rather remarkable. For that reason we comment briefly on the proof. Throughout the following discussion, the parameter T is as described in Theorem 4.8.

The natural first question is whether the representations β_T are faithful. Remarkably, one finds the following relation holds for all parameter values T, where $X = \beta_T(x)$ and $Y = \beta_T(y)$:

$$X^{-1}YX^{-1}YX^{-1}X^{-1}YYYXYYXY^{-1}X$$
$$= XY^{-1}XYYXYYYX^{-1}X^{-1}YX^{-1}YX^{-1}.$$

Now it is not difficult to check that the groups $\beta_T(\Gamma)$ contain unipotent elements — for example $\beta_T(y^2)$ is a unipotent. That $\beta_T(\Gamma)$ has finite index in $\mathrm{SL}(3, \mathbb{Z})$ relies on the following result:

Theorem 4.9 [27, Theorem 3.7]. *Suppose that $n \geq 3$ and $x \in \mathrm{SL}(n, \mathbb{Z})$ is a unipotent matrix such that $x - 1$ has matrix rank 1. Suppose that $y \in \mathrm{SL}(n, \mathbb{Z})$ is another unipotent such that x and y generate a free abelian group N of rank 2. Then any Zariski dense subgroup of $\mathrm{SL}(n, \mathbb{Z})$ containing N virtually, is of finite index in $\mathrm{SL}(n, \mathbb{Z})$.*

Given this result, in [20] we proceed to exhibit unipotent matrices b_1 and b_2 in $\beta_T(F)$ such that $b_1 - 1$ has rank 1 and $\langle b_1, b_2\rangle \cong \mathbb{Z} \oplus \mathbb{Z}$. In terms of X and Y the matrices are

$$b_1 = X^{-1}YYYXYYXY^{-1}X,$$
$$b_2 = XY^{-1}XYYXYYYX^{-1}.$$

It is shown in [20] that there is a conjugation of b_1 and b_2 so that they have the form

$$P^{-1}b_1P = \begin{pmatrix} 1 & 0 & -T^2(-1+2T)(-5+3T^3) \\ 0 & 1 & -T(-1+2T)(-2+3T^3) \\ 0 & 0 & 1 \end{pmatrix},$$

$$P^{-1}b_2P = \begin{pmatrix} 1 & 0 & -3T^2(-1+2T) \\ 0 & 1 & -T(-1+2T)(-2+3T^3) \\ 0 & 0 & 1 \end{pmatrix}.$$

It follows from Venkataramana's result that $\beta_T(\Gamma)$ has finite index.

By contrast, the representations ρ_k remain largely mysterious. Apart from a few small values of k, namely $\{0, 2, 3, 4, 5\}$ (where we can follow the idea described for the β_T representations described above), at present the following remains open.

Question 4.10. For k an integer and $k \neq 0, 2, 3, 4, 5$ is $\rho_k(\Gamma) < \mathrm{SL}(3, \mathbb{Z})$ a thin subgroup?

Experimentation suggests that $\rho_1(\Gamma)$ is virtually free (and so will be an infinite index subgroup of $\mathrm{SL}(3, \mathbb{Z})$).

5. Nonuniform lattices in SL(n, $\mathbb{R}$)

For convenience we recall a construction of nonuniform arithmetic lattices in $\mathrm{SL}(n, \mathbb{R})$ for $n \geq 3$. We refer the reader to [28, Propositions 6.42 and 6.55] for more details.

Let L be a real quadratic number field with ring of integers $\mathcal{O}_L$ and nontrivial Galois automorphism τ. For a matrix $A \in \mathrm{SL}(n, L)$, denote by A^* the matrix obtained by taking the transpose of the matrix obtained from A by applying τ (the nontrivial Galois automorphism) to all its entries.

Theorem 5.1. *Let L, $\mathcal{O}_L$ and τ be as above, and let $b_1, \ldots, b_n$ be nonzero elements of $\mathbb{Z}$. Setting $J = \mathrm{diag}(b_1, \ldots, b_n)$, then the group*

$$\mathrm{SU}(J, \mathcal{O}_L, \tau) = \{A \in \mathrm{SL}(n, \mathcal{O}_L) \mid A^* J A = J\}$$

is a nonuniform lattice in $\mathrm{SL}(n, \mathbb{R})$.

Proposition 6.42 of [28] deals with the case of a form that is not diagonal. Following the lines of Witte's argument, one can prove the following.

Let J be any matrix of the form

$$J = \begin{pmatrix} 0 & 0 & a \\ 0 & b & 0 \\ a & 0 & 0 \end{pmatrix},$$

where $a, b \in \mathbb{Z}$, $ab \neq 0$.

As above, let

$$\mathrm{SU}(J, \mathcal{O}_k, \tau) = \{A \in \mathrm{SL}(3, \mathcal{O}_k) \mid A^* J A = J\}.$$

Theorem 5.2. $\mathrm{SU}(J, \mathcal{O}_k, \tau)$ *is a (nonuniform) lattice in* $\mathrm{SL}(3, \mathbb{R})$.

We also note that Proposition 6.46 of [28] shows that up to conjugacy the lattices constructed in Theorem 5.1 together with $\mathrm{SL}(3, \mathbb{Z})$ represent the totality of commensurability classes of nonuniform lattices in $\mathrm{SL}(3, \mathbb{R})$. Indeed, as is discussed in Section 6 of [28], it suffices to take $a = b = 1$ in the matrix J above to describe the totality of commensurability classes of nonuniform lattices in $\mathrm{SL}(3, \mathbb{R})$ (up to conjugacy).

References

[1] H. Bass, *Groups of integral representation type*, Pacific J. Math. **86** (1980), no. 1, 15–51.

[2] Y. Benoist, *Convexes divisibles, II*, Duke Math. J. **120** (2003), 97–120.

[3] ———, *Convexes divisibles, III*, Ann. Sci. École Norm. Sup. (4) **38** (2005), no. 5, 793–832.

[4] J. S. Birman, *Braids, links, and mapping class groups*, Annals of Mathematics Studies, no. 82, Princeton University Press, 1975.

[5] J. Bourgain, A. Gamburd, and P. Sarnak, *Affine linear sieve, expanders, and sum-product*, Invent. Math. **179** (2010), no. 3, 559–644.

[6] S. Boyer, C. McA Gordon, and L. Watson, *On L-spaces and left-orderable fundamental groups*, Math. Ann. **356** (2013), no. 4, 1213–1245.

[7] M. R. Bridson and H. Wilton, *On the difficulty of presenting finitely presentable groups*, Groups Geom. Dyn. **5** (2011), no. 2, 301–325.

[8] S. Choi and W. M. Goldman, *The deformation spaces of convex* $\mathbb{RP}^2$*–structures on 2–orbifolds*, Amer. J. Math. **127** (2005), no. 5, 1019–1102.

[9] A. Clay and L. Watson, *Left-orderable fundamental groups and Dehn surgery*, Int. Math. Res. Not. (2013), no. 12, 2862–2890 (rns129).

[10] D. Cooper, D. Long, and M. Thistlethwaite, *Computing varieties of representations of hyperbolic 3–manifolds into* SL(4, $\mathbb{R}$), Experiment. Math. **15** (2006), 291–305.

[11] D. Cooper, D. D. Long, and A. W. Reid, *Essential closed surfaces in bounded 3–manifolds*, J. Amer. Math. Soc. **10** (1997), no. 3, 553–563.

[12] M. K. Dąbkowski, J. H. Przytycki, and A. A. Togha, *Non-left-orderable 3–manifold groups*, Canad. Math. Bull. **48** (2005), no. 1, 32–40.

[13] J. S. Ellenberg, C. Hall, and E. Kowalski, *Expander graphs, gonality, and variation of Galois representations*, Duke Math. J. **161** (2012), no. 7, 1233–1275.

[14] E. Fuchs, *The ubiquity of thin groups*, Thin groups and superstrong approximation (Hee Oh and Emmanuel Breuillard, eds.), Math. Sci. Res. Inst. Publ., no. 61, Cambridge, New York, 2014, pp. 73–92.

[15] E. Fuchs, C. Meiri, and P. Sarnak, *Hyperbolic monodromy groups for the hypergeometric equation and Cartan involutions*, preprint, 2013. arXiv 1305.0729

[16] W. M. Goldman, *Convex real projective structures on compact surfaces*, J. Differential Geom. **31** (1990), no. 3, 791–845.

[17] J. Kahn and V. Markovic, *Immersing almost geodesic surfaces in a closed hyperbolic three manifold*, Ann. of Math. (2) **175** (2012), no. 3, 1127–1190.

[18] J.-L. Koszul, *Déformations de connexions localement plates*, Ann. Inst. Fourier (Grenoble) **18** (1968), no. 1, 103–114.

[19] F. Labourie, *Anosov flows, surface groups and curves in projective space*, Invent. Math. **165** (2006), no. 1, 51–114.

[20] D. D. Long and A. W. Reid, *Small subgroups of* SL(3, $\mathbb{Z}$), Experiment. Math. **20** (2011), no. 4, 412–425.

[21] ———, *Constructing thin subgroups in* SL(4, $\mathbb{R}$), Int. Math. Res. Not. (2013), rns281.

[22] D. D. Long, A. W. Reid, and M. Thistlethwaite, *Zariski dense surface subgroups in* SL(3, $\mathbb{Z}$), Geom. Topol. **15** (2011), no. 1, 1–9.

[23] G. A. Margulis, *Discrete subgroups of semisimple Lie groups*, Ergebnisse der Math. und ihrer Grenzgebiete (3), no. 17, Springer, Berlin, 1991.

[24] A. W. Reid, *A non-Haken hyperbolic 3–manifold covered by a surface bundle*, Pacific J. Math. **167** (1995), no. 1, 163–182.

[25] P. Sarnak, *Notes on thin matrix groups*, Thin groups and superstrong approximation (Hee Oh and Emmanuel Breuillard, eds.), Math. Sci. Res. Inst. Publ., no. 61, Cambridge, New York, 2014, pp. 343–362.

[26] C. Squier, *The Burau representation is unitary*, Proc. Amer. Math. Soc. **90** (1984), no. 2, 199–202.

[27] T. N. Venkataramana, *Zariski dense subgroups of arithmetic groups*, J. Algebra **108** (1987), no. 2, 325–339.

[28] D. Witte Morris, *Introduction to arithmetic groups*, preprint, 2008. arXiv 0106063v3

[29] ———, *Can lattices in* $\mathrm{SL}(n, \mathbb{R})$ *act on the circle*?, Geometry, rigidity, and group actions (Benson Farb and David Fisher, eds.), University of Chicago Press, Chicago, 2011, pp. 158–207. arXiv 0811.0051

long@math.ucsb.edu *Department of Mathematics, University of California, Santa Barbara, Santa Barbara, CA 93106, United States*

areid@math.utexas.edu *Department of Mathematics, University of Texas, Austin, TX 78712, United States*

Thin groups and superstrong approximation
MSRI Publications
Volume **61**, 2013

Ergodic properties of the Burger–Roblin measure

AMIR MOHAMMADI

1. Introduction

In this article we describe some dynamical properties of one-parameter unipotent flows on the frame bundle of a convex cocompact hyperbolic 3-manifold.

Much effort and study have been spent in the case of manifolds with finite volume, and quite a rich theory has developed in this case. The case of infinite volume manifolds, however, is far less understood. The goal here is to highlight some of the difficulties one faces, and possible modifications, in extending techniques developed in the finite volume case to the case of infinite volume manifolds.

Throughout, $G = \mathrm{PSL}_2(\mathbb{C})$, the group of orientation-preserving isometries of the hyperbolic space $\mathbb{H}^3$. We let Γ be a nonelementary discrete subgroup of G which is *convex cocompact*, that is, the convex hull of the limit set of Γ is compact modulo Γ. Equivalently, $\Gamma\backslash\mathbb{H}^3$ admits a finite sided fundamental domain with no cusps.

The frame bundle of the manifold $\Gamma\backslash\mathbb{H}^3$ is identified with the homogeneous space $X = \Gamma\backslash G$. Certain subgroups of G will be of particular importance in the sequel. Let $K = \mathrm{PSU}_2$, $A = \{a_s : s \in \mathbb{R}\}$, and $N = \{n_z : z \in \mathbb{C}\}$, where

$$a_s = \begin{pmatrix} e^{s/2} & 0 \\ 0 & e^{-s/2} \end{pmatrix} \quad \text{and} \quad n_z = \begin{pmatrix} 1 & 0 \\ z & 1 \end{pmatrix}.$$

Any one-parameter unipotent subgroup of G is conjugate to

$$U = \left\{ u_t := \begin{pmatrix} 1 & 0 \\ t & 1 \end{pmatrix} : t \in \mathbb{R} \right\}. \tag{1}$$

The flow considered here is indeed the right action of U on X.

One possible starting point is to study ergodicity[1] of this action with respect to natural measures on X. If X has finite volume, then Moore's ergodicity theorem [1966] implies that this flow is ergodic with respect to the volume measure, i.e.,

The author is supported in part by NSF grant 1200388.

[1] Any invariant Borel subset is either null or conull.

the G-invariant measure. If Γ is not a lattice, however, the volume measure is not ergodic for the action of N. It turns out, in this case the interesting measure to consider is the so-called BR measure. This is an infinite measure when Γ is not a lattice, and is the volume measure when Γ is a lattice.

We now describe a construction of the BR measure. Let Γ be as above, and let δ denote the critical exponent of Γ. Fix $o \in \mathbb{H}^3$ stabilized by K, and denote by ν_o the Patterson–Sullivan measure on the boundary $\partial(\mathbb{H}^3)$ associated to o [Patterson 1976; Sullivan 1979], we will refer to it as the PS measure. Using the transitive action of K on $\partial(\mathbb{H}^3)$, we lift ν_o to a measure on K trivially, and continue to denote this extension by ν_o. Using the Iwasawa decomposition, $G = KAN$, we define a measure on G:

Burger–Roblin (BR) measure. Define the measure $\tilde{m}^{\mathrm{BR}}$ on G as follows: for $\psi \in C_c(G)$,

$$\tilde{m}^{\mathrm{BR}}(\psi) = \int_G \psi(ka_s n_z) e^{-\delta s} \, d\nu_o(k) \, ds \, dz,$$

where ds and dz are Lebesgue measures on $\mathbb{R}$ and $\mathbb{C}$ respectively. It is left Γ-equivariant and right N-invariant. The BR measure, m^{BR}, is a locally finite measure on X induced by $\tilde{m}^{\mathrm{BR}}$. It is an infinite measure except when $\delta = 2$, in which case it is the G-invariant measure on X. It is shown in [Burger 1990], for surfaces, and in [Roblin 2003], in general, that the BR measure is the unique N-invariant ergodic measure on X, not supported on a closed N-orbit.[2]

The question we are interested in is: whether the BR measure is ergodic for the action of U. The answer to this question turns out to depend on δ:

Theorem 1.1 [Mohammadi and Oh 2011]. *Let Γ be a nonelementary convex cocompact subgroup. The action of U on X is ergodic with respect to m^{BR} if and only if $\delta > 1$.*[3]

The rest of this article is devoted to describing ideas involved in the proof of Theorem 1.1.

We close this section by mentioning that since in the interesting case, that is, $\delta < 2$, the BR measure is an infinite measure, the usual approach, using the study of $L^2(X, m^{\mathrm{BR}})$, falls short in proving ergodicity. It is also worth mentioning that a priori it is not even clear that the action of U is conservative, that is: for any subset B of positive measure the $\{u_t\}$-orbit of almost every point in B spends infinite amount of time in B. Indeed one of the ingredients in the proof is to show that when $\delta > 1$ the flow is conservative and in the case $\delta \leq 1$ the flow does not have certain recurrence properties which are necessary for ergodicity.

[2] The result in [Roblin 2003] is much more general.

[3] Understanding all U-invariant ergodic Radon measures seems to be a very interesting and difficult problem.

2. Properties of the BR measure

The proof of Theorem 1.1 is based on a careful study of conditional measures of m^{BR} and its push-forward, $(a_{-s})_* m^{\mathrm{BR}}$, along N-leaves. It turns out that a lot of the interesting dynamical properties of the action of A, and that of N, on (X, m^{BR}) are in fact governed by a probability measure on X, called the BMS measure. In this section we describe local structure of the BR and the BMS measures, and state an important result of Roblin relating the two.

Retain the notation from the introduction. Further, let M denote the centralizer of A in K. Indeed the unit tangent bundle, $\mathrm{T}^1(\mathbb{H}^3)$, is identified with G/M. We let $\pi : \mathrm{T}^1\,\mathbb{H}^3 \to \mathbb{H}^3$ denote the natural projection.

Recall that a collection of nonzero finite Borel measures, $\{\mu_x : x \in \partial\mathbb{H}^3\}$, on $\partial\mathbb{H}^3$ is called *Γ-invariant conformal density of dimension δ_ν* if for any $x, y \in \mathbb{H}^3$, $\gamma \in \Gamma$, and $\xi \in \partial\mathbb{H}^3$ we have

$$\gamma_*\mu_x = \mu_{\gamma x} \quad \text{and} \quad \frac{d\mu_y}{d\mu_x}(\xi) = e^{-\delta_\mu \beta_\xi(y,x)}.$$

Fix two Γ-invariant conformal densities μ and μ' of dimension δ_μ and $\delta_{\mu'}$. Following Roblin, we define a left Γ-equivariant right M-invariant measure $\tilde{m}^{\mu,\mu'}$ on G as follows: fix $o \in \mathbb{H}^3$ and identify $\mathrm{T}^1(\mathbb{H}^3)$ with

$$\big(\partial(\mathbb{H}^3) \times \partial(\mathbb{H}^3) - \{(\xi,\xi) : \xi \in \partial(\mathbb{H}^3)\}\big) \times \mathbb{R},$$

via the map $x \mapsto (x^+, x^-, \beta_{x^-}(o, x))$. Here $x^\pm \in \partial\mathbb{H}^3$ denotes the end points of the corresponding geodesic, and for any $\xi \in \partial\mathbb{H}^3$, β_ξ is the Busemann function based at ξ. Define

$$\tilde{m}^{\mu,\mu'}(x) = e^{\delta_\mu \beta_{x^+}(o,\pi(x))}\, e^{\delta_{\mu'} \beta_{x^-}(o,\pi(x))}\, d\mu_o(x^+) d\mu'_o(x^-)\, dt,$$ [4]

Since μ and μ' are Γ-invariant, this defines a left Γ-equivariant measure on $\mathrm{T}^1\,\mathbb{H}^3$. We lift this to an M-invariant measure on G. Let $m^{\mu,\mu'}$ denote the measure induced by this measure on $X = \Gamma\backslash G$. This measure is quasiinvariant under the action of a_s. Indeed

$$a_s m^{\mu,\mu'} = e^{(\delta_\mu - \delta_{\mu'})s}\, m^{\mu,\mu'}. \tag{2}$$

With this notation, the Haar measure on X corresponds to $\mu = \mu' = m$, the (2-dimensional) G-invariant density on $\partial\mathbb{H}^3$. Let ν denote the unique (δ-dimensional) Γ-invariant geometric density on $\partial\mathbb{H}^3$ supported on the limit set, $\Lambda(\Gamma)$, of Γ. This will be referred to as the Patterson–Sullivan (PS) density; see

[4] See [Oh 2013] in this volume for further discussion of this definition.

[Patterson 1976; Sullivan 1979]. With our choice of Γ, the PS measure coincides with the δ-dimensional Hausdorff measure of the limit set of Γ. Put

$$m^{\mathrm{BR}} = \mu^{m,\nu} \quad \text{and} \quad m^{\mathrm{BMS}} = \mu^{\nu,\nu}.$$

We will refer to these measures as the BMS and the BR measures. The above definition coincides with the definition of the BR measure given in the introduction. It follows from (2) that m^{BR} is a finite measure if and only if $\delta = 2$, that is, if and only if Γ is a lattice [Yau 1975; Sullivan 1979]. Similarly we get that the BMS measure is A-invariant.

For later use, let us record here that $\mathrm{supp}(m^{\mathrm{BR}})$ is $\Gamma\backslash\{g \in G : g^- \in \Lambda(\Gamma)\}$, and $\mathrm{supp}(m^{\mathrm{BMS}})$ is $\Gamma\backslash\{g \in G : g^\pm \in \Lambda(\Gamma)\}$. In particular when Γ is convex cocompact the BMS measure has compact support, and is a finite measure.[5]

The important properties of the BR and the BMS measure that will be used here are listed in the following theorem. Of particular interest to us is (iii) in this theorem, which is [Roblin 2003, Theorem 3.4]. This relates the BR measure, which is an infinite measure and $\{a_s\}$ quasi invariant, to the BMS measure which is a finite measure and $\{a_s\}$ invariant. Indeed we loose the N-invariance in this transition, but since the conditional measures of the BMS measure along the N-leaves are δ-dimensional Hausdorff measures, they still carries quite a lot of information.

Theorem 2.1 [Flaminio and Spatzier 1990; Roblin 2003]. *Let Γ be as before and normalize so that m^{BMS} is a probability measure.*

(i) *The action of $\{a_s\}$ on X is mixing with respect to m^{BMS}, that is, for any $\psi_1, \psi_2 \in L^2(X, m^{\mathrm{BMS}})$, as $s \to \pm\infty$,*

$$\int_X \psi_1(ga_s)\psi_2(g)\, dm^{\mathrm{BMS}}(g) \to m^{\mathrm{BMS}}(\psi_1)m^{\mathrm{BMS}}(\psi_2).$$

(ii) *m^{BR} is the unique N-ergodic measure on X which is not supported on a closed N-orbit.*

(iii) *For any $\psi \in C_c(X)$ or for $\psi_i = \chi_{E_i}$ for bounded Borel subsets $E_i \subset X$ with $m^{\mathrm{BMS}}(\partial(E_i)) = 0$,*

$$\int_X \psi_1(ga_{-s})\psi_2(g)\, dm^{\mathrm{BR}}(g) \to m^{\mathrm{BMS}}(\psi_1)m^{\mathrm{BR}}(\psi_2) \quad \textit{as } s \to +\infty.$$

In the sequel, especially when applying this theorem, we will need to take "nice" local neighborhoods. We will use this definition: A subset $E \subseteq X$ is called a *BMS box* if $E = x_0 N_\rho^- A_\rho N_\rho M$, where $x_0 \in \mathrm{supp}(m^{\mathrm{BMS}})$, $\rho > 0$ is at most

[5] The BMS measure is a finite measure for any geometrically finite group [Sullivan 1984].

the injectivity radius at x_0 and S_ρ means the ρ-neighborhood of e in S for any subset S of G.

3. Recurrence properties of the action of U and Theorem 1.1

In this section we describe the proof of Theorem 1.1, modulo the "window theorem", which will be stated and used as a black box. The proof of the window theorem will be explained in the next section.

As was mentioned in the introduction for an infinite measure, L^2 considerations are not enough in proving ergodicity. Our proof is based on the polynomial divergence of unipotent flows. We investigate the intermediate range for two orbits of the unipotent flow and produce extra invariance.[6] The study of unipotent orbits in the intermediate range is not new, by any means, and has been used successfully in several prior works, for example, [Margulis 1975; 1989; Ratner 1990; 1991; Lindenstrauss 2006; Einsiedler et al. 2006].

Let us explain the proof in the case of a finite measure. Suppose μ is an N-invariant and ergodic probability measure on X. Further, assume that μ is an "interesting" measure, one not supported on a closed N-orbit. In fact, for simplicity, we will assume that we can find x_n and $y_n = x_n g_n$, generic for the action of U in the sense of the Birkhoff ergodic theorem, and satisfying the following conditions:

(g1) $g_n \to e$.

(g2) The $(1,2)$ entry of g_n, which will be denoted by $(g_n)_{12}$, is nonzero.

(g3) The real and the imaginary parts of $(g_n)_{12}$ have comparable sizes.[7]

Now for any $f \in C_c(X)$ we have

$$\frac{1}{T}\int_0^T f(\bullet u_t) \to \int_X f d\mu_\bullet \quad \text{for } \bullet = x_n, y_n, \tag{3}$$

where $\mu_\bullet$ denotes the corresponding ergodic components of μ.

We now compare the two orbits $x_n u_t$ and $y_n u_t$. Properties (g2) and (g3) imply that the divergence, in the transversal direction to U, is given by $u_{-t} g_n u_t$. The matrix entries of $u_{-t} g_n u_t$ are polynomials of degree at most 2, and the fastest divergence is along the $(2,1)$-entry, which is a polynomial of degree 2. This polynomial is (i) nonconstant, (ii) has size roughly 1, certainly nonzero, if

[6] The fact that polynomial divergence can be used to prove ergodicity of unipotent flows is due to Margulis.

[7] This last condition is admittedly stating quite precise information about the structure of the "generic set" and in general is not easy to guarantee without having some information about the A-action on this measure space.

$t \in [(1-\epsilon)T, T]$ for small ϵ and T of size $1/\sqrt{|(g_n)_{12}|}$, and (iii) $u_{-T} g_n u_T \in N - U$, when T is of size $1/\sqrt{|(g_n)_{12}|}$.

This special behavior of unipotent orbits in the intermediate range, and the fact that (3) *holds not only for* $[0, T]$ *but also for short intervals like* $[(1-\epsilon)T, T]$, imply that: if x_n and y_n are chosen from a "good" set in the sense that

(i) the convergence in (3) is uniform on this set, and

(ii) $\{x_n\}, \{y_n\}$ are in a compact set where the map $\bullet \mapsto \mu_\bullet$ is continuous on this set,[8]

and if we suppose $x_n \to x$, then μ_x is invariant under an element in $N - U$. This and N-ergodicity of μ, using commutativity of N and standard arguments, will imply that μ is U-ergodic.

This argument is indeed an important step in Ratner's proof of measure classification theorem; arguments of this kind, in topological context, are also crucial in Margulis' proof of the Oppenheim conjecture.

The proof of Theorem 1.1 is along the lines of the above argument. However there are some rather serious difficulties in carrying out this idea, which we now discuss. The first difficulty is that in order for this argument to even start, one needs to have an ergodic theorem for the action of U. The suitable ergodic theorem is the Hopf ratio ergodic theorem [1937]. The form that we need (see Theorem 7.4 of [Mohammadi and Oh 2011] and references there) is this: Let μ be a U-invariant and conservative measure on X, and let $\psi \in L^1(X, \mu)$ be nonnegative. Then for μ-a.e. x such that $\int_0^T \psi(xu_t)\,dt \to \infty$, we have

$$\frac{\int_0^T f(xu_t)}{\int_0^T \psi(xu_t)\,dt} \to \frac{\mu_x(f)}{\mu_x(\psi)} \quad \text{for any } f \in L^1(X, \mu).$$

Note that the Hopf ergodic theorem only kicks in if we have already established that the action of U is conservative, which, by Poincaré recurrence, is immediate in the case of a probability measure.

Theorem 3.1 [Mohammadi and Oh 2011, Theorems 6.6 and 9.4]. *Retain the above notation.*

(i) *If* $\delta > 1$, *then the action of U is conservative for* m^{BR}.

(ii) *Suppose either that* $0 < \delta < 1$ *or that* $\Lambda(\Gamma)$ *is a purely unrectifiable* 1*-set. Then the action of U on X is not strongly recurrent for* m^{BR}. *In particular* m^{BR} *is not U-ergodic.*

[8] Indeed by Egorov's and Lusin's theorems, these hold on a set of "almost" full measure.

Let us recall from [Mohammadi and Oh 2011, Definition 9.3] that a measure-preserving flow $\{u_t\}$, on a σ-finite measure space (X, μ), is called *strongly recurrent* if for any two nonnull measurable subsets B_1, B_2, we have $\{t : xu_t \in B_2\}$ is unbounded for μ-a.e. $x \in B_1$. It follows from the Hopf ratio ergodic theorem that any nontransitive ergodic action of U is strongly recurrent.

The proofs of the above statements are based on the general philosophy that the support of the BMS measure controls a lot of the dynamics of the action of U, and the action of A. The proofs also use some standard facts from geometric measure theory regarding dimension of projections, slices of certain measures, and some properties of *condensation points*.

Thanks to this theorem, we need to prove ergodicity in the case of $\delta > 1$. Thus we assume $\delta > 1$ for the rest of the argument. The algebraic part of the above argument goes through without change. That is: if we can find x_n, and $y_n = x_n g_n$, where the g_n satisfy properties (g1)–(g3), then we can construct a nontrivial polynomial of degree 2 in the (2,1) matrix entry which governs the fastest divergence of the two orbits $x_n U$ and $y_n U$.

Let us continue to denote points by x_n, and $y_n = x_n g_n$. In order to get an element in $N - U$, which is the goal of the argument, it is essential to consider times in the intermediate range, that is, when the two orbits are roughly size one apart from each other. Indeed a simple matrix multiplication, considering $u_{-t} g_n u_t$, implies: this holds for $t \in [rT, T]$, where $0 < r < 1$ and T of order of $1/\sqrt{|(g_n)_{12}|}$.

In order to be able to use this algebraic fact, however, one needs to have some information regarding dynamics of this piece of the orbit. In the finite measure case it follows from the Birkhoff Ergodic Theorem that for a typical point x, the piece of the orbit, $\{xu_t : t \in [rT, T]\}$, is equidistributed. In the case of an infinite measure, on the other hand, not only is this not free of charge, it seems to even be wrong in general.[9] This is the main technical difficulty in carrying out the above outline to the case in hand.

As we will see later our treatment of the main difficulty above makes it crucial that we are able to find g_n satisfying (g1)–(g3) in "all scales". The precise formulation is this:

Proposition 3.2 [Mohammadi and Oh 2011, Proposition 4.4]. *Let $\delta > 1$. Fix some* BMS *box E and some $0 < r < 1$. There exist positive numbers $d_0 = d_0(r) > 1$ and $s_0 \gg 1$ such that for any Borel subset $F \subset E$ with $m^{\mathrm{BR}}(F) > r \cdot m^{\mathrm{BR}}(E)$ and any $s \geq s_0$, there exists a pair of elements $x_s, y_s \in F$ satisfying*

(i) $x_s = y_s n^-_{w_s}$ *for* $n^-_{w_s} \in N^-$,

(ii) $1/d_0 s \leq |w_s| \leq d_0/s$ *and*

[9] The author, however, does not know of an example.

(iii) $|\Im(w_s)| \geq |\Re(w_s)|/d_0$.

This proposition gives a rather precise description of how two generic points approach each other along contracting leaves. In the case of a probability measure, conditions similar to this proposition, but weaker, can be proved using positive entropy. Our proof uses similar tools: we are able to guarantee this, because the conditional measures of BR along N^-, transpose of N, carry quite a lot of information. In particular, the conditional measure of a ball of radius r in N^-, centered at a point in the support of this conditional, is of order r^δ. This fact in view of $\delta > 1$, and a covering argument implies the proposition.

As we mentioned above the major difficulty in the proof is to get equidistribution of the orbits in the intermediate range of time. The following is a partial resolution of this problem and is the main technical result used in the proof of Theorem 1.1:

Theorem 3.3 (window theorem [Mohammadi and Oh 2011, Theorem 1.3]). *Retain the above notation and assumptions, in particular $\delta > 1$. Let $E \subset X$ be a* BMS *box, and suppose $\psi \in C_c(X)$ be a nonnegative function with $\psi|_E > 0$. Then there exist $0 < r < 1$ and $T_0 > 1$ such that for any $T \geq T_0$,*

$$m^{\mathrm{BR}}\left\{x \in E : \int_{-rT}^{rT} \psi(xu_t)\,dt \leq (1-r)\int_{-T}^{T} \psi(xu_t)\,dt\right\} > \frac{r}{2} \cdot m^{\mathrm{BR}}(E). \quad (4)$$

This theorem is used to control the dynamics when two orbits have moderately diverged. Let us however mention two difficulties one faces in applying this theorem. First (and more importantly) is that the set in (4), where one has a "doubling" property of return times for the action of U, depends on the time parameter T. Hence in order to be able to apply this theorem successfully, one needs to be able to find "good points", as above, which are close to each other in all scales. This is why Proposition 3.2 is essential to our analysis. In working with an unknown measure a statement of this form will be difficult to utilize.

Secondly, as in the case of probability measure, we really need equidistribution of $[(1-\epsilon)T, T]$, for small ϵ. In order to achieve this, for any fixed n, we use the window theorem above and a simple covering argument to find a subinterval, I say, with length ϵT, where we have the equidistribution for x_n. We then use the fact that the two pieces of orbits $\{\bullet u_t : t \in I\}$, for $\bullet = x_n, y_n$, stay $O(1)$ of each other for $t \in [0, T]$, to show that $\{y_n u_t : t \in I\}$ is also equidistributed.

4. Proof of the window theorem

In this section we describe the ingredients involved in the proof of Theorem 3.3. The proof has elements similar to the low entropy method introduced in [Lindenstrauss 2006].

The rough idea is this: The BMS measure is an A-invariant probability measure, and the leafwise measures of BMS along N are δ-dimensional Hausdorff measure. This, and the M-invariance of the BMS measure, imply: when $\delta > 1$ the entropy of a_s along U is nontrivial. Then, facts relating entropy and leafwise measures (see [Lindenstrauss 2006; Einsiedler et al. 2006]) imply that the leafwise measures of BMS measure along U restricted to $[-1, 1]$ satisfy a non-concentration property on a set of "almost" full measure; see [Lindenstrauss 2006; Einsiedler et al. 2006]. This is restated in [Mohammadi and Oh 2011, Theorem 6.11]. Now by (iii) in Theorem 2.1 we have, after renormalization, $a_{-s}m^{\mathrm{BR}}|_E$ weakly converges to the BMS measure, as $s \to \infty$. Here E is a BMS box fixed once and for all. The goal now is to use these to prove such non-concentration properties for the return times of the action of U with respect to the BR measure.

It is worth mentioning, however, that usually such information cannot be extracted — for example, due to discontinuity of the entropy with respect to weak star topology. We succeed, essentially, for two reasons:

(i) The BR and BMS measures have the same transversal measures, that is, locally, up to normalization, they only differ along N-leaves. This follows from the definition of these two measure.

(ii) The measures in question, that is, m^{BMS} and $a_{-s}m^{\mathrm{BR}}|_E$, are quite regular; see Propositions 4.1, 4.2 below.

Let us now fix some notation to be used in the course of the proof. For each $s > 0$, define a Borel measure $\mu_{E,s}^{\mathrm{BR}}$ on X to be the normalization of $a_{-s}m^{\mathrm{BR}}|_E$: for $\psi \in C_c(X)$,

$$\mu_{E,s}^{\mathrm{BR}}(\psi) := \frac{1}{m^{\mathrm{BR}}(E)} \int_E \psi(ga_{-s})\, dm^{\mathrm{BR}}(g).$$

It follows from (iii) in Theorem 2.1 that $\mu_{E,s} \to m^{\mathrm{BMS}}$ in the weak star topology. We have the following; see [Mohammadi and Oh 2011, Theorem 3.3].

Proposition 4.1. *Suppose that $x^- \in \Lambda(\Gamma)$. For all small enough $\rho > 0$, let $\lambda_{E,x,s}$ denote the conditional measure of $\mu_{E,s}^{\mathrm{BR}}$ along xN_ρ. We have,*

$$\lim_{s\to\infty} \lambda_{E,x,s}(\psi) = \mu_x^{\mathrm{PS}}(\psi) \quad \text{for any } \psi \in C_c(xN_\rho),$$

The proof of the above proposition draws from the convergence of $\mu_{E,s}^{\mathrm{BR}}$ to m^{BMS}, and a rather special feature of $\mu_{E,s}^{\mathrm{BR}}$: the conditionals of $\mu_{E,s}^{\mathrm{BR}}$ along N-leaves are obtained from a measure on the boundary, and in particular they change regularly as we move in the transversal direction.

We will now fix some x with $x^{\pm} \in \Lambda(\Gamma)$, and some small $\rho > 0$, and investigate $\lambda_{E,x,s}$. Given some $\theta \in [0, 2\pi]$ we let $m_\theta \in M$ denote the corresponding element. Also we will put

$$U_\theta^\rho := \{t \exp(2\pi i \theta) : t \in [-\rho, \rho]\},$$
$$V_\theta^\rho := \{it \exp(2\pi \theta) : t \in [-\rho, \rho]\}.$$

For any $0 < \tau \leq \rho$, and $\theta \in [0, 2\pi]$ we let $\sigma_{x,\theta,s}^\tau$ (resp. $\sigma_{x,\theta}^\tau$) denote the projection to V_θ^τ of $\lambda_{E,x,s}|_{xU_\theta^\tau V_\theta^\tau}$ (resp. $\mu_x^{\mathrm{PS}}|_{xU_\theta^\tau V_\theta^\tau}$). Furthermore, for any measure $\bullet$ on $\mathbb{R}$, we let $D(\bullet)$ denote the Radon–Nikodym derivative of $\bullet$ with respect to the Lebesgue measure.

We unfortunately cannot show a statement quite as strong as Proposition 4.1 for the disintegration of $\lambda_{E,x,s}$ along U-directions. The following is our replacement, which is [Mohammadi and Oh 2011, Proposition 5.10].

Proposition 4.2. *Let $s_i \to +\infty$ be a fixed sequence. For every $\epsilon > 0$ and every finite subset $\{\tau_1, \ldots, \tau_n\}$ of $(0, \rho]$, there exists a Borel subset $\Theta(x) \subset M$ of measure at least $1 - \epsilon$ such that for any $\theta \in \Theta(x)$ we have:*

(i) *For all $1 \leq \ell \leq n$, the measure $\sigma_{x,\theta}^{\tau_\ell}$ is absolutely continuous with respect to the Lebesgue measure on $xV_\theta^{\tau_\ell}$. Furthermore*

$$D(\sigma_{x,\theta}^{\tau_\ell}) \in H^r(xV_\theta^{\tau_\ell}), \quad \textit{for } r = \frac{\delta - 1}{4}.$$

(ii) *There is a subsequence $\{s_{i_j}\}$, depending on (x, θ), such that*

$$D(\sigma_{x,\theta,s_{i_j}}^{\tau_\ell}) \xrightarrow{L^2(xV_\theta)} D(\sigma_{x,\theta}^{\tau_\ell}) \quad \textit{for each} \quad 1 \leq \ell \leq n.$$

The proof of this proposition uses some techniques from geometric measure theory. Indeed an essential ingredient in the proof is a uniform bound for the α-dimensional energy of the $\lambda_{E,x,s}$, for some $1 < \alpha$; see [Mohammadi and Oh 2011, Theorem 5.7]. This gives uniform control on fractional derivatives of the Radon–Nikodym derivative of the projection of these measures with respect to the Lebesgue measure; see [Peres and Schlag 2000, Proposition 2.2] or [Mattila 2004, Theorem 4.5]. The L^2-convergence in the above then follows from compact embedding theorem of Sobolev spaces.

The desired energy estimate is proved using (i) the nonfocusing property of the PS measure, which was also used in the proof of Proposition 3.2, (ii) the fact that support of μ_s is contained in $O(e^{-s})$-thickening of the support of BMS [Mohammadi and Oh 2011, Lemma 5.6], and (iii) some covering argument.

We can now complete the proof of the window theorem. As we mentioned above, for the BMS measure the contribution of U to the entropy of a_s is nontrivial. Thus there is a set, Ω', of almost full BMS-measure and some $\beta > 0$, such that

$$(m^{\mathrm{BMS}})_x^U[-\beta,\beta] < \tfrac{1}{2}(m^{\mathrm{BMS}})_x^U[-1,1] \quad \text{for all } x \in \Omega', \tag{5}$$

where $(m^{\mathrm{BMS}})_x^U$ denotes the U-leafwise measure. Now suppose the window theorem fails, then there is a sequence $r_i \to 0$ and a sequence $T_i \to \infty$ such that

$$m^{\mathrm{BR}}\left\{x \in E : \int_{-r_iT_i}^{r_iT_i} \psi(xu_t)\,dt \ge (1-r_i)\int_{-T_i}^{T_i} \psi(xu_t)\,dt\right\} > (1-r_i)m^{\mathrm{BR}}(E).$$

Let $s_i = \log T_i$ and denote the set on the left side of the above by $E(s_i, r_i)$. If we flow this by a_{-s_i}, and use the definition of μ_{E,s_i}, we get

$$(\mu_{E,s_i})_x^U([-r_i, r_i]) > 1 - r_i,$$

for all x in a subset $E_{s_i}(r_i)$ of E, with $\mu_{E,x,s_i}(E_{s_i}(r_i)) > 1 - r_i$.

Now by Fubini's theorem there is a subset in the transversal direction to N of measure $1 - r_i$, such that if the transversal component of x is in this set, then $\lambda_{E,x,s_i}(xN \cap E_{s_i}(r_i)) > 1 - \sqrt{r_i}$. This set, in the transversal direction, however depends on i. To find a set which works for all i, we pass to a subsequence and assume that $\sum \sqrt{r_i} < \epsilon$, and replace these sets by their intersection. Altogether: there is some x such that $\lambda_{E,x,s_i}(xN \cap E_{s_i}(r_i)) > 1 - \sqrt{r_i}$ simultaneously for all i.

Indeed this condition is essentially to say that the disintegration of $\lambda_i = \lambda_{E,x,s_i}$, as a measure on N, along the direction of U is like a Dirac mass. Note also that since $\mu_{E,s}$ is M-invariant the above could be done so that the same holds not only for x, but also for xm_θ for "most" θ. Hence we have the slices of λ_i along U_θ is almost a dirac measure for many directions θ.

It is more convenient for us, however, to work with the projections. Fix some small $\tau \ll \beta$. We take projection of the set $\mathrm{Bad}_{\bullet,i} = xN_\bullet \cap E_{s_i}(r_i)$, for $\bullet = 1, \tau$, onto xV_θ. We have $\sigma_i^1(\mathrm{Bad}_{\tau,i}) > 1 - 2r_i\sigma_i^1(\mathrm{Bad}_{1,i})$, where σ denotes the projection of λ_i onto xV_θ. It is worth mentioning that one needs to be careful here, since these sets change as i changes; we need to show that $\mathrm{Bad}_{\bullet,i}$ has almost full PS measure in $xN_\bullet$. If this is established then (5) would give a contradiction. This thankfully follows from the L^2-convergence statement in Proposition 4.2; see [Mohammadi and Oh 2011, Lemma 5.11]. This completes the proof.

Acknowledgement

This article is based on joint work with Hee Oh, to whom the author is grateful.

References

[Burger 1990] M. Burger, "Horocycle flow on geometrically finite surfaces", *Duke Math. J.* **61**:3 (1990), 779–803.

[Einsiedler et al. 2006] M. Einsiedler, A. Katok, and E. Lindenstrauss, "Invariant measures and the set of exceptions to Littlewood's conjecture", *Ann. of Math.* (2) **164**:2 (2006), 513–560.

[Flaminio and Spatzier 1990] L. Flaminio and R. J. Spatzier, "Geometrically finite groups, Patterson–Sullivan measures and Ratner's rigidity theorem", *Invent. Math.* **99**:3 (1990), 601–626.

[Hopf 1937] E. Hopf, *Ergodentheorie*, Ergebnisse der Math. **5**, Springer, 1937.

[Lindenstrauss 2006] E. Lindenstrauss, "Invariant measures and arithmetic quantum unique ergodicity", *Ann. of Math.* (2) **163**:1 (2006), 165–219.

[Margulis 1975] G. A. Margulis, "On the action of unipotent groups in the space of lattices", pp. 365–370 in *Lie groups and their representations* (Budapest, 1971), edited by I. M. Gelfand, Halsted, New York, 1975.

[Margulis 1989] G. A. Margulis, "Indefinite quadratic forms and unipotent flows on homogeneous spaces", pp. 399–409 in *Dynamical systems and ergodic theory* (Warsaw, 1986), edited by K. Krzyżewski, Banach Center Publ. **23**, PWN, Warsaw, 1989.

[Mattila 2004] P. Mattila, "Hausdorff dimension, projections, and the Fourier transform", *Publ. Mat.* **48**:1 (2004), 3–48.

[Mohammadi and Oh 2011] A. Mohammadi and H. Oh, "Ergodicity of unipotent flows and Kleinian groups", preprint, 2011. arXiv 1112.4024

[Moore 1966] C. C. Moore, "Ergodicity of flows on homogeneous spaces", *Amer. J. Math.* **88** (1966), 154–178.

[Oh 2013] H. Oh, "Harmonic analysis, ergodic theory and counting for thin groups", Math. Sci. Res. Inst. Publ. **61**, Cambridge Univ. Press, Cambridge, 2013.

[Patterson 1976] S. J. Patterson, "The limit set of a Fuchsian group", *Acta Math.* **136**:3-4 (1976), 241–273.

[Peres and Schlag 2000] Y. Peres and W. Schlag, "Smoothness of projections, Bernoulli convolutions, and the dimension of exceptions", *Duke Math. J.* **102**:2 (2000), 193–251.

[Ratner 1990] M. Ratner, "On measure rigidity of unipotent subgroups of semisimple groups", *Acta Math.* **165**:3-4 (1990), 229–309.

[Ratner 1991] M. Ratner, "On Raghunathan's measure conjecture", *Ann. of Math.* (2) **134**:3 (1991), 545–607.

[Roblin 2003] T. Roblin, *Ergodicité et équidistribution en courbure négative*, Mém. Soc. Math. Fr. (N.S.) **95**, 2003.

[Sullivan 1979] D. Sullivan, "The density at infinity of a discrete group of hyperbolic motions", *Inst. Hautes Études Sci. Publ. Math.* 50 (1979), 171–202.

[Sullivan 1984] D. Sullivan, "Entropy, Hausdorff measures old and new, and limit sets of geometrically finite Kleinian groups", *Acta Math.* **153**:3-4 (1984), 259–277.

[Yau 1975] S. T. Yau, "Harmonic functions on complete Riemannian manifolds", *Comm. Pure Appl. Math.* **28** (1975), 201–228.

amir@math.utexas.edu *Department of Mathematics, The University of Texas, Austin, TX 78750, United States*

Thin groups and superstrong approximation
MSRI Publications
Volume **61**, 2013

Harmonic analysis, ergodic theory and counting for thin groups

HEE OH

For a geometrically finite group Γ of $G = \mathrm{SO}(n, 1)$, we survey recent developments on counting and equidistribution problems for orbits of Γ in a homogeneous space $H\backslash G$ where H is trivial, symmetric or horospherical. Main applications are found in an affine sieve on orbits of thin groups as well as in sphere counting problems for sphere packings invariant under a geometrically finite group. In our sphere counting problems, spheres can be ordered with respect to a general conformal metric.

1. Introduction

In this article we discuss counting and equidistribution problems for orbits of thin groups in homogeneous spaces.

Let G be a connected semisimple Lie group and H a closed subgroup. We consider the homogeneous space $V = H\backslash G$ and fix the identity coset $x_0 = [e]$. Let Γ be a discrete subgroup of G such that the orbit $x_0\Gamma$ is discrete, and let $\{B_T : T > 1\}$ be a family of compact subsets of V whose volume tends to infinity as $T \to \infty$. Understanding the asymptotic of $\#(x_0\Gamma \cap B_T)$ is a fundamental problem which bears many applications in number theory and geometry. We refer to this type of counting problem as an archimedean counting as opposed to a combinatorial counting where the elements in $x_0\Gamma$ are ordered with respect to a word metric on Γ; both have been used in applications to sieves; see [Kowalski 2010; 2014].

When Γ is a lattice in G, i.e., when $\Gamma\backslash G$ admits a finite invariant measure, this problem is well understood for a large class of subgroups H, e.g., when H is a maximal subgroup: assuming that the boundaries of the B_T are sufficiently regular and that $H\cap\Gamma$ is a lattice in H, the value of $\#(x_0\Gamma\cap B_T)$ is asymptotically proportional to the volume of B_T, computed with respect to a suitably normalized G-invariant measure in V [Duke et al. 1993; Eskin and McMullen 1993; Eskin et al. 1996; Benoist and Oh 2012]. See also the survey paper [Oh 2010].

When Γ is a thin group, i.e., a Zariski dense subgroup which is not a lattice in G, it is far less understood in a general setting. In this article we focus on the case when G is the special orthogonal group $\mathrm{SO}(n, 1)$ and H is either a symmetric subgroup or a horospherical subgroup (the case of H being the trivial subgroup will be treated as well). In this case, we have a more or less satisfactory understanding for the counting problem for groups Γ equipped with a certain finiteness property, called the geometric finiteness (see Definition 2.1).

By the fundamental observation of Duke, Rudnick and Sarnak [Duke et al. 1993], the counting problem for $x_0\Gamma\cap B_T$ can be well approached via the following equidistribution problem:

Describe the asymptotic distribution of $\Gamma\backslash\Gamma Hg$ *in* $\Gamma\backslash G$ *as* $g\to\infty$.

The assumption that H is either symmetric or horospherical is made so that we can approximate the translate $\Gamma\backslash\Gamma Hg$ locally by the matrix coefficient function in the quasiregular representation space $L^2(\Gamma\backslash G)$. This idea goes back to Margulis's 1970 thesis (translated in [Margulis 2004]; also see [Kleinbock and Margulis 1996]), but a more systematic formulation in our setting is due to Eskin and McMullen [1993].

The fact that the trivial representation is contained in $L^2(\Gamma\backslash G)$ for Γ lattice is directly related to the phenomenon that, when $\Gamma\backslash\Gamma H$ is of finite volume, the translate $\Gamma\backslash\Gamma Hg$ becomes equidistributed in $\Gamma\backslash G$ with respect to a G-*invariant* measure as $g\to\infty$ in $H\backslash G$. When Γ is not a lattice, the minimal subrepresentation, that is, the subrepresentation with the slowest decay of matrix coefficients of $L^2(\Gamma\backslash G)$, is infinite-dimensional and this makes the distribution of $\Gamma\backslash\Gamma Hg$ much more intricate, and understanding it requires introducing several singular measures in $\Gamma\backslash G$. A key input is the work of Roblin [2003] on the asymptotic of matrix coefficients for $L^2(\Gamma\backslash G)$, which he proves using ergodic theoretic methods.

There is a finer distinction among geometrically finite groups depending on the size of their critical exponents. When the critical exponent δ of Γ exceeds $\frac{n-1}{2}$, work of Lax and Phillips [1982] implies a spectral gap for $L^2(\Gamma\backslash\mathbb{H}^n)$; which we call a *spherical* spectral gap for $L^2(\Gamma\backslash G)$ as it concerns only the spherical part of $L^2(\Gamma\backslash G)$. We formulate a notion of a spectral gap for $L^2(\Gamma\backslash G)$ which deals with both spherical and nonspherical parts, based on the knowledge of the unitary dual of G. Under the hypothesis that $L^2(\Gamma\backslash G)$ admits a spectral gap (this is known to be true if $\delta > n-2$), developing Harish-Chandra's work on harmonic analysis on G in combination with Roblin's work on ergodic theory, we obtain an effective version of the asymptotic of matrix coefficients for $L^2(\Gamma\backslash G)$ in [Mohammadi and Oh 2012b]. This enables us to state an effective equidistribution of $\Gamma\backslash\Gamma Hg$ in $\Gamma\backslash G$. Similar to the condition that $\Gamma\backslash\Gamma H$ is of finite volume in the case of Γ lattice, there is also a certain restriction on the size of the orbit $\Gamma\backslash\Gamma H$ in order to deduce such an equidistribution. A precise condition is that the skinning measure μ_H^{PS} of $\Gamma\backslash\Gamma H$, introduced in [Oh and Shah 2013], is finite; roughly speaking $|\mu_H^{\mathrm{PS}}|$ measures asymptotically the portion of $\Gamma\backslash\Gamma H$ which returns to a compact subset after flowed by the geodesic flow. When the skinning measure μ_H^{PS} is compactly supported, the passage from the asymptotic of the matrix coefficient to the equidistribution of $\Gamma\backslash\Gamma Hg$ can be done by the so-called usual thickening methods. However when μ_H^{PS} is not compactly supported, this step requires a genuinely different strategy from the lattice case via the study of the transversal intersections, carefully done in [Oh and Shah 2013].

The error term in our effective equidistribution result of $\Gamma\backslash\Gamma Hg$ depends only on the spectral gap data of Γ. This enables us to state the asymptotic of $\#(x_0\Gamma_d\gamma \cap B_T)$ effectively in a uniform manner for all $\gamma \in \Gamma$ and for any family $\{\Gamma_d < \Gamma\}$ of subgroups of finite index which has a uniform spectral gap, if they satisfy $\Gamma_d \cap H = \Gamma \cap H$. When Γ is a subgroup of an arithmetic subgroup of G with $\delta > n-2$, the work of Salehi Golsefidy and Varjú [2012], extending an earlier work of Bourgain, Gamburd and Sarnak [Bourgain et al. 2010a], provides a certain congruence family $\{\Gamma_d\}$ satisfying this condition.

A recent development on an affine sieve [Bourgain et al. 2011] then tells us that such a uniform effective counting statement can be used to describe the distribution of almost prime vectors, as well as to give a sharp upper bound for primes (see Theorem 6.6).

One of the most beautiful applications of the study of thin orbital counting problem can be found in Apollonian circle packings. We will describe this application as well as its higher-dimensional analogues. The ordering in counting circles can be done not only in the Euclidean metric, but also in general conformal metrics. It is due to this flexibility that we can also describe the asymptotic number of circles in the ideal triangle of the hyperbolic plane ordered by the hyperbolic area in the last section.

2. Γ-invariant conformal densities and measures on $\Gamma\backslash G$

In the whole article, let $(\mathbb{H}^n, d)$ be the n-dimensional real hyperbolic space with constant curvature -1 and $\partial(\mathbb{H}^n)$ its geometric boundary. Set $G := \mathrm{Isom}^+(\mathbb{H}^n) \simeq \mathrm{SO}(n,1)^\circ$. Let Γ be a discrete torsion-free subgroup of G. We assume that Γ is nonelementary, or equivalently, Γ has no abelian subgroup of finite index.

We review some basic geometric and measure theoretic concepts for Γ and define several locally finite Borel measures (Radon measures) on $\Gamma\backslash G$ associated to Γ-invariant conformal densities on $\partial(\mathbb{H}^n)$. When Γ is a lattice, these measures all coincide with each other, being simply a G-invariant measure. But for a thin subgroup Γ, they are all different and singular, and appear in our equidistribution and counting statements. General references for this section are [Ratcliffe 2006; Bowditch 1993; Patterson 1976; Sullivan 1979; 1984; Roblin 2003].

Limit set and geometric finiteness. We denote by $\delta_\Gamma = \delta$ the critical exponent of Γ, i.e., the abscissa of convergence of the Poincaré series $\sum_{\gamma\in\Gamma} e^{-sd(o,\gamma(o))}$ for $o \in \mathbb{H}^n$. We have $0 < \delta \le n-1$. The limit set $\Lambda(\Gamma)$ is defined to be the set of all accumulation points of $\Gamma(z)$ in the compactification $\overline{\mathbb{H}^n} = \mathbb{H}^n \cup \partial(\mathbb{H}^n)$, $z \in \mathbb{H}^n$. As Γ is discrete, $\Lambda(\Gamma)$ is contained in the boundary $\partial(\mathbb{H}^n)$.

Definition 2.1. (1) The convex core $C(\Gamma)$ of Γ is the quotient by Γ of the smallest convex subset of $\mathbb{H}^n$ containing all geodesics connecting points in $\Lambda(\Gamma)$.

(2) Γ is called geometrically finite (resp. convex cocompact) if the unit neighborhood of the convex core of Γ has finite volume (resp. compact).

A lattice is clearly a geometrically finite group and so is a discrete group admitting a finite sided convex fundamental domain in $\mathbb{H}^n$. An important characterization of a geometrically finite group is given in terms of its limit set. For this, we need to define: a point $\xi \in \Lambda(\Gamma)$ is called a parabolic limit point for Γ if ξ is a unique fixed point in $\partial\mathbb{H}^n$ for an element of Γ and a radial limit point if the projection of a geodesic ray ξ_t toward ξ in $\Gamma\backslash\mathbb{H}^n$ meets a compact subset for an unbounded sequence of time t.

Now Γ is geometrically finite if and only if $\Lambda(\Gamma)$ consists only of parabolic and radial limit points [Bowditch 1993]. For Γ geometrically finite, its critical exponent δ is equal to the Hausdorff dimension of $\Lambda(\Gamma)$, and is $n-1$ only when Γ is a lattice in G [Sullivan 1979].

Conformal densities. To define a conformal density, we first recall the Busemann function $\beta_\xi(x,y)$ for $x, y \in \mathbb{H}^n$ and $\xi \in \partial(\mathbb{H}^n)$:

$$\beta_\xi(x,y) = \lim_{t\to\infty} d(x,\xi_t) - d(y,\xi_t)$$

where ξ_t is a geodesic toward ξ. Hence $\beta_\xi(x, y)$ measures a signed distance between horospheres based at ξ passing through x and y (a horosphere based at ξ is a Euclidean sphere in $\mathbb{H}^n$ tangent at ξ).

Definition 2.2. A Γ-invariant conformal density of dimension $\delta_\mu > 0$ is a family $\{\mu_x : x \in \mathbb{H}^n\}$ of finite positive measures on $\partial(\mathbb{H}^n)$ satisfying

(1) $\gamma_* \mu_x = \mu_{\gamma(x)}$ for any $\gamma \in \Gamma$, and

(2) $\dfrac{d\mu_x}{d\mu_y}(\xi) = e^{\delta_\mu \beta_\xi(y,x)}$ for all $x, y \in \mathbb{H}^n$ and $\xi \in \partial(\mathbb{H}^n)$.

It is easy to construct such a density of dimension $n-1$, as we simply need to set m_x to be the $\mathrm{Stab}_G(x)$-invariant probability measure on $\mathbb{H}^n$. This is a unique up to scaling, and does not depend on Γ. We call it the Lebesgue density.

How about in other dimensions? A fundamental work of Patterson [1976], generalized by Sullivan [1979], shows the following by an explicit construction:

Theorem 2.3. *There exists a Γ-invariant conformal density of dimension $\delta_\Gamma = \delta$.*

Assuming that Γ is of divergence type, i.e., its Poincaré series diverges at $s = \delta$, Patterson's construction can be summarized as follows: Fixing $o \in \mathbb{H}^n$, for each $x \in \mathbb{H}^n$, consider the finite measure on $\overline{\mathbb{H}^n}$ given by

$$\nu_{x,s} = \frac{1}{\sum_{\gamma\in\Gamma} e^{-sd(o,\gamma(o))}} \sum_{\gamma\in\Gamma} e^{-sd(x,\gamma(o))} \delta_{\gamma(o)}.$$

Then ν_x is the (unique) weak-limit of $\nu_{x,s}$ as $s \to \delta^+$, and $\{\nu_x : x \in \mathbb{H}^n\}$ is the desired density of dimension δ.

In the following, we fix a Γ-invariant conformal density $\{\nu_x\}$ of dimension δ, and call it the Patterson–Sullivan density (or simply the PS density). It is known to be unique up to scaling, when Γ is of divergence type, e.g., geometrically finite groups.

Denoting by Δ the hyperbolic Laplacian on $\mathbb{H}^n$, the PS density is closely related to the bottom of the spectrum of Δ for its action on smooth functions on $\Gamma\backslash\mathbb{H}^n$. If we set $\phi_0(x) := |\nu_x|$ for each $x \in \mathbb{H}^n$, then

$$\phi_0(x) = \int_{\xi\in\Lambda(\Gamma)} d\nu_x(\xi) = \int_{\xi\in\Lambda(\Gamma)} \frac{d\nu_x}{d\nu_o}(\xi)\, d\nu_o(\xi) = \int_{\xi\in\Lambda(\Gamma)} e^{\delta\beta_\xi(o,x)} d\nu_o(\xi).$$

Since $\gamma_*\nu_x = \nu_{\gamma(x)}$, we note that $\phi_0(\gamma(x)) = \phi_0(x)$, i.e., ϕ_0 is a function on $\Gamma\backslash\mathbb{H}^n$, which is *positive* everywhere! Furthermore:

(1) $\Delta(\phi_0) = \delta(n-1-\delta)\phi_0$.

(2) If Γ is geometrically finite, then $\phi_0 \in L^2(\Gamma\backslash\mathbb{H}^n)$ if and only if $\delta > \frac{n-1}{2}$.

Measures on $\Gamma\backslash G$ associated to a pair of conformal densities. For $v \in \mathrm{T}^1(\mathbb{H}^n)$, we denote by v^+ and v^- the forward and backward endpoints of the geodesic determined by v. Fixing $o \in \mathbb{H}^n$, the map $v \mapsto (v^+, v^-, s = \beta_{v^-}(o, v))$ gives a homeomorphism between $\mathrm{T}^1(\mathbb{H}^n)$ and $(\partial(\mathbb{H}^n) \times \partial(\mathbb{H}^n) - \text{diagonal}) \times \mathbb{R}$. Therefore we may use the coordinates $(v^+, v^-, s = \beta_{v^-}(o, v))$ of v in order to define measures on $\mathrm{T}^1(\mathbb{H}^n)$.

Let $\{\mu_x\}$ and $\{\mu'_x\}$ be Γ-invariant conformal densities on $\partial(\mathbb{H}^n)$ of dimensions δ_μ and $\delta_{\mu'}$ respectively. After [Roblin 2003], we define a measure $\tilde{m}^{\mu,\mu'}$ on $\mathrm{T}^1(\mathbb{H}^n)$ associated to $\{\mu_x\}$ and $\{\mu'_x\}$ by

$$d\tilde{m}^{\mu,\mu'}(v) = e^{\delta_\mu \beta_{v^+}(o,v)}\, e^{\delta_{\mu'} \beta_{v^-}(o,v)}\, d\mu_o(v^+)\, d\mu'_o(v^-)\, ds. \tag{2.4}$$

It follows from the Γ-invariant conformal properties of $\{\mu_x\}$ and $\{\mu'_x\}$ that the definition of $\tilde{m}^{\mu,\mu'}$ is independent of the choice of $o \in \mathbb{H}^n$ and that $\tilde{m}^{\mu,\mu'}$ is left Γ-invariant. Hence it induces a Radon measure $m^{\mu,\mu'}$ on the quotient space $\mathrm{T}^1(\Gamma\backslash\mathbb{H}^n) = \Gamma\backslash \mathrm{T}^1(\mathbb{H}^n)$.

We will lift the measure $m^{\mu,\mu'}$ to $\Gamma\backslash G$. This lift depends on the choice of subgroups K, M and $A = \{a_t\}$ of G. Here K is a maximal compact subgroup of G and M is the stabilizer of a vector $X_0 \in \mathrm{T}^1(\mathbb{H}^n)$ based at $o \in \mathbb{H}^n$ with $K = G_o$. Via the isometric action of G, we may identify the quotient spaces G/K and G/M with $\mathbb{H}^n$ and $\mathrm{T}^1(\mathbb{H}^n)$ respectively. Let $A = \{a_t\}$ be the one parameter subgroup of diagonalizable elements of G such that the right multiplication by a_t on G/M corresponds to the geodesic flow on $\mathrm{T}^1(\mathbb{H}^n)$ for time t.

By abuse of notation, we use the same notation $m^{\mu,\mu'}$ for the M-invariant extension of $m^{\mu,\mu'}$ on $\Gamma\backslash G/M = \Gamma\backslash \mathrm{T}^1(\mathbb{H}^n)$ to $\Gamma\backslash G$, that is, for $\Psi \in C_c(\Gamma\backslash G)$,

$$m^{\mu,\mu'}(\Psi) = \int_{x \in \Gamma\backslash G/M} \Psi^M(x)\, dm^{\mu,\mu'}(x)$$

where $\Psi^M(x) = \int_M \Psi(xm)\, dm$ for the probability M-invariant measure dm on M.

The measures $m^{\mu,\mu'}$ on $\Gamma\backslash G$ where μ and μ' are the PS-density $\{\nu_x\}$ or the Lebesgue density $\{m_x\}$ are of special importance. We name them as follows:

- Bowen–Margulis–Sullivan measure: $m^{\mathrm{BMS}} := m^{\nu,\nu}$
- Burger–Roblin measure: $m^{\mathrm{BR}} := m^{m,\nu}$
- Burger–Roblin $*$-measure: $m^{\mathrm{BR}}_* := m^{\nu,m}$
- Haar measure: $m^{\mathrm{Haar}} := m^{m,m}$.

For brevity, we refer to these as BMS, BR, BR_*, and Haar measures, respectively. As the naming indicates, m^{Haar} turns out to be a G-invariant measure. For $g \in G$, we use the notation $g^\pm$ for $(gM)^\pm$ with gM considered as a vector in $\mathrm{T}^1(\mathbb{H}^n)$.

It is clear from the definition that the supports of BMS, BR, BR_* measures are respectively given by $\{g \in \Gamma\backslash G : g^{\pm} \in \Lambda(\Gamma)\}$, $\{g \in \Gamma\backslash G : g^{-} \in \Lambda(\Gamma)\}$, and $\{g \in \Gamma\backslash G : g^{+} \in \Lambda(\Gamma)\}$. In particular, the support of BMS measure is contained in the convex core of Γ. Sullivan showed that $|m^{\mathrm{BMS}}| < \infty$ if Γ is geometrically finite.

The BMS, BR and BR_* measures are respectively invariant under A, N^+ and N^- where N^+ and N^- denote the expanding and contracting horospherical subgroups of G for a_t:

$$N^{\pm} = \{g \in G : a_t g a_{-t} \to e \text{ as } t \to \pm\infty\}.$$

The finiteness of m^{BMS} turns out to be a critical condition for the ergodic theory on $\Gamma\backslash G$.

Theorem 2.5. *Suppose that $|m^{\mathrm{BMS}}| < \infty$ and that Γ is Zariski dense.*

(1) m^{BMS} *is A-mixing: for any $\Psi_1, \Psi_2 \in L^2(\Gamma\backslash G)$,*

$$\lim_{t\to\infty} \int_{\Gamma\backslash G} \Psi_1(ga_t)\Psi_2(g)\, dm^{\mathrm{BMS}}(g) = \frac{1}{|m^{\mathrm{BMS}}|} m^{\mathrm{BMS}}(\Psi_1) \cdot m^{\mathrm{BMS}}(\Psi_2).$$

(2) *Any locally finite N^+-ergodic invariant measure on $\Gamma\backslash G$ is either supported on a closed N^+M_0-orbit where M_0 is an abelian closed subgroup of M or m^{BR}.*

(3) m^{BR} *is a finite measure if and only if Γ is a lattice in G.*

Claim (1) was first made in [Flaminio and Spatzier 1990]. However there is a small gap in their proof which is now fixed in [Winter $\geq$ 2012]. Winter also obtained Claim (1) in a general rank one symmetric space. For M-invariant functions, this claim was earlier proved by Babillot [2002] and in this case the Zariski density assumption is not needed. Claim (2) was first proved by Burger [1990] for a convex cocompact surface with critical exponent bigger than 1/2. In a general case, Winter obtained Claim (2) from Roblin's work [2003] and Claim (1).

Claim (3) is proved in [Oh and Shah 2013], using Ratner's measure classification [1991] of finite measures invariant under unipotent flows. Namely, we show that m^{BR} is not one of those homogeneous measures that her classification theorem lists for finite invariant measures.

In the spirit of Ratner's measure classification theorem, we pose the following question:

Problem 2.6. Under the assumption of Theorem 2.5, let U be a connected unipotent subgroup of G or more generally a connected subgroup generated by unipotent one-parameter subgroups.

(1) Classify all locally finite U-invariant ergodic measures in $\Gamma\backslash G$.

(2) Describe the closures of U-orbits in $\Gamma\backslash G$.

The emphasis here is that we want to understand not only finite measures but all Radon measures. In general, this seems to be a very challenging question. We mention a recent related result from [Mohammadi and Oh 2012a]: if Γ is a convex cocompact subgroup of G and U is a connected unipotent subgroup of N^+ of dimension k, m^{BR} is U-ergodic if $\delta > (n-1)-k$. Precisely speaking, this is proved only for $n=3$ in [Mohammadi and Oh 2012a], but the methods of proof works for a general $n \geq 3$ as well.

3. Matrix coefficients for $L^2(\Gamma\backslash G)$

Let Γ be a discrete, torsion-free, nonelementary subgroup of $G = \mathrm{SO}(n,1)^\circ$. The right translation action of G on $L^2(\Gamma\backslash G, m^{\mathrm{Haar}})$ gives rise to a unitary representation, as m^{Haar} is G-invariant. For $\Psi_1, \Psi_2 \in L^2(\Gamma\backslash G)$, the matrix coefficient function is a smooth function on G defined by

$$g \mapsto \langle g.\Psi_1, \Psi_2\rangle := \int_{\Gamma\backslash G} \Psi_1(xg)\overline{\Psi_2(x)}\,dm^{\mathrm{Haar}}(x).$$

Understanding the asymptotic expansion of $\langle a_t.\Psi_1, \Psi_2\rangle$ (as $t \to \infty$) is a basic problem in harmonic analysis as well as a main tool in our approach to the counting problem.

The quality of the error term in this type of the asymptotic expansion usually depends on Sobolev norms of Ψ_i's. For $\Psi \in C^\infty(\Gamma\backslash G)$ and $d \in \mathbb{N}$, the d-th Sobolev norm of Ψ is given by $\mathcal{S}_d(\Psi) = \sum \|X(\Psi)\|_2$ where the sum is taken over all monomials X in some fixed basis of the Lie algebra of G of order at most d and $\|X(\Psi)\|_2$ denotes the L^2-norm of Ψ.

In order to describe our results as well as a conjecture, we begin by describing the unitary dual of G, i.e., the set of equivalence classes of all irreducible unitary representations of G.

The unitary dual $\hat{G}$. Let K be a maximal compact subgroup of G and fix a Cartan decomposition $G = KA^+K$. We parametrize $A^+ = \{a_t : t \geq 0\}$ so that a_t corresponds to the geodesic flow on $\mathrm{T}^1(\mathbb{H}^n) = G/M$ where $M := C_K(A^+)$ is the centralizer of A^+ in K.

A representation $\pi \in \hat{G}$ is said to be *tempered* if for any K-finite vectors v_1, v_2 of π, the matrix coefficient function $g \mapsto \langle \pi(g)v_1, v_2\rangle$ belongs to $L^{2+\epsilon}(G)$ for any $\epsilon > 0$. We write $\hat{G} = \hat{G}_{\mathrm{temp}} \cup \hat{G}_{\mathrm{nontemp}}$ as the disjoint union of tempered representations and nontempered representations.

The work of Hirai [1962] on the classification of $\hat{G}$ implies that nontempered part of the unitary dual $\hat{G}$ consists of the trivial representation, and complementary series representations $\mathcal{U}(\upsilon, s-n+1)$ parameterized by $\upsilon \in \hat{M}$ and $s \in I_\upsilon$, where $\hat{M}$ is the unitary dual of M and I_υ is an interval contained in $(\frac{n-1}{2}, n-1)$, depending on υ (see also [Knapp and Stein 1971, Propositions 49, 50]). Moreover $\mathcal{U}(\upsilon, s-n+1)$ is spherical if and only if υ is the trivial representation 1 of M. By choosing a Casimir operator $\mathcal{C}$ of the Lie algebra of G normalized so that it acts on $C^\infty(G)^K = C^\infty(\mathbb{H}^n)$ by the negative Laplacian, the normalization is made so that $\mathcal{C}$ acts on $\mathcal{U}(1, s-n+1)^\infty$ by the scalar $s(s-n+1)$.

***Lattice case:* $|m^{\mathrm{Haar}}| < \infty$.** Set

$$L^2_0(\Gamma\backslash G) := \{\Psi \in L^2(\Gamma\backslash G) : \int \Psi \, dm^{\mathrm{Haar}} = 0\}.$$

When Γ is a lattice, we have $L^2(\Gamma\backslash G) = \mathbb{C} \oplus L^2_0(\Gamma\backslash G)$. It is well-known that there exists $\frac{n-1}{2} < s_0 < (n-1)$ such that $L^2_0(\Gamma\backslash G)$ does not contain any complementary series representation of parameter $s \geq s_0$ [Borel and Garland 1983]. This implies:

Theorem 3.1 [Shalom 2000; Kontorovich and Oh 2011, Proposition 5.3]. *Suppose Γ is a lattice in G, that is, $|m^{\mathrm{Haar}}| < \infty$. There exists $\ell \in \mathbb{N}$ such that for any $\Psi_1, \Psi_2 \in L^2(\Gamma\backslash G) \cap C(\Gamma\backslash G)^\infty$, we have, as $t \to \infty$,*

$$\langle a_t.\Psi_1, \Psi_2\rangle = \frac{1}{|m^{\mathrm{Haar}}|} m^{\mathrm{Haar}}(\Psi_1) m^{\mathrm{Haar}}(\Psi_2) + O(\mathcal{S}_\ell(\Psi_1)\mathcal{S}_\ell(\Psi_2) e^{(s_0-n+1)t})$$

where $m^{\mathrm{Haar}}(\Psi_i) = \int_{\Gamma\backslash G} \Psi_i(x)\, dm^{\mathrm{Haar}}(x)$ for $i = 1, 2$.

We note that the constant function $1/\sqrt{|m^{\mathrm{Haar}}|}$ is a unit vector in $L^2(\Gamma\backslash G)$ and the main term $\frac{1}{|m^{\mathrm{Haar}}|} m^{\mathrm{Haar}}(\Psi_1) m^{\mathrm{Haar}}(\Psi_2)$ is simply the product of the projections of Ψ_1 and Ψ_2 to the minimal subrepresentation space, which is $\mathbb{C}$, of $L^2(\Gamma\backslash G)$.

***Discrete groups with* $|m^{\mathrm{BMS}}| < \infty$.** When Γ is not a lattice, we have $L^2(\Gamma\backslash G) = L^2_0(\Gamma\backslash G)$. By the well-known decay of the matrix coefficients of unitary representations with no G-invariant vectors due to Howe and Moore [Howe and Moore 1979], we have, for any $\Psi_1, \Psi_2 \in L^2(\Gamma\backslash G)$,

$$\lim_{t\to\infty} \langle a_t.\Psi_1, \Psi_2\rangle = 0.$$

However we have a much more precise description on the decay of $\langle a_t.\Psi_1, \Psi_2\rangle$ due to Roblin:

Theorem 3.2 [Roblin 2003]. *Let Γ be Zariski dense with $|m^{\text{BMS}}| < \infty$. For any $\Psi_1, \Psi_2 \in C_c(\Gamma\backslash G)$,*

$$\lim_{t\to\infty} e^{(n-1-\delta)t}\langle a_t.\Psi_1, \Psi_2\rangle = \frac{m^{\text{BR}}(\Psi_1)\cdot m_*^{\text{BR}}(\Psi_2)}{|m^{\text{BMS}}|}.$$

Roblin proved this theorem for M-invariant functions using the mixing of the geodesic flow due to Babillot [2002]. His proof extends without difficulty to general functions, based on the A-mixing in $\Gamma\backslash G$ stated as in Theorem 2.5.

Geometrically finite groups with $\delta > \frac{n-1}{2}$. In this subsection, we assume that Γ is a geometrically finite, Zariski dense, discrete subgroup of G with $\delta > \frac{n-1}{2}$. Under this assumption, the works of Lax and Phillips [Lax and Phillips 1982] and Sullivan [Sullivan 1984] together imply that there exist only finitely many $(n-1) \ge s_0 > s_1 \ge \cdots \ge s_\ell > \frac{n-1}{2}$ such that the spherical complementary series representation $\mathcal{U}(1, s-n+1)$ occurs as a subrepresentation of $L^2(\Gamma\backslash G)$ and $s_0 = \delta$. In particular, there is no spherical complementary representation $\mathcal{U}(1, s-n+1)$ contained in $L^2(\Gamma\backslash G)$ for $\delta < s < s_1$; hence we have a *spherical* spectral gap for $L^2(\Gamma\backslash G)$.

Using the classification of $\hat{G}_{\text{nontemp}}$, we formulate the notion of a spectral gap. Recall that a unitary representation π is said to be weakly contained in a unitary representation π' if any diagonal matrix coefficients of π can be approximated, uniformly on compact subsets, by convex combinations of diagonal matrix coefficients of π'.

Definition 3.3. We say that $L^2(\Gamma\backslash G)$ has a *strong spectral gap* if

(1) $L^2(\Gamma\backslash G)$ does not contain any $\mathcal{U}(\upsilon, \delta-n+1)$ with $\upsilon \ne 1$;

(2) there exists $s_0(\Gamma)$ with $\frac{n-1}{2} < s_0(\Gamma) < \delta$ and such that $L^2(\Gamma\backslash G)$ does not weakly contain any $\mathcal{U}(\upsilon, s-n+1)$ with $s \in (s_0(\Gamma), \delta)$ and $\upsilon \in \hat{M}$.

For $\delta \le \frac{n-1}{2}$, the Laplacian spectrum of $L^2(\Gamma\backslash\mathbb{H}^n)$ is continuous; this implies that there is no spectral gap for $L^2(\Gamma\backslash G)$.

Conjecture 3.4 (spectral gap conjecture, [Mohammadi and Oh 2012b]). *If Γ is a geometrically finite and Zariski dense subgroup of G with $\delta > \frac{n-1}{2}$, $L^2(\Gamma\backslash G)$ has a strong spectral gap.*

If $\delta > \frac{n-1}{2}$ for $n = 2, 3$, or if $\delta > (n-2)$ for $n \ge 4$, then $L^2(\Gamma\backslash G)$ has a strong spectral gap. This observation follows from the classification of $\hat{G}_{\text{nontemp}}$ which says that there is no nonspherical complementary series representation $\mathcal{U}(\upsilon, s-n+1)$ of parameter $n-2 < s < n-1$ [Hirai 1962].

Our theorems are proved under the following slightly weaker spectral gap property assumption:

Definition 3.5. [Mohammadi and Oh 2012b] We say that $L^2(\Gamma\backslash G)$ has a *spectral gap* if there exist $\frac{n-1}{2} < s_0 = s_0(\Gamma) < \delta$ and $n_0 = n_0(\Gamma) \in \mathbb{N}$ such that

(1) the multiplicity of $\mathcal{U}(\upsilon, \delta-n+1)$ contained in $L^2(\Gamma\backslash G)$ is at most $\dim(\upsilon)^{n_0}$ for any $\upsilon \in \hat{M}$;

(2) $L^2(\Gamma\backslash G)$ does not weakly contain any $\mathcal{U}(\upsilon, s-n+1)$ with $s \in (s_0, \delta)$ and $\upsilon \in \hat{M}$.

The pair $(s_0(\Gamma), n_0(\Gamma))$ will be referred to as the spectral gap data for Γ.

The spectral gap hypothesis implies that for $\Psi_1, \Psi_2 \in L^2(\Gamma\backslash G)$, the leading term of the asymptotic expansion of the matrix coefficient $\langle a_t.\Psi_1, \Psi_2\rangle$ is determined by $\langle a_t.P_\delta(\Psi_1), P_\delta(\Psi_2)\rangle$ where P_δ is the projection operator from $L^2(\Gamma\backslash G)$ to $\mathcal{H}_\delta^\dagger$, which is the sum of all complementary series representations $\mathcal{U}(\upsilon, \delta-n+1)$, $\upsilon \in \hat{M}$ occurring in $L^2(\Gamma\backslash G)$ as subrepresentations.

Building up on the work of Harish-Chandra on the asymptotic behavior of the Eisenstein integrals (see [Warner 1972a; 1972b]), we obtain an asymptotic formula for $\langle a_t \upsilon, w\rangle$ for all K-isotypic vectors $\upsilon, w \in \mathcal{H}_\delta^\dagger$. This extension alone does not quite explain the leading term of $\langle a_t P_\delta(\Psi_1), P_\delta(\Psi_2)\rangle$ in terms of Ψ_1 and Ψ_2; however, with the help of Theorem 3.2, we are able to prove this:

Theorem 3.6 [Mohammadi and Oh 2012b]. *Suppose that $L^2(\Gamma\backslash G)$ possesses a spectral gap. There exist $\eta_0 > 0$ and $\ell \in \mathbb{N}$ such that for any $\Psi_1, \Psi_2 \in C_c^\infty(\Gamma\backslash G)$, as $t \to \infty$,*

$$e^{(n-1-\delta)t}\langle a_t\Psi_1, \Psi_2\rangle = \frac{m^{\mathrm{BR}}(\Psi_1)\cdot m_*^{\mathrm{BR}}(\Psi_2)}{|m^{\mathrm{BMS}}|} + O(\mathcal{S}_\ell(\Psi_1)\mathcal{S}_\ell(\Psi_2)e^{-\eta_0 t}).$$

We reiterate that the leading term in Theorem 3.6 is from the ergodic theory and the error term is from the harmonic analysis and the spectral gap.

Effective mixing for m^{BMS}. Theorem 3.6 can be used to obtain an effective mixing for the BMS measure (i.e., an effective version of Theorem 2.5(1)) for geometrically finite, Zariski dense subgroups with a spectral gap [Mohammadi and Oh 2012b].

For geometrically finite groups with $\delta \le \frac{n-1}{2}$, we cannot expect Theorem 3.6 to hold for such groups. However Guillarmou and Mazzeo [2012] have established meromorphic extensions of the resolvents of the Laplacian and of the Poincaré series.

For Γ convex cocompact, Stoyanov obtained, via the spectral properties of Ruelle transfer operators, an effective mixing of the geodesic flow for the BMS measure, regardless of the size of the critical exponent (see [Stoyanov 2011]). We remark that the idea of using spectral estimates of Ruelle transfer operators in obtaining an effective mixing originated in [Dolgopyat 1998].

It will be interesting to see if the results in [Guillarmou and Mazzeo 2012] can be used to answer the question:

Problem 3.7. Prove an effective mixing of the geodesic flow for the BMS measure for all geometrically finite groups.

We close this section with an open problem:

Problem 3.8. Find a suitable analogue of Theorem 3.6 for a higher rank simple Lie group G such as $\mathrm{SL}_2(\mathbb{R}) \times \mathrm{SL}_2(\mathbb{R})$ or $\mathrm{SL}_3(\mathbb{R})$.

4. Distribution of Γ in G

For a family $\{B_T : T > 1\}$ of compact subsets in G, the study of the asymptotic of $\#(\Gamma \cap B_T)$ can be approached directly by Theorems 3.2 and 3.6. The link is given by the following function on $\Gamma\backslash G \times \Gamma\backslash G$:

$$F_T(g,h) := \sum_{\gamma\in\Gamma} \chi_{B_T}(g^{-1}\gamma h)$$

where χ_{B_T} denotes the characteristic function of B_T. Observing that $F_T(e,e) = \#(\Gamma \cap B_T)$, we will explain how the asymptotic behavior of $F_T(e,e)$ is related to the matrix coefficient function for $L^2(\Gamma\backslash G)$. For $g = k_1 a_t k_2 \in KA^+K$, we have $dm^{\mathrm{Haar}}(g) = \xi(t)\,dt\,dk_1\,dk_2$ where $\xi(t) = e^{(n-1)t}(1 + O(e^{-\beta t}))$ for some $\beta > 0$.

Now assume that Γ admits a spectral gap, so that Theorem 3.6 holds. For any real-valued $\Psi_1, \Psi_2 \in C_c^\infty(\Gamma\backslash G)$, if we set $\Psi_i^k(g) := \Psi_i(gk)$, then we can deduce from Theorem 3.6 that

$$\begin{aligned}
&\langle F_T, \Psi_1 \otimes \Psi_2\rangle_{\Gamma\backslash G\times\Gamma\backslash G} \\
&= \int_{g\in B_T} \langle \Psi_1, g.\Psi_2\rangle_{L^2(\Gamma\backslash G)}\, dm^{\mathrm{Haar}}(g) \\
&= \int_{k_1 a_t k_2 \in B_T} \langle \Psi_1^{k_1^{-1}}, a_t.\Psi_2^{k_2}\rangle \cdot \xi(t)\,dt\,dk_1\,dk_2 \\
&= \frac{1}{|m^{\mathrm{BMS}}|}\int_{k_1 a_t k_2 \in B_T} \big(m_*^{\mathrm{BR}}(\Psi_1^{k_1^{-1}}) m^{\mathrm{BR}}(\Psi_2^{k_2}) e^{\delta t} + O(e^{(\delta-\eta)t})\big)\,dt\,dk_1\,dk_2 \qquad (4.1)
\end{aligned}$$

for some $\eta > 0$, with the implied constant depending only on the Sobolev norms of Ψ_i's.

Therefore, if $F_T(e,e)$ can be *effectively approximated* by $\langle F_T, \Psi_\epsilon \otimes \Psi_\epsilon\rangle$ for an approximation of identity Ψ_ϵ in $\Gamma\backslash G$, a condition which depends on the regularity of the boundary of B_T relative to the Patterson–Sullivan density, then $F_T(e,e)$ can be computed by evaluating the integral (4.1) for $\Psi_1 = \Psi_2 = \Psi_\epsilon$ and by taking ϵ to be a suitable power of e^{-t}.

To state our counting theorem more precisely, we need:

Definition 4.2. Define a Borel measure $\mathcal{M}_G$ on G as follows: for $\psi \in C_c(G)$,

$$\mathcal{M}_G(\psi) = \frac{1}{|m^{\text{BMS}}|} \int_{k_1 a_t k_2 \in KA^+K} \psi(k_1 a_t k_2) e^{\delta t} \, d\nu_o(k_1) \, dt \, d\nu_o(k_2^{-1})$$

here ν_o is the M-invariant lift to K of the PS measure on $\partial(\mathbb{H}^n) = K/M$ viewed from $o \in \mathbb{H}^n$ with $K = \text{Stab}_G(o)$.

Definition 4.3. For a family $\{B_T \subset G\}$ of compact subsets with $\mathcal{M}_G(B_T)$ tending to infinity as $T \to \infty$, we say that $\{B_T\}$ is *effectively well-rounded with respect to* Γ if there exists $p > 0$ such that for all small $\epsilon > 0$ and $T \gg 1$:

$$\mathcal{M}_G(B_{T,\epsilon}^+ - B_{T,\epsilon}^-) = O(\epsilon^p \cdot \mathcal{M}_G(B_T))$$

where $B_{T,\epsilon}^+ = G_\epsilon B_T G_\epsilon$ and $B_{T,\epsilon}^- = \bigcap_{g_1, g_2 \in G_\epsilon} g_1 B_T g_2$. Here G_ϵ denotes a symmetric ϵ-neighborhood of e in G.

For the next two theorems, we assume that $\Gamma' < \Gamma$ is a subgroup of finite index and both Γ and Γ' have spectral gaps.

Theorem 4.4 [Mohammadi and Oh 2012b]. *If $\{B_T\}$ is effectively well-rounded with respect to Γ, then there exists $\eta_0 > 0$ such that for any $\gamma \in \Gamma$,*

$$\#(\Gamma'\gamma \cap B_T) = \frac{1}{[\Gamma : \Gamma']} \mathcal{M}_G(B_T) + O(\mathcal{M}_G(B_T)^{1-\eta_0})$$

where $\eta_0 > 0$ depends only on a uniform spectral gap for Γ and Γ', and the implied constant is independent of Γ' and γ.

If $\|\cdot\|$ is a norm on the space $M_{n+1}(\mathbb{R})$ of $(n+1) \times (n+1)$ matrices, then the norm ball $B_T = \{g \in \text{SO}(n, 1) : \|g\| \le T\}$ is effectively well-rounded with respect to Γ, and hence Theorem 4.4 applies.

Consider the set $B_T = \Omega_1 A_T \Omega_2$ where $\Omega_i \subset K$ and $A_T = \{a_t : 0 \le t \le T\}$. Then $\{B_T\}$ is effectively well-rounded with respect to Γ if there exists $\beta' > 0$ such that the PS-measures of the ϵ-neighborhoods of the boundaries of $\Omega_1 M/M$ and $\Omega_2^{-1} M/M$ are at most of order $\epsilon^{\beta'}$ for all small $\epsilon > 0$. These conditions on Ω_i are satisfied for instance if their boundaries in K/M are disjoint from the limit set $\Lambda(\Gamma)$. But also many (but not all) compact subsets with piecewise smooth boundary also satisfy this condition (see Section 7 of [Mohammadi and Oh 2012b]).

Hence one can deduce the following:

Theorem 4.5 [Mohammadi and Oh 2012b]. *If $\{B_T = \Omega_1 A_T \Omega_2\}$ is effectively well-rounded with respect to Γ, then, as $T \to \infty$,*

$$\#(\Gamma'\gamma \cap \Omega_1 A_T \Omega_2) = \frac{\Xi(\Gamma, \Omega_1, \Omega_2)}{[\Gamma : \Gamma']} e^{\delta T} + O(e^{(\delta - \eta_0)T}),$$

where $\Xi(\Gamma, \Omega_1, \Omega_2) := \dfrac{\nu_o(\Omega_1)\nu_o(\Omega_2^{-1})}{\delta \cdot |m^{\mathrm{BMS}}|}$ *and the implied constant is independent of* Γ' *and* $\gamma \in \Gamma$.

The fact that the only dependence of Γ' on the right hand side of the above formula is on the index $[\Gamma : \Gamma']$ is of crucial importance for our intended applications to an affine sieve (page 197). Bourgain, Kontorovich and Sarnak [Bourgain et al. 2010b] showed Theorem 4.5 for the case $G = \mathrm{SO}(2, 1)$ via an explicit computation of matrix coefficients using the Gauss hypergeometric functions, and Vinogradov [2012] generalized their method to $G = \mathrm{SO}(3, 1)$. We note that in view of Theorem 3.2, the noneffective versions of Theorems 4.4 and 4.5 hold for any Zariski dense Γ with $|m^{\mathrm{BMS}}| < \infty$; this was obtained in [Roblin 2003] (see also [Oh and Shah 2013]) with a different proof. For Γ lattice, Theorem 4.5 is known in much greater generality [Gorodnik and Oh 2007].

5. Asymptotic distribution of $\Gamma\backslash\Gamma Ha_t$

The counting problem in $H\backslash G$ for H a nontrivial subgroup can be approached via studying the asymptotic distribution of translates $\Gamma\backslash\Gamma Ha_t$ as mentioned in the introduction. We will assume in this section that H is either a symmetric subgroup or a horospherical subgroup of G. Recall that H is symmetric means that H is the subgroup of fixed points under an involution of G. In this case, we can relate the distribution of $\Gamma\backslash\Gamma Ha_t$ to the matrix coefficient function of $L^2(\Gamma\backslash G)$.

We fix a generalized Cartan decomposition $G = HAK$ (for H horospherical, it is just an Iwasawa decomposition) where K is a maximal compact subgroup and A is a one-dimensional subgroup of diagonalizable elements. As before, we parametrize $A = \{a_t : t \in \mathbb{R}\}$ so that the right multiplication by a_t on $G/M = \mathrm{T}^1(\mathbb{H}^n)$ corresponds to the geodesic flow for time t.

When H is horospherical, we will assume that $H = N^+$, i.e, the expanding horospherical subgroup with respect to a_t.

Any symmetric subgroup H of G is known to be locally isomorphic to $\mathrm{SO}(k, 1) \times \mathrm{SO}(n-k)$ for some $0 \le k \le n-1$, and the H-orbit of the identity coset in G/K (resp. G/M) is (resp. the unit normal bundle to) a complete totally geodesic subspace $\mathbb{H}^k$ of $\mathbb{H}^n$ of dimension k. The right multiplication by a_t on G/M corresponds to the geodesic flow and the image of Ha_t in G/M represents the expansion of the totally geodesic subspace of dimension k by distance t.

Measures on $\Gamma\backslash\Gamma H$ ***associated to a conformal density.*** The leading term in the description of the asymptotic distribution of $\Gamma\backslash\Gamma Ha_t$ turns out to be a new measure on $\Gamma\backslash\Gamma H$ associated to the PS density (see [Oh and Shah 2013]). We assume that $\Gamma\backslash\Gamma H$ is closed in $\Gamma\backslash G$ in the rest of this section.

For a Γ-invariant conformal density $\{\mu_x\}$ on $\partial(\mathbb{H}^n)$ of dimension δ_μ, define a measure $\tilde{\mu}_H$ on $H/(H \cap M)$ by

$$d\tilde{\mu}_H([h]) = e^{\delta_\mu \beta_{[h]^+}(o,[h])}\, d\mu_o([h]^+)$$

where $o \in \mathbb{H}^n$ and $[h] \in H/(H \cap M)$ is considered as an element of $G/M = \mathrm{T}^1(\mathbb{H}^n)$ under the injective map $H/(H \cap M) \to G/M$.

By abuse of notation, we use the same notation $\tilde{\mu}_H$ for the $H \cap M$-invariant lift of $\tilde{\mu}_H$ to H: for $\psi \in C_c(\Gamma\backslash G)$,

$$\tilde{\mu}_H(\psi) = \int_{H/(H\cap M)} \psi^{H\cap M}(x)\, d\tilde{\mu}_H(x)$$

where $\psi^{H\cap M}(x) = \int_{H\cap M} \psi(xm)\, d_{H\cap M}(m)$ for the $H \cap M$-invariant probability measure $d_{H\cap M}$. This definition is independent of the choice of $o \in \mathbb{H}^n$ and the measure $\tilde{\mu}_H$ is $H \cap \Gamma$-invariant from the left, and hence induces a measure μ_H on $(H \cap \Gamma)\backslash H$ or equivalently on $\Gamma\backslash\Gamma H$.

For the PS density $\{\nu_x\}$ and the Lebesgue density $\{m_x\}$, the following two locally finite measures on $\Gamma\backslash\Gamma H$ are of special importance:

- Skinning measure: $\mu_H^{\mathrm{PS}} = \nu_H$.
- H-invariant measure: $\mu_H^{\mathrm{Leb}} = m_H$.

We remark that μ_H^{PS} is different from μ_H^{Leb} in general even when $H \cap \Gamma$ is a lattice in H.

Finiteness of the skinning measure μ_H^{PS}. The finiteness of the skinning measure μ_H^{PS} turns out to be the precise replacement for the finiteness of the volume measure μ_H^{Leb}, in extending the equidistribution statement from Γ lattices to thin subgroups.

When is the skinning measure μ_H^{PS} finite? This question is completely answered in [Oh and Shah 2013] for Γ geometrically finite. First, when H is a horospherical subgroup, the support of μ_H^{PS} is compact. When H is a symmetric subgroup, i.e., isomorphic to $\mathrm{SO}(k,1) \times \mathrm{SO}(n-k)$ locally, the answer to this question depends on the notion of the parabolic corank of $\Gamma \cap H$: let $\Lambda_p(\Gamma)$ denote the set of all parabolic limit points of Γ. For $\xi \in \Lambda_p(\Gamma)$, the stabilizer Γ_ξ has a free abelian subgroup of finite index, whose rank is defined to be the rank of ξ (or the rank of Γ_ξ).

Definition 5.1. The parabolic corank of $\Gamma \cap H$ in Γ is defined to be the maximum of the difference $\mathrm{rank}(\Gamma_\xi) - \mathrm{rank}(\Gamma \cap H)_\xi$ over all $\xi \in \Lambda_p(\Gamma) \cap \partial(\mathbb{H}^k)$.

Theorem 5.2 [Oh and Shah 2013]. *Let Γ be geometrically finite.*

(1) *μ_H^{PS} is compactly supported if and only if the parabolic corank of $\Gamma \cap H$ is zero.*

(2) *$|\mu_H^{\mathrm{PS}}| < \infty$ if and only if δ is bigger than the parabolic corank of $\Gamma \cap H$.*

As we show $\text{rank}(\Gamma_\xi) - \text{rank}(\Gamma \cap H)_\xi \le n-k$ for all $\xi \in \Lambda_p(\Gamma) \cap \partial(\mathbb{H}^k)$, we have:

Corollary 5.3 [Oh and Shah 2013]. *If $\delta > (n-k)$, then $|\mu_H^{\text{PS}}| < \infty$.*

For instance, if we assume $\delta > \frac{n-1}{2}$, then $|\mu_H^{\text{PS}}| < \infty$ whenever $k \ge (n+1)/2$.

Distribution of $\Gamma \backslash \Gamma H a_t$. We first recall:

Theorem 5.4. *Suppose that Γ is a lattice in G and that $H \cap \Gamma$ is a lattice in H. In other words, $|m^{\text{Haar}}| < \infty$ and $|m_H^{\text{Leb}}| < \infty$. Then there exist $\eta_0 > 0$ (depending only on the spectral gap for Γ) and $\ell \in \mathbb{N}$ such that for any $\Psi \in C_c^\infty(\Gamma \backslash G)$, as $t \to \infty$,*

$$\int_{\Gamma \backslash \Gamma H} \Psi(ha_t)\, d\mu_H^{\text{Leb}}(h) = \frac{|\mu_H^{\text{Leb}}|}{|m^{\text{Haar}}|} \cdot m^{\text{Haar}}(\Psi) + O(\mathcal{S}_\ell(\Psi) \cdot e^{-\eta_0 t}).$$

In fact, this theorem holds in much greater generality of any connected semisimple Lie group: the noneffective statement is due to Duke et al. [1993] (also see [Eskin and McMullen 1993]). For the effective statement, see [Duke et al. 1993] for the case when $H \cap \Gamma \backslash H$ is compact and [Benoist and Oh 2012] in general.

An analogue of Theorem 5.4 for discrete groups that are not necessarily lattices is this:

Theorem 5.5 [Oh and Shah 2013]. *Let Γ be Zariski dense with $|m^{\text{BMS}}| < \infty$. Suppose $|\mu_H^{\text{PS}}| < \infty$. Then for any $\Psi \in C_c(\Gamma \backslash G)$,*

$$\lim_{t\to\infty} e^{(n-1-\delta)t} \int_{h \in \Gamma \backslash \Gamma H} \Psi(ha_t) d\mu_H^{\text{Leb}}(h) = \frac{|\mu_H^{\text{PS}}|}{|m^{\text{BMS}}|} m^{\text{BR}}(\Psi).$$

Theorem 5.6 [Mohammadi and Oh 2012b]. *Let Γ be a geometrically finite Zariski dense subgroup of G with a spectral gap (e.g., $\delta > n-2$). Suppose $|\mu_H^{\text{PS}}| < \infty$. Then there exist $\eta_0 > 0$ and $\ell \in \mathbb{N}$ such that for any $\Psi \in C_c^\infty(\Gamma \backslash G)$, as $t \to \infty$,*

$$e^{(n-1-\delta)t} \int_{h \in \Gamma \backslash \Gamma H} \Psi(ha_t)\, d\mu_H^{\text{Leb}}(h) = \frac{|\mu_H^{\text{PS}}|}{|m^{\text{BMS}}|} m^{\text{BR}}(\Psi) + O(\mathcal{S}_\ell(\Psi) \cdot e^{-\eta_0 t}).$$

In the case when μ_H^{PS} is compactly supported, we can show that there exists a compact subset $\mathcal{O}_H$ (depending on Ψ) of $\Gamma \backslash \Gamma H$ such that for any $t \in \mathbb{R}$, $\Psi(ha_t) = 0$ for all $h \notin \mathcal{O}_H$. Therefore

$$\int \Psi(ha_t)\, d\mu_H^{\text{Leb}} = \int_{\mathcal{O}_H} \Psi(ha_t)\, d\mu_H^{\text{Leb}}.$$

Now the assumption on H being either symmetric or horospherical ensures the wave front property of [Eskin and McMullen 1993], which can be used to establish, as $t \to \infty$,

$$\int_{\mathcal{O}_H} \Psi(ha_t)\, d\mu_H^{\mathrm{Leb}} \approx \langle a_t \Psi, \rho_{\mathcal{O}_H,\epsilon} \rangle_{L^2(\Gamma\backslash G)} \tag{5.7}$$

where $\rho_{\mathcal{O}_H,\epsilon} \in C_c^\infty(\Gamma\backslash G)$ is an ϵ-approximation of $\mathcal{O}_H$. Therefore the estimates on the matrix coefficients in Theorems 3.2 and 3.6 can be used to establish Theorems 5.5 and 5.6.

The case when μ_H^{PS} is not compactly supported turns out to be much more intricate, the main reason being that we are taking the integral with respect to μ_H^{Leb} as well as multiplying the weight factor $e^{(n-1-\delta)t}$ in the left hand side of Theorem 5.5, whereas the finiteness assumption is made on the skinning measure μ_H^{PS}. In this case, we first develop a version of thick-thin decomposition of the nonwandering set, that is, $\mathcal{W} := \{h \in \Gamma\backslash\Gamma H : ha_t \in \mathrm{supp}(\Psi)\}$, which resembles that of the support of μ_H^{PS}. This together with (5.7) takes care of the integral $\int \Psi_{ha_t}\, d\mu_H^{\mathrm{Leb}}(h)$ over a thick part as well as a very thin part of $\mathcal{W}$. What is left is the integration over an intermediate range, which is investigated by comparing the two measures $(a_t)_*\mu_H^{\mathrm{PS}}$ and $(a_t)_*\mu_H^{\mathrm{Leb}}$ via the transversal intersections of the orbits $\Gamma\backslash\Gamma H a_t$ with the weak-stable horospherical foliations . A key reason that this approach works is that these transversal intersections are governed by the topological properties of the orbit $\Gamma\backslash\Gamma H a_t$, independent of the measures put on $\Gamma\backslash\Gamma H$.

In the special case of $n = 2, 3$ and H horospherical, Theorem 5.6 was proved in [Lee and Oh 2013] by a different method.

6. Distribution of Γ orbits in $H\backslash G$ and affine sieve

Distribution of Γ orbits in $H\backslash G$. Let H be as in the previous section (i.e., symmetric or horospherical subgroup) and let $\{B_T\}$ be a family of compact subsets in $H\backslash G$ which is getting larger and larger as $T \to \infty$. We assume that $[e]\Gamma$ is discrete in $H\backslash G$. The study of the asymptotic of $\#([e]\Gamma \cap B_T)$ can be now approached by Theorems 5.5 and 5.6 via the following counting function on $\Gamma\backslash G$:

$$F_T(g) := \sum_{\gamma \in H\cap\Gamma\backslash\Gamma} \chi_{B_T}(\gamma g)$$

as $F_T(e) = \#([e]\Gamma \cap B_T)$. Depending on the subgroup H, we have $G = HA^+K$ or $G = HA^+K \cup HA^-K$. For the sake of simplicity, we will consider the sets B_T contained in HA^+K. We have for $g = ha_t k$, $dm^{\mathrm{Haar}}(g) = \rho(t)\, d\mu^{\mathrm{Leb}}(h)\, dt\, dk$ where $\rho(t) = e^{(n-1)t}(1 + O(e^{-\beta t}))$ for some $\beta > 0$. Let μ be a G-invariant measure on $H\backslash G$ normalized so that $dm^{\mathrm{Haar}} = d\mu_H^{\mathrm{Leb}} \otimes d\mu$ locally. Then, under the assumption that Γ has a spectral gap and $|\mu_H^{\mathrm{PS}}| < \infty$, we deduce from

Theorem 5.6 that

$$
\begin{aligned}
\langle F_T, \Psi\rangle &= \int_{g\in B_T}\int_{h\in\Gamma\backslash\Gamma H} \Psi(hg)\, d\mu_H^{\mathrm{Leb}}(h)\, d\mu(g) \\
&= \int_{[e]a_t k\in B_T}\int_{h\in\Gamma\backslash\Gamma H} \Psi^k(ha_t) d\mu_H^{\mathrm{Leb}}(h)\rho(t)\, dt\, dk \\
&= \int_{[e]a_t k\in B_T}\left(\frac{|\mu_H^{\mathrm{PS}}|}{|m^{\mathrm{BMS}}|} m^{\mathrm{BR}}(\Psi^k)e^{\delta t} + O(e^{(\delta-\eta_0)t})\right) dt\, dk \qquad (6.1)
\end{aligned}
$$

for some $\eta_0 > 0$.

Therefore, similarly to the discussion in Section 4, if $F_T(e)$ can be *effectively approximated* by $\langle F_T, \Psi_\epsilon\rangle$ for an approximation of identity Ψ_ϵ in $\Gamma\backslash G$, a condition which depends on the regularity of the boundary of B_T relative to the PS density, then $F_T(e)$ can be computed by evaluating the integral (6.1) for $\Psi = \Psi_\epsilon$ and by taking ϵ a suitable power of e^{-t}.

For H horospherical or symmetric, we have

$$G = HA^+K$$

or

$$G = HA^+K \cup HA^-K$$

(as a disjoint union except for the identity element), where $A^\pm = \{a_{\pm t} : t \geq 0\}$.

Definition 6.2. Define a Borel measure $\mathcal{M}_{H\backslash G}$ on $H\backslash G$ as follows: for $\psi \in C_c(H\backslash G)$,

$$
\mathcal{M}_{H\backslash G}(\psi) = \begin{cases} \dfrac{|\mu_H^{\mathrm{PS}}|}{|m^{\mathrm{BMS}}|}\displaystyle\int_{a_t k\in A^+K} \psi([e]a_t k)e^{\delta t} dt\, d\nu_o(k^{-1}) & \text{if } G = HA^+K, \\ \displaystyle\sum \dfrac{|\mu_{H,\pm}^{\mathrm{PS}}|}{|m^{\mathrm{BMS}}|}\int_{a_{\pm t}k\in A^\pm K} \psi([e]a_{\pm t}k)e^{\delta t} dt\, d\nu_o(k^{-1}) & \text{otherwise,} \end{cases}
$$

where $o \in \mathbb{H}^n$ is the point fixed by K, ν_o is the right M-invariant measure on K, which projects to the PS-measure ν_o on $K/M = \partial(\mathbb{H}^n)$ and $\mu_{H,-}^{\mathrm{PS}}$ is the skinning measure on $\Gamma \cap H\backslash H$ in the negative direction.

Definition 6.3. For a family $\{B_T \subset H\backslash G\}$ of compact subsets with $\mathcal{M}_{H\backslash G}(B_T)$ tending to infinity as $T \to \infty$, we say that $\{B_T\}$ is *effectively well-rounded with respect to* Γ if there exists $p > 0$ such that for all small $\epsilon > 0$ and $T \gg 1$:

$$\mathcal{M}_{H\backslash G}(B_{T,\epsilon}^+ - B_{T,\epsilon}^-) = O(\epsilon^p \cdot \mathcal{M}_{H\backslash G}(B_T))$$

where $B_{T,\epsilon}^+ = B_T G_\epsilon$ and $B_{T,\epsilon}^- = \cap_{g\in G_\epsilon} B_T g$.

In the next two theorems 6.4 and 6.5, we assume that $\Gamma' < \Gamma$ is a subgroup of finite index with $H \cap \Gamma = H \cap \Gamma'$ and that both Γ and Γ' have spectral gaps.

Theorem 6.4 [Mohammadi and Oh 2012b]. *When H is symmetric, we assume that $|\mu_H^{\rm PS}| < \infty$. Suppose that $\{B_T\}$ is effectively well-rounded with respect to Γ. Then for any $\gamma \in \Gamma$, there exists $\eta_0 > 0$ (depending only on the uniform spectral gaps of Γ and Γ') such that*

$$\#([e]\Gamma'\gamma \cap B_T) = \frac{1}{[\Gamma : \Gamma']}\mathcal{M}_{H\backslash G}(B_T) + O(\mathcal{M}_{H\backslash G}(B_T)^{1-\eta_0})$$

with the implied constant independent of Γ' and $\gamma \in \Gamma$.

For a given family $\{B_T\}$, understanding its effective well-roundedness can be the hardest part of the work in general. However verifying the family of norm balls is effectively well-rounded is manageable (see Section 7 of [Mohammadi and Oh 2012b]). Consider an example of the family $\{B_T = [e]A_T\Omega\}$ where $\Omega \subset K$ and $A_T = \{a_t : 0 \le t \le T\}$. In this case, it is rather simple to formulate the effective well-rounded condition: there exists $\beta' > 0$ such that the PS-measure of the ϵ-neighborhood of $\partial(\Omega^{-1}M/M)$ is at most of order $\epsilon^{\beta'}$ for all small $\epsilon > 0$. As mentioned before, this holds when the boundary of $\Omega^{-1}M/M$ does not intersect the limit set $\Lambda(\Gamma)$ (see [loc. cit.] for a more general condition).

Then, setting

$$\Xi(\Gamma, \Omega) := \frac{|\mu_H^{\rm PS}| \cdot \nu_o(\Omega^{-1})}{\delta \cdot |m^{\rm BMS}|},$$

one can deduce this from (6.1):

Theorem 6.5 [Mohammadi and Oh 2012b]. *Under the same assumption on $\mu_H^{\rm PS}$ and Γ', there exists $\eta_0 > 0$ (depending only on the uniform spectral gaps of Γ and Γ') such that for any $\gamma \in \Gamma$*

$$\#([e]\Gamma'\gamma \cap [e]A_T\Omega) = \frac{\Xi(\Gamma, \Omega)}{[\Gamma : \Gamma']}e^{\delta T} + O(e^{(\delta-\eta)T}),$$

with implied constant independent of Γ' and γ.

We note that in view of (4.1), the noneffective version of this theorem holds for any Zariski dense Γ with $|m^{\rm BMS}| < \infty$ and $|\mu_H^{\rm PS}| < \infty$, as proved in [Oh and Shah 2013].

Affine sieve. For applications to an affine sieve, we consider the case when the homogeneous space $H\backslash G$ is defined over $\mathbb{Z}$. More precisely, we assume that G is defined over $\mathbb{Z}$ and acts linearly and irreducibly on a finite-dimensional vector space W defined over $\mathbb{Z}$ in such a way that $G(\mathbb{Z})$ preserves $W(\mathbb{Z})$. Let $w_0 \in W(\mathbb{Z})$ be a nonzero vector such the stabilizer of w_0 is a symmetric subgroup or the stabilizer of the line $\mathbb{R}w_0$ is a parabolic subgroup. We set $V := w_0 G$. Let Γ be a geometrically finite Zariski dense subgroup of G with a spectral gap, which is contained in $G(\mathbb{Z})$.

Let F be an integer-valued polynomial on the orbit $w_0\Gamma$. Salehi Golsefidy and Sarnak [2013], generalizing [Bourgain et al. 2010a], showed that for some $R > 1$, the set of $\boldsymbol{x} \in w_0\Gamma$ such that $F(\boldsymbol{x})$ has at most R prime factors is Zariski dense in V.

For a square-free integer d, let $\Gamma_d < \Gamma$ be a subgroup which contains $\{\gamma \in \Gamma : \gamma \equiv e \bmod d\}$ and satisfies $\mathrm{Stab}_{\Gamma_d}(w_0) = \mathrm{Stab}_{\Gamma}(w_0)$. For instance, we can set $\Gamma_d = \{\gamma \in \Gamma : w_0\gamma \equiv w_0 \bmod d\}$.

We say the family $\{\Gamma_d\}$ has a uniform spectral gap if $\sup_d s_0(\Gamma_d) < \delta$ and $\sup_d n_0(\Gamma_d) < \infty$ (see Definition 3.5 for notation).

Salehi Golsefidy and Varjú [2012], generalizing [Bourgain et al. 2010a], showed that the family of Cayley graphs of Γ/Γ_d's (with respect to the projections of a fixed symmetric generating set of Γ) forms expanders as d runs through square-free integers with large prime factors. If $\delta > \frac{n-1}{2}$, the transfer property from the combinatorial spectral gap to the archimedean one established in [Bourgain et al. 2011] (see also [Kim 2012]) implies that the family $\{\Gamma_d : d \text{ square-free}\}$ has a uniform *spherical* spectral gap. Together with the classification of $\hat{G}$, it follows that $\{\Gamma_d\}$ admits a uniform spectral gap if $\delta > \frac{n-1}{2}$ for $n = 2, 3$ or if $\delta > n-2$ for $n \geq 4$.

For the following discussion, we assume that there is a finite set of primes S such that the family Γ_d with d square-free with no prime factors in S admits a uniform spectral gap. Then we can apply Theorem 6.5 to Γ_d's. By a recent development on the affine linear sieve on homogeneous spaces ([Bourgain et al. 2010a], [Nevo and Sarnak 2010]), we are able to deduce: let F be an integer-valued polynomial on $w_0\Gamma$. Letting $F = F_1F_2\cdots F_r$ be a factorization into irreducible polynomials, assume that all F_j's are integral on $w_0\Gamma$ and distinct from each other. Let λ be the log of the largest eigenvalue of a_1 on the $\mathbb{R}$-span of w_0G.

Theorem 6.6 [Mohammadi and Oh 2012b]. *For any norm $\|\cdot\|$ on V,*

(1) $\#\{\boldsymbol{x} \in w_0\Gamma : \|\boldsymbol{x}\| < T,\ F_j(\boldsymbol{x}) \text{ is prime for } j = 1, \ldots, r\} \ll \dfrac{T^{\delta/\lambda}}{(\log T)^r}$;

(2) *there exists $R = R(F, w_0\Gamma) \geq 1$ such that*

$$\#\{\boldsymbol{x} \in w_0\Gamma : \|\boldsymbol{x}\| < T,\ F(\boldsymbol{x}) \text{ has at most } R \text{ prime factors}\} \gg \frac{T^{\delta/\lambda}}{(\log T)^r}.$$

The significance of $T^{\delta/\lambda}$ in the above theorem is that for $H = \mathrm{Stab}_G(w_0)$,

$$\mathcal{M}_{H\backslash G}\{w \in w_0G : \|w\| < T\} \asymp \#\{\boldsymbol{x} \in w_0\Gamma : \|\boldsymbol{x}\| < T\} \asymp T^{\delta/\lambda}.$$

In view of the above discussion, it is also possible to state Theorem 6.6 for more general sets B_T, instead of the norm balls, which then provides a certain uniform distribution of almost prime vectors.

When Γ is an arithmetic subgroup of a simply connected semi-simple algebraic $\mathbb{Q}$-group G, and H is a symmetric subgroup, the analogue of Theorem 6.6 was obtained in [Benoist and Oh 2012]. Strictly speaking, Theorem 1.3 of that reference is stated only for a fixed group; however it is clear from its proof that the statement also holds uniformly over its congruence subgroups with the correct main term. Based on this, one can use the combinatorial sieve to obtain an analogue of Theorem 6.6, as was done for a group variety in [Nevo and Sarnak 2010]. Theorem 6.6 on lower bound for arithmetic was obtained in [Gorodnik and Nevo 2012] further assuming that $H \cap \Gamma$ is cocompact in H.

7. Application to sphere packings

In this section we will discuss counting problems for sphere packings in $\mathbb{R}^n$ as an application of an orbital counting problem for thin subgroups of $\mathrm{SO}(n, 1)$. By a sphere packing in the Euclidean space $\mathbb{R}^n$ for $n \geq 1$, we simply mean a union of (possibly intersecting) $(n-1)$-dimensional spheres; here an $(n-1)$-dimensional plane is regarded as a sphere of infinite radius.

Fixing a sphere packing $\mathcal{P}$ in $\mathbb{R}^n$, a basic problem is to understand the asymptotic of the number $\#\{S \in \mathcal{P} : \mathrm{Radius}(S) > t\}$ or equivalently $\#\{S \in \mathcal{P} : \mathrm{Vol}(S) > t\}$ where $\mathrm{Vol}(S)$ means the volume of the ball enclosed by S. For the sake of brevity, we will simply refer $\mathrm{Vol}(S)$ to the volume of S. We will consider this problem in a more general setting, that is, allowing the volume of the sphere S to be computed in various conformal metrics.

Let U be an open subset of $\mathbb{R}^n$ and f a positive continuous function on U. A conformal metric associated to the pair (U, f) is a metric on U of the form $f(x)\,dx$, where dx is the Euclidean metric on $\mathbb{R}^n$. Given a conformal metric (U, f), we set $\mathrm{Vol}_f(S) := \int_B f(x)^n dx$ where B is the ball enclosed by S. When S is a plane, we put $\mathrm{Vol}_f(S) = \infty$.

For any compact subset E of U and $t > 0$, set

$$N_t(\mathcal{P}, f, E) := \#\{S \in \mathcal{P} : \mathrm{Vol}_f(S) > t,\ E \cap S \neq \varnothing\}.$$

For simplicity, we will omit $\mathcal{P}$ in the notation $N_t(\mathcal{P}, f, E)$. When $(U, f) = (\mathbb{R}^n, 1)$, i.e., the standard Euclidean metric, we simply write $N_t(E)$ instead of $N_t(1, E)$. We call $\mathcal{P}$ *locally finite* if $N_t(E) < \infty$ for all E bounded.

In order to approach the problem of the computation of the asymptotic of $N_t(f, E)$, a crucial condition is that $\mathcal{P}$ admits enough symmetries of Mobius transformations of $\mathbb{R}^n$. By the Poincaré extension, the group $\mathrm{MG}(\mathbb{R}^n)$ of Mobius transformations of $\mathbb{R}^n$ can be identified with the isometry group of the upper half space $\mathbb{H}^{n+1} = \{(z, r) : z \in \mathbb{R}^n, r > 0\}$. We assume that $\mathcal{P}$ is invariant under a nonelementary discrete subgroup Γ of $G := \mathrm{Isom}^+(\mathbb{H}^{n+1})$. As before, let δ

denote the critical exponent of Γ and $\{\nu_x : x \in \mathbb{H}^{n+1}\}$ a Patterson–Sullivan density for Γ.

We recall from [Oh and Shah 2012]:

Definition 7.1 (Γ-skinning size of $\mathcal{P}$). For a sphere packing $\mathcal{P}$ invariant under Γ, define $0 \le \mathrm{sk}_\Gamma(\mathcal{P}) \le \infty$ as follows:

$$\mathrm{sk}_\Gamma(\mathcal{P}) := \sum_{i \in I} \int_{s \in \mathrm{Stab}_\Gamma(S_i^\dagger) \backslash S_i^\dagger} e^{\delta \beta_{s^+}(o,s)} d\nu_o(s^+)$$

where $o \in \mathbb{H}^{n+1}$, $\{S_i : i \in I\}$ is a set of representatives of Γ-orbits in $\mathcal{P}$, and $S_i^\dagger \subset \mathrm{T}^1(\mathbb{H}^{n+1})$ is the set of unit normal vectors to the convex hull of S_i.

Definition 7.2. For the pair (U, f), we define a Borel measure $\omega_{\Gamma,f}$ on U: for $\psi \in C_c(U)$ and for $o \in \mathbb{H}^{n+1}$,

$$\omega_{\Gamma,f}(\psi) = \int_{z \in U} \psi(z) f(z)^\delta e^{\delta \beta_z(o,(z,1))} \, d\nu_o(z).$$

Alternatively, we have the simple formula

$$d\omega_{\Gamma,f} = f(z)^\delta (|z|^2 + 1)^\delta d\nu_{e_{n+1}},$$

where $e_{n+1} = (0, \dots, 0, 1) \in \mathbb{H}^{n+1}$.

Example 7.3. (1) For the spherical metric $(\mathbb{R}^n, 2/(1 + |z|^2))$ (also called the chordal metric) on $\mathbb{R}^n$, $\mathrm{Vol}_f(S)$ is the spherical volume of the ball enclosed by S and $d\omega_{\Gamma,f} = 2^\delta \cdot d\nu_{e_{n+1}}$.

(2) For the hyperbolic metric $(\mathbb{H}^n = \{z \in \mathbb{R}^n : z_n > 0\}, 1/z_n)$, $\mathrm{Vol}_f(S)$ is the hyperbolic volume of the ball enclosed by S and

$$d\omega_{\Gamma,f} = \frac{(1 + |z|^2)^\delta}{z_n^\delta} d\nu_{e_{n+1}}.$$

Definition 7.4. By an *infinite bouquet of tangent spheres glued at a point* $\xi \in \mathbb{R}^n \cup \{\infty\}$, we mean a union of two collections, each consisting of infinitely many pairwise internally tangent spheres with the common tangent point ξ and their radii tending to 0, such that the spheres in each collection are externally tangent to the spheres in the other at ξ.

We denote by $v_n := \mathrm{Vol}(B(0, 1))$ the Euclidean volume of the unit ball $B(0, 1) := \{x \in \mathbb{R}^n : \|x\| < 1\}$; v_n is equal to $(2\pi)^{n/2}/(2 \cdot 4 \cdots n)$ if n is even and $2(2\pi)^{(n-1)/2}/(1 \cdot 3 \cdots n)$ if n is odd.

Theorem 7.5. *Let $\mathcal{P}$ be a locally finite sphere packing invariant under a geometrically finite group Γ with finitely many Γ-orbits. In the case of $\delta \le 1$, we also* *assume that $\mathcal{P}$ has no infinite bouquet of spheres glued at a parabolic fixed point of Γ.*

Then for any conformal metric (U, f) and for any compact subset E of U whose boundary has zero Patterson–Sullivan density, as $t \to 0$,

$$\lim_{t\to 0} N_t(f, E)t^{\delta/n} = \frac{\mathrm{sk}_\Gamma(\mathcal{P}) \cdot v_n^{\delta/n} \cdot \omega_{\Gamma,f}(E)}{\delta \cdot |m^{\mathrm{BMS}}|}.$$

The assumption on the nonexistence of an infinite bouquet in $\mathcal{P}$ is to ensure that the Γ-skinning size for $\mathcal{P}$ is finite.

In the case when $(U, f) = (\mathbb{R}^n, 1)$, Theorem 7.5 was proved in [Oh and Shah 2012] for the case of $n = 2$ and and the proof given there extends easily for any $n \geq 2$ using the general equidistribution result in [Oh and Shah 2013]. See also [Oh and Shah 2010] for the case when f defines the spherical metric. To give an interpretation of Theorem 7.5 as a special case of the orbital counting problem discussed at the beginning of Section 6, fixing H to be the stabilizer of a sphere S_0 in $\mathcal{P}$, we may think of $H\backslash G$ as the space of all totally geodesic planes in $\mathbb{H}^{n+1}$. Then a key point is to describe a particular subset $B_t(E)$ in $H\backslash G$ such that $N_t(E)$ is same as $\#[e]\Gamma \cap B_t(E)$.

The extension to a general conformal metric (U, f) is possible basically due to the uniform continuity of f on a compact subset E and a covering argument. We give a brief sketch as follows (the argument below was established in a discussion with Shah): denote by $Q_z(\eta)$ the cube $\{z' \in \mathbb{R}^n : \max_{1\leq i\leq n} |z_i - z_i'| \leq \eta\}$ centered at $z \in \mathbb{R}^n$ with radius η.

First, Theorem 7.5 for $f = 1$, together with the uniform continuity of f on E, implies that for any $\epsilon > 0$, there exists $\eta = \eta(\epsilon) > 0$ (depending only on E and f) such that for any cube $Q_z(\eta)$ centered at $z \in E$

$$\frac{N_t(f, Q_z(\eta))t^{\delta/n}}{v_n^{\delta/n}} = (1 + O(\epsilon))\frac{\mathrm{sk}_\Gamma(\mathcal{P})}{\delta \cdot |m^{\mathrm{BMS}}|} f(z)^\delta \omega_\Gamma(Q_z(\eta)). \qquad (7.6)$$

Let k be the minimal integer such that a k-dimensional sphere, say, P charges a positive PS density. As the PS density is atom-free, we have $k > 0$ and the limit set $\Lambda(\Gamma)$ is contained in P (see [Roblin 2003, Proposition 3.1]). We cover $E \cap P$ with cubes $\{Q_z(\eta) : z \in I_\eta\}$ with disjoint interiors for a finite subset I_η of $E \cap P$. As each cube is centered at a point of P which is a sphere, the intersection of its boundary with P is contained in a $(k-1)$-dimensional sphere for all small $\eta > 0$. It follows that the boundary of each cube has zero PS density by the minimality assumption on k. Let $\mathscr{C}(\eta) := \{Q_z(\eta) : z \in \tilde{I}_\eta\}$ be a covering of E with $\tilde{I}_\eta \supset I_\eta$. Note that the boundary of each cube in $\mathscr{C}(\eta)$ has zero PS density. We can find compact subsets $E_\epsilon^\pm$ of U with $E_\epsilon^- \subset E \subset E_\epsilon^+$ and a positive integer m_ϵ so that

$\omega_{\Gamma,f}(E_\epsilon^+ - E_\epsilon^-) < \epsilon$ and E_ϵ^+ (resp. E) contains all cubes centered at E (resp. E_ϵ^-) and of size less than η possibly except at most m_ϵ number of such cubes.

We may also assume that $f(z) = (1 + O(\epsilon))f(z')$ for all $z' \in Q_z(\eta) \in \mathcal{C}(\eta)$ where the implied constant is uniform for all $z \in E$. We may now apply Equation (7.6) for $f = 1$ to each cube in the covering of $\mathcal{C}(\eta)$ to obtain that

$$\begin{aligned}\frac{N_t(f,E)t^{\delta/n}}{v_n^{\delta/n}} &= (1+O(\epsilon))\frac{\mathrm{sk}_\Gamma(\mathcal{P})}{\delta\cdot|m^{\mathrm{BMS}}|}\sum_{z\in\tilde{I}_\eta} f(z)^\delta \omega_\Gamma(Q_z(\eta)) \\ &= (1+O(\epsilon))\frac{\mathrm{sk}_\Gamma(\mathcal{P})}{\delta\cdot|m^{\mathrm{BMS}}|}\omega_{\Gamma,f}(E_\epsilon^+) + O(\epsilon) \\ &= (1+O(\epsilon))\frac{\mathrm{sk}_\Gamma(\mathcal{P})}{\delta\cdot|m^{\mathrm{BMS}}|}\omega_{\Gamma,f}(E) + O(\epsilon).\end{aligned}$$

As $\epsilon > 0$ is arbitrary, this proves Theorem 7.5.

8. On Apollonian circle packings

Construction. In the case of Apollonian circle packings in the plane $\mathbb{R}^2 = \mathbb{C}$, Theorem 7.5 can be made more explicit as the measure $\omega_{\Gamma,f}$ turns out to be the δ-dimensional Hausdorff measure with respect to (U, f), restricted to $\Lambda(\Gamma)$.

We begin by recalling Apollonian circle packings, whose construction is based on the following theorem of Apollonius of Perga:

Theorem 8.1 (Apollonius, 200 BC). *Given* 3 *mutually tangent circles in the plane (with distinct tangent points), there exist precisely two circles tangent to all three circles.*

Consider four mutually tangent circles in the plane with distinct points of tangency. By Apollonius' theorem, one can add four new circles each of which is tangent to three of the given ones. Continuing to repeatedly add new circles tangent to three of the previous circles, we obtain an infinite circle packing, called an *Apollonian circle packing*. Figure 1 shows the first three generations of this procedure where each circle is labeled with its curvature (the reciprocal of its radius).

Apollonian packing and Hausdorff measure. We start by recalling:

Definition 8.2. For $s > 0$, the s-dimensional Hausdorff (also known as covering) measure $\mathcal{H}^s$ of a closed subset E of $\mathbb{R}^2$ is defined as follows:

$$\mathcal{H}^s(E) := \lim_{\epsilon\to 0} \inf\left\{\sum_{i\in I} \mathrm{diam}(D_i)^s : E \subset \bigcup_{i\in I} D_i,\ \mathrm{diam}(D_i) \le \epsilon\right\}.$$

The Hausdorff dimension of E is then given as

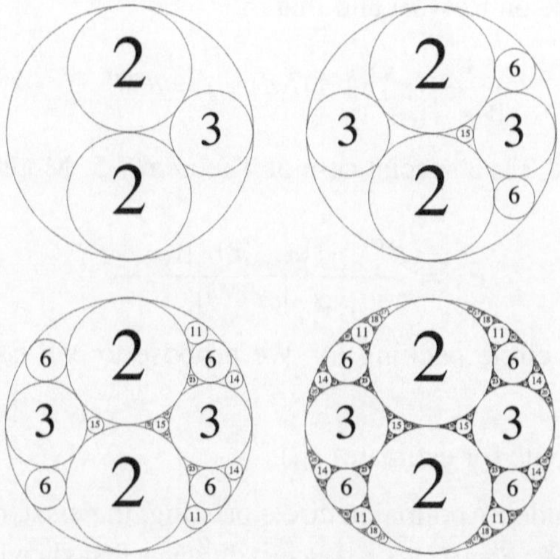

Figure 1. An Apollonian circle packing.

$$\dim_{\mathcal{H}}(E) := \sup\{s : \mathcal{H}^s(E) = \infty\} = \inf\{s \geq 0 : \mathcal{H}^s(E) = 0\}.$$

For s a positive integer, the s-dimensional Hausdorff measure is proportional to the usual Lebesgue measure on $\mathbb{R}^s$.

For an Apollonian circle packing $\mathcal{P}$, the residual set $\mathrm{Res}(\mathcal{P})$ is defined to be the closure of the union of all circles in $\mathcal{P}$. Its Hausdorff dimension, say α, is independent of $\mathcal{P}$ and known to be approximately 1.30568(8) [McMullen 1998].

Theorem 8.3. *Let $\mathcal{P}$ be any Apollonian circle packing. For any conformal metric (U, f) and for any compact subset $E \subset U$ with smooth boundary, we have*

$$\lim_{t\to 0} t^{\alpha/2} \cdot \#\{C \in \mathcal{P} : \mathrm{area}_f(C) > t, C \cap E \neq \varnothing\} = c_A \cdot \mathcal{H}^{\alpha}_f(\mathrm{Res}(\mathcal{P}) \cap E)$$

where $c_A > 0$ is independent of $\mathcal{P}$ and $d\mathcal{H}^{\alpha}_f(z) = f(z)^{\alpha} \cdot d\mathcal{H}^{\alpha}(z)$.

The symmetry group $\Gamma_{\mathcal{P}} := \{g \in \mathrm{PSL}_2(\mathbb{C}) : g(\mathcal{P}) = \mathcal{P}\}$ satisfies the following:

(1) $\Gamma_{\mathcal{P}}$ is geometrically finite.

(2) The limit set of $\Gamma_{\mathcal{P}}$ coincides with $\mathrm{Res}(\mathcal{P})$; in particular, its critical exponent is α.

(3) There are only finitely many $\Gamma_{\mathcal{P}}$-orbits of circles in $\mathcal{P}$.

Let $\nu_{\mathcal{P},j}$ denote the PS measure viewed from $j = (0, 0, 1) \in \mathbb{H}^3$ for the group $\Gamma_{\mathcal{P}}$. As $\Gamma_{\mathcal{P}}$ has no rank 2 parabolic limit points and $\alpha > 1$, Sullivan's work [Sullivan 1984] implies that the α-dimensional Hausdorff measure $\mathcal{H}^{\alpha}$ is a locally finite measure on $\mathrm{Res}(\mathcal{P})$ and that

$$\frac{1}{|\nu_{\mathcal{P},j}|}(|z|^2+1)^\alpha \, d\nu_{\mathcal{P},j} = d\mathcal{H}^\alpha.$$

Therefore Theorem 8.3 is a special case of Theorem 7.5. Moreover the constant c_A is given by

$$c_A = \frac{\pi^{\alpha/2} \cdot \mathrm{sk}_{\Gamma_{\mathcal{P}}}(\mathcal{P}) \cdot |\nu_{\mathcal{P},j}|}{\alpha \cdot |m^{\mathrm{BMS}}|}$$

for any Apollonian circle packing $\mathcal{P}$. We propose to call c_A the Apollonian constant.

Problem 8.4. Compute (or estimate) c_A!

When $\mathcal{P}$ is a bounded Apollonian circle packing, the existence of the asymptotic formula $\#\{C \in \mathcal{P} : \mathrm{area}(C) > t, \} \sim c_{\mathcal{P}} \cdot t^{\alpha/2}$ was first shown in [Kontorovich and Oh 2011] without an error term, and later in [Lee and Oh 2013] with an error term (see also [Vinogradov 2012]). In view of [Oh and Shah 2012], Theorem 5.6 can be used to prove an effective circle count in a compact region E for general Apollonian packings, provided the boundary of E satisfies a regularity property.

There is also a beautiful arithmetic aspect of Apollonian circle packings which is entirely omitted in this article; see [Sarnak 2011; Oh 2011; Bourgain and Fuchs 2011; Bourgain and Kontorovich 2012].

Apollonian sphere packing for n = 3. Given $n+1$ mutually tangent spheres in $\mathbb{R}^n$ with disjoint interiors, it is known that there is a unique sphere, called a *dual sphere*, passing through their points of tangency and orthogonal to all $n+1$ spheres [Graham et al. 2006, Theorem 7.1]. Hence for $n+2$ mutually tangent spheres with disjoint interiors $S_1, \dots, S_{n+2}$ in $\mathbb{R}^n$, there are $n+2$ dual spheres, say, $\tilde{S}_1, \dots, \tilde{S}_{n+2}$. The Apollonian group $\mathcal{A} = \mathcal{A}(S_1, \dots, S_{n+2})$ is generated by the inversions with respect to $\tilde{S}_i$, $1 \le i \le n+2$.

Only for $n = 2$ or 3, the Apollonian group $\mathcal{A}$ is a discrete subgroup of $\mathrm{MG}(\mathbb{R}^n)$ [Graham et al. 2006, Theorem 4.1] and in this case its orbit $\mathcal{P} := \bigcup_{i=1}^{n+2} \mathcal{A}(S_i)$ consists of spheres with disjoint interiors. For $n = 2$, $\mathcal{P}$ is an Apollonian circle packing. For $n = 3$, $\mathcal{P}$ is called an Apollonian sphere packing. Note that $\mathcal{A}$ is geometrically finite and that $\mathcal{P}$ is locally finite, as the spheres in $\mathcal{P}$ have disjoint interiors, Hence Theorem 7.5 applies to $\mathcal{P}$. The critical exponent of $\mathcal{A}$ for $n = 3$ has been estimated to be 2.473946(5) in [Borkovec et al. 1994].

Dual apollonian cluster ensemble for any n ≥ 2. Given $n+2$ mutually tangent spheres with disjoint interiors $S_1, \dots, S_{n+2}$ in $\mathbb{R}^n$, let $\mathcal{A}^*$ denote the group generated by the inversions with respect to S_i, $1 \le i \le n+2$. The dual Apollonian group $\mathcal{A}^*$ is a discrete geometrically finite subgroup of $\mathrm{MG}(\mathbb{R}^n)$ for all $n \ge 2$

and the orbit $\mathcal{P} := \bigcup_{i=1}^{n+2} \mathcal{A}^*(S_i)$ is a sphere packing, in our sense, consisting of spheres nested in S_i's. We note that $\mathcal{P}$ is locally finite, as nested spheres are getting smaller and smaller and hence Theorem 7.5 applies to $\mathcal{P}$.

9. Packing circles of the ideal triangle in $\mathbb{H}^2$

Consider an ideal triangle $\mathcal{T}$ in $\mathbb{H}^2$ i.e., a triangle whose sides are hyperbolic lines connecting vertices on the boundary of $\mathbb{H}^2$. An ideal triangle exists uniquely up to hyperbolic congruences. Consider $\mathcal{P}(\mathcal{T})$ the circle packing of an ideal triangle by filling in largest inner circles. The notation $\overline{\mathcal{P}(\mathcal{T})}$ denotes the closure of $\mathcal{P}(\mathcal{T})$ and $\operatorname{area}_{\mathrm{Hyp}}(C)$ is the hyperbolic area of the disk enclosed by C.

Theorem 9.1 (packing circles of the ideal triangle). *Let $\mathcal{T}$ be the ideal triangle of $\mathbb{H}^2$. Then*

$$\lim_{t\to 0} t^{\alpha/2} \cdot \#\{C \in \mathcal{P}(\mathcal{T}) : \operatorname{area}_{\mathrm{Hyp}}(C) > t\} = c_A \cdot \int_{\overline{\mathcal{P}(\mathcal{T})}} y^{-\alpha}\, d\mathcal{H}^\alpha(z)$$

where c_A denotes the Apollonian constant.

Fix the Apollonian circle packing $\mathcal{P}_0$ generated by two vertical lines $x = \pm 1$ and the unit circle $\{|z| = 1\}$. The corresponding Apollonian group $\Gamma_0 = \Gamma(\mathcal{P}_0)$ is generated by the inversions with respect to horizontal lines $y = 0$ and $y = -2i$ and the circles $\{|z - (\pm 1 - i)| = 1\}$. We set $\mathbb{R}^2_+ := \{x + iy : y > 0\}$. Now for the conformal metric $(U, f) = (\mathbb{R}^2_+, 1/y) = \mathbb{H}^2$, we note that $\{C \in \mathcal{P}(\mathcal{T}) : \operatorname{area}_{\mathrm{Hyp}}(C) > t\} = \{C \in \mathcal{P}_0 : \operatorname{area}_f(C) > t, C \cap \mathcal{T} \neq \varnothing\}$.

However Theorem 9.1 does not immediately follow from Theorem 8.3 since the ideal triangle $\mathcal{T}$ is not a compact subset of $\mathbb{H}^2$.

We need to understand the $\mathcal{H}^\alpha_f$-measure of neighborhoods of cusps in the triangle for $f = 1/y$. For the next two theorems, consider a conformal metric $(\mathbb{R}^2_+, f)$.

Theorem 9.2. *If $f(x+iy) \ll y^{-k}$ for some real number $k > \alpha^{-1}$ with implied constant independent of $|x| \le 1$, then for any $\eta > 0$,*

$$\mathcal{H}^\alpha_f\{|x| \le 1, y > \eta\} < \infty.$$

Moreover for any Borel subset $E \subset \{|x| \le 1, y > \eta\}$ (not necessarily compact) with smooth boundary,

$$\lim_{t\to\infty} t^{\alpha/2} \cdot \#\{C \in \mathcal{P}_0 : \operatorname{area}_f(C) > t, C \cap E \neq \varnothing\} \sim c_A \cdot \mathcal{H}^\alpha_f(E).$$

Proof. It suffices to show the claim for $\eta = 1$, since $\{|x| \le 1, \eta \le y \le 1\}$ is a compact subset. So we put $\eta = 1$ and set $U_R := \{|x| \le 1, y > R\}$. Define $F_t(E) := \{C \in \mathcal{P}_0 : \operatorname{area}_f(C) > t, C \cap E \neq \varnothing\}$ and $E_n := \{|x| \le 1, n \le y < n+1\}$. Then $F_t(U_1) = \bigcup_{n \ge 1} F_t(E_n)$.

For $C \in F_t(E_n)$, $C-(n-1)i \in F_t(E_1)$ and

$$\begin{aligned}\text{area}(C-(n-1)i) &= \int_{C-(n-1)i} f(z)^2 dz \\ &= \int_C f(z+(n-1)i))^2 dz \ll n^{-2k}\text{area}(C).\end{aligned}$$

Hence we get an injective map $F_t(E_n)$ to $F_{tn^{-2k}}(E_1)$ and hence for $R_0 \geq 1$,

$$\#F_t(U_{R_0}) = \sum_{n \geq R_0} \#F_{tn^{-2k}}(E_1).$$

By Theorem 8.3, for $\epsilon > 0$, there exists t_ϵ such that for all $t < t_\epsilon$,

$$\#F_t(E_1) \leq t^{-\alpha/2} c_A(\mathcal{H}_f^\alpha(E_1) + \epsilon)$$

and hence there exists $N_\epsilon > 1$ such that for all $t \leq 1$ and $n > N_\epsilon$,

$$\#F_{tn^{-2k}}(E_1) \leq t^{-\alpha/2} n^{-k\alpha} c_A(\mathcal{H}_f^\alpha(E_1) + \epsilon)$$

and hence

$$\#F_t(U_{N_\epsilon})t^{\alpha/2} \leq \left(\sum_{n \geq N_\epsilon} n^{-k\alpha}\right) c_A(\mathcal{H}_f^\alpha(E_1) + \epsilon).$$

Since $\left(\sum_{n \geq N_\epsilon} n^{-k\alpha}\right) < \infty$ as $k\alpha > 1$, if $N_\epsilon \gg 1$ is sufficiently large, we can make $\#F_t(U_{N_\epsilon})t^{\alpha/2} \leq \epsilon$ and $\mathcal{H}_f^\alpha(U_{N_\epsilon}) < \epsilon$. This implies $\mathcal{H}_f^\alpha(U_{N_\epsilon}) < \infty$. Moreover,

$$\limsup \#F_t(U_1)t^{\alpha/2} = \limsup \# \sum_{1 \leq n < N_\epsilon} F_t(E_n)t^{\alpha/2} + O(\epsilon) = c_A \cdot \mathcal{H}_f^\alpha(U_1) + O(\epsilon).$$

As $\epsilon > 0$ is arbitrary, we have

$$\limsup \#F_t(U_1)t^{\alpha/2} = c_A \cdot \mathcal{H}_f^\alpha(U_1).$$

Similarly we have $\liminf \#F_t(U_1)t^{\alpha/2} = c_A \cdot \mathcal{H}_f^\alpha(U_1)$. Hence

$$\lim \#F_t(U_1)t^{\alpha/2} = c_A \cdot \mathcal{H}_f^\alpha(U_1).$$

In the same way, we can deduce the claim for any Borel set E using the fact that $\mathcal{H}_f^\alpha(E \cap U_{N_\epsilon}) = O(\epsilon)$. □

We note that the boundary of $\mathbb{R}_+^2$ meets with $\text{Res}(\mathcal{P}_0)$ at three points $1, -1, \infty$ and these three points are in one Γ_0-orbit. Note that Γ_0 contains an element γ_0: $\gamma_0(x+yi) = x + (y+2)i$.

Theorem 9.3. *Let $(\mathbb{R}_+^2, f)$ be a conformal metric such that $f(x+iy) \asymp y^{-k}$ for $k \in \mathbb{R}$.*

(1) *If $\alpha^{-1} < k < 2 - \alpha^{-1}$, we have $\mathcal{H}_f^\alpha(\mathbb{R}_+^2) < \infty$.*

(2) *If either $k < \alpha^{-1}$ or $k > 2 - \alpha^{-1}$, we have $\mathcal{H}_f^\alpha(\mathbb{R}_+^2) = \infty$.*

Proof. Let $\gamma(z) = \frac{z-i}{z-1-i}$. Then $\gamma \in \Gamma_0$ and $\gamma(\infty) = 1$. Hence if we set $U_\eta := \{|x| \le 1, y > \eta\}$, $\gamma(U_\eta)$ is a neighborhood of 1 for all large $\eta > 1$. We will show that $\mathcal{H}^\alpha_f(\gamma(U_\eta)) < \infty$ if $k < 2 - \alpha^{-1}$.

We have

$$\mathcal{H}^\alpha_f(\gamma(U_\eta)) = \int_{z \in \gamma(U_\eta)} f(z)^\alpha \, d\mathcal{H}^\alpha(z) = \int_{w \in U_\eta} f(\gamma(w))^\alpha |\gamma'(w)|^\alpha \, d\mathcal{H}^\alpha(w).$$

Since $|\gamma'(w)| = |w - 1 + i|^{-2}$ and $\Im(\gamma(w)) \asymp y^{-1}$, we have

$$f(\gamma(w))|\gamma'(w)| \asymp y^{k-2}.$$

Hence by Theorem 9.2, if $2 - k > \alpha^{-1}$,

$$\mathcal{H}^\alpha_f(\gamma(U_\eta)) < \infty.$$

The remaining cases can be proved similarly and we leave them to the reader. □

Note that $\mathcal{T} := \mathcal{P}_0 \cap \mathbb{R}^2_+$ is a circle packing of the curvilinear triangle made by largest inner circles.

As $\alpha > 1$, Theorem 9.1 is a special case of the following:

Theorem 9.4. *Let $(\mathbb{R}^2_+, f)$ be a conformal metric such that $f(x + iy) \asymp y^{-k}$ where $\alpha^{-1} < k < 2 - \alpha^{-1}$. Then, for any (not necessarily compact) Borel subset $E \subset \mathbb{R}^2_+$ with smooth boundary, we have*

$$\lim_{t \to \infty} t^{\alpha/2} \cdot \#\{C \in \mathcal{T} : \operatorname{area}_f(C) > t,\ C \cap E \ne \varnothing\} \sim c_A \cdot \mathcal{H}^\alpha_f(E).$$

Acknowledgments

This article are mostly based on the papers [Oh and Shah 2013] and [Mohammadi and Oh 2012b]. I am grateful to Nimish Shah and Amir Mohammadi for their collaborations. I would also like to thank Tim Austin for useful discussions concerning the last section.

References

[Babillot 2002] M. Babillot, "On the mixing property for hyperbolic systems", *Israel J. Math.* **129** (2002), 61–76.

[Benoist and Oh 2012] Y. Benoist and H. Oh, "Effective equidistribution of S–integral points on symmetric varieties", *Ann. Inst. Fourier (Grenoble)* **62**:5 (2012), 1889–1942.

[Borel and Garland 1983] A. Borel and H. Garland, "Laplacian and the discrete spectrum of an arithmetic group", *Amer. J. Math.* **105**:2 (1983), 309–335.

[Borkovec et al. 1994] M. Borkovec, W. de Paris, and R. Peikert, "The fractal dimension of the Apollonian sphere packing", *Fractals* **2**:4 (1994), 521–526.

[Bourgain and Fuchs 2011] J. Bourgain and E. Fuchs, "A proof of the positive density conjecture for integer Apollonian circle packings", *J. Amer. Math. Soc.* **24**:4 (2011), 945–967.

[Bourgain and Kontorovich 2012] J. Bourgain and A. Kontorovich, "On the strong density conjecture of Apollonian packings", preprint, 2012. arXiv 1205.4416

[Bourgain et al. 2010a] J. Bourgain, A. Gamburd, and P. Sarnak, "Affine linear sieve, expanders, and sum-product", *Invent. Math.* **179**:3 (2010), 559–644.

[Bourgain et al. 2010b] J. Bourgain, A. Kontorovich, and P. Sarnak, "Sector estimates for hyperbolic isometries", *Geom. Funct. Anal.* **20**:5 (2010), 1175–1200.

[Bourgain et al. 2011] J. Bourgain, A. Gamburd, and P. Sarnak, "Generalization of Selberg's $\frac{3}{16}$ theorem and affine sieve", *Acta Math.* **207**:2 (2011), 255–290.

[Bowditch 1993] B. H. Bowditch, "Geometrical finiteness for hyperbolic groups", *J. Funct. Anal.* **113**:2 (1993), 245–317.

[Burger 1990] M. Burger, "Horocycle flow on geometrically finite surfaces", *Duke Math. J.* **61**:3 (1990), 779–803.

[Dolgopyat 1998] D. Dolgopyat, "On decay of correlations in Anosov flows", *Ann. of Math.* (2) **147**:2 (1998), 357–390.

[Duke et al. 1993] W. Duke, Z. Rudnick, and P. Sarnak, "Density of integer points on affine homogeneous varieties", *Duke Math. J.* **71**:1 (1993), 143–179.

[Eskin and McMullen 1993] A. Eskin and C. McMullen, "Mixing, counting, and equidistribution in Lie groups", *Duke Math. J.* **71**:1 (1993), 181–209.

[Eskin et al. 1996] A. Eskin, S. Mozes, and N. Shah, "Unipotent flows and counting lattice points on homogeneous varieties", *Ann. of Math.* (2) **143**:2 (1996), 253–299.

[Flaminio and Spatzier 1990] L. Flaminio and R. J. Spatzier, "Geometrically finite groups, Patterson–Sullivan measures and Ratner's rigidity theorem", *Invent. Math.* **99**:3 (1990), 601–626.

[Gorodnik and Nevo 2012] A. Gorodnik and A. Nevo, "Lifting, restricting and sifting integral points on affine homogeneous varieties", *Compos. Math.* **148**:6 (2012), 1695–1716.

[Gorodnik and Oh 2007] A. Gorodnik and H. Oh, "Orbits of discrete subgroups on a symmetric space and the Furstenberg boundary", *Duke Math. J.* **139**:3 (2007), 483–525.

[Graham et al. 2006] R. L. Graham, J. C. Lagarias, C. L. Mallows, A. R. Wilks, and C. H. Yan, "Apollonian circle packings: Geometry and group theory, III: Higher dimensions", *Discrete Comput. Geom.* **35**:1 (2006), 37–72.

[Guillarmou and Mazzeo 2012] C. Guillarmou and R. Mazzeo, "Resolvent of the Laplacian on geometrically finite hyperbolic manifolds", *Invent. Math.* **187**:1 (2012), 99–144.

[Hirai 1962] T. Hirai, "On irreducible representations of the Lorentz group of n – th order", *Proc. Japan Acad.* **38** (1962), 258–262.

[Howe and Moore 1979] R. E. Howe and C. C. Moore, "Asymptotic properties of unitary representations", *J. Funct. Anal.* **32**:1 (1979), 72–96.

[Kim 2012] I. Kim, "Counting, mixing and equidistribution of horospheres in geometrically finite rank one locally symmetric manifolds", (2012). arXiv 1103.5003

[Kleinbock and Margulis 1996] D. Y. Kleinbock and G. A. Margulis, "Bounded orbits of nonquasiunipotent flows on homogeneous spaces", pp. 141–172 in *Sinaĭ's Moscow Seminar on Dynamical Systems*, edited by L. A. Bunimovich et al., Amer. Math. Soc. Transl. Ser. 2 **171**, Amer. Math. Soc., Providence, RI, 1996.

[Knapp and Stein 1971] A. W. Knapp and E. M. Stein, "Intertwining operators for semisimple groups", *Ann. of Math.* (2) **93** (1971), 489–578.

[Kontorovich and Oh 2011] A. Kontorovich and H. Oh, "Apollonian circle packings and closed horospheres on hyperbolic 3–manifolds", *J. Amer. Math. Soc.* **24**:3 (2011), 603–648.

[Kowalski 2010] E. Kowalski, "Sieve in expansion", preprint, 2010. arXiv 1012.2793

[Kowalski 2014] E. Kowalski, "Sieves in discrete groups, especially sparse", pp. 107–140 in *Thin groups and superstrong approximation*, edited by H. Oh and E. Breuillard, Math. Sci. Res. Inst. Publ. **61**, Cambridge, New York, 2014.

[Lax and Phillips 1982] P. D. Lax and R. S. Phillips, "The asymptotic distribution of lattice points in Euclidean and non-Euclidean spaces", *J. Funct. Anal.* **46**:3 (1982), 280–350.

[Lee and Oh 2013] M. Lee and H. Oh, "Effective circle count for Apollonian packings and closed horospheres", *Geom. Funct. Anal.* **23**:2 (2013), 580–621.

[Margulis 2004] G. A. Margulis, *On some aspects of the theory of Anosov systems*, Springer, Berlin, 2004.

[McMullen 1998] C. T. McMullen, "Hausdorff dimension and conformal dynamics, III: Computation of dimension", *Amer. J. Math.* **120**:4 (1998), 691–721.

[Mohammadi and Oh 2012a] A. Mohammadi and H. Oh, "Ergodicity of unipotent flows and Kleinian groups", preprint, 2012. arXiv 1112.4024

[Mohammadi and Oh 2012b] A. Mohammadi and H. Oh, "Matrix coefficients, counting and primes for orbits of geometrically finite groups", preprint, 2012. To appear in *J. European Math. Soc.* arXiv 1208.4139

[Nevo and Sarnak 2010] A. Nevo and P. Sarnak, "Prime and almost prime integral points on principal homogeneous spaces", *Acta Math.* **205**:2 (2010), 361–402.

[Oh 2010] H. Oh, "Orbital counting via mixing and unipotent flows", pp. 339–375 in *Homogeneous flows, moduli spaces and arithmetic* (Pisa, 2007), edited by M. L. Einsiedler et al., Clay Math. Proc. **10**, Amer. Math. Soc., Providence, RI, 2010.

[Oh 2011] H. Oh, "Dynamics on geometrically finite hyperbolic manifolds with applications to Apollonian circle packings and beyond", pp. 1308–1331 in *Proceedings of the International Congress of Mathematicians, III* (Hyperabad, 2010), edited by R. Bhatia et al., Hindustan Book Agency, New Delhi, 2011.

[Oh and Shah 2010] H. Oh and N. Shah, "Counting visible circles on the sphere and Kleinian groups", preprint, 2010. arXiv 1004.2129

[Oh and Shah 2012] H. Oh and N. Shah, "The asymptotic distribution of circles in the orbits of Kleinian groups", *Invent. Math.* **187**:1 (2012), 1–35.

[Oh and Shah 2013] H. Oh and N. A. Shah, "Equidistribution and counting for orbits of geometrically finite hyperbolic groups", *J. Amer. Math. Soc.* **26**:2 (2013), 511–562.

[Patterson 1976] S. J. Patterson, "The limit set of a Fuchsian group", *Acta Math.* **136**:3-4 (1976), 241–273.

[Ratcliffe 2006] J. G. Ratcliffe, *Foundations of hyperbolic manifolds*, 2nd ed., Graduate Texts in Mathematics **149**, Springer, New York, 2006.

[Ratner 1991] M. Ratner, "On Raghunathan's measure conjecture", *Ann. of Math.* (2) **134**:3 (1991), 545–607.

[Roblin 2003] T. Roblin, *Ergodicité et équidistribution en courbure négative*, Mém. Soc. Math. Fr. (N.S.) **95**, Société Mathématique de France, Paris, 2003.

[Salehi Golsefidy and Sarnak 2013] A. Salehi Golsefidy and P. Sarnak, "The affine sieve", *J. Amer. Math. Soc.* **26**:4 (2013), 1085–1105.

[Salehi Golsefidy and Varjú 2012] A. Salehi Golsefidy and P. P. Varjú, "Expansion in perfect groups", *Geom. Funct. Anal.* **22**:6 (2012), 1832–1891.

[Sarnak 2011] P. Sarnak, "Integral Apollonian packings", *Amer. Math. Monthly* **118**:4 (2011), 291–306.

[Shalom 2000] Y. Shalom, "Rigidity, unitary representations of semisimple groups, and fundamental groups of manifolds with rank one transformation group", *Ann. of Math.* (2) **152**:1 (2000), 113–182.

[Stoyanov 2011] L. Stoyanov, "Spectra of Ruelle transfer operators for axiom A flows", *Nonlinearity* **24**:4 (2011), 1089–1120.

[Sullivan 1979] D. Sullivan, "The density at infinity of a discrete group of hyperbolic motions", *Inst. Hautes Études Sci. Publ. Math.* 50 (1979), 171–202.

[Sullivan 1984] D. Sullivan, "Entropy, Hausdorff measures old and new, and limit sets of geometrically finite Kleinian groups", *Acta Math.* **153**:3-4 (1984), 259–277.

[Vinogradov 2012] I. Vinogradov, "Effective bisector estimate with applications to Apollonian circle packings", preprint, 2012. arXiv 1204.5498

[Warner 1972a] G. Warner, *Harmonic analysis on semi-simple Lie groups, I*, Die Grundlehren der mathematischen Wissenschaften **188**, Springer, New York, 1972.

[Warner 1972b] G. Warner, *Harmonic analysis on semi-simple Lie groups, II*, Die Grundlehren der mathematischen Wissenschaften **189**, Springer, New York, 1972.

[Winter $\geq$ 2012] D. Winter, In preparation.

hee.oh@yale.edu

Mathematics Department, Yale University,
New Haven, CT 06520, United States

Korea Institute for Advanced Study, Seoul, Korea

Thin groups and superstrong approximation
MSRI Publications
Volume **61**, 2013

Generic elements in Zariski-dense subgroups and isospectral locally symmetric spaces

GOPAL PRASAD AND ANDREI S. RAPINCHUK

The article contains a survey of our results on length-commensurable and isospectral locally symmetric spaces and of related problems in the theory of semisimple algebraic groups. We discuss some of the techniques involved in this work (in particular, the existence of generic tori in semisimple algebraic groups over finitely generated fields and of generic elements in finitely generated Zariski-dense subgroups) and some open problems.

1. Introduction

This article contains an exposition of recent results on isospectral and length-commensurable locally symmetric spaces associated with simple real algebraic groups [Prasad and Rapinchuk 2009; 2013] and related problems in the theory of semisimple algebraic groups [Garibaldi 2012; Garibaldi and Rapinchuk 2013; Prasad and Rapinchuk 2010a]. One of the goals of [Prasad and Rapinchuk 2009] was to study the problem beautifully formulated by Mark Kac in [1966] as "*Can one hear the shape of a drum*?" for the quotients of symmetric spaces of the groups of real points of absolutely simple real algebraic groups by cocompact arithmetic subgroups. A precise mathematical formulation of Kac's question is *whether two compact Riemannian manifolds which are isospectral* (i.e., have equal spectra — eigenvalues and multiplicities — for the Laplace–Beltrami operator) *are necessarily isometric*. In general, the answer to this question is in the negative as was shown by John Milnor already in [1964] by constructing two nonisometric isospectral flat tori of dimension 16. Later M.-F. Vignéras [1980] used arithmetic properties of quaternion algebras to produce examples of arithmetically defined isospectral, but not isometric, Riemann surfaces. On the other hand, T. Sunada [1985], inspired by a construction of nonisomorphic number fields with the same Dedekind zeta-function, proposed a general and basically purely group-theoretic method of producing nonisometric isospectral Riemannian manifolds which has since then been used in various ways. It is important to note, however, that the nonisometric isospectral manifolds constructed by Vignéras and Sunada are *commensurable*, that is, have a common finite-sheeted cover. This suggests that one should probably settle for the following weaker version of Kac's original question: *Are any two isospectral compact Riemannian manifolds necessarily commensurable?* The answer to this modified question is still negative in the general case: Lubotzky, Samuels and Vishne [Lubotzky et al. 2006], using the Langlands correspondence, have constructed examples of noncommensurable isospectral locally symmetric spaces associated with absolutely simple real groups of type A_n (see Problem 10.7). Nevertheless, it turned out that the answer is actually in the affirmative for several classes of locally symmetric spaces. Prior to our paper [Prasad and Rapinchuk 2009], this was known to be the case only for the following two classes: arithmetically defined Riemann surfaces [Reid 1992] and arithmetically defined hyperbolic 3-manifolds [Chinburg et al. 2008].

In [Prasad and Rapinchuk 2009], we used Schanuel's conjecture from transcendental number theory (for more about this conjecture, and how it comes up in our work, see below) and the results of [Garibaldi 2012; Prasad and Rapinchuk 2010a] to prove that any two compact *isospectral* arithmetically defined locally symmetric spaces associated with absolutely simple real algebraic groups of type other than A_n $(n > 1)$, D_{2n+1} $(n > 1)$, or E_6 are *necessarily commensurable*. One of the important ingredients of the proof is the connection between isospectrality and another property of Riemannian manifolds called *isolength spectrality*.

More precisely, for a Riemannian manifold M we let $L(M)$ denote the *weak length spectrum* of M, that is, the collection of the lengths of all closed geodesics in M (note that for the existence of a "nice" Laplace spectrum, M is required to be compact, but the weak length spectrum $L(M)$ can be considered for *any* M —i.e., we do not need to assume that M is compact). Then two Riemannian manifolds M_1 and M_2 are called *isolength spectral* if $L(M_1) = L(M_2)$. Any two compact isospectral locally symmetric spaces are isolength spectral; this was first proved in [Gangolli 1977] in the rank-one case, then in [Duistermaat and Guillemin 1975] and [Duistermaat et al. 1979] in the general case; see [Prasad and Rapinchuk 2009, Theorem 10.1]. So, the emphasis in this last paper is really on the analysis of isolength spectral locally symmetric spaces M_1 and M_2.

In fact, we prove our results under the much weaker assumption of *length-commensurability*, which means that $\mathbb{Q}\cdot L(M_1) = \mathbb{Q}\cdot L(M_2)$. (The set $\mathbb{Q}\cdot L(M)$ is sometimes called the *rational length spectrum* of M; its advantage, particularly in the analysis of questions involving commensurable manifolds, is that it is invariant under passing to a finite-sheeted cover—this property fails for the Laplace spectrum or the length spectrum. At the same time, $\mathbb{Q} \cdot L(M)$ can actually be computed in at least some cases, while precise computation of $L(M)$ or the Laplace spectrum is not available for any compact locally symmetric space at this point.) The notion of length-commensurability was introduced in [Prasad and Rapinchuk 2009], and the investigation of its qualitative and quantitative consequences for general locally symmetric spaces is an ongoing project. For arithmetically defined spaces, however, the main questions were answered in the same paper, and we would like to complete this introduction by showcasing the results for arithmetic hyperbolic spaces.

Let $\mathbb{H}^n$ be the real hyperbolic n-space. By an arithmetically defined real hyperbolic n-manifold we mean the quotient $\mathbb{H}^n/\Gamma$, where Γ is an arithmetic subgroup of $\mathrm{PSO}(n,1)$ (which is the isometry group of $\mathbb{H}^n$); see Section 3 regarding the notion of arithmeticity.

Theorem 1.1 [Prasad and Rapinchuk 2009, Corollary 8.17 and Remark 8.18]. *Let M_1 and M_2 be arithmetically defined real hyperbolic n-manifolds.*

If $n \not\equiv 1 \pmod 4$, then in case M_1 and M_2 are not commensurable, after a possible interchange of M_1 and M_2, there exists $\lambda_1 \in L(M_1)$ such that for any $\lambda_2 \in L(M_2)$, the ratio λ_1/λ_2 is transcendental over $\mathbb{Q}$. (Thus, for such n the length-commensurability, and hence isospectrality, of M_1 and M_2 implies their commensurability.)

On the contrary, for any $n \equiv 1 \pmod 4$, there exist M_1 and M_2 as above that are length-commensurable, but not commensurable.

What is noteworthy is that there is no apparent geometric reason for this dramatic distinction between the length-commensurability of hyperbolic n-manifolds when $n \not\equiv 1 \pmod 4$ and $n \equiv 1 \pmod 4$ — in our argument the difference comes from considerations involving Galois cohomology; see Theorem 4.2 and subsequent comments.

Our general results for arithmetically defined length-commensurable locally symmetric spaces (Section 5) imply similar (but not identical!) assertions for complex and quaternionic hyperbolic manifolds. At the same time, one can ask about possible relations between $L(M_1)$ and $L(M_2)$ (or between $\mathbb{Q}\cdot L(M_1)$ and $\mathbb{Q}\cdot L(M_2)$) if M_1 and M_2 are *not* length-commensurable. The results we will describe in Section 8 assert that if $\mathbb{Q}\cdot L(M_1) \neq \mathbb{Q}\cdot L(M_2)$, then no polynomial-type relation between $L(M_1)$ and $L(M_2)$ can ever exist; in other words, these sets are *very* different. This is, for example, the case if M_1 and M_2 are hyperbolic manifolds of finite volume having different dimensions!

2. Length-commensurable locally symmetric spaces and weakly commensurable subgroups

2.1. ***Riemann surfaces.*** Our analysis of length-commensurability of locally symmetric spaces relies on a purely algebraic relation between their fundamental groups which we termed *weak commensurability*. It is easiest to motivate this notion by looking at the length-commensurability of Riemann surfaces. In this discussion we will be using the realization of $\mathbb{H}^2$ as the complex upper half-plane with the standard hyperbolic metric $ds^2 = y^{-2}(dx^2 + dy^2)$. The usual action of $\mathrm{SL}_2(\mathbb{R})$ on $\mathbb{H}^2$ by fractional linear transformations is isometric and allows us to identify $\mathbb{H}^2$ with the symmetric space $\mathrm{SO}_2(\mathbb{R})\backslash\mathrm{SL}_2(\mathbb{R})$. It is well known that any compact Riemann surface M of genus > 1 can be obtained as a quotient of $\mathbb{H}^2$ by a discrete subgroup $\Gamma \subset \mathrm{SL}_2(\mathbb{R})$ with torsion-free image in $\mathrm{PSL}_2(\mathbb{R})$. Now, given any such subgroup Γ, we let $\pi\colon \mathbb{H}^2 \to \mathbb{H}^2/\Gamma =: M$ denote the canonical projection. It is easy to see that

$$t \mapsto e^t i = i \cdot \begin{pmatrix} e^{t/2} & 0 \\ 0 & e^{-t/2} \end{pmatrix}, \quad t \in \mathbb{R},$$

is a unit-velocity parametrization of a geodesic c in $\mathbb{H}^2$. So, if $\gamma = \mathrm{diag}(t_\gamma, t_\gamma^{-1}) \in \Gamma$, then the image $\pi(c)$ is a *closed geodesic* c_γ in M, whose length is given by the formula

$$\ell_\Gamma(c_\gamma) = \frac{2}{n_\gamma} \cdot \log t_\gamma \tag{1}$$

(assuming that $t_\gamma > 1$), where n_γ is an integer $\geqslant 1$ (winding number in case c_γ is not primitive). Generalizing this construction, one shows that every semisimple element $\gamma \in \Gamma \setminus \{\pm 1\}$ gives rise to a closed geodesic c_γ in M whose length is given by (1) where t_γ is the eigenvalue of $\pm\gamma$ which is > 1, and conversely, any closed geodesic in M is obtained this way. As a result,

$$\mathbb{Q} \cdot L(M) = \mathbb{Q} \cdot \{\log t_\gamma \mid \gamma \in \Gamma\backslash\{\pm 1\} \text{ semisimple}\}.$$

Now, suppose we have two quotients $M_1 = \mathbb{H}^2/\Gamma_1$ and $M_2 = \mathbb{H}^2/\Gamma_2$ as above, and let c_{γ_i} be a closed geodesic in M_i for $i = 1, 2$. Then

$$\ell_{\Gamma_1}(c_{\gamma_1})/\ell_{\Gamma_2}(c_{\gamma_2}) \in \mathbb{Q} \iff t_{\gamma_1}^m = t_{\gamma_2}^n \text{ for some } m, n \in \mathbb{N},$$

or equivalently, the subgroups generated by the eigenvalues of γ_1 and γ_2 have nontrivial intersection. This leads us to the following.

Definition 2.2. Let $G_1 \subset \mathrm{GL}_{N_1}$ and $G_2 \subset \mathrm{GL}_{N_2}$ be two semisimple algebraic groups defined over a field F of characteristic zero.

(a) Two semisimple elements $\gamma_1 \in G_1(F)$ and $\gamma_2 \in G_2(F)$ are said to be *weakly commensurable* if the subgroups of $\bar{F}^\times$ generated by their eigenvalues intersect nontrivially.

(b) Two (Zariski-dense) subgroups $\Gamma_1 \subset G_1(F)$ and $\Gamma_2 \subset G_2(F)$ are *weakly commensurable* if every semisimple element $\gamma_1 \in \Gamma_1$ of infinite order is weakly commensurable to some semisimple element $\gamma_2 \in \Gamma_2$ of infinite order, and vice versa.

It should be noted that in [Prasad and Rapinchuk 2009] we gave a more technical, but equivalent, definition of weakly commensurable elements, viz. we required that there should exist maximal F-tori T_i of G_i for $i = 1, 2$ such that $\gamma_i \in T_i(F)$ and for some characters $\chi_i \in X(T_i)$ we have

$$\chi_1(\gamma_1) = \chi_2(\gamma_2) \neq 1.$$

This (equivalent) reformulation of (a) immediately demonstrates that the notion of weak commensurability does not depend on the choice of matrix realizations of the G_i, and more importantly, is more convenient for the proofs of our results.

The above discussion of Riemann surfaces implies that if two Riemann surfaces $M_1 = \mathbb{H}^2/\Gamma_1$ and $M_2 = \mathbb{H}^2/\Gamma_2$ are length-commensurable, then the corresponding fundamental groups Γ_1 and Γ_2 are weakly commensurable. Our next goal is to explain why this implication remains valid for general locally symmetric spaces.

2.3. *Length-commensurability and weak commensurability*: *The general case.* First, we need to fix some notations related to general locally symmetric spaces. Let G be a connected adjoint real semisimple algebraic group, let $\mathcal{G} = G(\mathbb{R})$ considered as a real Lie group, and let $\mathfrak{X} = \mathcal{K}\backslash\mathcal{G}$, where $\mathcal{K}$ is a maximal compact subgroup of $\mathcal{G}$, be the associated symmetric space endowed with the Riemannian metric coming from the Killing form on the Lie algebra $\mathfrak{g}$ of $\mathcal{G}$. Furthermore, given a torsion-free discrete subgroup Γ of $\mathcal{G}$, we let $\mathfrak{X}_\Gamma = \mathfrak{X}/\Gamma$ denote the corresponding locally symmetric space. Just as in the case of Riemann surfaces, to any nontrivial semisimple element $\gamma \in \Gamma$ there corresponds a closed geodesic c_γ whose length is given by

$$\ell_\Gamma(c_\gamma) = \frac{1}{n_\gamma} \cdot \lambda_\Gamma(\gamma),$$

where n_γ is an integer $\geqslant 1$ and

$$\lambda_\Gamma(\gamma)^2 := \sum_\alpha (\log|\alpha(\gamma)|)^2, \tag{2}$$

with the summation running over all roots of G with respect to a fixed maximal $\mathbb{R}$-torus T of G whose group of $\mathbb{R}$-points contains γ. This formula looks much more intimidating than (1), so in order to make it more manageable we first make the following observation. Of course, the $\mathbb{R}$-torus T may not be $\mathbb{R}$-split, so not every root α may be defined over $\mathbb{R}$. However,

$$|\alpha(\gamma)|^2 = \chi(\gamma),$$

where $\chi = \alpha + \overline{\alpha}$ (with $\overline{\alpha}$ being the conjugate character in terms of the natural action of $\mathrm{Gal}(\mathbb{C}/\mathbb{R})$ on $X(T)$, and, as usual, $X(T)$ is viewed as an additive group) is a character defined over $\mathbb{R}$ and which takes positive values on $T(\mathbb{R})$. Such characters will be called *positive*. So, we can now rewrite (2) in the form

$$\lambda_\Gamma(\gamma)^2 = \sum_{i=1}^{p} s_i(\log \chi_i(\gamma))^2, \tag{3}$$

where $\chi_1, \ldots, \chi_p$ are certain positive characters of T and $s_1, \ldots, s_p$ are positive rational numbers whose denominators are divisors of 4. The point to be made here is that the subgroup $P(T) \subset X(T)$ of positive characters may be rather small. More precisely, T is an almost direct product of an $\mathbb{R}$-anisotropic subtorus A and an $\mathbb{R}$-split subtorus S. Then any character of T which is defined over $\mathbb{R}$ vanishes on A. This easily implies that the restriction map yields an embedding $P(T) \hookrightarrow X(S)_{\mathbb{R}} = X(S)$ with finite cokernel; in particular, the rank of $P(T)$ as an abelian group coincides with the $\mathbb{R}$-rank $\mathrm{rk}_{\mathbb{R}}\, T$ of T.

Before formulating our results, we define the following property. Let $G \subset \mathrm{GL}_N$ be a semisimple algebraic group defined over a field F of characteristic zero. We say that a (Zariski-dense) subgroup $\Gamma \subset G(F)$ ***has property*** (*A*) if for any semisimple element $\gamma \in \Gamma$, all the eigenvalues of γ lie in the field of algebraic numbers $\overline{\mathbb{Q}}$ (note that the latter is equivalent to the fact that for any maximal F-torus T of G containing γ and any character $\chi \in X(T)$, we have $\chi(\gamma) \in \overline{\mathbb{Q}}^{\times}$ — this reformulation shows, in particular, that this property does not depend on the choice of a matrix realization of G). Of course, this property automatically holds if Γ is arithmetic, or more generally, if Γ can be conjugated into $\mathrm{SL}_N(K)$ for some number field K.

Let us now consider the rank-one case first.

The rank-one case. Suppose $\mathrm{rk}_{\mathbb{R}}\, G = 1$ (the examples include the adjoint groups of $\mathrm{SO}(n, 1)$, $\mathrm{SU}(n, 1)$ and $\mathrm{Sp}(n, 1)$; the corresponding symmetric spaces are respectively the real, complex and quaternionic hyperbolic n-spaces). Then given a nontrivial semisimple element $\gamma \in \Gamma$, for any maximal $\mathbb{R}$-torus T of G containing γ we have $\mathrm{rk}_{\mathbb{R}}\, T = 1$, which implies that the group $P(T)$ of positive characters is cyclic and is generated, say, by χ. Then it follows from (3) that

$$\lambda_\Gamma(\gamma) = \frac{\sqrt{m}}{2} \cdot |\log \chi(t)|, \tag{4}$$

where m is some integer $\geqslant 1$ depending only on G; note that this formula is still in the spirit of (1), but potentially involves some irrationality which can complicate the analysis of length-commensurability.

Now, suppose that G_1 and G_2 are two simple algebraic $\mathbb{R}$-groups of $\mathbb{R}$-rank one. For $i = 1, 2$, let $\Gamma_i \subset G_i(\mathbb{R}) = \mathcal{G}_i$ be a discrete torsion-free subgroup having property (A). Given a nontrivial semisimple element $\gamma_i \in \Gamma_i$, we pick a maximal $\mathbb{R}$-torus T_i of G_i whose group of $\mathbb{R}$-points contains γ_i and let χ_i be a generator of the group of positive characters $P(T_i)$. Then according to (4),

$$\lambda_{\Gamma_1}(\gamma_1) = \frac{\sqrt{m_1}}{2} \cdot |\log \chi_1(\gamma_1)| \quad \text{and} \quad \lambda_{\Gamma_2}(\gamma_2) = \frac{\sqrt{m_2}}{2} \cdot |\log \chi_2(\gamma_2)|,$$

for some integers $m_1, m_2 \geqslant 1$. By a theorem proved independently by Gelfond and Schneider in 1934 (which settled Hilbert's seventh problem; see [Baker 1990]), the ratio

$$\frac{\log \chi_1(\gamma_1)}{\log \chi_2(\gamma_2)}$$

is either rational or transcendental. This implies that the $\ell_{\Gamma_1}(\gamma_1)/\ell_{\Gamma_2}(\gamma_2)$, or equivalently the ratio $\lambda_{\Gamma_1}(\gamma_1)/\lambda_{\Gamma_2}(\gamma_2)$, can be rational only if

$$\chi_1(\gamma_1)^{n_1} = \chi_2(\gamma_2)^{n_2},$$

for some nonzero integers n_1, n_2, which makes the elements γ_1 and γ_2 weakly commensurable. (Of course, we get this conclusion without using the theorem of Gelfond–Schneider if $G_1 = G_2$, hence $m_1 = m_2$.) This argument shows that the length-commensurability of $\mathfrak{X}_{\Gamma_1}$ and $\mathfrak{X}_{\Gamma_2}$ implies the weak commensurability of Γ_1 and Γ_2.

Finally, we recall that if $G \subset \mathrm{GL}_N$ is an absolutely simple real algebraic group not isomorphic to PGL_2 then any lattice $\Gamma \subset G(\mathbb{R})$ can be conjugated into $\mathrm{SL}_N(K)$ for some number field K (see [Raghunathan 1972, 7.67 and 7.68]), hence possesses property (A). This implies that if $\mathfrak{X}_{\Gamma_1}$ and $\mathfrak{X}_{\Gamma_2}$ are rank one locally symmetric spaces of finite volume then their length-commensurability always implies the weak commensurability of Γ_1 and Γ_2 except possibly in the following situation: $G_1 = \mathrm{PGL}_2$ and Γ_1 cannot be conjugated into $\mathrm{PGL}_2(K)$ for any number field $K \subset \mathbb{R}$ while $G_2 \neq \mathrm{PGL}_2$ (in [Prasad and Rapinchuk 2009] this was called the exceptional case $(\mathscr{E})$). Nevertheless, the conclusion remains valid also in this case if one assumes the truth of Schanuel's conjecture (see below)—this follows from our recent results [Prasad and Rapinchuk 2013], which we will discuss in Section 8 (see Theorem 8.1).

The general case. If $\mathrm{rk}_{\mathbb{R}}\, G > 1$ then p may be > 1 in (3), hence $\lambda_\Gamma(\gamma)$, generally speaking, is *not* a multiple of the logarithm of the value of a positive character. Consequently, the fact that the ratio $\lambda_{\Gamma_1}(\gamma_1)/\lambda_{\Gamma_2}(\gamma_2)$ is a rational number does not imply *directly* that γ_1 and γ_2 are weakly commensurable. While the implication nevertheless *is* valid (under some natural technical assumptions), it is hardly surprising now that the proof requires some nontrivial information about the logarithms of the character values. More precisely, our arguments in [Prasad and Rapinchuk 2009; 2013] assume the truth of the following famous conjecture in transcendental number theory (see [Ax 1971], for example).

Conjecture 2.4 (Schanuel's conjecture). *If $z_1, \dots, z_n \in \mathbb{C}$ are linearly independent over $\mathbb{Q}$, the transcendence degree (over $\mathbb{Q}$) of the field generated by $z_1, \dots, z_n$ and $e^{z_1}, \dots, e^{z_n}$ is at least n.*

In fact, we will only need the consequence of this conjecture that for nonzero algebraic numbers $a_1, \dots, a_n$, (any values of) their logarithms $\log a_1, \dots, \log a_n$ are algebraically independent once they are linearly independent (over $\mathbb{Q}$). In order to apply this statement in our situation, we first prove the following elementary lemma.

Lemma 2.5. *Let G_1 and G_2 be two connected semisimple real algebraic groups. For $i = 1, 2$, let T_i be a maximal $\mathbb{R}$-torus of G_i, $\gamma_i \in T_i(\mathbb{R})$ and let $\chi_1^{(i)}, \dots, \chi_{d_i}^{(i)}$ be positive characters of T_i such that the set*

$$S_i = \{\log \chi_1^{(i)}(\gamma_i), \dots, \log \chi_{d_i}^{(i)}(\gamma_i)\} \subset \mathbb{R}$$

is linearly independent over $\mathbb{Q}$. If γ_1 and γ_2 are not *weakly commensurable then the set $S_1 \cup S_2$ is also linearly independent.*

Proof. Assume the contrary. Then there exist integers $s_1, \dots, s_{d_1}, t_1, \dots, t_{d_2}$, not all zero, such that

$$s_1 \log \chi_1^{(1)}(\gamma_1) + \dots + s_{d_1} \log \chi_{d_1}^{(1)}(\gamma_1) - t_1 \log \chi_1^{(2)}(\gamma_2) - \dots - t_{d_2} \log \chi_{d_2}^{(2)}(\gamma_2) = 0.$$

Consider the characters

$$\psi^{(1)} := s_1 \chi_1^{(1)} + \dots + s_{d_1} \chi_{d_1}^{(1)} \quad \text{and} \quad \psi^{(2)} := t_1 \chi_1^{(2)} + \dots + t_{d_2} \chi_{d_2}^{(2)}$$

of T_1 and T_2 respectively. Then $\psi^{(1)}(\gamma_1) = \psi^{(2)}(\gamma_2)$, and hence

$$\psi^{(1)}(\gamma_1) = 1 = \psi^{(2)}(\gamma_2),$$

because γ_1 and γ_2 are not weakly commensurable. This means that in

$$s_1 \log \chi_1^{(1)}(\gamma_1) + \dots + s_{d_1} \log \chi_{d_1}^{(1)}(\gamma_1) = t_1 \log \chi_1^{(2)}(\gamma_2) + \dots + t_{d_2} \log \chi_{d_2}^{(2)}(\gamma_2)$$

both sides vanish. But the sets S_1 and S_2 are linearly independent, so all the coefficients are zero, a contradiction. □

We are now ready to connect length-commensurability with weak commensurability.

Proposition 2.6. *Let G_1 and G_2 be two connected semisimple real algebraic groups. For $i = 1, 2$, let $\Gamma_i \subset G_i(\mathbb{R})$ be a subgroup satisfying property* (A). *Assume that Schanuel's conjecture holds. If semisimple elements $\gamma_1 \in \Gamma_1$ and $\gamma_2 \in \Gamma_2$ are not weakly commensurable then $\lambda_{\Gamma_1}(\gamma_1)$ and $\lambda_{\Gamma_2}(\gamma_2)$ are algebraically independent over $\mathbb{Q}$.*

Proof. It follows from (3) that

$$\lambda_{\Gamma_1}(\gamma_1)^2 = \sum_{i=1}^{p} s_i (\log \chi_i^{(1)}(\gamma_1))^2 \quad \text{and} \quad \lambda_{\Gamma_2}(\gamma_2)^2 = \sum_{i=1}^{q} t_i (\log \chi_i^{(2)}(\gamma_2))^2, \quad (5)$$

where s_i and t_i are positive rational numbers, and $\chi_i^{(1)}$ and $\chi_i^{(2)}$ are positive characters on maximal $\mathbb{R}$-tori T_1 and T_2 of G_1 and G_2 whose groups of $\mathbb{R}$-points contain the elements γ_1 and γ_2, respectively. After renumbering the characters, we can assume that

$$a_1 := \log \chi_1^{(1)}(\gamma_1), \quad \ldots, \quad a_m := \log \chi_m^{(1)}(\gamma_1)$$

(resp., $b_1 := \log \chi_1^{(2)}(\gamma_2), \ldots, b_n = \log \chi_n^{(2)}(\gamma_2)$) form a basis of the $\mathbb{Q}$-subspace of $\mathbb{R}$ spanned by $\log \chi_i^{(1)}(\gamma_1)$ for $i \leqslant p$ (resp., by $\log \chi_i^{(2)}(\gamma_2)$ for $i \leqslant q$). It follows from Lemma 2.5 that the numbers

$$a_1, \ldots, a_m \quad \text{and} \quad b_1, \ldots, b_n, \quad (6)$$

are linearly independent. By our assumption, Γ_1 and Γ_2 possess property (A), so the character values $\chi_i^{(j)}(\gamma_j)$ are all algebraic numbers. So, it follows from Schanuel's conjecture that the numbers in (6) are algebraically independent over $\mathbb{Q}$. As is seen from (5), $\lambda_{\Gamma_1}(\gamma_1)$ and $\lambda_{\Gamma_2}(\gamma_2)$ are represented by nonzero homogeneous polynomials of degree two, with rational coefficients, in $a_1, \ldots, a_m$ and $b_1, \ldots, b_n$, respectively, and therefore they are algebraically independent. □

This proposition leads us to the following.

Theorem 2.7. *Let G_1 and G_2 be two connected semisimple real algebraic groups. For $i = 1, 2$, let $\Gamma_i \subset G_i(\mathbb{R})$ be a discrete torsion-free subgroup having property (A). Assume that Schanuel's conjecture holds. If Γ_1 and Γ_2 are not weakly commensurable, then, possibly after reindexing, we can find $\lambda_1 \in L(\mathfrak{X}_{\Gamma_1})$ which is algebraically independent from* any *$\lambda_2 \in L(\mathfrak{X}_{\Gamma_2})$. In particular, $\mathfrak{X}_{\Gamma_1}$ and $\mathfrak{X}_{\Gamma_2}$ are not length-commensurable.*

Combining this with the discussion above of property (A) for lattices and of the exceptional case ($\mathscr{E}$), we obtain the following.

Corollary 2.8. *Let G_1 and G_2 be two absolutely simple real algebraic groups, and for $i = 1, 2$ let Γ_i be a lattice in $G_i(\mathbb{R})$ (so that the locally symmetric space $\mathfrak{X}_{\Gamma_i}$ has finite volume). If $\mathfrak{X}_{\Gamma_1}$ and $\mathfrak{X}_{\Gamma_2}$ are length-commensurable then Γ_1 and Γ_2 are weakly commensurable.*

The results we discussed in this section shift the focus in the analysis of length-commensurability and/or isospectrality of locally symmetric spaces to that of weak commensurability of finitely generated Zariski-dense subgroups of simple (or semisimple) algebraic groups. In Section 3, we will first present some basic results dealing with the weak commensurability of such subgroups in a completely general situation, one of which states that the mere existence of such subgroups implies that the ambient algebraic groups either are of the same type, or one of them is of type B_n and the other of type C_n for some $n \geqslant 3$ (Theorem 3.1). We then turn to much more precise results in the case where the algebraic groups are of the same type and the subgroups are S-arithmetic (see Section 4), and finally derive some geometric consequences of these results (see Section 5). Next, Section 7 contains an exposition of the recent results of [Garibaldi and Rapinchuk 2013], which completely characterize weakly commensurable S-arithmetic subgroups in the case where one of the two groups is of type B_n and the other is of type C_n ($n \geqslant 3$). In Section 8 we discuss a more technical version of the notion of weak commensurability, which enabled us to show in [Prasad and Rapinchuk 2013] (under mild technical assumptions) that if two arithmetically defined locally symmetric spaces $M_1 = \mathfrak{X}_{\Gamma_1}$ and $M_2 = \mathfrak{X}_{\Gamma_2}$ are not length-commensurable then the sets $L(M_1)$ and $L(M_2)$ (or $\mathbb{Q} \cdot L(M_1)$ and $\mathbb{Q} \cdot (M_2)$) are *very* different. The proofs of all these results use the existence (first established in [Prasad and Rapinchuk 2003]) of special elements, which we call *generic elements*, in arbitrary finitely generated Zariski-dense subgroups; we briefly review these and more recent results in this direction in Section 9 along with the results that relate the analysis of weak commensurability with a problem of independent interest in the theory of semisimple algebraic groups of characterizing simple K-groups having the same isomorphism classes of maximal K-tori (Section 6). Finally, in Section 10 we discuss some open problems.

3. Two basic results implied by weak commensurability and the definition of arithmeticity

Our next goal is to give an account of the results from [Prasad and Rapinchuk 2009] concerning weakly commensurable subgroups of semisimple algebraic groups. We begin with the following two theorems that provide the basic results about weak commensurability of arbitrary finitely generated Zariski-dense subgroups of semisimple groups.

Theorem 3.1. *Let G_1 and G_2 be two connected absolutely almost simple algebraic groups defined over a field F of characteristic zero. Assume that there exist finitely generated Zariski-dense subgroups Γ_i of $G_i(F)$ which are weakly commensurable. Then either G_1 and G_2 are of the same Killing–Cartan type, or one of them is of type B_n and the other is of type C_n for some $n \geqslant 3$.*

The way we prove this theorem is by showing that the Weyl groups of G_1 and G_2 have the same order, as it is well known that the order of the Weyl group uniquely determines the type of the root system, except for the ambiguity between B_n and C_n. On the other hand, groups G_1 and G_2 of types B_n and C_n with $n > 2$ respectively, may indeed contain weakly commensurable subgroups. This was first shown in [Prasad and Rapinchuk 2009, Example 6.7] using a cohomological construction which we will briefly recall in Section 7. Recently in [Garibaldi and Rapinchuk 2013] another explanation was given using commutative étale subalgebras of simple algebras with involution. We refer the reader to Section 7 for this argument as well as a complete characterization of weakly commensurable S-arithmetic subgroups in the algebraic groups of types B_n and C_n (see Theorem 7.2).

Theorem 3.2. *Let G_1 and G_2 be two connected absolutely almost simple algebraic groups defined over a field F of characteristic zero. For $i = 1, 2$, let Γ_i be a finitely generated Zariski-dense subgroup of $G_i(F)$, and K_{Γ_i} be the subfield of F generated by the traces $\operatorname{Tr}\operatorname{Ad}\gamma$, in the adjoint representation, of $\gamma \in \Gamma_i$. If Γ_1 and Γ_2 are weakly commensurable, then $K_{\Gamma_1} = K_{\Gamma_2}$.*

We now turn to the results concerning weakly commensurable Zariski-dense S-arithmetic subgroups, which are surprisingly strong. In Section 4 we will discuss the weak commensurability of S-arithmetic subgroups in absolutely almost simple algebraic groups G_1 and G_2 of the same type, postponing the case where one of the groups is of type B_n and the other of type C_n to Section 7. Since our results rely on a specific way of describing S-arithmetic subgroups in absolutely almost simple groups, we will discuss this issue first.

3.3. ***The definition of arithmeticity.*** Let G be an algebraic group defined over a number field K, and let S be a finite subset of the set V^K of all places of K containing the set V_∞^K of archimedean places. Fix a K-embedding $G \subset \mathrm{GL}_N$, and consider the group of S-integral points

$$G(\mathcal{O}_K(S)) := G \cap \mathrm{GL}_N(\mathcal{O}_K(S)).$$

Then, for any field extension F/K, the subgroups of $G(F)$ that are commensurable[1] with $G(\mathcal{O}_K(S))$ are called *S-arithmetic*, and in the case where $S = V_\infty^K$ simply *arithmetic* (note that $\mathcal{O}_K(V_\infty^K) = \mathcal{O}_K$, the ring of algebraic integers in K). It is well known (see, for example, [Platonov and Rapinchuk 1994]) that the resulting class of S-arithmetic subgroups does not depend on the choice of K-embedding $G \subset \mathrm{GL}_N$. The question, however, is what we should mean by an arithmetic subgroup of $G(F)$ when G is an algebraic group defined over a field F of characteristic zero that is not equipped with a structure of K-group over some

[1] We recall that two subgroups $\mathcal{H}_1$ and $\mathcal{H}_2$ of an abstract group $\mathcal{G}$ are called *commensurable* if their intersection $\mathcal{H}_1 \cap \mathcal{H}_2$ is of finite index in each of the subgroups.

number field $K \subset F$. For example, what is an arithmetic subgroup of $G(\mathbb{R})$ where $G = \mathrm{SO}_3(f)$ and $f = x^2 + ey^2 - \pi z^2$? For absolutely almost simple groups the "right" concept that we will formalize below is given in terms of the forms of G over the subfields $K \subset F$ that are number fields. In our example, we can consider the following rational quadratic forms that are equivalent to f over $\mathbb{R}$:

$$f_1 = x^2 + y^2 - 3z^2 \quad \text{and} \quad f_2 = x^2 + 2y^2 - 7z^2,$$

and set $G_i = \mathrm{SO}_3(f_i)$. Then for each $i = 1, 2$, we have an $\mathbb{R}$-isomorphism $G_i \simeq G$, so the natural arithmetic subgroup $G_i(\mathbb{Z}) \subset G_i(\mathbb{R})$ can be thought of as an "arithmetic" subgroup of $G(\mathbb{R})$. Furthermore, one can consider quadratic forms over other number subfields $K \subset \mathbb{R}$ that again become equivalent to f over $\mathbb{R}$; for example,

$$K = \mathbb{Q}(\sqrt{2}) \quad \text{and} \quad f_3 = x^2 + y^2 - \sqrt{2}z^2.$$

Then for $G_3 = \mathrm{SO}_3(f_3)$, there is an $\mathbb{R}$-isomorphism $G_3 \simeq G$ which allows us to view the natural arithmetic subgroup $G_3(\mathcal{O}_K) \subset G_3(\mathbb{R})$, where $\mathcal{O}_K = \mathbb{Z}[\sqrt{2}]$, as an "arithmetic" subgroup of $G(\mathbb{R})$. One can easily generalize such constructions from arithmetic to S-arithmetic groups by replacing the rings of integers with the rings of S-integers. So, generally speaking, by an S-arithmetic subgroup of $G(\mathbb{R})$ we mean a subgroup which is commensurable to one of the subgroups obtained through this construction for some choice of a number subfield $K \subset \mathbb{R}$, a finite set S of places of K containing all the archimedean ones, and a quadratic form $\tilde{f}$ over K that becomes equivalent to f over $\mathbb{R}$. The technical definition is as follows.

Let G be a connected absolutely almost simple algebraic group defined over a field F of characteristic zero, $\overline{G}$ be its adjoint group, and $\pi\colon G \to \overline{G}$ be the natural isogeny. Suppose we are given the following data:

- a *number field* K with a *fixed* embedding $K \hookrightarrow F$;
- an F/K-form $\mathcal{G}$ of $\overline{G}$, which is an algebraic K-group such that there exists an F-isomorphism ${}_F\mathcal{G} \simeq \overline{G}$, where ${}_F\mathcal{G}$ is the group obtained from $\mathcal{G}$ by the extension of scalars from K to F;
- a finite set S of places of K containing V_∞^K but not containing any nonarchimedean places v such that $\mathcal{G}$ is K_v-anisotropic.[2]

We then have an embedding $\iota\colon \mathcal{G}(K) \hookrightarrow \overline{G}(F)$ which is well defined up to an F-automorphism of $\overline{G}$ (note that we do *not* fix an isomorphism ${}_F\mathcal{G} \simeq \overline{G}$). A subgroup Γ of $G(F)$ such that $\pi(\Gamma)$ is commensurable with $\sigma(\iota(\mathcal{G}(\mathcal{O}_K(S))))$, for

[2] We note that if $\mathcal{G}$ is K_v-anisotropic then $\mathcal{G}(\mathcal{O}_K(S))$ and $\mathcal{G}(\mathcal{O}_K(S \cup \{v\}))$ are commensurable, and therefore the classes of S- and $(S \cup \{v\})$-arithmetic subgroups coincide. Thus, this assumption on S is necessary if we want to recover it from a given S-arithmetic subgroup.

some F-automorphism σ of $\overline{G}$, will be called a *$(\mathscr{G}, K, S)$-arithmetic subgroup*,[3] or an S-arithmetic subgroup described in terms of the triple $(\mathscr{G}, K, S)$. As usual, $(\mathscr{G}, K, V_\infty^K)$-arithmetic subgroups will simply be called $(\mathscr{G}, K)$-arithmetic.

We also need to introduce a more general notion of commensurability. The point is that since weak commensurability is defined in terms of eigenvalues, a subgroup $\Gamma \subset G(F)$ is weakly commensurable with any conjugate subgroup, while the latter may not be commensurable with the former in the usual sense. So, to make theorems asserting that in certain situations *weak commensurability implies commensurability* possible (and such theorems are in fact one of the goals of our analysis) one definitely needs to modify the notion of commensurability. The following notion works well in geometric applications. Let G_i, for $i = 1, 2$, be a connected absolutely almost simple F-group, and let $\pi_i \colon G_i \to \overline{G}_i$ be the isogeny onto the corresponding adjoint group. We will say that the subgroups Γ_i of $G_i(F)$ are *commensurable up to an F-isomorphism between $\overline{G}_1$ and $\overline{G}_2$* if there exists an F-isomorphism $\sigma \colon \overline{G}_1 \to \overline{G}_2$ such that $\sigma(\pi_1(\Gamma_1))$ is commensurable with $\pi(\Gamma_2)$ in the usual sense. The key observation is that the description of S-arithmetic subgroups in terms of triples $(\mathscr{G}, K, S)$ is very convenient for determining when two such subgroups are commensurable in the new generalized sense.

Proposition 3.4. *Let G_1 and G_2 be connected absolutely almost simple algebraic groups defined over a field F of characteristic zero, and for $i = 1, 2$, let Γ_i be a Zariski-dense $(\mathscr{G}_i, K_i, S_i)$-arithmetic subgroup of $G_i(F)$. Then Γ_1 and Γ_2 are commensurable up to an F-isomorphism between $\overline{G}_1$ and $\overline{G}_2$ if and only if $K_1 = K_2 =: K$, $S_1 = S_2$, and $\mathscr{G}_1$ and $\mathscr{G}_2$ are K-isomorphic.*

It follows from the above proposition that the arithmetic subgroups Γ_1, Γ_2, and Γ_3 constructed above, of $G(\mathbb{R})$, where $G = \mathrm{SO}_3(f)$, are pairwise noncommensurable: indeed, Γ_3, being defined over $\mathbb{Q}(\sqrt{2})$, cannot possibly be commensurable to Γ_1 or Γ_2 as these two groups are defined over $\mathbb{Q}$; at the same time, the noncommensurability of Γ_1 and Γ_2 is a consequence of the fact that $\mathrm{SO}_3(f_1)$ and $\mathrm{SO}_3(f_2)$ are not $\mathbb{Q}$-isomorphic since the quadratic forms f_1 and f_2 are not equivalent over $\mathbb{Q}$.

4. Results on weakly commensurable S-arithmetic subgroups

In view of Proposition 3.4, the central question in the analysis of weak commensurability of S-arithmetic subgroups is the following: *Suppose we are given two Zariski-dense S-arithmetic subgroups that are described in terms of triples. Which components of these triples coincide given the fact that the subgroups*

[3]This notion of arithmetic subgroups coincides with that in Margulis' book [1991] for absolutely simple adjoint groups.

are weakly commensurable? As the following result demonstrates, two of these components *must* coincide.

Theorem 4.1. *Let G_1 and G_2 be two connected absolutely almost simple algebraic groups defined over a field F of characteristic zero. If Zariski-dense $(\mathscr{G}_i, K_i, S_i)$-arithmetic subgroups Γ_i of $G_i(F)$, where $i = 1, 2$, are weakly commensurable for $i = 1, 2$, then $K_1 = K_2$ and $S_1 = S_2$.*

In general, the forms $\mathscr{G}_1$ and $\mathscr{G}_2$ do not have to be K-isomorphic (see [Prasad and Rapinchuk 2009], Examples 6.5 and 6.6, as well as the general construction in Section 9). In the next theorem we list the cases where it can nevertheless be asserted that $\mathscr{G}_1$ and $\mathscr{G}_2$ are necessarily K-isomorphic, and then give a general finiteness result for the number of K-isomorphism classes.

Theorem 4.2. *Let G_1 and G_2 be two connected absolutely almost simple algebraic groups defined over a field F of characteristic zero, of the* same *type different from A_n, D_{2n+1}, with $n > 1$, or E_6. If for $i = 1, 2$, $G_i(F)$ contain Zariski-dense weakly commensurable $(\mathscr{G}_i, K, S)$-arithmetic subgroups Γ_i, then $\mathscr{G}_1 \simeq \mathscr{G}_2$ over K, and hence Γ_1 and Γ_2 are commensurable up to an F-isomorphism between $\overline{G}_1$ and $\overline{G}_2$.*

In this theorem, type D_{2n} ($n \geqslant 2$) required special consideration. The case $n > 2$ was settled in [Prasad and Rapinchuk 2010a] using the techniques of [Prasad and Rapinchuk 2009] in conjunction with results on embeddings of fields with involutive automorphisms into simple algebras with involution. The remaining case of type D_4 was treated by Skip Garibaldi [2012], whose argument actually applies to all n and explains the result from the perspective of Galois cohomology, providing thereby a cohomological insight into the difference between the types D_{2n} and D_{2n+1}. We note that the types excluded in the theorem are precisely the types for which the automorphism $\alpha \mapsto -\alpha$ of the corresponding root system is not in the Weyl group. More importantly, all these types are honest exceptions to the theorem — a general Galois-cohomological construction of weakly commensurable, but not commensurable, Zariski-dense S-arithmetic subgroups for all of these types is given in [Prasad and Rapinchuk 2009, Section 9].

Theorem 4.3. *Let G_1 and G_2 be two connected absolutely almost simple groups defined over a field F of characteristic zero. Let Γ_1 be a Zariski-dense $(\mathscr{G}_1, K, S)$-arithmetic subgroup of $G_1(F)$. Then the set of K-isomorphism classes of K-forms $\mathscr{G}_2$ of $\overline{G}_2$ such that $G_2(F)$ contains a Zariski-dense $(\mathscr{G}_2, K, S)$-arithmetic subgroup weakly commensurable to Γ_1 is finite.*

In other words, the set of all Zariski-dense (K, S)-arithmetic subgroups of $G_2(F)$ which are weakly commensurable to a given Zariski-dense (K, S)-arithmetic subgroup is a union of finitely many commensurability classes.

A noteworthy fact about weak commensurability is that it has the following implication for the existence of unipotent elements in arithmetic subgroups (even though it is formulated entirely in terms of semisimple ones). We recall that a semisimple K-group is called *K-isotropic* if $\mathrm{rk}_K\, G > 0$; in characteristic zero, this is equivalent to the existence of nontrivial unipotent elements in $G(K)$. Moreover, if K is a number field then G is K-isotropic if and only if every S-arithmetic subgroup contains unipotent elements, for any S.

Theorem 4.4. *Let G_1 and G_2 be two connected absolutely almost simple algebraic groups defined over a field F of characteristic zero. For $i = 1, 2$, let Γ_i be a Zariski-dense $(\mathcal{G}_i, K, S)$-arithmetic subgroup of $G_i(F)$. If Γ_1 and Γ_2 are weakly commensurable then $\mathrm{rk}_K \mathcal{G}_1 = \mathrm{rk}_K \mathcal{G}_2$; in particular, if $\mathcal{G}_1$ is K-isotropic, then so is $\mathcal{G}_2$.*

We note that in [Prasad and Rapinchuk 2009, Section 7] we prove a somewhat more precise result, viz. that if G_1 and G_2 are of the same type, then the Tits indices of $\mathcal{G}_1/K$ and $\mathcal{G}_2/K$ are isomorphic, but we will not get into these technical details here.

The following result asserts that a lattice[4] which is weakly commensurable with an S-arithmetic group is itself S-arithmetic.

Theorem 4.5. *Let G_1 and G_2 be two connected absolutely almost simple algebraic groups defined over a nondiscrete locally compact field F of characteristic zero, and for $i = 1, 2$, let Γ_i be a Zariski-dense lattice in $G_i(F)$. Assume that Γ_1 is a (K, S)-arithmetic subgroup of $G_1(F)$. If Γ_1 and Γ_2 are weakly commensurable, then Γ_2 is a (K, S)-arithmetic subgroup of $G_2(F)$.*

5. Geometric applications

We are now in a position to give the precise statements of our results on isospectral and length-commensurable locally symmetric spaces. Throughout this subsection, for $i = 1, 2$, G_i will denote an absolutely simple real algebraic group and $\mathfrak{X}_i$ the symmetric space of $\mathcal{G}_i = G_i(\mathbb{R})$. Furthermore, given a discrete torsion-free subgroup $\Gamma_i \subset \mathcal{G}_i$, we let $\mathfrak{X}_{\Gamma_i} = \mathfrak{X}_i/\Gamma_i$ denote the corresponding locally symmetric space. The geometric results are basically obtained by combining Theorem 2.7 and Corollary 2.8 with the results on weakly commensurable subgroups from the previous section. It should be emphasized that when $\mathfrak{X}_{\Gamma_1}$ and $\mathfrak{X}_{\Gamma_2}$ are both rank.one spaces and we are not in the exceptional case $(\mathcal{E})$ (which is the case, for example, for all hyperbolic n-manifolds with $n \geqslant 4$) our

[4]A discrete subgroup Γ of a locally compact topological group $\mathcal{G}$ is said to be a lattice in $\mathcal{G}$ if $\mathcal{G}/\Gamma$ carries a *finite* $\mathcal{G}$-invariant Borel measure.

results are *unconditional*, while in all other cases they depend on the validity of Schanuel's conjecture.

Now, applying Theorems 3.1 and 3.2 we obtain the following.

Theorem 5.1. *Let G_1 and G_2 be connected absolutely simple real algebraic groups, and let $\mathfrak{X}_{\Gamma_i}$ be a locally symmetric space of finite volume, of $\mathcal{G}_i$, for $i = 1, 2$. If $\mathfrak{X}_{\Gamma_1}$ and $\mathfrak{X}_{\Gamma_2}$ are length-commensurable, then:*

(i) *either G_1 and G_2 are of same Killing–Cartan type, or one of them is of type B_n and the other is of type C_n for some $n \geqslant 3$;*

(ii) $K_{\Gamma_1} = K_{\Gamma_2}$.

It should be pointed out that assuming Schanuel's conjecture in all cases, one can prove this theorem (in fact, a much stronger statement; see Theorem 8.1) assuming only that Γ_1 and Γ_2 are finitely generated and Zariski-dense.

Next, using Theorems 4.2 and 4.3 we obtain

Theorem 5.2. *Let G_1 and G_2 be connected absolutely simple real algebraic groups, and let $\mathcal{G}_i = G_i(\mathbb{R})$, for $i = 1, 2$. Then the set of arithmetically defined locally symmetric spaces $\mathfrak{X}_{\Gamma_2}$ of $\mathcal{G}_2$, which are length-commensurable to a given arithmetically defined locally symmetric space $\mathfrak{X}_{\Gamma_1}$ of $\mathcal{G}_1$, is a union of finitely many commensurability classes. It in fact consists of a single commensurability class if G_1 and G_2 have the same type different from A_n, D_{2n+1}, with $n > 1$, or E_6.*

Furthermore, Theorems 4.4 and 4.5 imply the following rather surprising result which has so far defied all attempts of a purely geometric proof.

Theorem 5.3. *Let G_1 and G_2 be connected absolutely simple real algebraic groups, and let $\mathfrak{X}_{\Gamma_1}$ and $\mathfrak{X}_{\Gamma_2}$ be length-commensurable locally symmetric spaces of $\mathcal{G}_1$ and $\mathcal{G}_2$ respectively, of finite volume. Assume that at least one of the spaces is arithmetically defined. Then the other space is also arithmetically defined, and the compactness of one of the spaces implies the compactness of the other.*

In fact, if one of the spaces is compact and the other is not, the weak length spectra $L(\mathfrak{X}_{\Gamma_1})$ and $L(\mathfrak{X}_{\Gamma_2})$ are quite different. See Theorem 8.6 for a precise statement. (We note that the proof of this result uses Schanuel's conjecture in all cases.)

Finally, we will describe some applications to isospectral compact locally symmetric spaces. So, in the remainder of this section, the locally symmetric spaces $\mathfrak{X}_{\Gamma_1}$ and $\mathfrak{X}_{\Gamma_2}$ as above will be assumed to be *compact*. Then, as we discussed in Section 1, the fact that $\mathfrak{X}_{\Gamma_1}$ and $\mathfrak{X}_{\Gamma_2}$ are isospectral implies that $L(\mathfrak{X}_{\Gamma_1}) = L(\mathfrak{X}_{\Gamma_2})$, so we can use our results on length-commensurable spaces. Thus, in particular we obtain the following.

Theorem 5.4. *Assume $\mathfrak{X}_{\Gamma_1}$ and $\mathfrak{X}_{\Gamma_2}$ are isospectral. If Γ_1 is arithmetic, then so is Γ_2.*

Thus, the Laplace spectrum can see if the fundamental group is arithmetic or not — to our knowledge, no results of this kind, particularly for general locally symmetric spaces, were previously known in spectral theory.

The following theorem settles the question "Can one hear the shape of a drum?" for arithmetically defined compact locally symmetric spaces.

Theorem 5.5. *Let $\mathfrak{X}_{\Gamma_1}$ and $\mathfrak{X}_{\Gamma_2}$ be compact locally symmetric spaces associated with absolutely simple real algebraic groups G_1 and G_2, and assume that at least one of the spaces is arithmetically defined. If $\mathfrak{X}_{\Gamma_1}$ and $\mathfrak{X}_{\Gamma_2}$ are isospectral then $G_1 = G_2 := G$. Moreover, unless G is of type A_n, D_{2n+1} $(n > 1)$, or E_6, the spaces $\mathfrak{X}_{\Gamma_1}$ and $\mathfrak{X}_{\Gamma_2}$ are commensurable.*

It should be noted that our methods based on length-commensurability or weak commensurability leave room for the following ambiguity in the proof of Theorem 5.5: either $G_1 = G_2$ or G_1 and G_2 are $\mathbb{R}$-split forms of types B_n and C_n for some $n \geqslant 3$ - and this ambiguity is unavoidable; see the end of Section 7. The fact that in the latter case the locally symmetric spaces cannot be isospectral was shown by Sai-Kee Yeung [2011] by comparing the traces of the heat operator (without using Schanuel's conjecture), which leads to the statement of the theorem given above.

6. Absolutely almost simple algebraic groups having the same maximal tori

The analysis of weak commensurability is related to another natural problem in the theory of algebraic groups of characterizing absolutely almost simple K-groups having the same isomorphism/isogeny classes of maximal K-tori — the exact nature of this connection will be clarified in Theorem 9.8 and the subsequent discussion. Some aspects of this problem over local and global fields were considered in [Garge 2005] and [Kariyama 1989]. Another direction of research, which has already generated a number of results — we mention [Bayer-Fluckiger 2011; Garibaldi 2012; Lee 2012; Prasad and Rapinchuk 2010a] — is the investigation of local-global principles for embedding tori into absolutely almost simple algebraic groups as maximal tori (in particular, for embedding of commutative étale algebras with involutive automorphisms into simple algebras with involution); some number-theoretic applications of these results can be found, for example, in [Fiori 2012]. A detailed discussion of these issues would be an independent undertaking, so we will limit ourselves here to the following theorem:

Theorem 6.1 (Prasad and Rapinchuk 2009, Theorem 7.5; Garibaldi and Rapinchuk 2013, Proposition 1.3). (1) *Let G_1 and G_2 be connected absolutely almost simple algebraic groups defined over a number field K, and let L_i be the smallest Galois extension of K over which G_i becomes an inner form of a split group. If G_1 and G_2 have the same K-isogeny classes of maximal K-tori then either G_1 and G_2 are of the same Killing–Cartan type, or one of them is of type B_n and the other is of type C_n, and moreover, $L_1 = L_2$.*

(2) *Fix an absolutely almost simple K-group G. Then the set of isomorphism classes of all absolutely almost simple K-groups G' having the same K-isogeny classes of maximal K-tori is finite.*

(3) *Fix an absolutely almost simple simply connected K-group G whose Killing–Cartan type is different from A_n, D_{2n+1} $(n > 1)$ or E_6. Then any K-form G' of G (in other words, any absolutely almost simple simply connected K-group G' of the* same *type as G) that has the same K-isogeny classes of maximal K-tori as G, is isomorphic to G.*

The construction described in [Prasad and Rapinchuk 2009, Section 9] shows that the types excluded in Theorem 6.1(3) are honest exceptions; that is, for each of those types one can construct nonisomorphic absolutely almost simple simply connected K-groups G_1 and G_2 of this type over a number field K that have the same isomorphism classes of maximal K-tori. Furthermore, the analysis of the situation where G_1 and G_2 are of types B_n and C_n, respectively, over a number field K and have the same isomorphism/isogeny classes of maximal K-tori is given in Theorem 7.3 below (see [Garibaldi and Rapinchuk 2013, Theorem 1.4 and 1.5]).

Of course, the question about determining absolutely almost simple algebraic K-groups by their maximal K-tori makes sense over general fields. It is particularly interesting for division algebra where it can be reformulated as the following question which is somewhat reminiscent of Amitsur's famous theorem on generic splitting fields [Amitsur 1955; Gille and Szamuely 2006]: *What can one say about two finite-dimensional central division algebras D_1 and D_2 over the same field K given the fact that they have the same isomorphism classes of maximal subfields?* For recent results on this problem, see [Chernousov et al. 2012; Garibaldi and Saltman 2010; Krashen and McKinnie 2011; Rapinchuk and Rapinchuk 2010].

7. Weakly commensurable subgroups in groups of types B and C

Let G_1 and G_2 be absolutely almost simple algebraic groups over a field K of characteristic zero. According to Theorem 3.1, finitely generated weakly commensurable Zariski-dense subgroups $\Gamma_1 \subset G_1(K)$ and $\Gamma_2 \subset G_2(K)$ can exist only if G_1 and G_2 are of the same Killing–Cartan type or one of them is of type B_n and the other is of type C_n for some $n \geqslant 3$. Moreover, the results we described in Section 4 provide virtually complete answers to the key questions about weakly commensurable S-arithmetic subgroups in the case where G_1 and G_2 are *of the same type*. In this section, we will discuss recent results [Garibaldi and Rapinchuk 2013] that determine weakly commensurable arithmetic subgroups when G_1 is *of type* B_n and G_2 is *of type* C_n ($n \geqslant 3$).

First of all, it should be pointed out that S-arithmetic subgroups in groups of types B_n and C_n can indeed be weakly commensurable. The underlying reason is that if G_1 is a split adjoint group of type B_n and G_2 is a split simply connected group of type C_n ($n \geqslant 2$) over any field K of characteristic $\neq 2$, then G_1 and G_2 have the *same isomorphism classes of maximal K-tori*. For the reader's convenience we briefly recall the Galois-cohomological proof of this fact given in [Prasad and Rapinchuk 2009, Example 6.7].

It is well known that for any semisimple K-group G there is a natural bijection between the set of $G(K)$-conjugacy classes of maximal K-tori of G and the set

$$\mathscr{C}_K := \mathrm{Ker}\big(H^1(K,N) \to H^1(K,G)\big),$$

where T is a maximal K-torus of G and N is the normalizer of T in G (see [Prasad and Rapinchuk 2009, Lemma 9.1]). Let $W = N/T$ be the corresponding Weyl group and introduce the following natural maps in Galois cohomology:

$$\theta_K \colon H^1(K,N) \to H^1(K,W) \quad \text{and} \quad \nu_K \colon H^1(K,W) \to H^1(K, \mathrm{Aut}\, T).$$

To apply these considerations to the groups G_1 and G_2, we will denote by T_i a fixed maximal K-split torus of G_i and let N_i, W_i, $\mathscr{C}_K^{(i)}$, $\theta_K^{(i)}$ and $\nu_K^{(i)}$ be the corresponding objects attached to G_i. It follows from an explicit description of the root systems of types B_n and C_n that there exist K-isomorphisms $\varphi \colon T_1 \to T_2$ and $\psi \colon W_1 \to W_2$ such that for the natural action of W_i on T_i we have

$$\varphi(w \cdot t) = \psi(w) \cdot \varphi(t), \quad \text{for all } t \in T_1, w \in W_1.$$

Since G_i is K-split, we have $\theta_K^{(i)}(\mathscr{C}_K^{(i)}) = H^1(K, W_i)$ (see [Gille 2004; Kottwitz 1982; Raghunathan 2004]). So, ψ induces a natural bijection between $\theta_K^{(1)}(\mathscr{C}_K^{(1)})$ and $\theta_K^{(2)}(\mathscr{C}_K^{(2)})$. Finally, we observe that if S_i is a maximal K-torus of G_i in the $G_i(K)$-conjugacy class corresponding to $c_i \in \mathscr{C}_K^{(i)}$, then the K-isomorphism class of S_i is determined by $\nu_K^{(i)}(\theta_K^{(i)}(c_i))$, and if $\psi(\theta_K^{(1)}(c_1)) = \theta_K^{(2)}(c_2)$ then S_1 and S_2 are K-isomorphic. It follows that G_1 and G_2 have the same isomorphism classes of maximal K-tori, as required.

Subsequently, in [Garibaldi and Rapinchuk 2013] a more explicit explanation of this fact was given. More precisely, let A be a central simple algebra over K with a K-linear involution τ (involution of the first kind). We recall that τ is called *orthogonal* if $\dim_K A^\tau = n(n+1)/2$ and *symplectic* if $\dim_K A^\tau = n(n-1)/2$. Furthermore, if τ is orthogonal and $n = 2m+1$ ($m \geqslant 2$) then $A = M_n(K)$ and the corresponding algebraic group $G = \mathrm{SU}(A, \tau)$ coincides with the orthogonal group $\mathrm{SO}_n(q)$ of a nondegenerate n-dimensional quadratic form $q = q_\tau$ over K, hence is a simple adjoint algebraic K-group of type B_m (note that the K-rank of G equals the Witt index of q). If τ is symplectic then necessarily $n = 2m$ and $G = \mathrm{SU}(A, \tau)$ is an almost simple simply connected K-group of type C_m; moreover G is K-split if and only if $A = M_n(K)$, in which case G is of course isomorphic to Sp_{2m}. Next, in all cases, any maximal K-torus T of G has the form $T = \mathrm{SU}(E, \sigma)$ where E is a τ-invariant n-dimensional commutative étale K-subalgebra of A such that for $\sigma = \tau | E$ we have

$$\dim E^\sigma = \left[\frac{n+1}{2}\right]. \tag{7}$$

So, the question whether $G = \mathrm{SU}(A, \tau)$, with A and τ as above, has a maximal K-torus of a specific type can be reformulated as follows: Let (E, σ) be an n-dimensional commutative étale K-algebra with an involutive K-automorphism σ satisfying (7). When does there exist an embedding $(E, \sigma) \hookrightarrow (A, \tau)$ as algebras with involution? While in the general case this question is nontrivial (see [Prasad and Rapinchuk 2010a]), the answer in the case where the group G splits over K is quite straightforward.

Proposition 7.1 [Garibaldi and Rapinchuk 2013, 2.3 and 2.5]. *Let $A = M_n(K)$ with a K-linear involution τ, and let (E, σ) be an n-dimensional commutative étale K-algebra with involution satisfying* (7). *In each of the following situations, there exists a K-embedding $(E, \sigma) \hookrightarrow (A, \tau)$:*

(1) *τ is symplectic.*

(2) *$n = 2m+1$ and τ is orthogonal such that the corresponding quadratic form q_τ has Witt index m.*

(This proposition should be viewed as an analogue for algebras with involution of the following result of Steinberg [1965]: Let G_0 be a *quasisplit* simply connected almost simple algebraic group over a field K. Then given an inner form G of G_0, *any maximal K-torus T of G admits a K-embedding into G_0*. While the proof of this result is rather technical, the proof of Proposition 7.1, as well as of the corresponding assertion for the algebras with involution arising in the description of algebraic groups of type A and D is completely elementary; see [Garibaldi and Rapinchuk 2013, Section 2].)

Now, fix $n \geqslant 2$ and let $A_1 = M_{n_1}(K)$, where $n_1 = 2n+1$, with an orthogonal involution τ_1 such that the Witt index of the corresponding quadratic form q_{τ_1} is n, and let $A_2 = M_{n_2}(K)$, where $n_2 = 2n$, with a symplectic involution τ_2. According to Proposition 7.1, for $i = 1, 2$, the maximal K-tori of $G_i = \mathrm{SU}(A_i, \tau_i)$ are of the form $T_i = \mathrm{SU}(E_i, \sigma_i)$ for a commutative étale K-algebra E_i of dimension n_i with an involution σ_i satisfying (7). On the other hand, the correspondence

$$(E_2, \sigma_2) \mapsto (E_1, \sigma_1) := (E_2 \times K\,,\, \sigma_2 \times \mathrm{id}_K)$$

defines a natural bijection between the isomorphism classes of commutative étale K-algebras with involution satisfying (7), of dimension n_2 and n_1, respectively. Since $\mathrm{SU}(E_1, \sigma_1) = \mathrm{SU}(E_2, \sigma_2)$ in these notations, we again obtain that G_1 and G_2 have the same isomorphism classes of maximal K-tori (see [Garibaldi and Rapinchuk 2013, Remark 2.6]).

Now, let K be a number field and S be any finite set of places of K containing the set V_∞^K of archimedean places. Furthermore, let G_1 be a split adjoint K-group of type B_n, and G_2 be a split simply connected K-group of type C_n $(n \geqslant 2)$. Then the fact, discussed above, that G_1 and G_2 have the same isomorphism classes of maximal K-tori immediately implies that the S-arithmetic subgroups in G_1 and G_2 are weakly commensurable (see [Prasad and Rapinchuk 2009, Examples 6.5 and 6.7]).

A complete determination of weakly commensurable S-arithmetic subgroups in algebraic groups G_1 and G_2 of types B_n and C_n $(n \geqslant 3)$ respectively was recently obtained by Skip Garibaldi and the second-named author [Garibaldi and Rapinchuk 2013]. To formulate the result we need the following definition. Let $\mathscr{G}_1$ and $\mathscr{G}_2$ be absolutely almost simple algebraic groups of types B_n and C_n with $n \geqslant 2$, respectively, over a number field K. We say that $\mathscr{G}_1$ and $\mathscr{G}_2$ are *twins* (over K) if for each place v of K, either both groups are split or both are anisotropic over the completion K_v. (We note that since groups of these types cannot be anisotropic over K_v when v is nonarchimedean, our condition effectively says that $\mathscr{G}_1$ and $\mathscr{G}_2$ must be K_v-split for *all* nonarchimedean v.)

Theorem 7.2 [Garibaldi and Rapinchuk 2013, Theorem 1.2]. *Let G_1 and G_2 be absolutely almost simple algebraic groups over a field F of characteristic zero having Killing–Cartan types B_n and C_n $(n \geqslant 3)$ respectively, and let Γ_i be a Zariski-dense $(\mathscr{G}_i, K, S)$-arithmetic subgroup of $G_i(F)$ for $i = 1, 2$. Then Γ_1 and Γ_2 are weakly commensurable if and only if the groups $\mathscr{G}_1$ and $\mathscr{G}_2$ are twins.*

(We recall that according to Theorem 4.1, if Zariski-dense $(\mathscr{G}_1, K_1, S_1)$- and $(\mathscr{G}_2, K_2, S_2)$-arithmetic subgroups are weakly commensurable then necessarily $K_1 = K_2$ and $S_1 = S_2$, so Theorem 7.2 in fact treats the most general situation.)

The necessity is proved using generic tori (Section 9) in conjunction with the analysis of maximal tori in real groups of types B_n and C_n (which can also be found in [Đoković and Thăńg 1994]). The proof of sufficiency is obtained using the above description of maximal K-tori in terms of commutative étale K-subalgebras with involution and the local-global results for the existence of an embedding of commutative étale algebras with involution into simple algebras with involution established in [Prasad and Rapinchuk 2010a]; an alternative argument along the lines outlined in the beginning of this section can be given using Galois cohomology of algebraic groups (see [Garibaldi and Rapinchuk 2013, Section 9]).

As we already mentioned in Section 6, the analysis of weak commensurability involved in the proof of Theorem 7.2 leads to, and at the same time depends on, the following result describing when groups of types B_n and C_n have the same isogeny/isomorphism classes of maximal K-tori.

Theorem 7.3 [Garibaldi and Rapinchuk 2013, Theorem 1.4]. *Let G_1 and G_2 be absolutely almost simple algebraic groups over a number field K of types B_n and C_n respectively for some $n \geqslant 3$.*

(1) *The groups G_1 and G_2 have the same* isogeny *classes of maximal K-tori if and only if they are twins.*

(2) *The groups G_1 and G_2 have the same* isomorphism *classes of maximal K-tori if and only if they are twins, G_1 is adjoint, and G_2 is simply connected.*

Theorem 7.2 has the following interesting geometric applications [Prasad and Rapinchuk 2013]. Again, let G_1 and G_2 be simple real algebraic groups of types B_n and C_n respectively. For $i = 1, 2$, let Γ_i be a discrete torsion-free $(\mathcal{G}_i, K)$-arithmetic subgroup of $\mathcal{G}_i = G_i(\mathbb{R})$, and let $\mathfrak{X}_{\Gamma_i}$ be the corresponding locally symmetric space of $\mathcal{G}_i$. Then if $\mathcal{G}_1$ and $\mathcal{G}_2$ are not twins, the locally symmetric spaces $\mathfrak{X}_{\Gamma_1}$ and $\mathfrak{X}_{\Gamma_2}$ are *not* length-commensurable. As one application of this result, we would like to point out the following assertion: *Let M_1 be an arithmetic quotient of the real hyperbolic space $\mathbb{H}^p$ ($p \geqslant 5$) and M_2 be an arithmetic quotient of the quaternionic hyperbolic space $\mathbb{H}^q_{\mathbf{H}}$ ($q \geqslant 2$). Then M_1 and M_2 are not length-commensurable.* The results of [Garibaldi and Rapinchuk 2013] are used to handle the case $p = 2n$ and $q = n - 1$ for some $n \geqslant 3$; for other values of p and q, the claim follows from Theorem 5.1.

On the other hand, suppose $G_1 = \mathrm{SO}(n+1, n)$ and $G_2 = \mathrm{Sp}_{2n}$ over $\mathbb{R}$ (i.e., G_1 and G_2 are the $\mathbb{R}$-split forms of types B_n and C_n, respectively) for some $n \geqslant 3$. Furthermore, let Γ_i be a discrete torsion-free $(\mathcal{G}_i, K)$-arithmetic subgroup of $\mathcal{G}_i$ for $i = 1, 2$, and let $M_i = \mathfrak{X}_{\Gamma_i}$. If $\mathcal{G}_1$ and $\mathcal{G}_2$ are twins then

$$\mathbb{Q} \cdot L(M_2) = \lambda \cdot \mathbb{Q} \cdot L(M_1), \quad \text{where } \lambda = \sqrt{\frac{2n+2}{2n-1}}.$$

Thus, despite the fact that they are associated with groups of different types, the locally symmetric spaces M_1 and M_2 can be made length-commensurable by scaling the metric on one of them; this, however, will *never* make them isospectral [Yeung 2011]. What is interesting is that so far this is the *only* situation in our analysis commensurability of isospectral and length-commensurable locally symmetric spaces where isospectrality manifests itself as an essentially stronger condition.

8. On the fields generated by the lengths of closed geodesics

In Section 5 (and also at the end of Section 7) we discussed the consequences of length-commensurability of two locally symmetric spaces M_1 and M_2; our focus in this section will be on the consequences of *nonlength-commensurability* of M_1 and M_2. More precisely, we will explore how different in this case the sets $L(M_1)$ and $L(M_2)$ (or $\mathbb{Q} \cdot L(M_1)$ and $\mathbb{Q} \cdot L(M_2)$) are and whether they can in fact be related in *any* reasonable way? Of course, one can ask a number of specific questions that fit this general perspective: for example, can $L(M_1)$ and $L(M_2)$ differ only in a finite number of elements, in other words, can the symmetric difference $L(M_1) \bigtriangleup L(M_2)$ be finite? Or can it happen that $\overline{\mathbb{Q}} \cdot L(M_1) = \overline{\mathbb{Q}} \cdot L(M_2)$, where $\overline{\mathbb{Q}}$ is the field of all algebraic numbers; in other words, can the use of the field $\overline{\mathbb{Q}}$ in place of $\mathbb{Q}$ in the definition of length-commensurability essentially change this relation? One relation between $L(M_1)$ and $L(M_2)$ that would make a lot of sense geometrically is that of similarity, requiring that there be a real number $\alpha > 0$ such that

$$L(M_2) = \alpha \cdot L(M_1) \quad (\text{or } \mathbb{Q} \cdot L(M_2) = \alpha \cdot \mathbb{Q} \cdot L(M_1)),$$

which means that M_1 and M_2 can be made isolength spectral (or length-commensurable) by scaling the metric on one of them. From the algebraic standpoint, one can generalize this relation by considering arbitrary polynomial relations between $L(M_1)$ and $L(M_2)$ instead of just linear relations although this perhaps does not have a clear geometric interpretation. To formalize this general idea, we need to introduce some additional notations and definitions.

For a Riemannian manifold M, we let $\mathcal{F}(M)$ denote the subfield of $\mathbb{R}$ generated by the set $L(M)$. Given two Riemannian manifolds M_1 and M_2, we set $\mathcal{F}_i = \mathcal{F}(M_i)$ for $i \in \{1, 2\}$ and consider the following condition:

(T_i) *The compositum $\mathcal{F}_1\mathcal{F}_2$ has infinite transcendence degree over the field $\mathcal{F}_{3-i}$.*

Informally, this condition means that $L(M_i)$ contains "many" elements which are algebraically independent from all the elements of $L(M_{3-i})$, implying the nonexistence of any nontrivial polynomial dependence between $L(M_1)$ and $L(M_2)$. In particular, (T_i) implies the following condition:

(N_i) $L(M_i) \not\subset A \cdot \mathbb{Q} \cdot L(M_{3-i})$ *for any finite set* A *of real numbers.*

In [Prasad and Rapinchuk 2013], we have proved a series of results asserting that if $M_i = \mathfrak{X}_{\Gamma_i}$ for $i = 1, 2$, are the quotients of symmetric spaces $\mathfrak{X}_i$ associated with absolutely simple real algebraic groups G_i by Zariski-dense discrete torsion-free subgroups $\Gamma_i \subset G_i(\mathbb{R})$, then in many situations the fact that M_1 and M_2 are not length-commensurable implies that conditions (T_i) and (N_i) hold for at least one $i \in \{1, 2\}$. To give precise formulations, in addition to the standard notations used earlier, we let w_i denote the order of the (absolute) Weyl group of G_i. We also need to emphasize that *all geometric results in* [Prasad and Rapinchuk 2013] *assume the validity of Shanuel's conjecture.* This assumption, however, enables one to establish results that are somewhat stronger than the corresponding results in Section 5 and do not require that Γ_1 and Γ_2 have property (A) (see Theorem 2.7 and Corollary 2.8). We begin with the following result which strengthens Theorem 5.1.

Theorem 8.1. *Assume that the Zariski-dense subgroups* Γ_1 *and* Γ_2 *are finitely generated (which is automatically the case if these subgroups are lattices).*

(1) *If* $w_1 > w_2$ *then* (T_1) *holds.*

(2) *If* $w_1 = w_2$ *but* $K_{\Gamma_1} \not\subset K_{\Gamma_2}$ *then again* (T_1) *holds.*

Thus, unless $w_1 = w_2$ *and* $K_{\Gamma_1} = K_{\Gamma_2}$, *the condition* (T_i) *holds for at least one* $i \in \{1, 2\}$; *in particular,* M_1 *and* M_2 *are not length-commensurable.*

As follows from Theorem 8.1, we only need to consider the case where $w_1 = w_2$, which we will assume — recall that this entails that either G_1 and G_2 are of the same Killing–Cartan type, or one of them is of type B_n and the other is of type C_n ($n \geqslant 3$). Then it is convenient to divide our results for *arithmetic* subgroups Γ_1 and Γ_2 into three theorems: the first one will treat the case where G_1 and G_2 are of the *same type* which is different from A_n, D_{2n+1} ($n > 1$) and E_6, the second one — the case where both G_1 and G_2 are one of the types A_n, D_{2n+1} ($n > 1$) and E_6, and the third one – the case where G_1 is of type B_n and G_2 is of type C_n for some $n \geqslant 3$.

Theorem 8.2. *With notations as above, assume that* G_1 *and* G_2 *are of the same Killing–Cartan type which is different from* A_n, D_{2n+1} ($n > 1$) *and* E_6 *and that the subgroups* Γ_1 *and* Γ_2 *are arithmetic. Then either* $M_1 = \mathfrak{X}_{\Gamma_1}$ *and* $M_2 = \mathfrak{X}_{\Gamma_2}$ *are commensurable (hence also length-commensurable), or conditions* (T_i) *and* (N_i) *hold for at least one* $i \in \{1, 2\}$.

(This theorem strengthens part of Theorem 5.2. We also note that (T_i) and (N_i) may not hold for *both* $i = 1$ and 2; in fact, it is possible that one of $L(M_1)$ and $L(M_2)$ is contained in the other.)

Theorem 8.3. *Again, keep the above notations and assume that the common Killing–Cartan type of G_1 and G_2 is one of the following: A_n, D_{2n+1} $(n > 1)$ or E_6 and that the subgroups Γ_1 and Γ_2 are arithmetic. Assume in addition that $K_{\Gamma_i} \neq \mathbb{Q}$ for at least one $i \in \{1, 2\}$. Then either M_1 and M_2 are length-commensurable (although not necessarily commensurable), or conditions (T_i) and (N_i) hold for at least one $i \in \{1, 2\}$.*

To illustrate possible applications of these theorems, we will give in Theorem 1.1 explicit statements for real hyperbolic manifolds; similar results are available for complex and quaternionic hyperbolic spaces.

Corollary 8.4. *Let M_i $(i = 1, 2)$ be the quotients of the real hyperbolic space $\mathbb{H}^{d_i}$ with $d_i \neq 3$ by a torsion-free Zariski-dense discrete subgroup Γ_i of $G_i(\mathbb{R})$, where $G_i = \mathrm{PSO}(d_i, 1)$.*

(i) *If $d_1 > d_2$ then conditions (T_1) and (N_1) hold.*

(ii) *If $d_1 = d_2$ but $K_{\Gamma_1} \not\subset K_{\Gamma_2}$ then again conditions (T_1) and (N_1) hold.*

Thus, unless $d_1 = d_2$ and $K_{\Gamma_1} = K_{\Gamma_2}$, conditions (T_i) and (N_i) hold for at least one $i \in \{1, 2\}$.

Assume now that $d_1 = d_2 =: d$ and the subgroups Γ_1 and Γ_2 are arithmetic.

(iii) *If d is either even or is congruent to* 3 (mod 4), *then either M_1 and M_2 are commensurable, hence length-commensurable, or (T_i) and (N_i) hold for at least one $i \in \{1, 2\}$.*

(iv) *If $d \equiv 1$ (mod 4) and in addition $K_{\Gamma_i} \neq \mathbb{Q}$ for at least one $i \in \{1, 2\}$ then either M_1 and M_2 are length-commensurable (although not necessarily commensurable), or conditions (T_i) and (N_i) hold for at least one $i \in \{1, 2\}$.*

Now, we consider the case where one of the groups is of type B_n and the other of type C_n $(n \geqslant 3)$. The theorem below strengthens the results of Section 7.

Theorem 8.5. *Notations as above, assume that G_1 is of type B_n and G_2 is of type C_n for some $n \geqslant 3$ and the subgroups Γ_1 and Γ_2 are arithmetic. Then either (T_i) and (N_i) hold for at least one $i \in \{1, 2\}$, or*

$$\mathbb{Q} \cdot L(M_2) = \lambda \cdot \mathbb{Q} \cdot L(M_1), \quad \textit{where } \lambda = \sqrt{\frac{2n+2}{2n-1}}.$$

The following interesting result holds for all types (cf. Theorem 5.3):

Theorem 8.6. *For $i = 1, 2$, let $M_i = \mathfrak{X}_{\Gamma_i}$ be an arithmetically defined locally symmetric space, and assume that $w_1 = w_2$. If M_2 is compact and M_1 is not, then conditions (T_1) and (N_1) hold.*

Finally, we have the following result, which shows that the notion of "length-similarity" for arithmetically defined locally symmetric spaces is redundant if G_1 and G_2 are of the same type (but compare Theorem 8.5 for the case where G_1 and G_2 are of types B_n and C_n).

Corollary 8.7. *Let $M_i = \mathfrak{X}_{\Gamma_i}$, for $i = 1, 2$, be arithmetically defined locally symmetric spaces. Assume that there exists $\lambda \in \mathbb{R}_{>0}$ such that*

$$\mathbb{Q} \cdot L(M_1) = \lambda \cdot \mathbb{Q} \cdot L(M_2).$$

(1) *If G_1 and G_2 are of the same type which is different from A_n, D_{2n+1} $(n > 1)$ and E_6, then M_1 and M_2 are commensurable, hence length-commensurable.*

(2) *If G_1 and G_2 are of the same type which is one of the following: A_n, D_{2n+1} $(n > 1)$ or E_6, then, provided that $K_{\Gamma_i} \neq \mathbb{Q}$ for at least one $i \in \{1, 2\}$, the spaces M_1 and M_2 are length-commensurable* (*although not necessarily commensurable*).

The proofs of the results in this section use a generalization of the notion of weak commensurability which we termed *weak containment.* To give a precise definition, we temporarily return to the general set-up where G_1 and G_2 are semisimple algebraic groups defined over a field F of characteristic zero, and Γ_i is a Zariski-dense subgroup of $G_i(F)$ for $i = 1, 2$.

Definition 8.8. (a) Semisimple elements

$$\gamma_1^{(1)}, \ldots, \gamma_{m_1}^{(1)} \in \Gamma_1$$

are *weakly contained* in Γ_2 if there are semisimple elements $\gamma_1^{(2)}, \ldots, \gamma_{m_2}^{(2)}$ such that

$$\chi_1^{(1)}(\gamma_1^{(1)}) \cdots \chi_{m_1}^{(1)}(\gamma_{m_1}^{(1)}) = \chi_1^{(2)}(\gamma_1^{(2)}) \cdots \chi_{m_2}^{(2)}(\gamma_{m_2}^{(2)}) \neq 1, \tag{8}$$

for some maximal F-tori $T_k^{(j)}$ of G_j whose group of F-rational points contains elements $\gamma_k^{(j)}$ and some characters $\chi_k^{(j)}$ of $T_k^{(j)}$ for $j \in \{1, 2\}$ and $k \leqslant m_j$.[5]

(b) Semisimple elements $\gamma_1^{(1)}, \ldots, \gamma_{m_1}^{(1)} \in \Gamma_1$ are *multiplicatively independent* if for some (equivalently, any) choice of maximal F-tori $T_i^{(1)}$ of G_1 such that $\gamma_i^{(1)} \in T_i(F)$ for $i \leqslant m_1$, a relation of the form

$$\chi_1^{(1)}(\gamma_1^{(1)}) \cdots \chi_{m_1}^{(1)}(\gamma_{m_1}^{(1)}) = 1,$$

[5]Note that (8) means that the subgroups of $\overline{F}^\times$ generated by the eigenvalues of $\gamma_1^{(1)}, \ldots, \gamma_{m_1}^{(1)}$ and by those of $\gamma_1^{(2)}, \ldots, \gamma_{m_2}^{(2)}$ for some (equivalently, any) matrix realizations of $G_1 \subset \mathrm{GL}_{N_1}$ and $G_2 \subset \mathrm{GL}_{N_2}$, intersect nontrivially.

where $\chi_i \in X(T_i)$ implies that

$$\chi_1^{(1)}(\gamma_1^{(1)}) = \cdots = \chi_{m_1}^{(1)}(\gamma_{m_1}) = 1.$$

(c) We say that Γ_1 and Γ_2 as above satisfy *property* (C_i), where $i = 1$ or 2, if for any $m \geqslant 1$ there exist semisimple elements $\gamma_1^{(i)}, \ldots, \gamma_m^{(i)} \in \Gamma_i$ of infinite order that are multiplicatively independent and are *not* weakly contained in Γ_{3-i}.

Using Schanuel's conjecture, we proved in [Prasad and Rapinchuk 2013, Corollary 7.3] that *if $\mathfrak{X}_{\Gamma_1}$ and $\mathfrak{X}_{\Gamma_2}$ are locally symmetric spaces as above with finitely generated Zariski-dense fundamental groups Γ_1 and Γ_2, then the fact that these groups satisfy property (C_i) for some $i \in \{1, 2\}$ implies that the locally symmetric spaces satisfy conditions (T_i) and (N_i) for the same i.* So, the way we prove Theorems 8.1–8.3 and 8.5–8.6 is by showing that condition (C_i) holds in the respective situations. For example, Theorem 8.1 is a consequence of the following algebraic result.

Theorem 8.9. *Assume that Γ_1 and Γ_2 are finitely generated (and Zariski-dense).*

(i) *If $w_1 > w_2$ then condition (C_1) holds.*

(ii) *If $w_1 = w_2$ but $K_{\Gamma_1} \not\subset K_{\Gamma_2}$ then again (C_1) hods.*

Thus, unless $w_1 = w_2$ and $K_{\Gamma_1} = K_{\Gamma_2}$, condition (C_i) holds for at least one $i \in \{1, 2\}$.

In [Prasad and Rapinchuk 2013] we prove much more precise results in the case where the Γ_i are arithmetic, which leads to the geometric applications described above. We refer the interested reader to [Prasad and Rapinchuk 2013] for the technical formulations of these results; a point we would like to make here, however, is that our "algebraic" results (i.e., those asserting that condition (C_i) holds in certain situations) do not depend on Schanuel's conjecture.

9. Generic elements and tori

The analysis of weak commensurability and its variations in [Garibaldi and Rapinchuk 2013; Prasad and Rapinchuk 2009; 2013] relies on the remarkable fact, first established in [Prasad and Rapinchuk 2003], that any Zariski-dense subgroup of the group of rational points of a semisimple group over a finitely generated field of characteristic zero contains special elements, to be called *generic elements* here. It is convenient to begin our discussion of these elements with the definition of *generic tori*.

Let G be a connected semisimple algebraic group defined over an infinite field K. Fix a maximal K-torus T of G, and, as usual, let $\Phi = \Phi(G, T)$ denote the corresponding root system, and let $W(G, T)$ be its Weyl group. Furthermore, we let K_T denote the (minimal) splitting field of T in a fixed separable closure $\overline{K}$

of K. Then the natural action of the Galois group $\mathrm{Gal}(K_T/K)$ on the character group $X(T)$ of T induces an injective homomorphism

$$\theta_T : \mathrm{Gal}(K_T/K) \to \mathrm{Aut}(\Phi(G,T)).$$

We say that T is *generic* (over K) if

$$\theta_T(\mathrm{Gal}(K_T/K)) \supset W(G,T). \tag{9}$$

For example, any maximal K-torus of $G = \mathrm{SL}_n/K$ is of the form

$$T = \mathrm{R}^{(1)}_{E/K}(\mathrm{GL}_1),$$

for some n-dimensional commutative étale K-algebra E. Then such a torus is generic over K if and only if E is a separable field extension of K and the Galois group of the normal closure L of E over K is isomorphic to the symmetric group S_n. It is well known that for each $n \geqslant 2$ one can write down a system of congruences such that any monic polynomial $f(t) \in \mathbb{Z}[t]$ satisfying this system of congruences has Galois group S_n. It turns out that one can prove a similar statement for maximal tori of an arbitrary semisimple algebraic group G over a finitely generated field K of characteristic zero (see [Prasad and Rapinchuk 2003, Theorem 3]). In order to avoid technical details, we will restrict ourselves here to the case of absolutely almost simple groups.

Theorem 9.1 [Prasad and Rapinchuk 2009, Theorem 3.1]. *Let G be a connected absolutely almost simple algebraic group over a finitely generated field K of characteristic zero, and let r be the number of nontrivial conjugacy classes of the Weyl group of G.*

(1) *There exist r inequivalent nontrivial discrete valuations $v_1, \dots, v_r$ of K such that the completion K_{v_i} is locally compact and G splits over K_{v_i} for all $i = 1, \dots, r$.*

(2) *For any choice of discrete valuations $v_1, \dots, v_r$ as in* (1), *one can find maximal K_{v_i}-tori $T(v_i)$ of G, one for each $i \in \{1, \dots, r\}$, with the property that any maximal K-torus T of G which is conjugate to $T(v_i)$ by an element of $G(K_{v_i})$, for all $i = 1, \dots, r$, is generic (i.e., the inclusion* (9) *holds).*

The first assertion is an immediate consequence of the following, which actually shows that we can find the v_j's so that $K_{v_j} = \mathbb{Q}_{p_j}$, where $p_1, \dots, p_r$ are distinct primes.

Proposition 9.2 [Prasad and Rapinchuk 2002; 2003]. *Let $\mathcal{K}$ be a finitely generated field of characteristic zero and $\mathcal{R} \subset \mathcal{K}$ a finitely generated subring. Then there exists an infinite set Π of primes such that for each $p \in \Pi$ there exists an embedding $\varepsilon_p : \mathcal{K} \hookrightarrow \mathbb{Q}_p$ with the property $\varepsilon_p(\mathcal{R}) \subset \mathbb{Z}_p$.*

To sketch a proof of the second assertion of Theorem 9.1, we fix a maximal K-torus T_0 of G. Given any other maximal torus T of G defined over an extension F of K there exists $g \in G(\bar{F})$ such that $T = \iota_g(T_0)$, where $\iota_g(x) = gxg^{-1}$. Then ι_g induces an isomorphism between the Weyl groups $W(G, T_0)$ and $W(G, T)$. A different choice of g will change this isomorphism by an inner automorphism of the Weyl group, implying that there is a *canonical bijection* between the sets $[W(G, T_0)]$ and $[W(G, T)]$ of conjugacy classes in the respective groups; we will denote this bijection by $\iota_{T_0,T}$.

Now, let v be a nontrivial discrete valuation of K such that the completion K_v is locally compact and splits T_0. Using the Frobenius automorphism of the maximal unramified extension K_v^{ur} in conjunction with the fact that $H^1(K_v, \widetilde{G})$, where $\widetilde{G}$ is the simply connected cover of G, vanishes (see [Bruhat and Tits 1987; Kneser 1965a; 1965b]),[6] one shows that given a nontrivial conjugacy class $c \in [W(G, T_0)]$, one can find a maximal K_v-torus $T(v, c)$ such that given any maximal K_v-torus T of G that is conjugate to $T(v, c)$ by an element of $G(K_v)$, for its splitting field K_{vT} we have

$$\theta_T(\mathrm{Gal}(K_{vT}/K_v)) \cap \iota_{T_0,T}(c) \neq \varnothing. \tag{10}$$

Now, if $v_1, \ldots, v_r$ are as in part (1), then using the weak approximation property of the variety of maximal tori (see [Platonov and Rapinchuk 1994, Corollary 7.3]), one can pick a maximal K-torus T_0 which splits over K_{v_i} for all $i = 1, \ldots, r$. Let $c_1, \ldots, c_r$ be the nontrivial conjugacy classes of $W(G, T_0)$. Set $T(v_i) = T(v_i, c_i)$ for $i = 1, \ldots, r$ in the above notation. Then it is not difficult to show that the tori $T(v_1), \ldots, T(v_r)$ are as required.

The method described above enables one to construct generic tori with various additional properties, in particular, having prescribed local behavior.

Corollary 9.3 [Prasad and Rapinchuk 2009, Corollary 3.2]. *Let G and K be as in Theorem 9.1, and let V be a finite set of inequivalent nontrivial rank* 1 *valuations of K. Suppose that for each $v \in V$ we are given a maximal K_v-torus $T(v)$ of G. Then there exists a maximal K-torus T of G for which* (9) *holds and which is conjugate to $T(v)$ by an element of $G(K_v)$, for all $v \in V$.*

It should be noted that the method of p-adic embeddings that we used in the proof of Theorem 9.1, and which is based on Proposition 9.2, has many other applications; see [Prasad and Rapinchuk 2010b].

---

[6] One can alternatively use the fact that if we endow $\widetilde{G}$ with the structure of a group scheme over $\mathbb{O}_v$ as a Chevalley group, then $H^1(K_v^{\mathrm{ur}}/K_v, \widetilde{G}(\mathbb{O}_v^{\mathrm{ur}}))$, where K_v^{ur} is the maximal unramified extension of K_v with the valuation ring $\mathbb{O}_v^{\mathrm{ur}}$, vanishes, which follows from Lang's theorem [1956] or its generalization due to Steinberg [1965]; see [Platonov and Rapinchuk 1994, Theorem 6.8].

We are now prepared to discuss *generic elements* whose existence in any finitely generated Zariski-dense subgroup is the core issue in this section.

Definition 9.4. Let G be a connected semisimple algebraic group defined over a field K. A regular semisimple element $g \in G(K)$ is called *generic* (over K) if the maximal torus $T := Z_G(g)^\circ$ is generic (over K). We shall refer to T as the torus associated with g.

Before proceeding with the discussion of generic elements, we would like to point out that some authors adopt a slight variant of this definition by requiring that the extension K_g of K generated by the eigenvalues of g be "generic", which is more consistent with the notion of a "generic polynomial" in Galois theory. We note that $K_g \subset K_T$ making the Galois group $\mathrm{Gal}(K_g/K)$ a *quotient* of the group $\mathrm{Gal}(K_T/K)$. Then the requirement that the order of $\mathrm{Gal}(K_g/K)$ be divisible by $|W(G,T)|$ (which is probably one of the most natural ways to express the "genericity" of K_g/K) a priori may not imply the inclusion (9), which is most commonly used in applications (although, by Lemma 4.1 of [Prasad and Rapinchuk 2009], the former does imply the latter if G is an inner form of a split group — in particular, G is absolutely almost simple of type different from A_n $(n > 1)$, D_n $(n \geqslant 4)$ and E_6). On the other hand, even for a regular element $g \in T(K)$, where T is a maximal generic K-torus of G, the field K_g may be strictly smaller than K_T. (Example: Let $G = \mathrm{PSL}_2$ over K, and let T be a maximal K-torus of G of the form $\mathrm{R}^{(1)}_{L/K}(\mathrm{GL}_1)$ where L/K is a quadratic extension; then an element $g \in T(K)$ of order two is regular with $K_g = K$, while $K_T = L$). This problem, however, does not arise if G is absolutely simple, T is generic and $g \in T(K)$ has infinite order. Indeed, then the K-torus T is *irreducible*; that is, it does not contain any proper K-subtori since $W(G,T)$ acts irreducibly on $X(T) \otimes_{\mathbb{Z}} \mathbb{Q}$ (cf. [Prasad and Rapinchuk 2001]). It follows that every element $g \in T(K)$ of infinite order generates a Zariski-dense subgroup of T, hence $K_g = K_T$, so the order of $\mathrm{Gal}(K_g/K)$ is divisible by $|W(G,T)|$. Recall that according to a famous result of Selberg [1972, Theorem 6.11], any finitely generated subgroup Γ of $G(K)$ contains a torsion-free subgroup Γ' of finite index, which therefore is also Zariski-dense. So, the following theorem (Theorem 9.6) that asserts the existence of generic elements in an arbitrary Zariski-dense subgroup in the sense of our definition also implies the existence of generic elements in the sense of the other definition.

9.5. Let K be a field and G a connected absolutely almost simple K-group. Let $g \in G(K)$ be a generic element and let $T = Z_G(g)^\circ$. Then $g \in T(K)$ (see [Borel 1991, Corollary 11.12]). As T does not contain proper K-subtori, the cyclic group generated by any t $(\in T(K))$ of infinite order is Zariski-dense in T. So $Z_G(t) = Z_G(T) = T$; which implies that t is generic (over K) and $Z_G(t)$ is connected; in particular, if g is of infinite order, then $Z_G(g)$ $(= T)$ is connected. Moreover, if $n \in G(K)$ is such that ntn^{-1} commutes with t, then n lies in the normalizer $N_G(T)(K)$ of T in $G(K)$. If x is an element of $G(K)$ of infinite order such that for some nonzero integer a, $t := x^a$ lies in $T(K)$, then as x commutes with t, it commutes with T and hence it lies in $T(K)$.

It is also clear that the tori associated to two generic elements are equal if and only if the elements commute.

It is easily seen that the natural action of $N_G(T)(K)$ on the character group $X(T)$ commutes with the natural action of $\mathrm{Gal}(K_T/K)$. Now since T is generic, $\theta_T(\mathrm{Gal}(K_T/K)) \supset W(G,T)$, and as $W(G,T)$ acts irreducibly on $X(T) \otimes_{\mathbb{Z}} \mathbb{C}$, we see that the elements of $N_G(T)(K)$ act by $\pm I$ on $X(T)$. Therefore, for $n \in N_G(T)(K)$, n^2 commutes with T and hence lies in $T(K)$. Now if $n \in G(K)$ is such that ngn^{-1} commutes with g, then n belongs to $N_G(T)(K)$ and $n^2 \in T(K)$. So, if, moreover, n is of infinite order, then it actually lies in $T(K)$. Thus an element of $G(K)$ of infinite order which does not belong to $T(K)$ cannot normalize T.

Theorem 9.6. *Let G be a connected absolutely almost simple algebraic group over a finitely generated field K, and let Γ be a finitely generated Zariski-dense subgroup of $G(K)$. Then Γ contains a generic element (over K) of infinite order.*

(It is not difficult to show, e.g., using Burnside's characterization of absolutely irreducible linear groups, that any Zariski-dense subgroup $\Gamma \subset G(K)$ contains a *finitely generated* Zariski-dense subgroup, so the assumption in the theorem that Γ be finitely generated can actually be omitted.)

Sketch of the proof. Fix a matrix K-realization $G \subset \mathrm{GL}_N$, and pick a finitely generated subring $R \subset K$ so that $\Gamma \subset G(R) := G \cap \mathrm{GL}_N(R)$. Let r be the number of nontrivial conjugacy classes in the Weyl group $W(G,T)$. Using Proposition 9.2, we can find r distinct primes $p_1, \dots, p_r$ such that for each $i \leqslant r$ there exists an embedding $\varepsilon_i \colon K \hookrightarrow \mathbb{Q}_{p_i}$ such that $\varepsilon_i(R) \subset \mathbb{Z}_{p_i}$ and G splits over $\mathbb{Q}_{p_i}$. Let v_i be the discrete valuation of K obtained as the pullback of the p_i-adic valuation of $\mathbb{Q}_{p_i}$ so that $K_{v_i} = \mathbb{Q}_{p_i}$. Pick maximal K_{v_i}-tori $T(v_1), \dots, T(v_r)$ as in part (2) of Theorem 9.1. Let Σ_i be the Zariski-open K_{v_i}-subvariety of regular elements in $T(v_i)$. It follows from the implicit function theorem that the image Ω_i of the map

$$G(K_{v_i}) \times \Sigma_i(K_{v_i}) \to G(K_{v_i}), \quad (g,t) \mapsto gtg^{-1},$$

is open in $G(K_{v_i})$ and intersects every open subgroup of the latter. On the other hand, as explained in [Rapinchuk 2014, Section 3], the closure of the image of the diagonal embedding $\Gamma \hookrightarrow \prod_{i=1}^r G(K_{v_i})$ is open, hence contains some $U = \prod_{i=1}^r U_i$, where $U_i \subset G(K_{v_i})$ is a Zariski-open torsion-free subgroup.

Then

$$U_0 := \prod_{i=1}^{r} (U_i \cap \Omega_i)$$

is an open set that intersects Γ, and it follows from our construction that any element $g \in \Gamma \cap U_0$ is a generic element of infinite order. □

Basically, our proof shows that given a finitely generated Zariski-dense subgroup Γ of $G(K)$, one can produce a finite system of congruences (defined in terms of suitable valuations of K) such that the set of elements $\gamma \in \Gamma$ satisfying this system of congruences consists entirely of generic elements (and additionally this set is in fact a coset of a finite index subgroup in Γ, in particular, it is Zariski-dense in G). Recently, Jouve, Kowalski and Zywina [Jouve et al. 2013], Gorodnik and Nevo [2011] and Lubotzky and Rosenzweig [2012] developed different *quantitative* ways of showing that generic elements exist in abundance (in fact, these results demonstrate that "most" elements in Γ are generic). More precisely, the result of [Gorodnik and Nevo 2011] gives the asymptotics of the number of generic elements of a given height in an arithmetic group, while the results of [Lubotzky and Rosenzweig 2012], generalizing earlier results of [Jouve et al. 2013], are formulated in terms of random walks on groups and apply to arbitrary Zariski-dense subgroups in not necessarily connected semisimple groups. These papers introduce several new ideas and techniques, but at the same time employ the elements of the argument from [Prasad and Rapinchuk 2003] we outlined above.

The proofs of the results in [Prasad and Rapinchuk 2009; 2013] use not only Theorem 9.6 itself but also its different variants that provide generic elements with additional properties — e.g., having prescribed local behavior (Corollary 9.3). We refer the interested reader to these papers for precise formulations (which are rather technical), and will only indicate here the basic "multidimensional" version of Theorem 9.6.

Theorem 9.7 [Prasad and Rapinchuk 2013, Theorem 3.4]. *Let G, K and $\Gamma \subset G(K)$ be as in Theorem 9.6. Then for any $m \geqslant 1$ one can find generic semisimple elements $\gamma_1, \dots, \gamma_m \in \Gamma$ of infinite order that are multiplicatively independent.*

Finally, we would like to formulate a result that enables one to pass from the weak commensurability of two generic semisimple elements to an isogeny, and in most cases even to an isomorphism, of the ambient tori. This result relates the analysis of weak commensurability to the problem of characterizing algebraic group having the same isomorphism/isogeny classes of maximal tori.

Theorem 9.8 (isogeny theorem [Prasad and Rapinchuk 2009, Theorem 4.2]). *Let G_1 and G_2 be two connected absolutely almost simple algebraic groups defined over an infinite field K, and let L_i be the minimal Galois extension of K*

over which G_i becomes an inner form of a split group. Suppose that for $i = 1, 2$, we are given a semisimple element $\gamma_i \in G_i(K)$ contained in a maximal K-torus T_i of G_i. Assume that:

(i) *G_1 and G_2 are either of the same Killing–Cartan type, or one of them is of type B_n and the other is of type C_n;*

(ii) *γ_1 has infinite order;*

(iii) *T_1 is K-irreducible; and*

(iv) *γ_1 and γ_2 are weakly commensurable.*

Then:

(1) *There exists a K-isogeny $\pi\colon T_2 \to T_1$ which carries $\gamma_2^{m_2}$ to $\gamma_1^{m_1}$ for some integers $m_1, m_2 \geqslant 1$.*

(2) *If $L_1 = L_2 =: L$ and $\theta_{T_1}(\mathrm{Gal}(L_{T_1}/L)) \supset W(G_1, T_1)$, then*
$$\pi^*\colon X(T_1) \otimes_{\mathbb{Z}} \mathbb{Q} \to X(T_2) \otimes_{\mathbb{Z}} \mathbb{Q}$$
has the property that $\pi^*(\mathbb{Q} \cdot \Phi(G_1, T_1)) = \mathbb{Q} \cdot \Phi(G_2, T_2)$. Moreover, if G_1 and G_2 are of the same Killing–Cartan type different from $B_2 = C_2$, F_4 or G_2, then a suitable rational multiple of π^* maps $\Phi(G_1, T_1)$ onto $\Phi(G_2, T_2)$, and if G_1 is of type B_n and G_2 is of type C_n, with $n > 2$, then a suitable rational multiple λ of π^* takes the long roots in $\Phi(G_1, T_1)$ to the short roots in $\Phi(G_2, T_2)$ while 2λ takes the short roots in $\Phi(G_1, T_1)$ to the long roots in $\Phi(G_2, T_2)$.

It follows that in the situations where π^* can be, and has been, scaled so that $\pi^*(\Phi(G_1, T_1)) = \Phi(G_2, T_2)$, it induces K-isomorphisms $\tilde{\pi}\colon \tilde{T}_2 \to \tilde{T}_1$ and $\overline{\pi}\colon \overline{T}_2 \to \overline{T}_1$ between the corresponding tori in the simply connected and adjoint groups $\tilde{G}_i$ and $\overline{G}_i$, respectively, that extend to $\overline{K}$-isomorphisms $\tilde{G}_2 \to \tilde{G}_1$ and $\overline{G}_2 \to \overline{G}_1$. Thus, the fact that Zariski-dense torsion-free subgroups $\Gamma_1 \subset G_1(K)$ and $\Gamma_2 \subset G_2(K)$ are weakly commensurable implies (under some minor technical assumptions) that G_1 and G_2 have the same K-isogeny classes (and under some additional assumptions, even the same K-isomorphism classes) of generic maximal K-tori that nontrivially intersect Γ_1 and Γ_2, respectively.

For a "multidimensional" version of Theorem 9.8, which is formulated using the notion of weak containment (see Section 8) in place of weak commensurability, see [Prasad and Rapinchuk 2013, Theorem 2.3].

9.9. We conclude this section with one new observation (Theorem 9.10) which is directly related to the main theme of the workshop—thin groups. This observation was inspired by a conversation of the first-named author with Igor Rivin at the Institute for Advanced Study.

Let G be a connected absolutely almost simple algebraic group over a field K of characteristic zero, and let T be a maximal K-torus of G. We let $\Phi_>(G,T)$ (resp., $\Phi_<(G,T)$) denote the set of all long (resp., short) roots in the root system $\Phi(G,T)$; by convention,

$$\Phi_>(G,T) = \Phi_<(G,T) = \Phi(G,T)$$

if all roots have the same length. Furthermore, we let $G_T^>$ denote the K-subgroup of G generated by T and the one-parameter unipotent subgroups U_a for a in $\Phi_>(G,T)$. Then $G_T^>$ is a connected semisimple subgroup of G of maximal absolute rank (so, in fact, just the U_a's for $a \in \Phi_>(G,T)$ generate $G_T^>$). By direct inspection, one verifies that $G_T^> \neq G$ precisely when $\Phi(G,T)$ has roots of different lengths, and then $G_T^>$ is a semisimple group of type $(\mathsf{A}_1)^n$ if G is of type C_n, and an absolutely almost simple group of type D_n, D_4 and A_2 if G is of type B_n, F_4 and G_2, respectively. On the other hand, the subgroups U_a for $a \in \Phi_<(G,T)$ generate G in all cases. Finally, for any connected subgroup of G containing T there exists a subset $\Psi \subset \Phi(G,T)$ such that G is generated by T and U_a for all $a \in \Psi$.

Theorem 9.10. *Let g be a generic element of infinite order and $T := Z_G(g)$ be the associated maximal torus. Let $x \in G(K)$ be any element of infinite order not contained in $T(K)$. Furthermore, let Γ be the (abstract) subgroup of $G(K)$ generated by g and x, and let H be the identity component of the Zariski-closure of Γ. Then either $H = G$ or $H = G_T^>$. Consequently, g and x generate a Zariski-dense subgroup of G if all roots in the root system $\Phi(G,T)$ are of same length.*

Proof. As g is a generic element of infinite order, the cyclic group generated by it is Zariski-dense in T, and so the cyclic group generated by xgx^{-1} is Zariski-dense in the conjugate torus xTx^{-1}. Since $x \notin T(K)$ and is of infinite order, it cannot normalize T (see 9.5). Thus H contains at least two different (generic) maximal K-tori, namely T and xTx^{-1}. Assume that $H \neq G$. Since H is connected and properly contains T, it must contain a one-parameter subgroup U_a for some $a \in \Phi(G,T)$. Then being defined over K, H also contains U_b for all b of the form $b = \sigma(a)$ with $\sigma \in \mathrm{Gal}(K_T/K)$. Now since T is generic, using the fact that the Weyl group $W(G,T)$ acts transitively on the subsets of roots of same length (by [Bourbaki 1968, Chapter VI, Proposition 11]), we see that H contains U_b for all roots $b \in \Phi(G,T)$ of same length as a. If a were a short root then the above remarks would imply that $H = G$, which is not the case. Thus, a must be long, and therefore H contains $G_T^>$ but does not contain U_b for any short root b. This clearly implies that $H = G_T^>$. □

Remark 9.11. It is worth noting that the types with roots of different lengths are honest exceptions in Theorem 9.10 in the sense that for any absolutely almost simple algebraic group G of one of those types over a finitely generated field K one can find two generic elements $\gamma_1, \gamma_2 \in G(K)$ that generate $G_T^{>} \neq G$ for a generic maximal K-torus T. To see this, we first pick an arbitrary generic element $\gamma_1 \in G(K)$ of infinite order provided by Theorem 9.6, and let $T = Z_G(\gamma_1)$ be the corresponding torus. Since $H := G_T^{>}$ is semisimple, the group $H(K)$ is Zariski-dense in H (by [Borel 1991, Corollary 18.3]). So, there exists $h \in H(K)$ such that $\gamma_2 := h\gamma_1 h^{-1} \notin T(K)$. Then $\gamma_2 \in H(K)$ is also generic over K, and the Zariski-closure of the subgroup generated by γ_1 and γ_2 is contained in (actually, is equal to) H.

10. Some open problems

The analysis of weak commensurability has led to a number of interesting problems in the theory of algebraic and Lie groups (see Section 6, for example) and its applications to locally symmetric spaces, and we would like to conclude this article with a brief discussion of some of these problems.

According to Theorem 4.5, if two lattices in the groups of rational points of connected absolutely almost simple groups over a nondiscrete locally compact field are weakly commensurable and one of the lattices is arithmetic, then so is the other. At the same time, it has been shown by means of an example (see [Prasad and Rapinchuk 2009, Remark 5.5]) that a Zariski-dense subgroup weakly commensurable to a *rank-one* arithmetic subgroup need not be arithmetic. It would be interesting, however, to understand what happens with *higher-rank* S-arithmetic subgroups.

Problem 10.1. Let G_1 and G_2 be two connected absolutely almost simple algebraic groups defined over a field F of characteristic zero, and let $\Gamma_1 \subset G_1(F)$ be a Zariski-dense (K, S)-arithmetic subgroup whose S-rank[7] is at least 2. If $\Gamma_2 \subset G_2(F)$ is a Zariski-dense subgroup weakly commensurable to Γ_1, is Γ_2 necessarily S-arithmetic?

This problem appears to be very challenging; the answer is not known even in the cases where Γ_1 is $\mathrm{SL}_3(\mathbb{Z})$ or $\mathrm{SL}_2(\mathbb{Z}[1/p])$. One should probably start by considering Problem 10.1 in a more specialized situation, e.g., assuming that F is a nondiscrete locally compact field, $\Gamma_1 \subset G_1(F)$ is a *discrete* Zariski-dense (higher-rank) S-arithmetic subgroup, and $\Gamma_2 \subset G_2(F)$ is a (finitely generated) *discrete* Zariski-dense subgroup weakly commensurable to Γ_1 (these restrictions would eliminate $\mathrm{SL}_2(\mathbb{Z}[1/p])$ as a possibility for Γ_1, but many interesting groups

[7]We recall that if Γ is $(\mathscr{G}, K, S)$-arithmetic, then the S-rank of Γ is defined to be $\sum_{v \in S} \mathrm{rk}_{K_v} \mathscr{G}$, where $\mathrm{rk}_F \mathscr{G}$ denotes the rank of $\mathscr{G}$ over a field $F \supset K$.

such as $SL_3(\mathbb{Z})$ would still be included). The nature of these assumptions brings up another question of independent interest.

Problem 10.2. Let G_1 and G_2 be connected absolutely almost simple algebraic groups over a nondiscrete locally compact field F, and let Γ_i be a finitely generated Zariski-dense subgroup of $G_i(F)$ for $i = 1, 2$. Assume that Γ_1 and Γ_2 are weakly commensurable. Does the discreteness of Γ_1 imply the discreteness of Γ_2?

An affirmative answer to Problem 10.2 was given in [Prasad and Rapinchuk 2009, Propisition 5.6] for the case where F is a *nonarchimedean* local field, but the case $F = \mathbb{R}$ or $\mathbb{C}$ remains open. Another interesting question is whether weak commensurability preserves cocompactness of lattices.

Problem 10.3. Let G_1 and G_2 be connected absolutely almost simple algebraic groups over $F = \mathbb{R}$ or $\mathbb{C}$, and let $\Gamma_i \subset G_i(F)$ be a lattice for $i = 1, 2$. Assume that Γ_1 and Γ_2 are weakly commensurable. Does the compactness of $G_1(F)/\Gamma_1$ imply the compactness of $G_2(F)/\Gamma_2$?[8]

We recall that the cocompactness of a lattice in a semisimple real Lie group is equivalent to the absence of nontrivial unipotents in it; see [Raghunathan 1972, Corollary 11.13]. So, Problem 10.3 can be rephrased as the question whether for two weakly commensurable lattices Γ_1 and Γ_2, the existence of nontrivial unipotent elements in one of them implies their existence in the other; in this form the question is meaningful for arbitrary Zariski-dense subgroups (not necessarily discrete or of finite covolume). The combination of Theorems 4.4 and 4.5 implies the affirmative answer to Problem 10.3 in the case where one of the lattices is arithmetic, but no other cases have been considered so far.

From the general perspective, one important problem is to try to generalize our results on length-commensurable and/or isospectral arithmetically defined locally symmetric spaces of absolutely simple real Lie groups to arithmetically defined locally symmetric spaces of arbitrary semisimple Lie groups, or at least those of $\mathbb{R}$-simple Lie groups. To highlight the difficulty, we will make some comments about the latter case. An $\mathbb{R}$-simple adjoint group G can be written in the form $G = \mathrm{R}_{\mathbb{C}/\mathbb{R}}(H)$ (restriction of scalars) where H is an absolutely simple complex algebraic group. Arithmetic lattices in $G(\mathbb{R}) \simeq H(\mathbb{C})$ come from the forms of H over a number field admitting exactly one complex embedding. The analysis of weak commensurability of even arithmetic lattices $\Gamma_1 \subset G_1(\mathbb{R})$ and $\Gamma_2 \subset G_2(\mathbb{R})$, where $G_i = \mathrm{R}_{\mathbb{C}/\mathbb{R}}(H_i)$ for $i = 1, 2$, cannot be implemented via the

[8] It is well known that for a semisimple algebraic group G over a nondiscrete nonarchimedean locally compact field F of characteristic zero and a discrete subgroup $\Gamma \subset G(F)$, the quotient $G(F)/\Gamma$ has finite measure if and only if it is compact, so the problem in this case becomes vacuous.

study of the forms of the G_i's, forcing us to study directly the forms of the H_i's. But the relation of weak commensurability of semisimple elements $\gamma_1 \in \Gamma_1$ and $\gamma_2 \in \Gamma_2$ in terms of G_1 and G_2—i.e., the fact that $\chi_1(\gamma_1) = \chi_2(\gamma_2) \neq 1$ for some characters χ_i of maximal $\mathbb{R}$-tori T_i of G_i such that $\gamma_i \in T_i(\mathbb{R})$—translates into a significantly more complicated relation in terms of H_1 and H_2. Indeed, pick maximal $\mathbb{C}$-tori S_i of H_i so that $T_i = \mathrm{R}_{\mathbb{C}/\mathbb{R}}(S_i)$, and let $\delta_i \in S_i(\mathbb{C})$ be the element corresponding to γ_i under the identification $T_i(\mathbb{R}) \simeq S_i(\mathbb{C})$. Then there exist characters χ_i', χ_i'' of S_i such that $\chi_i(\gamma_i) = \chi_i'(\delta_i)\overline{\chi_i''(\delta_i)}$. So, the relation of weak commensurability of γ_1 and γ_2 assumes the following form in terms of δ_1 and δ_2:

$$\chi_1'(\delta_1)\overline{\chi_1''(\delta_1)} = \chi_2'(\delta_2)\overline{\chi_2''(\delta_2)}.$$

It is not clear if this type of relation would lead to the results similar to those we described in this article for the weakly commensurable arithmetic subgroups of absolutely almost simple groups. So, the general problem at this stage is to formulate for general semisimple groups (or at least $\mathbb{R}$-simple groups) the "right" notion of weak commensurability and explore its consequences. We will now formulate a particular case of this general program that would be interesting for geometric applications.

Problem 10.4. Let G_1 and G_2 be almost simple complex algebraic groups. Two semisimple elements $\gamma_i \in G_i(\mathbb{C})$ are called $\mathbb{R}$-weakly commensurable if there exist complex maximal tori T_i of G_i for $i = 1, 2$ such that $\gamma_i \in T_i(\mathbb{C})$ and for suitable characters χ_i of T_i we have

$$|\chi_1(\gamma_1)| = |\chi_2(\gamma_2)| \neq 1.$$

Furthermore, Zariski-dense (discrete) subgroups $\Gamma_i \subset G_i(\mathbb{C})$ are $\mathbb{R}$-weakly commensurable if every semisimple element $\gamma_1 \in \Gamma_1$ of infinite order is $\mathbb{R}$-weakly commensurable to some semisimple element $\gamma_2 \in \Gamma_2$ of infinite order, and vice versa. Under what conditions does the $\mathbb{R}$-weak commensurability of Zariski-dense (arithmetic) lattices $\Gamma_i \subset G_i(\mathbb{C})$ $(i = 1, 2)$ imply their commensurability?

The result of [Chinburg et al. 2008] seems to imply that the $\mathbb{R}$-weak commensurability of arithmetic lattices in $\mathrm{SL}_2(\mathbb{C})$ does imply their commensurability, but no other results in this direction are available.

Turning now to the geometric aspect, we would like to reiterate that most of our results deal with the analysis of the new relation of length-commensurability, which eventually implies the results about isospectral locally symmetric spaces. At the same time, the general consequences of isospectrality and isolength spectrality are much better understood than those of length-commensurability. So, as an overarching problem, we would like to propose the following.

Problem 10.5. Understand consequences (qualitative and quantitative) of length-commensurability for locally symmetric spaces.

(Here by *quantitative consequences* we mean results stating that in certain situations a family of length-commensurable locally symmetric spaces consists either of a single commensurability class or of a certain bounded number of commensurability classes, and by *qualitative consequences* - results guaranteeing that the number of commensurability classes in a given class of length-commensurable locally symmetric spaces is finite.)

There are various concrete questions within the framework provided by Problem 10.5 that were resolved in [Prasad and Rapinchuk 2009] for arithmetically defined locally symmetric spaces but remain open for locally symmetric spaces which are not arithmetically defined. For example, according to Theorem 5.3, if $\mathfrak{X}_{\Gamma_1}$ and $\mathfrak{X}_{\Gamma_2}$ are length-commensurable and at least one of the spaces is arithmetically defined, then the compactness of one of them implies the compactness of the other. It is natural to ask if this can be proved for locally symmetric spaces which are not arithmetically defined.

Problem 10.6 (geometric version of Problem 10.3). Let $\mathfrak{X}_{\Gamma_1}$ and $\mathfrak{X}_{\Gamma_2}$ be length-commensurable locally symmetric spaces of finite volume. Does the compactness of $\mathfrak{X}_{\Gamma_1}$ always imply the compactness of $\mathfrak{X}_{\Gamma_2}$?

In [Prasad and Rapinchuk 2009, Section 9], for each of the exceptional types A_n, D_{2n+1} $(n > 1)$ and E_6, we have constructed examples of length-commensurable, but not commensurable, compact arithmetically defined locally symmetric spaces associated with a simple real algebraic group of this type. It would be interesting to see if this construction can be refined to provide examples of isolength spectral or even isospectral compact arithmetically defined locally symmetric spaces that are not commensurable.

Problem 10.7. For inner and outer types A_n $(n > 1)$, D_{2n+1} $(n > 1)$ and E_6, construct examples of isospectral compact arithmetically defined locally symmetric spaces that are not commensurable.

Currently, such a construction is known only for *inner forms* of type A_n (see [Lubotzky et al. 2006]); it relies on some delicate results from the theory of automorphic forms [Harris and Taylor 2001], the analogues of which are not yet available for groups of other types.

As we already mentioned, in [Prasad and Rapinchuk 2009] we focused on the case where G_1 and G_2 are absolutely (almost) simple real algebraic groups. From the geometric perspective, however, it would be desirable to consider a more general situation where G_1 and G_2 are allowed to be either arbitrary real semisimple groups (without compact factors), or at least arbitrary $\mathbb{R}$-simple groups. This problem is intimately related to the problem, discussed above, of generalizing our results on weak commensurability from absolutely almost simple to arbitrary semisimple groups. In particular, a successful resolution of Problem 10.4 would enable us to extend our results to the (arithmetically defined) locally symmetric spaces associated with $\mathbb{R}$-simple groups providing thereby a significant generalization of the result of [Chinburg et al. 2008] where the case $G_1 = G_2 = \mathrm{R}_{\mathbb{C}/\mathbb{R}}(\mathrm{SL}_2)$ (that leads to arithmetically defined hyperbolic 3-manifolds) was considered.

Finally, the proof of the result that connects the length-commensurability of $\mathfrak{X}_{\Gamma_1}$ and $\mathfrak{X}_{\Gamma_2}$ to the weak commensurability of Γ_1 and Γ_2 relies (at least in the higher-rank case) on Schanuel's conjecture. It would be interesting to see if our geometric results can be made independent of Schanuel's conjecture. Some results in this direction were obtained in [Bhagwat et al. 2012].

Acknowledgements

Both authors were partially supported by the NSF (grants DMS-1001748 and DMS-0965758) and the Humboldt Foundation. The first-named author thanks the Institute for Advanced Study (Princeton) for its hospitality and support during 2012. The hospitality of the MSRI during the workshop "Hot Topics: Thin Groups and Superstrong Approximation" (February 6–10, 2012) is also thankfully acknowledged. We thank Skip Garibaldi and Igor Rapinchuk for carefully reading the manuscript and for their comments and corrections. We also thank Igor Rivin for useful discussions related to Theorem 9.10.

References

[Amitsur 1955] S. A. Amitsur, "Generic splitting fields of central simple algebras", *Ann. of Math.* (2) **62** (1955), 8–43.

[Ax 1971] J. Ax, "On Schanuel's conjectures", *Ann. of Math.* (2) **93** (1971), 252–268.

[Baker 1990] A. Baker, *Transcendental number theory*, 2nd ed., Cambridge University Press, 1990.

[Bayer-Fluckiger 2011] E. Bayer-Fluckiger, "Isometries of quadratic spaces over global fields", preprint, 2011, http://www.mathematik.uni-bielefeld.de/LAG/man/453.pdf.

[Bhagwat et al. 2012] C. Bhagwat, S. Pisolkar, and C. S. Rajan, "Commensurability and representation equivalent arithmetic lattices", preprint, 2012. To appear in *Int. Math. Res. Not.* arXiv 1207.3891

[Borel 1991] A. Borel, *Linear algebraic groups*, 2nd ed., Graduate Texts in Mathematics **126**, Springer, New York, 1991.

[Bourbaki 1968] N. Bourbaki, *Groupes et algèbres de Lie, Chapitres IV–VI*, Actualités Scientifiques et Industrielles **1337**, Hermann, Paris, 1968.

[Bruhat and Tits 1987] F. Bruhat and J. Tits, "Groupes algébriques sur un corps local, III: Compléments et applications à la cohomologie galoisienne", *J. Fac. Sci. Univ. Tokyo Sect. IA Math.* **34**:3 (1987), 671–698.

[Chernousov et al. 2012] V. I. Chernousov, A. S. Rapinchuk, and I. A. Rapinchuk, "On the genus of a division algebra", *C. R. Math. Acad. Sci. Paris* **350**:17-18 (2012), 807–812.

[Chinburg et al. 2008] T. Chinburg, E. Hamilton, D. D. Long, and A. W. Reid, "Geodesics and commensurability classes of arithmetic hyperbolic 3–manifolds", *Duke Math. J.* **145**:1 (2008), 25–44.

[Đoković and Thǎńg 1994] D. Ž. Đoković and N. Q. Thǎńg, "Conjugacy classes of maximal tori in simple real algebraic groups and applications", *Canad. J. Math.* **46**:4 (1994), 699–717.

[Duistermaat and Guillemin 1975] J. J. Duistermaat and V. W. Guillemin, "The spectrum of positive elliptic operators and periodic bicharacteristics", *Invent. Math.* **29**:1 (1975), 39–79.

[Duistermaat et al. 1979] J. J. Duistermaat, J. A. C. Kolk, and V. S. Varadarajan, "Spectra of compact locally symmetric manifolds of negative curvature", *Invent. Math.* **52**:1 (1979), 27–93.

[Fiori 2012] A. Fiori, "Characterization of special points of orthogonal symmetric spaces", *J. Algebra* **372** (2012), 397–419.

[Gangolli 1977] R. Gangolli, "The length spectra of some compact manifolds of negative curvature", *J. Differential Geom.* **12**:3 (1977), 403–424.

[Garge 2005] S. M. Garge, "Maximal tori determining the algebraic groups", *Pacific J. Math.* **220**:1 (2005), 69–85.

[Garibaldi 2012] S. Garibaldi, "Outer automorphisms of algebraic groups and determining groups by their maximal tori", *Michigan Math. J.* **61**:2 (2012), 227–237.

[Garibaldi and Rapinchuk 2013] S. Garibaldi and A. Rapinchuk, "Weakly commensurable S-arithmetic subgroups in almost simple algebraic groups of types B and C", *Algebra Number Theory* **7**:5 (2013), 1147–1178.

[Garibaldi and Saltman 2010] S. Garibaldi and D. J. Saltman, "Quaternion algebras with the same subfields", pp. 225–238 in *Quadratic forms, linear algebraic groups, and cohomology*, edited by J.-L. Colliot-Thélène et al., Dev. Math. **18**, Springer, New York, 2010.

[Gille 2004] P. Gille, "Type des tores maximaux des groupes semi-simples", *J. Ramanujan Math. Soc.* **19**:3 (2004), 213–230.

[Gille and Szamuely 2006] P. Gille and T. Szamuely, *Central simple algebras and Galois cohomology*, Cambridge Studies in Advanced Mathematics **101**, Cambridge University Press, 2006.

[Gorodnik and Nevo 2011] A. Gorodnik and A. Nevo, "Splitting fields of elements in arithmetic groups", *Math. Res. Lett.* **18**:6 (2011), 1281–1288.

[Harris and Taylor 2001] M. Harris and R. Taylor, *The geometry and cohomology of some simple Shimura varieties*, Annals of Mathematics Studies **151**, Princeton University Press, 2001.

[Jouve et al. 2013] F. Jouve, E. Kowalski, and D. Zywina, "Splitting fields of characteristic polynomials of random elements in arithmetic groups", *Israel J. Math.* **193**:1 (2013), 263–307.

[Kac 1966] M. Kac, "Can one hear the shape of a drum?", *Amer. Math. Monthly* **73**:4-2 (1966), 1–23.

[Kariyama 1989] K. Kariyama, "On conjugacy classes of maximal tori in classical groups", *J. Algebra* **125**:1 (1989), 133–149.

[Kneser 1965a] M. Kneser, "Galois–Kohomologie halbeinfacher algebraischer Gruppen über $\mathfrak{p}$–adischen Körpern, I", *Math. Z.* **88** (1965), 40–47.

[Kneser 1965b] M. Kneser, "Galois–Kohomologie halbeinfacher algebraischer Gruppen über $\mathfrak{p}$–adischen Körpern, II", *Math. Z.* **89** (1965), 250–272.

[Kottwitz 1982] R. E. Kottwitz, "Rational conjugacy classes in reductive groups", *Duke Math. J.* **49**:4 (1982), 785–806.

[Krashen and McKinnie 2011] D. Krashen and K. McKinnie, "Distinguishing division algebras by finite splitting fields", *Manuscripta Math.* **134**:1-2 (2011), 171–182.

[Lang 1956] S. Lang, "Algebraic groups over finite fields", *Amer. J. Math.* **78** (1956), 555–563.

[Lee 2012] T.-Y. Lee, "Embedding functors and their arithmetic properties", preprint, 2012. arXiv 1211.3564

[Lubotzky and Rosenzweig 2012] A. Lubotzky and L. Rosenzweig, "The Galois group of random elements of linear groups", preprint, 2012. arXiv 1205.5290

[Lubotzky et al. 2006] A. Lubotzky, B. Samuels, and U. Vishne, "Division algebras and noncommensurable isospectral manifolds", *Duke Math. J.* **135**:2 (2006), 361–379.

[Margulis 1991] G. A. Margulis, *Discrete subgroups of semisimple Lie groups*, Ergebnisse der Mathematik und ihrer Grenzgebiete (3) **17**, Springer, Berlin, 1991.

[Milnor 1964] J. Milnor, "Eigenvalues of the Laplace operator on certain manifolds", *Proc. Nat. Acad. Sci. U.S.A.* **51** (1964), 542.

[Platonov and Rapinchuk 1994] V. Platonov and A. Rapinchuk, *Algebraic groups and number theory*, Pure and Applied Mathematics **139**, Academic Press, Boston, 1994.

[Prasad and Rapinchuk 2001] G. Prasad and A. S. Rapinchuk, "Irreducible tori in semisimple groups", *Int. Math. Res. Not.* **2001**:23 (2001), 1229–1242.

[Prasad and Rapinchuk 2002] G. Prasad and A. S. Rapinchuk, "Subnormal subgroups of the groups of rational points of reductive algebraic groups", *Proc. Amer. Math. Soc.* **130**:8 (2002), 2219–2227.

[Prasad and Rapinchuk 2003] G. Prasad and A. S. Rapinchuk, "Existence of irreducible $\mathbb{R}$-regular elements in Zariski-dense subgroups", *Math. Res. Lett.* **10**:1 (2003), 21–32.

[Prasad and Rapinchuk 2009] G. Prasad and A. S. Rapinchuk, "Weakly commensurable arithmetic groups and isospectral locally symmetric spaces", *Inst. Hautes Études Sci. Publ. Math.* 109 (2009), 113–184.

[Prasad and Rapinchuk 2010a] G. Prasad and A. S. Rapinchuk, "Local-global principles for embedding of fields with involution into simple algebras with involution", *Comment. Math. Helv.* **85**:3 (2010), 583–645.

[Prasad and Rapinchuk 2010b] G. Prasad and A. S. Rapinchuk, "Number-theoretic techniques in the theory of Lie groups and differential geometry", pp. 231–250 in *Fourth International Congress of Chinese Mathematicians*, edited by L. Ji et al., AMS/IP Stud. Adv. Math. **48**, Amer. Math. Soc., Providence, RI, 2010.

[Prasad and Rapinchuk 2013] G. Prasad and A. S. Rapinchuk, "On the fields generated by the lengths of closed geodesics in locally symmetric spaces", *Geometriae Dedicata* (2013).

[Raghunathan 1972] M. S. Raghunathan, *Discrete subgroups of Lie groups*, Ergebnisse der Mathematik und ihrer Grenzgebiete **68**, Springer, New York, 1972.

[Raghunathan 2004] M. S. Raghunathan, "Tori in quasi-split-groups", *J. Ramanujan Math. Soc.* **19**:4 (2004), 281–287.

[Rapinchuk 2014] A. S. Rapinchuk, "Strong approximation for algebraic groups", pp. 269–298 in *Thin groups and superstrong approximation*, edited by H. Oh and E. Breuillard, Math. Sci. Res. Inst. Publ. **61**, Cambridge, New York, 2014.

[Rapinchuk and Rapinchuk 2010] A. S. Rapinchuk and I. A. Rapinchuk, "On division algebras having the same maximal subfields", *Manuscripta Math.* **132**:3-4 (2010), 273–293.

[Reid 1992] A. W. Reid, "Isospectrality and commensurability of arithmetic hyperbolic 2– and 3–manifolds", *Duke Math. J.* **65**:2 (1992), 215–228.

[Steinberg 1965] R. Steinberg, "Regular elements of semisimple algebraic groups", *Inst. Hautes Études Sci. Publ. Math.* 25 (1965), 49–80.

[Sunada 1985] T. Sunada, "Riemannian coverings and isospectral manifolds", *Ann. of Math.* (2) **121**:1 (1985), 169–186.

[Vignéras 1980] M.-F. Vignéras, "Variétés riemanniennes isospectrales et non isométriques", *Ann. of Math.* (2) **112**:1 (1980), 21–32.

[Yeung 2011] S.-K. Yeung, "Isospectral problem of locally symmetric spaces", *Int. Math. Res. Not.* **2011**:12 (2011), 2810–2824.

gprasad@umich.edu *Department of Mathematics, University of Michigan, Ann Arbor, MI 48109, United States*

asr3x@virginia.edu *Department of Mathematics, University of Virginia, Charlottesville, VA 22904, United States*

Thin groups and superstrong approximation
MSRI Publications
Volume **61**, 2013

Growth in linear groups

LÁSZLÓ PYBER AND ENDRE SZABÓ

We give a description of nongrowing subsets in linear groups over arbitrary fields, which extends the product theorem for simple groups of Lie type. We also give an account of various related aspects of growth in linear groups.

1. A polynomial inverse theorem in linear groups

The following theorem was proved independently in 2010 by Breuillard, Green and Tao [Breuillard et al. 2011a] and the authors [Pyber and Szabó 2010].

Theorem 1 (product theorem). *Let L be a finite simple group of Lie type of rank r and A a generating set of L. Then either $A^3 = L$ or*

$$|A^3| > |A|^{1+\varepsilon},$$

where ε depends only on r.

For $G = \mathrm{PSL}(2, p)$, p prime, this is a famous result of Helfgott [2008]. Other special cases that preceded the general proof include $G = \mathrm{PSL}(3, p)$ [Helfgott 2011] and $G = \mathrm{PSL}(2, q)$, q a prime power [Dinai 2011; Varjú 2012].

In [Pyber and Szabó 2010] we give various examples which show that in the above result the dependence of ε on r is necessary. In particular we construct generating sets of $\mathrm{SL}(n, 3)$ of size $2^{n-1} + 4$ with $|A^3| < 100|A|$ for $n \geq 3$.

For the groups $G = \mathrm{PSL}(n, q)$ (which are simple groups of Lie type of rank $n - 1$) the product theorem can be reformulated as follows: If A is a generating set of G, such that $|A^3| < K|A|$ for some number $K \geq 1$, then A has at most $K^{c(n)}$ (i.e., polynomially many) elements. This reformulation turns out to be quite useful when we seek an extension of the product theorem which describes nongrowing subsets of linear groups.

The product theorem has quickly become a central result of finite asymptotic group theory with many applications. To describe a few of these we introduce some graph-theoretic terminology. The *diameter*, diam X, of an undirected graph $X = (V, E)$ is the largest distance between two of its vertices. The *girth* of X is the length of the shortest cycle in X. Given a subset A of the vertex set V

László Pyber is supported in part by OTKA NK78439 and K84233. Endre Szabó is supported in part by OTKA NK81203 and K84233.

the *expansion* of A, $c(A)$, is defined to be the ratio $|\sigma(X)|/|X|$ where $\sigma(X)$ is the set of vertices at distance 1 from A. A graph is a C-*expander* for some $C > 0$ if for all subsets A of V with $|A| < |V|/2$ we have $c(A) \geq C$. A family of graphs is an *expander family* if all of its members are C-expanders for some fixed positive constant C.

A subset S of a group G is called *symmetric* if it is inverse-closed (throughout the paper S will denote a symmetric subset). The Cayley graph $\text{Cay}(G, S)$ is a graph whose vertices are the elements of G and which has an edge from x to y if and only if $x = sy$ for some $s \in S$. Then the diameter of $\text{Cay}(G, S)$ is the smallest number d such that $S^d = G$.

The following classical conjecture is due to Babai. (It first appeared in print in [Babai and Seress 1992].)

Conjecture 2. *For every nonabelian finite simple group L and every symmetric generating set S of L we have* $\text{diam}\,\text{Cay}(L, S) \leq C(\log|L|)^c$ *where c and C are absolute constants.*

The product theorem easily implies the following.

Corollary 3. *Conjecture 2 holds for simple groups of Lie type of bounded rank.*

The following conjecture of Liebeck, Nikolov, and Shalev is more recent:

Conjecture 4 [Liebeck et al. 2012]. *There exists an absolute constant c such that if L is a finite simple group and A is a subset of L of size at least two, then L is a product of N conjugates of A for some $N \leq c\log|L|/\log|A|$.*

By a deep (and widely applied) theorem of Liebeck and Shalev [2001] the conjecture holds when S is a conjugacy class or, more generally, a normal subset (that is a union of conjugacy classes).

Theorem 5. *Conjecture 4 holds for simple groups of Lie type of bounded rank.*

This is proved in [Gill et al. 2011] using the product theorem and an extra trick which handles small subsets S.

Breuillard, Green, Guralnick and Tao have announced a deep result of similar flavour:

Theorem 6 [Breuillard et al. 2013]. *Let G be a simple group of Lie type of rank r. Then for two random elements x, y the Cayley graph* $\text{Cay}(G, \{x, y\})$ *is an $\varepsilon(r)$ expander with probability going to* 1 *as* $|G| \to \infty$, *where $\varepsilon(r)$ depends only on r.*

As an open problem Lubotzky [2012] suggested deciding whether this holds for all nonabelian finite simple groups with an absolute constant ε.

The proof of Theorem 6 is based on the product theorem, some results in [Breuillard et al. 2012a] and the proof scheme (termed the Bourgain–Gamburd expansion machine in [Tao 2012]) first used in the proof of the following theorem.

Theorem 7 [Bourgain and Gamburd 2008]. *For any* $0 < \delta \in \mathbb{R}$ *there exists* $\varepsilon = \varepsilon(\delta) > 0$ *such that for every prime* p, *if* S *is a symmetric set of generators of* $G = \mathrm{SL}(2, p)$ *such that girth of the Cayley graph* $\mathrm{Cay}(G; S)$ *is at least* $\delta \log p$ *then* $\mathrm{Cay}(G; S)$ *is an* ε*-expander.*

It seems quite plausible that this result extends to all nonabelian finite simple groups. In view of the girth bounds in [Gamburd et al. 2009] this would imply Theorem 6. At present however such an extension is not even known to hold for the groups $\mathrm{SL}(2, q)$, q a prime power.

For the above applications of the product theorem it is essential that the size of a generating set A is bounded by a polynomial of the tripling constant $K = |A^3|/|A|$ (unless A is very large). For a discussion of related issues, see [Tao 2012]. Guided by this insight, and remarks of Helfgott [2011, page 764] and Breuillard, Green and Tao [Breuillard et al. 2012b, Remark 1.8], and various discussions on Tao's blog, we proved in 2012 the following polynomial inverse theorem, which significantly improves the results in [Pyber and Szabó 2010].

Theorem 8 [Pyber and Szabó $\geq$ 2012]. *Let* S *be a symmetric subset of* $\mathrm{GL}(n, \mathbb{F})$ *satisfying* $|S^3| \leq K|S|$ *for some* $K \geq 1$, *where* $\mathbb{F}$ *is an arbitrary field. Then* S *is contained in the union of polynomially many (more precisely* $K^{c(n)}$*) cosets of a finite-by-soluble subgroup* Γ *normalised by* S.

Moreover, Γ *has a finite subgroup* P *normalised by* S *such that* Γ/P *is soluble, and* S^3 *contains a coset of* P.

The theorem extends and unifies several earlier results. Most importantly (to us) it contains the product theorem (for symmetric sets) as a special case. For subsets of $\mathrm{SL}(2, p)$ and $\mathrm{SL}(3, p)$ similar results were obtained by Helfgott [2008; 2011].

In characteristic zero Theorem 8 was first proved in [Breuillard et al. 2011a]. The earliest result in this direction for subsets of $\mathrm{SL}(2, \mathbb{R})$ is due to Elekes and Király [2001]. The proof in [Breuillard et al. 2011a] uses the fact that a virtually soluble subgroup of $\mathrm{SL}(n, \mathbb{C})$ has a soluble subgroup of n-bounded index, which is no longer true in positive characteristic.

Finally the theorem implies a result of Hrushovski for linear groups over arbitrary fields obtained by model-theoretic tools [Hrushovski 2012]. In his theorem the structure of Γ is described in a less precise way and the number of covering cosets is only bounded by some large function of n and K.

It seems possible that a result similar to Theorem 8 can be proved with H nilpotent (possibly at the cost of loosing the normality of H and P in $\langle S \rangle$). Indeed in characteristic zero such a result follows by combining [Breuillard et al. 2011a] with [Breuillard and Green 2011], and for prime fields it follows from [Pyber and Szabó 2010] and the results of [Gill and Helfgott 2010] for soluble groups. In general this may be technically quite challenging even though soluble linear groups have a nilpotent-by-abelian subgroup of n-bounded index.

It may also be possible that a more general polynomial inverse theorem holds (see [Breuillard 2014]). We pose the following question which bypasses abelian and finite obstacles.

Question 9. *Let S be a finite symmetric subset of an arbitrary group G such that $|S^3| \le K|S|$ for some $K \ge 1$. Is it true that S is contained in $K^{c(G)}$ cosets of some virtually soluble subgroup of G?*

The above question may be viewed as a counterpart of the polynomial Freiman–Ruzsa Conjecture which asserts that (a variant of) Freiman's famous Inverse theorem holds with polynomial constants. The existence of some huge bound $f(K)$ for the number of covering cosets, as above, follows from the very general inverse theorem of Breuillard, Green and Tao [Breuillard et al. 2012b]. See [Breuillard 2014] for a detailed discussion of these issues.

Theorem 8 shows that the answer is positive for the groups $G = \mathrm{SL}(n, \mathbb{F})$. It would interesting to investigate this question for various groups of intermediate word growth, such as the Grigorchuk groups (see, e.g., [de la Harpe 2000]).

It would be extremely interesting if the number of cosets required would be bounded by K^c for some absolute constant c for all groups G. Obtaining such a result for all linear groups G (in which c does not depend on the dimension) already seems to require some essential new ideas.

2. Applications of the Sarnak–Xue–Gowers trick

A subset X of a group G is called product-free if there are no solutions to $xy = z$ with $x, y, z \in X$.

Answering a question in [Babai and Sós 1985], Gowers [2008] showed that the group $G = \mathrm{PSL}(2, p)$ has no product-free subset of size greater than $c|G|^{8/9}$ for some $c > 0$. Gowers' proof is closely related to an argument of Sarnak and Xue [1991], which also plays an important role in the proof of Theorem 7.

Notation. For any group G let $\deg_{\mathbb{C}}(G)$ denote the minimum degree of a nontrivial complex representation.

It was observed in [Nikolov and Pyber 2011], that as a byproduct of the results of [Gowers 2008] one obtains the following extremely useful corollary.

Corollary 10. *Let G be a group of order n such that $\deg_{\mathbb{C}}(G) = k$. If A is a subset of G such that $|A| > n/k^{1/3}$, then we have $A^3 = G$.*

For groups of Lie type rather strong lower bounds on the minimal degree of a complex representation are known [Landazuri and Seitz 1974]. Combining these bounds with Corollary 10 we obtain the following proposition, which of course implies the product theorem for very large subsets.

Proposition 11. *Let A be a subset of $G = \mathrm{PSL}(n,q)$ of size at least $2|G|/q^{\frac{n-1}{3}}$. Then we have $A^3 = G$.*

Another interesting application is the following "Waring-type" theorem:

Theorem 12 [Nikolov and Pyber 2011]. *Let w be a nontrivial group word. Let L be a finite simple group of Lie type of rank r over the field $\mathbb{F}_q$. Let W be any subset of $w(L)$ such that $|W| \geq |w(L)|/q^{r/13}$. Then there exists a positive integer N depending only on w such that if $|L| > N$ then*

$$W^3 = L.$$

Earlier, as the main result of a difficult paper, Shalev [2009] has obtained the same result in the case $W = w(L)$ (allowing L to be also an alternating group).

An advantage of the above sparse version is that one can impose further restrictions on W. For example one can require that no two elements of W are inverses each other, or images of each other under Frobenius automorphisms. It would be interesting to obtain a sparse "Waring-type" theorem for the alternating groups.

The proof of Theorem 12 is relatively short compared to the one in [Shalev 2009]. One only has to show that for simple groups of Lie type the sets $w(L)$ are large enough.

Larsen, Shalev and Tiep, in a major recent work [Larsen et al. 2011], have shown that in fact, for a nontrivial word w, $w(L)^2 = L$ holds for large enough finite simple groups.

In certain situations, in which Corollary 10 cannot be applied directly, the following extension due to Babai, Nikolov and Pyber has turned out to be quite useful.

Lemma 13 [Babai et al. 2008]. *Let G be a group of order n such that $\deg_{\mathbb{C}}(G) = k$. Let $A_1, \ldots, A_t$ $(t \geq 3)$ be nonempty subsets of G. If*

$$\prod_{i=1}^{t} |A_i| \geq \frac{n^t}{k^{t-2}},$$

then $\prod_{i=1}^{t} A_i = G$.

A major result of Nikolov and Segal [2007a; 2007b] states that finite index subgroups of finitely generated profinite groups are open, as conjectured by Serre. Improving their methods, the same authors later proved the following theorem, which also implies Serre's conjecture:

Theorem 14 [Nikolov and Segal 2011]. *Let $d, q \in \mathbb{N}$ and G a finite d-generated group. Let G^q be the subgroup of G generated by the q-th powers. There exists a function $f(d,q)$ such that every element of G^q is a product of at most $f(d,q)$ powers x^q.*

Recently Nikolov and Segal [2012] obtained much shorter proofs as well as vast generalisations of their results. Lemma 13 is a key tool in obtaining them.

Lemma 13 was also used to improve a result from [Liebeck and Pyber 2001]:

Theorem 15 [Babai et al. 2008]. *Let L be a finite simple group of Lie type in characteristic p. Then L is a product of five Sylow p-subgroups.*

As an illustration of the proof consider the Ree groups $L = {}^2F_4(q)$ (a difficult case in [Liebeck and Pyber 2001]). Take a pair of Sylow p-subgroups U and V in L such that $U \cap V = \{1\}$. Using the tables in [Kleidman and Liebeck 1990] it follows that

$$\frac{|L|}{|UV|} < \deg_{\mathbb{C}}(L)^{1/2}.$$

Hence Lemma 13 with $t = 4$ immediately gives

$$L = (UV)(VU)(UV)(VU) = UVUVU.$$

Combining Theorem 15 with an argument from [Liebeck and Pyber 2001] one obtains a similar improvement of another result in [Liebeck and Pyber 2001].

Theorem 16. *Let $\mathbb{F}$ be a field of characteristic p and G a finite subgroup of* $\mathrm{GL}(n, \mathbb{F})$ *generated by its Sylow p-subgroups. If p is sufficiently large compared to n then G is a product of* 5 *of its Sylow p-subgroups.*

This is a generalisation of the following curious result, originally obtained by model-theoretic tools.

Theorem 17 [Hrushovski and Pillay 1995]. *Let p be a prime, and G a subgroup of* $\mathrm{GL}(n, p)$ *generated by elements of order p. Then G can be written as a product $G = C_1C_2 \dots C_k$ where the C_i are cyclic subgroups of order p and k depends only on n.*

Returning to our main theme next we present a (previously unpublished) "baby version" of the product theorem. This result was in fact the starting point of the authors' joint work concerning growth in groups. It is proved by a "greedy argument" based on Corollary 10. The proof of the product theorem may be viewed as an analogous greedy argument based on certain Larsen–Pink type inequalities (see Section 3).

First we quote two combinatorial results which also play an important role in the proof of the product theorem. As noted in [Helfgott 2011], the first one is essentially due to Ruzsa (see [Ruzsa 2010; Ruzsa and Turjányi 1985]).

Proposition 18 [Helfgott 2011]. *Let $1 \in S$ be a symmetric finite subset of a group and $k \geq 3$ an integer. Then*

$$\frac{|S^k|}{|S|} \leq \left(\frac{|S^3|}{|S|}\right)^{k-2}.$$

That is, if a small power of S is much larger than S itself, then S^3 is already much larger than S.

The next lemma is a slight extension of a result from [Helfgott 2011].

Lemma 19 [Pyber and Szabó 2010]. *Let $1 \in S$ be a symmetric finite subset of a group G, and H a subgroup of G. Then for all integers $k > 0$ one has*

$$\frac{|S^k \cap H|}{|S^2 \cap H|} \leq \frac{|S^{k+1}|}{|S|}.$$

That is, growth in any subgroup H implies growth in G itself.

Here is the simplest version of our "baby product theorem".

Theorem 20. *Let S be a symmetric generating set of the group $G = \mathrm{SL}(n, q)$ (where $q \geq 4$ is a prime power) and H a subgroup of G isomorphic to $\mathrm{SL}(2, q)$. If S contains H then either $S^3 = G$ or*

$$|S^3| > |S| \cdot q^{1/1000}.$$

Proof. We have $\deg_{\mathbb{C}}(H) \geq (q-1)/2$. Assume first that S^6 does not contain all conjugates of H in G. Since S generates G, there exists a conjugate H_0 of H such that H_0 is contained in S^6, and an S-conjugate, say H_0^s, of H_0 is not contained in S^6. Then we have

$$|S^2 \cap H_0^s| \leq |H_0| \cdot \deg_{\mathbb{C}}(H)^{-1/3},$$

since otherwise S^6 would contain H_0^s by Corollary 10. It follows from the assumptions on H_0 that H_0^s is contained in S^8. Hence by Lemma 19 we have

$$\frac{|S^9|}{|S|} \geq \frac{|S^8 \cap H_0^s|}{|S^2 \cap H_0^s|} \geq \deg_{\mathbb{C}}(H)^{1/3}.$$

By Proposition 18 this implies

$$\left(\frac{|S^3|}{|S|}\right)^7 \geq \frac{|S^9|}{|S|} \geq \deg_{\mathbb{C}}(H)^{1/3}.$$

Therefore

$$|S^3| \geq |S| \cdot \deg_{\mathbb{C}}(H)^{1/27} = |S| \cdot \left(\frac{q-1}{2}\right)^{\frac{1}{27}},$$

and our statement follows in this case.

Assume now that S^6 contains all conjugates of H, but $S^3 \neq G$. In this case S^6 contains a noncentral conjugacy class C of G together with C^{-1}. By a result of Lawther and Liebeck [1998] (see also [Ellers et al. 1999]), if $\mathcal{K}$ is any noncentral conjugacy class of $\mathrm{SL}(n,q)$ ($q \geq 4$), then $(\mathcal{K} \cup \mathcal{K}^{-1})^{40n} = \mathrm{SL}(n,q)$.

This implies $S^{240n} = G$. On the other hand, since $S^3 \neq G$, by Proposition 11 (which also holds for $\mathrm{SL}(n,q)$), we have $|S| \leq 2|G|/q^{(n-1)/3}$. By Proposition 18,

$$\left(\frac{|S^3|}{|S|}\right)^{240n-2} \geq \frac{|S^{240n}|}{|S|} = \frac{|G|}{|S|} \geq \frac{1}{2} q^{(n-1)/3},$$

and our statement follows. □

Note that in the above result the exponent of q in the growth is independent of n.

Essentially the same result continues to hold when G is an arbitrary simple group of Lie type over $\mathbb{F}_q$. More significantly, if in Theorem 20 we replace H by any subgroup with $\deg_{\mathbb{C}}(H) \geq q^{O(1)}$, we still obtain the conclusion $|S^3| \geq |S| \cdot q^{O(1)}$ (unless $S^3 = G$). Here H could be, say, a Suzuki subgroup, or a subfield subgroup $\mathrm{SL}(n, q_0)$ for some $\mathbb{F}_{q_0} \leq \mathbb{F}_q$.

Definition 21. A family of finite groups G_i is an expanding family if, for some natural number k, there exist generating subsets S_i of G_i of size at most k such that the graphs $\mathrm{Cay}(G_i, S_i)$ form an expander family.

Kassabov, Lubotzky, and Nikolov proved that all finite simple groups, except possibly the Suzuki groups, form an expanding family (see the announcement [Kassabov et al. 2006] for references). The same is now known for the Suzuki groups as well by [Breuillard et al. 2011b], and more generally by Theorem 6.

As an important tool in the proof of this result Lubotzky [2011] has shown that a simple group of Lie type of rank r over the field $\mathbb{F}_q$ (not a Suzuki group) decomposes as a product of $f(r)$ subgroups isomorphic to $\mathrm{SL}(2,q)$ or $\mathrm{PSL}(2,q)$. The argument in [Lubotzky 2011] is based on model theory. Theorem 20, extended to simple groups of Lie type, immediately yields another proof of this result, which (assuming known results) is both elementary and short.

Another group-theoretic proof, that gives a quadratic bound for $f(r)$, which uses Lemma 13 in a different way, is given in [Liebeck et al. 2011]. That proof does not yield growth results, and does not apply to, say, Suzuki subgroups.

Somewhat surprisingly the following affine variant of Theorem 20 is also true.

Lemma 22. *Let H be a d-generated finite group with* $\deg_{\mathbb{C}}(H) > K^{21}$, *and A a $\mathbb{Z}H$-module. Let $0 \in S \subseteq A$ be a symmetric H-invariant set generating A. Then (with multiplicative notation) either $|S^3| > K \cdot |S|$ or S^{7d} contains the submodule $[H, A]$.*

This plays an essential role in the proof of the polynomial inverse theorem (Theorem 8) in the case of linear groups over finite fields.

We end this section with a question vaguely related to superstrong approximation (see, e.g., [Breuillard 2014]).

Question 23. *Let G_i be an expanding family of finite groups. Is it true that for all i and every symmetric generating set S_i of G_i we have*

$$\operatorname{diam} \operatorname{Cay}(G_i, S_i) \leq C(\log |G_i|)^c,$$

where the constants c and C depend only on the family?

Note that answering an earlier question of Lubotzky and Weiss [Lubotzky and Weiss 1993] in the negative, in [Alon et al. 2001] some expanding families of groups G_i are constructed together with bounded sized generating sets S_i such that the corresponding Cayley graphs do not form an expander family (see also [Meshulam and Wigderson 2004]). The alternating groups also have these properties, but the fact that they form an expanding family was proved somewhat later in the breakthrough paper of Kassabov [2007].

Note also that since nonabelian finite simple groups form an expanding family, a positive answer to Question 23 would imply Babai's conjecture.

3. On the proof of the product theorem

In this section we describe the main ideas in the proof of the product theorem for $\mathrm{SL}(n, q)$. Simple groups of Lie type can be handled by essentially the same argument. A far-reaching generalisation of the argument plays an important role in the proof of the polynomial inverse theorem.

In the course of the proof we obtain various results which say that if L is a "nice" subgroup of an algebraic group G, generated by a set A, then A grows in some sense. These were motivated by earlier results of Helfgott [2008; 2011], Hrushovski and Pillay [1995], and Theorem 20.

Assume for example that A generates $L = \mathrm{SL}(n, q)$ (q a power of a prime p), which is a subgroup of $G = \mathrm{SL}(n, \overline{\mathbb{F}}_p)$, and "$A$ does not grow", in the sense that $|A^3|$ is not much larger than $|A|$. Using an "escape from subvarieties" argument, Helfgott [2011] proved the following useful lemma: *If T is a maximal torus in G then $|T \cap A|$ is not much larger than $|A|^{1/(n+1)}$.* This is natural to expect for dimensional reasons since $\dim T / \dim G = (n-1)/(n^2-1) = 1/(n+1)$.

In particular we see that, if A contains a maximal torus of L, then A is a very large subset of L, hence $A^3 = L$ by Corollary 10. This consequence of Helfgott's lemma is very similar to our baby product theorem.

What is the right generalisation of these results? Our first answer to this question which is subsumed by [Pyber and Szabó 2010, Theorem 49], is the following inequality.

Theorem 24 [Pyber and Szabó 2009]. *Let* $\varepsilon > 0$ *be a fixed constant. Let* G *be a linear algebraic group defined over* $\mathbb{F}_q$ *which satisfies the following conditions*:

(a) *The centraliser of* $G(\mathbb{F}_q)$ *(the group of* $\mathbb{F}_q$*-points) in* G *is finite.*

(b) $G(\mathbb{F}_q)$ *does not normalise any closed subgroup* $H < G$ *with* $0 < \dim H < \dim G$.

Let $1 \in A$ *be a generating set of* $G(\mathbb{F}_q)$ *and* V *a subvariety of* G *of positive dimension. If* $A \cap V$ *is large enough (i.e., greater than some appropriate function of* ε, $\dim G$, *and the degrees of the varieties* G *and* V*) then*

$$|A^m| \geq |A \cap V|^{(1-\varepsilon)\dim G / \dim V},$$

for some m *which depends only on* ε *and* $\dim G$.

Taking G to be $\mathrm{SL}(n, \overline{\mathbb{F}}_p)$ and T a maximal torus in G we obtain Helfgott's lemma. Here we use the fact, essential to Helfgott's approach, that if A^3 is not much larger than A then A^m is not much larger either (see Proposition 18).

Breuillard, Green, and Tao arrived at a somewhat stronger inequality [Breuillard et al. 2011a], by taking hints from [Hrushovski 2012]. (Our inequality was obtained in March 2009, independently of [Hrushovski 2012].) For a formulation of their result, which they call a Larsen–Pink type inequality, see [Breuillard 2014] in this volume.

Anyway, the inequality says that if a generating set of $L = \mathrm{SL}(n, q)$ does not grow then it is distributed in a "balanced way" in L.

How do we break this balance? We have to start conjugating! More precisely, we have to answer another question. Which are the "interesting subvarieties" to which Theorem 24 should be applied to? Centralisers and conjugacy classes, as explained below.

For an element x consider the conjugation map $g \to x^g$. The image of this map is $cl(x)$, the conjugacy class of x in G, and the fibres are cosets of the centraliser $\mathscr{C}_G(x)$. For the closure $\overline{cl(x)}$ of the set $cl(x)$ this implies that

$$\dim \mathscr{C}_G(x) + \dim \overline{cl(x)} = \dim G.$$

It follows that an upper bound for $|A \cap \overline{cl(x)}|$ implies a lower bound for $|A \cap \mathscr{C}_G(x)|$ which matches the upper bound given by Theorem 24.

Maximal tori are the centralisers of their regular semisimple elements (that is elements with all eigenvalues distinct).

Let us say that A *covers* a maximal torus T if $|T \cap A|$ contains a regular semisimple element. We obtain a fundamental dichotomy:

Lemma [Pyber and Szabó 2010, Lemma 60]. *Assume that a generating set A does not grow.*

(i) *If A does not cover a maximal torus T then $|T \cap A|$ is not much larger than $|A|^{1/(n+1)-1/(n^2-1)}$.*

(ii) *If A covers T then $|T \cap AA^{-1}|$ is not much smaller than $|A|^{1/(n+1)}$.*

To break the balance when A is not very large, we only have to start conjugating tori with elements of A. That is, an argument analogous to the proof of our baby product theorem based on the above dichotomy completes the proof of the product theorem.

4. On the proof of the polynomial inverse theorem

A large part of the proof of the polynomial inverse theorem (Theorem 8), which is already contained in [Pyber and Szabó 2010], is to show that if S is a symmetric subset of $\mathrm{GL}(n, \overline{\mathbb{F}})$ such that $|S^3| \leq K|S|$, then S is contained in the union of polynomially many cosets of a soluble-by-finite subgroup.

For the proof of this we first establish a generalisation of Theorem 24, called the spreading theorem [Pyber and Szabó 2010, Theorem 49]. Roughly speaking it says the following.

For a subvariety V of positive dimension we define the *concentration* of a finite set S in V as $(\log|S \cap V|)/\dim V$.

Let S be a finite subset in a connected linear algebraic group G such that $\mathscr{C}_G(S)$ is finite. If G has a subvariety X in which S has much larger concentration than in G then we can find a connected closed subgroup $H \leq G$ normalised by S in which a small power of S has similarly large concentration. (When G is the simple algebraic group used to define a finite group of Lie type L and S generates L then H turns out to be G itself.)

Next we have to introduce certain "generalised tori". If G is a simple algebraic group then a maximal torus T can be obtained as the centraliser of a regular semisimple element of T. Moreover, most elements of T are regular semisimple. Since T is abelian, it coincides with its centraliser $\mathscr{C}_G(T)$.

In an arbitrary nonnilpotent linear algebraic group G we consider a class of subgroups we call CCC-subgroups with similar properties. Our CCC-subgroups are the connected centralisers of connected subgroups (which explains the name).

For a subvariety X denote by X^{gen} the set of all $(\dim G)$-tuples $\underline{g} \in \prod^{\dim G} X$ such that the connected centralisers $\mathscr{C}_G(\underline{g})^0$ and $\mathscr{C}_G(X)^0$ are equal.

It turns out that if X is a CCC-subgroup then X^{gen} is a dense open subset of $\prod^{\dim G} X$. Moreover, if Y is another CCC-subgroup, then $X^{\text{gen}} \cap Y^{\text{gen}} = \varnothing$. Finally we have $\mathscr{C}_G(\mathscr{C}_G(X)^0)^0 = X$.

If we replace the tori in the proof of the product theorem with CCC-subgroups, the proof goes through, and shows that a small power of S has very large concentration in some subgroup H of G normalised by S.

An induction argument then shows that S is contained in polynomially many cosets of some soluble-by-finite subgroup, as we wanted.

With various additional arguments this result can be upgraded as follows.

Theorem 25 [Pyber and Szabó 2010]. *Let $1 \in S$ be a finite symmetric subset of $\mathrm{GL}(n, \mathbb{F})$ such that $|S^3| \leq K|S|$ for some $K \geq 1$. Then S can be covered by polynomially many cosets of some soluble-by-finite subgroup Γ normalised by S. Moreover, Γ has a soluble normal subgroup H such that $\Gamma \subset S^6 H$.*

This is essentially a polynomial version of a result of Hrushovski [2012].

Since a soluble-by-finite subgroup of $\mathrm{GL}(n, \mathbb{C})$ has a soluble subgroup of bounded index, Theorem 25 implies the polynomial inverse theorem for fields of characteristic zero. As mentioned before, the characteristic zero case was first obtained in [Breuillard et al. 2011a].

In the rest of this section $\mathbb{F}$ denotes a field of characteristic $p > 0$.

To complete the proof of Theorem 8 next we have to settle the case of finite linear groups.

Definition 26. As usual, $O_p(G)$ denotes the maximal normal p-subgroup of a finite group G. A group is called *perfect*, if it has no abelian quotients. A finite group is called *quasisimple* if it is a perfect central extension of a finite simple group. We denote by $\mathrm{Lie}^*(p)$ the set of central products of quasisimple groups of Lie type of characteristic p.

The following deep result is essentially due to Weisfeiler [1984].

Proposition 27. *Let G be a finite subgroup of $\mathrm{GL}(n, \mathbb{F})$. Then G has a normal subgroup H of index at most $f(n)$ such that $H \geq O_p(G)$ and $H/O_p(G)$ is the central product of an abelian p'-group and quasisimple groups of Lie type of characteristic p, where the bound $f(n)$ depends on n.*

It was proved by Collins [2008] that for $n \geq 71$ one can take $f(n) = (n+2)!$. Remarkably a (noneffective) version of the above result was obtained by Larsen and Pink [2011] without relying on the classification of finite simple groups.

It is an easy consequence of Weisfeiler's theorem, that if $\deg_{\mathbb{C}}(G)$ is "large" then in fact $G/O_p(G)$ is in $\mathrm{Lie}^*(p)$. For such groups one can use Lemma 22 inductively to prove that if a symmetric set S with $|S^3| \leq K|S|$ projects onto $G/O_p(G)$, then in fact $S^3 = G$.

This result, and the product theorem are the main ingredients in the proof of the polynomial inverse theorem at least in the case finite linear groups.

The following classical theorem of Malcev (see, e.g., [Wehrfritz 1973]) makes it possible to use finite group theory to study properties of finitely generated linear groups.

Proposition 28. *Let* Γ *be a finitely generated subgroup of* $\mathrm{GL}(n, \mathbb{F})$. *For every finite set of elements* $g_1, \ldots, g_t$ *of* Γ *there exists a finite field* $\mathbb{K}$ *of the same characteristic and a homomorphism* $\phi : \Gamma \to \mathrm{GL}(n, \mathbb{K})$ *such that* $\phi(g_1), \ldots, \phi(g_t)$ *are all distinct.*

In the final part of the proof of the polynomial inverse theorem we have to further upgrade Theorem 25. Due to some lucky coincidences this can be done by combining the finite case with Malcev's theorem.

References

[Alon et al. 2001] N. Alon, A. Lubotzky, and A. Wigderson, "Semi-direct product in groups and zig-zag product in graphs: Connections and applications (extended abstract)", pp. 630–637 in *42nd IEEE Symposium on Foundations of Computer Science* (Las Vegas, 2001), IEEE Computer Soc., Los Alamitos, CA, 2001.

[Babai and Seress 1992] L. Babai and Á. Seress, "On the diameter of permutation groups", *European J. Combin.* **13**:4 (1992), 231–243.

[Babai and Sós 1985] L. Babai and V. T. Sós, "Sidon sets in groups and induced subgraphs of Cayley graphs", *European J. Combin.* **6**:2 (1985), 101–114.

[Babai et al. 2008] L. Babai, N. Nikolov, and L. Pyber, "Product growth and mixing in finite groups", pp. 248–257 in *Proceedings of the Nineteenth Annual ACM-SIAM Symposium on Discrete Algorithms*, ACM, New York, 2008.

[Bourgain and Gamburd 2008] J. Bourgain and A. Gamburd, "Uniform expansion bounds for Cayley graphs of $\mathrm{SL}_2(\mathbb{F}_p)$", *Ann. of Math.* (2) **167**:2 (2008), 625–642.

[Breuillard 2014] E. Breuillard, "A brief introduction to approximate groups", pp. 23–50 in *Thin groups and superstrong approximation*, edited by H. Oh and E. Breuillard, Math. Sci. Res. Inst. Publ. **61**, Cambridge Univ. Press, Cambridge, 2014.

[Breuillard and Green 2011] E. Breuillard and B. Green, "Approximate groups, II: The solvable linear case", *Q. J. Math.* **62**:3 (2011), 513–521.

[Breuillard et al. 2011a] E. Breuillard, B. Green, and T. Tao, "Approximate subgroups of linear groups", *Geom. Funct. Anal.* **21**:4 (2011), 774–819.

[Breuillard et al. 2011b] E. Breuillard, B. Green, and T. Tao, "Suzuki groups as expanders", *Groups Geom. Dyn.* **5**:2 (2011), 281–299.

[Breuillard et al. 2012a] E. Breuillard, B. Green, R. Guralnick, and T. Tao, "Strongly dense free subgroups of semisimple algebraic groups", *Israel J. Math.* **192**:1 (2012), 347–379.

[Breuillard et al. 2012b] E. Breuillard, B. Green, and T. Tao, "The structure of approximate groups", *Publ. Math. IHES* **116**:1 (2012), 115–221.

[Breuillard et al. 2013] E. Breuillard, B. Green, R. Guralnick, and T. Tao, "Expansion in finite simple groups of Lie type", preprint, 2013. arXiv 1309.1975

[Collins 2008] M. J. Collins, "Modular analogues of Jordan's theorem for finite linear groups", *J. Reine Angew. Math.* **624** (2008), 143–171.

[Dinai 2011] O. Dinai, "Growth in SL_2 over finite fields", *J. Group Theory* **14**:2 (2011), 273–297.

[Elekes and Király 2001] G. Elekes and Z. Király, "On the combinatorics of projective mappings", *J. Algebraic Combin.* **14**:3 (2001), 183–197.

[Ellers et al. 1999] E. W. Ellers, N. Gordeev, and M. Herzog, "Covering numbers for Chevalley groups", *Israel J. Math.* **111** (1999), 339–372.

[Gamburd et al. 2009] A. Gamburd, S. Hoory, M. Shahshahani, A. Shalev, and B. Virág, "On the girth of random Cayley graphs", *Random Structures Algorithms* **35**:1 (2009), 100–117.

[Gill and Helfgott 2010] N. Gill and H. Helfgott, "Growth in solvable subgroups of $\mathrm{GL}_r(\mathbb{Z}/p\mathbb{Z})$", preprint, 2010. arXiv 1008.5264

[Gill et al. 2011] N. Gill, L. Pyber, I. Short, and E. Szabó, "On the product decomposition conjecture for finite simple groups", preprint, 2011. Accepted in *Groups, Geometry, and Dynamics.* arXiv 1111.3497

[Gowers 2008] W. T. Gowers, "Quasirandom groups", *Combin. Probab. Comput.* **17**:3 (2008), 363–387.

[de la Harpe 2000] P. de la Harpe, *Topics in geometric group theory*, University of Chicago Press, Chicago, IL, 2000.

[Helfgott 2008] H. A. Helfgott, "Growth and generation in $\mathrm{SL}_2(\mathbb{Z}/p\mathbb{Z})$", *Ann. of Math.* (2) **167**:2 (2008), 601–623.

[Helfgott 2011] H. A. Helfgott, "Growth in $\mathrm{SL}_3(\mathbb{Z}/p\mathbb{Z})$", *J. Eur. Math. Soc.* **13**:3 (2011), 761–851.

[Hrushovski 2012] E. Hrushovski, "Stable group theory and approximate subgroups", *J. Amer. Math. Soc.* **25**:1 (2012), 189–243.

[Hrushovski and Pillay 1995] E. Hrushovski and A. Pillay, "Definable subgroups of algebraic groups over finite fields", *J. Reine Angew. Math.* **462** (1995), 69–91.

[Kassabov 2007] M. Kassabov, "Symmetric groups and expander graphs", *Invent. Math.* **170**:2 (2007), 327–354.

[Kassabov et al. 2006] M. Kassabov, A. Lubotzky, and N. Nikolov, "Finite simple groups as expanders", *Proc. Natl. Acad. Sci. USA* **103**:16 (2006), 6116–6119.

[Kleidman and Liebeck 1990] P. Kleidman and M. Liebeck, *The subgroup structure of the finite classical groups*, London Math. Soc. Lecture Note Series **129**, Cambridge University Press, 1990.

[Landazuri and Seitz 1974] V. Landazuri and G. M. Seitz, "On the minimal degrees of projective representations of the finite Chevalley groups", *J. Algebra* **32** (1974), 418–443.

[Larsen and Pink 2011] M. J. Larsen and R. Pink, "Finite subgroups of algebraic groups", *J. Amer. Math. Soc.* **24**:4 (2011), 1105–1158.

[Larsen et al. 2011] M. Larsen, A. Shalev, and P. H. Tiep, "The Waring problem for finite simple groups", *Ann. of Math.* (2) **174**:3 (2011), 1885–1950.

[Lawther and Liebeck 1998] R. Lawther and M. W. Liebeck, "On the diameter of a Cayley graph of a simple group of Lie type based on a conjugacy class", *J. Combin. Theory Ser. A* **83**:1 (1998), 118–137.

[Liebeck and Pyber 2001] M. W. Liebeck and L. Pyber, "Finite linear groups and bounded generation", *Duke Math. J.* **107**:1 (2001), 159–171.

[Liebeck and Shalev 2001] M. W. Liebeck and A. Shalev, "Diameters of finite simple groups: sharp bounds and applications", *Ann. of Math.* (2) **154**:2 (2001), 383–406.

[Liebeck et al. 2011] M. W. Liebeck, N. Nikolov, and A. Shalev, "Groups of Lie type as products of SL_2 subgroups", *J. Algebra* **326** (2011), 201–207.

[Liebeck et al. 2012] M. W. Liebeck, N. Nikolov, and A. Shalev, "Product decompositions in finite simple groups", *Bull. Lond. Math. Soc.* **44**:3 (2012), 469–472.

[Lubotzky 2011] A. Lubotzky, "Finite simple groups of Lie type as expanders", *J. Eur. Math. Soc.* **13**:5 (2011), 1331–1341.

[Lubotzky 2012] A. Lubotzky, "Expander graphs in pure and applied mathematics", *Bull. Amer. Math. Soc.* (*N.S.*) **49**:1 (2012), 113–162.

[Lubotzky and Weiss 1993] A. Lubotzky and B. Weiss, "Groups and expanders", pp. 95–109 in *Expanding graphs* (Princeton, NJ, 1992), edited by J. Friedman, DIMACS Ser. Discrete Math. Theoret. Comput. Sci. **10**, Amer. Math. Soc., Providence, RI, 1993.

[Meshulam and Wigderson 2004] R. Meshulam and A. Wigderson, "Expanders in group algebras", *Combinatorica* **24**:4 (2004), 659–680.

[Nikolov and Pyber 2011] N. Nikolov and L. Pyber, "Product decompositions of quasirandom groups and a Jordan type theorem", *J. Eur. Math. Soc.* **13**:4 (2011), 1063–1077.

[Nikolov and Segal 2007a] N. Nikolov and D. Segal, "On finitely generated profinite groups, I: Strong completeness and uniform bounds", *Ann. of Math.* (2) **165**:1 (2007), 171–238.

[Nikolov and Segal 2007b] N. Nikolov and D. Segal, "On finitely generated profinite groups, II: Products in quasisimple groups", *Ann. of Math.* (2) **165**:1 (2007), 239–273.

[Nikolov and Segal 2011] N. Nikolov and D. Segal, "Powers in finite groups", *Groups Geom. Dyn.* **5**:2 (2011), 501–507.

[Nikolov and Segal 2012] N. Nikolov and D. Segal, "Generators and commutators in finite groups; abstract quotients of compact groups", *Invent. Math.* **190**:3 (2012), 513–602.

[Pyber and Szabó 2009] L. Pyber and E. Szabó, "Generating simple groups", preprint, 2009, http://www.renyi.hu/~endre/growth.pdf.

[Pyber and Szabó 2010] L. Pyber and E. Szabó, "Growth in finite simple groups of Lie type of bounded rank", preprint, 2010. arXiv 1005.1858

[Pyber and Szabó ≥ 2012] L. Pyber and E. Szabó, "A polynomial inverse theorem in linear groups", in preparation.

[Ruzsa 2010] I. Z. Ruzsa, "Towards a noncommutative Plünnecke-type inequality", pp. 591–605 in *An irregular mind*, edited by I. Bárány and J. Solymosi, Bolyai Soc. Math. Stud. **21**, János Bolyai Math. Soc., Budapest, 2010.

[Ruzsa and Turjányi 1985] I. Z. Ruzsa and S. Turjányi, "A note on additive bases of integers", *Publ. Math. Debrecen* **32**:1-2 (1985), 101–104.

[Sarnak and Xue 1991] P. Sarnak and X. X. Xue, "Bounds for multiplicities of automorphic representations", *Duke Math. J.* **64**:1 (1991), 207–227.

[Shalev 2009] A. Shalev, "Word maps, conjugacy classes, and a noncommutative Waring-type theorem", *Ann. of Math.* (2) **170**:3 (2009), 1383–1416.

[Tao 2012] T. Tao, "The Bourgain–Gamburd expansion machine", Blog post, 2012, http://goo.gl/7lF86.

[Varjú 2012] P. P. Varjú, "Expansion in $\mathrm{SL}_d(\mathbb{O}_K/I)$, I square-free", *J. Eur. Math. Soc.* **14**:1 (2012), 273–305.

[Wehrfritz 1973] B. A. F. Wehrfritz, *Infinite linear groups. An account of the group-theoretic properties of infinite groups of matrices*, Ergebnisse der Mat. **76**, Springer, New York, 1973.

[Weisfeiler 1984] B. Weisfeiler, "Post-classification version of Jordan's theorem on finite linear groups", *Proc. Nat. Acad. Sci. U.S.A.* **81**:16, Phys. Sci. (1984), 5278–5279.

pyber@renyi.hu A. Rényi Institute of Mathematics, Hungarian Academy of Sciences, P.O. Box 127, H-1364 Budapest, Hungary

endre@renyi.hu A. Rényi Institute of Mathematics, Hungarian Academy of Sciences, P.O. Box 127, H-1364 Budapest, Hungary

Thin groups and superstrong approximation
MSRI Publications
Volume **61**, 2013

Strong approximation for algebraic groups

ANDREI S. RAPINCHUK

We survey results on strong approximation in algebraic groups, considering in detail the classical form of strong approximation as well as more recent results on strong approximation for arbitrary Zariski-dense subgroups. Some other topics, ranging from strong approximation in homogeneous spaces of algebraic groups to various applications of strong approximation, are also discussed.

1. Introduction

This article is a survey of known results related to strong approximation in algebraic groups. We focus primarily on two aspects: the classical form of strong approximation, which is really strong approximation for S-arithmetic groups (Section 2), and its more modern version for arbitrary Zariski-dense subgroups (Section 3). Along the way we will also mention results dealing with strong approximation in arbitrary varieties and particularly homogeneous spaces (which are probably not so well known to the general audience as some other results in the article) and some applications. The reader will find more applications of strong approximation for Zariski-dense subgroups in other articles in this volume.

1A. ***Strong approximation and congruences.*** The most elementary way to start thinking about strong approximation is in terms of lifting solutions of integer polynomial equations mod m for all $m \geqslant 1$, to integer solutions. So, suppose we have a family of polynomials

$$f_\alpha(x_1, \dots, x_d) \in \mathbb{Z}[x_1, \dots, x_d], \quad \alpha \in I,$$

and we let $X \subset \mathbb{A}^d_{\mathbb{Z}}$ denote the closed affine subscheme defined by these polynomials. Thus, for any $\mathbb{Z}$-algebra R, the scheme X has the following set of R-points:

$$X(R) = \{(a_1, \dots, a_d) \in R^d \mid f_\alpha(a_1, \dots, a_d) = 0 \text{ for all } \alpha \in I\}.$$

Then for any integer $m \geqslant 1$, we have a natural reduction modulo m map

$$\rho_m : X(\mathbb{Z}) \to X(\mathbb{Z}/m\mathbb{Z}),$$

and the question is whether these maps are *surjective* for all m. (Of course, this question is meaningful only if we assume that $X(\mathbb{Z}/m\mathbb{Z}) \neq \varnothing$ for all m.) Observe that for $m \mid n$, there is a canonical homomorphism $\mathbb{Z}/n\mathbb{Z} \to \mathbb{Z}/m\mathbb{Z}$, hence a natural map $\pi^n_m : X(\mathbb{Z}/n\mathbb{Z}) \to X(\mathbb{Z}/m\mathbb{Z})$. Clearly, $\{X(\mathbb{Z}/m\mathbb{Z}), \pi^n_m\}$ is an inverse system, so we can assemble all the $X(\mathbb{Z}/m\mathbb{Z})$'s together by taking the inverse limit:

$$\varprojlim X(\mathbb{Z}/m\mathbb{Z}) = X(\widehat{\mathbb{Z}}), \quad \text{where } \widehat{\mathbb{Z}} = \varprojlim \mathbb{Z}/m\mathbb{Z}.$$

Recall that the Chinese remainder theorem furnishes an isomorphism $\widehat{\mathbb{Z}} \simeq \prod_p \mathbb{Z}_p$, where $\mathbb{Z}_p$ is the ring of p-adic integers and the product is taken over all primes, which allows us to identify $X(\widehat{\mathbb{Z}})$ with $\prod_p X(\mathbb{Z}_p)$.

Just as above, for any integer $m \geqslant 1$, there is a natural map

$$\hat{\rho} : X(\widehat{\mathbb{Z}}) \longrightarrow X(\widehat{\mathbb{Z}}/m\widehat{\mathbb{Z}}) = X(\mathbb{Z}/m\mathbb{Z}).$$

The preimages of points under the $\hat{\rho}_m$ form a basis of a natural topology on $X(\widehat{\mathbb{Z}})$, which coincides with either of the following topologies:

- the topology of the inverse limit on $\varprojlim X(\mathbb{Z}/m\mathbb{Z})$ — cf. [Klopsch et al. 2011, Chapter I, 5.3];
- the topology induced by the embedding $X(\widehat{\mathbb{Z}}) \hookrightarrow \widehat{\mathbb{Z}}^d$, where $\widehat{\mathbb{Z}}$ is endowed with the inverse limit topology on $\varprojlim \mathbb{Z}/m\mathbb{Z}$;
- the direct product topology on $\prod_p X(\mathbb{Z}_p)$, where $X(\mathbb{Z}_p)$ gets its topology from the embedding $X(\mathbb{Z}_p) \hookrightarrow \mathbb{Z}_p^d$, and $\mathbb{Z}_p$ is endowed with the natural p-adic topology.

As an immediate consequence, we have:

Lemma 1.1. *The following conditions are equivalent*:

(1) $\rho_m : X(\mathbb{Z}) \to X(\mathbb{Z}/m\mathbb{Z})$ *is surjective for all integers* $m \geqslant 1$;

(2) *the natural embedding* $\iota : X(\mathbb{Z}) \hookrightarrow X(\widehat{\mathbb{Z}})$ *has a dense image in the above topology.*

In this situation, we say X has *strong approximation* if it satisfies the equivalent conditions of Lemma 1.1 (of course, this is only a first approximation to the precise definition(s) of strong approximation that will be given later; see Section 2A). Intuitively, strong approximation should not be very common as there are plentiful examples where $X(\widehat{\mathbb{Z}}) \neq \varnothing$ but $X(\mathbb{Z}) = \varnothing$ (i.e., the Hasse principle fails — note that here we actually omit the archimedean place of $\mathbb{Q}$), and also examples where $X(\mathbb{Z})$ is nonempty but so "small" that it cannot possibly be dense in $X(\widehat{\mathbb{Z}})$. A classical example of the second situation is a cubic hypersurface $X \subset \mathbb{A}^3$ given by the equation

$$3x^3 + 4y^3 + 5z^3 = 0;$$

it is known that $X(\mathbb{Z}) = \{(0,0,0)\}$ but $X(\mathbb{Z}_p) \neq \{(0,0,0)\}$ (hence infinite as any point on X other than the origin is smooth) for all prime p. In fact, very little appears to be known about strong approximation for schemes (varieties) lying outside some special classes such as homogeneous spaces — one can only give some *necessary conditions* (see Proposition 2.2 and subsequent remarks). So, in this article we will deal almost exclusively with algebraic groups.

1B. SL$_2$ *versus* GL$_2$. Let us start with two elementary examples: $G_1 = \mathrm{SL}_2$ and $G_2 = \mathrm{GL}_2$. One doesn't see much of a difference between these examples just by looking at the defining equations, which can be written as follows, with the obvious labeling of coordinates:

- G_1 can be realized as a hypersurface in $\mathbb{A}^4$, given by

$$x_{11}x_{22} - x_{12}x_{21} = 1.$$

- G_2 can be realized as a hypersurface in $\mathbb{A}^5$, given by

$$y(x_{11}x_{22} - x_{12}x_{21}) = 1.$$

However, G_1 has strong approximation, and G_2 does not.

Lemma 1.2. *For any $m > 1$, the reduction modulo m map*

$$\rho_m : \mathrm{SL}_2(\mathbb{Z}) \longrightarrow \mathrm{SL}_2(\mathbb{Z}/m\mathbb{Z})$$

is surjective.

Proof. The argument does not use equations (in fact, it is not a completely trivial task to prove strong approximation using equations in this case — see the discussion after Proposition 2.4). The crucial observation is that any $\bar{g} \in \mathrm{SL}_2(\mathbb{Z}/m\mathbb{Z})$ can be written as a product of elementary matrices:

$$\bar{g} = \prod_k e_{i_k j_k}(\bar{a}_k) \quad \text{with } (i_k, j_k) \in \{(1,2),(2,1)\} \text{ and } \bar{a}_k \in \mathbb{Z}/m\mathbb{Z}. \tag{1}$$

(As usual, for $i \neq j$, we let $e_{ij}(a)$ denote the elementary matrix having a as its ij-entry.) For this, one needs to observe that if $m = p_1^{\alpha_1} \cdots p_r^{\alpha_r}$ then it follows from the Chinese remainder theorem that

$$\mathrm{SL}_2(\mathbb{Z}/m\mathbb{Z}) = \mathrm{SL}_2(\mathbb{Z}/p_1^{\alpha_1}\mathbb{Z}) \times \cdots \times \mathrm{SL}_2(\mathbb{Z}/p_r^{\alpha_r}\mathbb{Z}),$$

which reduces the problem to the case where $m = p^\alpha$. Now, given

$$\bar{g} = \begin{pmatrix} x_{11} & x_{12} \\ x_{21} & x_{22} \end{pmatrix} \in \mathrm{SL}_2(\mathbb{Z}/p^\alpha\mathbb{Z}),$$

we see that either x_{11} or x_{12} is a unit mod p^α, so using Gaussian elimination one can easily write $\bar{g}$ as a product of elementaries over $\mathbb{Z}/p^\alpha\mathbb{Z}$.

Next, given an arbitrary $\bar{g} \in \mathrm{SL}_2(\mathbb{Z}/m\mathbb{Z})$, pick a factorization (1), and furthermore, for each k pick an integer a_k in the class $\bar{a}_k$ modulo m. Set

$$g = \prod_k e_{i_k j_k}(a_k) \in \mathrm{SL}_2(\mathbb{Z}).$$

Then $\rho_m(g) = \bar{g}$, proving the surjectivity of ρ_m. □

Note that the proof of Lemma 1.2 relies on the consideration of unipotent elements, so it is worth pointing out that, as we will see in the course of this article, unipotent elements are involved in one way or another in most known results on strong approximation (even when the group at hand does not contain any nontrivial rational unipotent elements, i.e., is anisotropic).

The fact that $G_2 = \mathrm{GL}_2$ does not have strong approximation is much easier: in fact, already the map

$$\rho_5 : \mathrm{GL}_2(\mathbb{Z}) \longrightarrow \mathrm{GL}_2(\mathbb{Z}/5\mathbb{Z})$$

fails to be surjective. (Indeed, since all matrices in $\mathrm{GL}_2(\mathbb{Z})$ have determinant ± 1, the matrices in $\rho_5(\mathrm{GL}_2(\mathbb{Z}))$ have determinant $\pm 1 \pmod 5$, and therefore, for example, $\left(\begin{smallmatrix} \bar{1} & \bar{0} \\ \bar{0} & \bar{2} \end{smallmatrix}\right)$ does not lie in the image of ρ_5.) One can conceptually articulate the obstruction that prevents GL_2 from having strong approximation in this case by saying that in order for an affine $\mathbb{Q}$-variety X to have strong approximation,

$$X(\mathbb{Z}) \text{ must be Zariski-dense in } X.$$

Indeed, let $Y = \overline{X(\mathbb{Z})}$ be the Zariski closure of $X(\mathbb{Z})$ in X, and assume that $Y \neq X$. Pick a point $a \in X(\bar{\mathbb{Q}}) \setminus Y(\bar{\mathbb{Q}})$, where $\bar{\mathbb{Q}}$ is an algebraic closure of $\mathbb{Q}$. Then one can find a polynomial $f \in \mathbb{Z}[x_1, \ldots, x_d]$ that vanishes on Y and such that $f(a) \neq 0$. It follows from Chebotarev's density theorem that for infinitely many primes p, we have $a \in X(\mathbb{Z}_p)$ and $f(a) \not\equiv 0 \pmod p$. Let $\bar{a} \in X(\mathbb{F}_p)$ be the reduction of a modulo p, where $\mathbb{F}_p = \mathbb{Z}/p\mathbb{Z} = \mathbb{Z}_p/p\mathbb{Z}_p$. (Note that it would be more appropriate to write $\underline{X}^{(p)}(\mathbb{F}_p)$ instead of $X(\mathbb{F}_p)$, where $\underline{X}^{(p)}$ denotes the reduction of X modulo p, but we will slightly abuse the notations in this introductory section in order to keep them simple.) Then clearly

$$\bar{a} \in X(\mathbb{F}_p) \setminus Y(\mathbb{F}_p),$$

and in particular, $X(\mathbb{F}_p) \neq Y(\mathbb{F}_p)$. On the other hand, the image of the reduction map $\rho_p : X(\mathbb{Z}) \to X(\mathbb{F}_p)$ is obviously contained in $Y(\mathbb{F}_p)$. Thus, if $X(\mathbb{Z})$ is not Zariski-dense in X then ρ_p fails to be surjective for infinitely many p, which certainly prevents X from having strong approximation. (Incidentally, this observation implies that if G is an algebraic $\mathbb{Q}$-group and $G(\mathbb{Z})$ is not Zariski-dense in G then the closure of $G(\mathbb{Z})$ in $G(\widehat{\mathbb{Z}})$ is of infinite index.)

In fact, the conclusion about the absence of strong approximation in X as above can be made even sharper. First, it is easy to show that X cannot possibly have strong approximation unless it is *absolutely irreducible* (see the remark after Proposition 2.2). So, assume that X is such. Then by the Lang–Weil estimates [1954] we have

$$|X(\mathbb{F}_p)| \approx p^{\dim X},$$

for p sufficiently large. Similarly, for any proper $\mathbb{Q}$-subvariety $Y \subset X$, the cardinality $|Y(\mathbb{F}_p)|$ is bounded above by an expression of the form $C \cdot p^{\dim Y}$, where C is a constant independent of p. It follows that $Y(\mathbb{F}_p) \neq X(\mathbb{F}_p)$ for almost all p, and therefore unless $X(\mathbb{Z})$ is Zariski-dense in X, the reduction map $\rho_p : X(\mathbb{Z}) \to X(\mathbb{F}_p)$ is not surjective for *almost all* p.

So, the fact that $\mathrm{GL}_2(\mathbb{Z})$ is not Zariski-dense in GL_2 (its Zariski closure is precisely the subgroup consisting of $g \in \mathrm{GL}_2$ that satisfy $(\det g)^2 - 1 = 0$), is definitely one of the factors that prevent GL_2 from having strong approximation; in fact, the reduction maps ρ_p are nonsurjective for all $p \geqslant 5$. Now, let us slightly change the set-up by replacing the ring of integers $\mathbb{Z}$ with some localization, for example, $\mathbb{Z}[\frac{1}{2}]$. Then $\mathrm{GL}_2\left(\mathbb{Z}[\frac{1}{2}]\right)$ is already Zariski-dense in GL_2, and in fact the map

$$\rho_5 : \mathrm{GL}_2(\mathbb{Z}[\tfrac{1}{2}]) \longrightarrow \mathrm{GL}_2(\mathbb{Z}/5\mathbb{Z})$$

is surjective, however the map

$$\rho_{17} : \mathrm{GL}_2(\mathbb{Z}[\tfrac{1}{2}]) \longrightarrow \mathrm{GL}_2(\mathbb{Z}/17\mathbb{Z})$$

is not. The reason is that the possible determinants of matrices in $\mathrm{GL}_2\left(\mathbb{Z}[\frac{1}{2}]\right)$ are of the form $\pm 2^{\ell}$ with $\ell \in \mathbb{Z}$, hence squares modulo $p = 17$ (in fact, this property will hold for any prime of the form $8k+1$, and by Dirichlet's theorem there are infinitely many such primes; compare Section 2B).

We see that Zariski density is definitely not sufficient for strong approximation in the general case. At the same time, let us consider the following example involving various subgroups of the group $\mathrm{SL}_2(\mathbb{Z})$. We have

$$\Gamma_0 := \mathrm{SL}_2(\mathbb{Z}) = \left\langle \left(\begin{smallmatrix} 1 & 1 \\ 0 & 1 \end{smallmatrix}\right), \left(\begin{smallmatrix} 1 & 0 \\ 1 & 1 \end{smallmatrix}\right) \right\rangle.$$

For $\ell \geqslant 1$, we define

$$\Gamma_\ell = \left\langle \begin{pmatrix} 1 & 2^\ell \\ 0 & 1 \end{pmatrix}, \begin{pmatrix} 1 & 0 \\ 2^\ell & 1 \end{pmatrix} \right\rangle.$$

Then we have the inclusions

$$\Gamma_0 \supset \Gamma_1 \supset \Gamma_2 \supset \cdots \supset \Gamma_\ell \supset \Gamma_{\ell+1} \supset \cdots,$$

with

$$[\Gamma_0 : \Gamma_1] = 12 \quad \text{and} \quad [\Gamma_\ell : \Gamma_{\ell+1}] = \infty, \quad \text{for } \ell \geqslant 1.$$

(The fastest way to verify both of these claims is to use the *virtual* Euler–Poincaré characteristic; cf. [Serre 1971]. It is known that the Euler–Poincaré characteristic $\chi(\Gamma_0) = -\frac{1}{12}$. On the other hand, for any $m \geqslant 2$ the matrices $\left(\begin{smallmatrix} 1 & m \\ 0 & 1 \end{smallmatrix}\right)$ and $\left(\begin{smallmatrix} 1 & 0 \\ m & 1 \end{smallmatrix}\right)$ generate a rank 2 free subgroup $\Delta_m \subset \Gamma_0$ (see [de la Harpe 2000, p. 26]), so $\chi(\Delta_m) = -1$. It is an elementary exercise to show that $\Gamma_1 = \Delta_2$ contains the congruence subgroup $\mathrm{SL}_2(\mathbb{Z}, 4)$ modulo 4, so the index $d = [\Gamma_0 : \Gamma_1]$ is finite. So we have

$$\chi(\Gamma_1) = d \cdot \chi(\Gamma_0),$$

whence $d = 12$, as claimed. On the other hand, the assumption that the index $[\Gamma_\ell : \Gamma_{\ell+1}]$ is finite — denote it by d, say — would imply that

$$-1 = \chi(\Gamma_{\ell+1}) = d \cdot \chi(\Gamma_\ell) = -d,$$

that is, $\Gamma_{\ell+1} = \Gamma_\ell$, which is clearly false (consider the reduction modulo $2^{\ell+1}$). Incidentally, the same argument shows that Δ_m is of infinite index in Γ_0 for *any* $m \geqslant 3$. Indeed, we can now assume that m is not a power of 2. If $[\Gamma_0 : \Delta_m] = d < \infty$ then

$$-1 = \chi(\Delta_m) = d \cdot \chi(\Gamma_0) = -\frac{d}{12},$$

implying that $d = 12$. But Δ_m is contained in the congruence subgroup $\mathrm{SL}_2(\mathbb{Z}, m)$, so if p is an odd prime divisor of m then

$$[\Gamma_0 : \Delta_m] \geqslant [\Gamma_0 : \mathrm{SL}_2(\mathbb{Z}, m)] \geqslant |\mathrm{SL}_2(\mathbb{F}_p)| = p(p^2 - 1) \geqslant 24,$$

a contradiction. We note that the group $\Delta_3 = \left\langle \left(\begin{smallmatrix} 1 & 3 \\ 0 & 1 \end{smallmatrix}\right), \left(\begin{smallmatrix} 1 & 0 \\ 3 & 1 \end{smallmatrix}\right) \right\rangle$ received a lot of attention during the workshop.)

So, for large ℓ, the subgroup Γ_ℓ is very "thin" in Γ_0, and essentially the only property it retains is Zariski density. Nevertheless, for all *odd* m we still have

$$\rho_m(\Gamma_\ell) = \rho_m(\Gamma_0) = \mathrm{SL}_2(\mathbb{Z}/m\mathbb{Z}).$$

So, if we ignore $p = 2$ (more precisely, the dyadic component $\mathbb{Z}_2$ of $\widehat{\mathbb{Z}}$), then we still have an analog of the property of strong approximation for Γ_ℓ, for *any* $\ell \geqslant 1$.

At the same time, the closure of Γ_ℓ in $\mathrm{SL}_2(\mathbb{Z}_2)$ is open (see Lemma 2.7 for a more general statement). Thus, we eventually obtain that the closure of Γ_ℓ in $\mathrm{SL}_2(\hat{\mathbb{Z}})$ is open — one should think of this property as being the next best thing to strong approximation. Note that for a general X as above, the openness of the closure of $X(\mathbb{Z})$ in $X(\hat{\mathbb{Z}})$ implies that the reduction maps $\rho_m : X(\mathbb{Z}) \to X(\mathbb{Z}/m\mathbb{Z})$ are surjective for all m coprime to some fixed exceptional number $N_0 = N_0(X)$.

To summarize, we see that generally speaking the idea that in certain situations Zariski density should (or may) imply some version of strong approximation, at least for subgroups, appears to be sound, but in order to make it more precise, we need to figure out what is wrong with GL_2 (compared to SL_2).

Before we do this, however, we would like to generalize our set-up and also describe a somewhat different (although closely related) approach to strong approximation. The issue is that typically an algebraic group does not come with a fixed geometric (or linear) realization $G \hookrightarrow \mathrm{GL}_n$, and different realizations may result in different groups of integral points. So, it makes sense to reformulate the property of strong approximation in terms of the *group of rational points*.

2. Strong approximation in algebraic groups and homogeneous spaces

2A. ***Adele groups and strong approximation.*** Let G be an algebraic group defined over a global field K, and let S be a set of places of K. For now, we fix a matrix realization $G \hookrightarrow \mathrm{GL}_n$, which enables us to define unambiguously the groups

$$G(\mathbb{O}_v) = G \cap \mathrm{GL}_n(\mathbb{O}_v),$$

for all nonarchimedean places v of K, where $\mathbb{O}_v$ is the valuation ring in the completion K_v. We let $\mathbb{A}_S$ denote the *ring of S-adeles* of K, and let

$$G(\mathbb{A}_S) = \left\{ g = (g_v) \in \prod_{v \notin S} G(K_v) \mid g_v \in G(\mathbb{O}_v) \text{ for almost all } v \notin S \right\}$$

be the *group of S-adeles* of G. We refer the reader to [Platonov and Rapinchuk 1994, Section 5.1] for a more detailed discussion of adeles, and in particular for the definition of the space of S-adeles $X(\mathbb{A}_S)$ for any affine algebraic K-variety X (we note that $\mathbb{A}_S$ is a K-algebra so we can in fact talk about the set of $\mathbb{A}_S$-points $X(\mathbb{A}_S)$ in an intrinsic way). Here we recall only that one endows $G(\mathbb{A}_S)$ with a natural topology (called the S-adelic topology) that makes it into a locally compact topological group. When S contains all archimedean places of K, this topology is obtained by taking the open subgroups of $\prod_{v \notin S} G(\mathbb{O}_v)$ for a fundamental system of neighborhoods of the identity — thus, the S-adelic topology on $G(\mathbb{A}_S)$ in this case is the "natural extension" of the product topology on $\prod_{v \notin S} G(\mathbb{O}_v)$. (We note that in the case $K = \mathbb{Q}$, $S = \{\infty\}$, the latter group coincides with $\prod_p G(\mathbb{Z}_p) = G(\widehat{\mathbb{Z}})$, so these adelic definitions are direct generalizations of the notions we discussed in Section 1.) One proves (see [Platonov and Rapinchuk 1994, Section 5.1]) that the topological group $G(\mathbb{A}_S)$ is independent of the choice of a K-realization $G \hookrightarrow \mathrm{GL}_n$. Furthermore, there is a canonical embedding $G(K) \hookrightarrow G(\mathbb{A}_S)$, so we can give the following.

Definition. An algebraic K-group G has *strong approximation* with respect to S if $G(K)$ is dense in $G(\mathbb{A}_S)$.

(Of course, one can give a similar definition for an arbitrary affine K-variety X. We note that if $S = \varnothing$ then $X(K)$ is a closed discrete subspace of $X(\mathbb{A}_S)$, so in discussing strong approximation one actually needs to assume from the outset that S is nonempty.)

Defined this way (in terms of rational points), the property of strong approximation does not depend on the choice of a matrix realization $G \hookrightarrow \mathrm{GL}_n$. On the other hand, in the case where S contains all nonarchimedean places, its validity implies that for *any* realization, the group $G(\mathbb{O}(S))$ of points over the ring of S-integers $\mathbb{O}(S)$, which can alternatively be described as

$$G(\mathbb{O}(S)) = G(K) \cap \prod_{v \notin S} G(\mathbb{O}_v),$$

is dense in $\prod_{v \notin S} G(\mathbb{O}_v)$ (thus, we have strong approximation in the sense discussed in Section 1 for any realization).

2B. *Absence of strong approximation in algebraic tori.* Our next goal is to explain why GL_2 has no chance to possess strong approximation. However, it is easiest to pin down the reason by working with the 1-dimensional $T = \mathbb{G}_m$: we will now show that it does not have strong approximation with respect to any finite set of places S, and will then demonstrate how the same phenomenon manifests itself in the case of GL_2 and other situations.

Let us start with the case $K = \mathbb{Q}$. If $S = \{\infty\}$ then $T(\mathbb{Z}) = \{\pm 1\}$ which is not even Zariski-dense. For $S = \{\infty, 2\}$, we have

$$T(\mathbb{Z}[\tfrac{1}{2}]) = \pm\langle 2\rangle,$$

which is already Zariski-dense, but nevertheless T still does not have strong approximation. Indeed, pick any prime p of the form $8k + 1$. Then -1 and 2 are squares modulo p, so the map

$$\pm\langle 2\rangle \to (\mathbb{Z}/p\mathbb{Z})^\times$$

is *not* surjective. What really happens here is that T possesses a 2-sheeted cover

$$\pi : T \to T, \quad t \mapsto t^2,$$

and for any prime $p \equiv 1 \pmod 8$ we have

$$T(\mathbb{Z}[\tfrac{1}{2}]) \subset \pi(T(\mathbb{Z}_p)) \subsetneqq T(\mathbb{Z}_p).$$

Since $\pi(T(\mathbb{Z}_p)) \subset T(\mathbb{Z}_p)$ is a closed subgroup, we obtain that $T(\mathbb{Z}[\tfrac{1}{2}])$ is *not* dense in $T(\mathbb{Z}_p)$ for any such p. Moreover, by Dirichlet's Prime Number Theorem, for any $r \geqslant 1$ we can find r distinct primes $p_1, \ldots, p_r$ congruent to $1 \pmod 8$. Then the image of the map

$$\pm\langle 2\rangle \to (\mathbb{Z}/p_1 \cdots p_r\mathbb{Z})^\times$$

is contained in $(\mathbb{Z}/p_1 \cdots p_r\mathbb{Z})^{\times 2}$, which has index 2^r in $(\mathbb{Z}/p_1 \cdots p_r\mathbb{Z})^\times$. It follows that the closure of $T(\mathbb{Z}[\tfrac{1}{2}])$ in $T(\widehat{\mathbb{Z}}) = \prod_p T(\mathbb{Z}_p)$ is *of infinite index.*

This approach easily generalizes. First, let $T = \mathbb{G}_m$ over an arbitrary number field K, and let S be an arbitrary finite set of places of K containing all archimedean ones. Then by Dirichlet's unit theorem [Lang 1994, p. 105], the group $T(\mathbb{O}(S))$ is generated by a finite collection of elements, say $t_1, \ldots, t_r$. Set $L = K(\sqrt{t_1}, \ldots, \sqrt{t_r})$. By Chebotarev's density theorem [Lang 1994, p. 169], there exist infinitely many places $v \notin S$ that totally split in L (i.e., $L \subset K_v$). Considering again the covering $\pi : T \to T$, $\pi(t) = t^2$, we see that for any such nondyadic v we have the inclusions

$$T(\mathbb{O}(S)) \subset \pi(T(\mathbb{O}_v)) \subsetneqq T(\mathbb{O}_v).$$

This implies that the closure of $T(\mathbb{O}(S))$ in $\prod_{v \notin S} T(\mathbb{O}_v)$ is of infinite index, and therefore the closure of $T(K)$ in $T(\mathbb{A}_S)$ is of infinite index as well.

Next, this argument can be extended to an arbitrary torus T over a global field K and any finite set S of places of K. Moreover, by considering coverings (isogenies) $\pi_m : T \to T$, $\pi_m(t) = t^m$ for various m prime to char K, one proves the following.

Proposition 2.1. *Let T be a nontrivial torus over a global field K, and S a finite set of places of K. If $\overline{T(K)}$ is the closure of $T(K)$ in $T(\mathbb{A}_S)$ then the quotient*

$$T(\mathbb{A}_S)/\overline{T(K)}$$

is a group of infinite exponent.

This proposition yields a strong version of the fact that a nontrivial torus over a global field always fails to have strong approximation with respect to any finite set of places S. Nevertheless, a torus may have strong approximation with respect to some infinite (and coinfinite) sets S; see Remark 3 after Theorem 2.3.

2C. ***Simply connectedness as a necessary condition.*** The discussion of tori in the previous subsection suggests that the existence of a nontrivial covering map for a given variety X over a global field K may prevent it from having strong approximation with respect to any finite set of places S. Indeed, as we will see soon, simply connectedness of a connected absolutely almost simple group G (i.e., the absence of nontrivial central isogenies $\pi : \widetilde{G} \to G$ with connected $\widetilde{G}$ — [Tits 1966] for a more detailed discussion) is one of the essential conditions in the strong approximation theorem for algebraic groups (see Theorem 2.3 below). But before we shift our focus entirely to algebraic groups, we would like to mention the following general result of Minchev [1989] which does not seem to be well-known to the general audience. (Note that we did not formally define adeles for arbitrary varieties, so the reader may want to assume that all varieties considered are actually affine, in which case the definitions are completely parallel to the above definitions for algebraic groups.)

Proposition 2.2 [Minchev 1989, Theorem 1]. *Let X be an irreducible normal variety over a number field K. If there exists a nontrivial connected unramified covering $f : Y \to X$ defined over an algebraic closure $\overline{K}$, then X does not have strong approximation with respect to any finite set S of places of K.*

Proof. Since this appeared only in Russian, in a journal with limited circulation, we sketch the argument here assuming X and Y to be affine and smooth and S to contain all archimedean places. We may assume that f is a Galois cover of degree $n > 1$, and pick a finite extension L/K such that Y and f are L-defined. For $x \in X(L)$, we let $L(f^{-1}(x))$ denote the extension of L generated by the coordinates of all preimages of x in $Y(\overline{K})$; note that $[L(f^{-1}(x)) : L] \leqslant n!$. Using the local version of the Chevalley–Weil theorem (cf. [Lang 1983, Chapter 2, Lemma 8.3]), for which we need f to be unramified, one shows that there exists a finite set of places S_1 of K containing S such that any $v \notin S_1$ is unramified in $L(f^{-1}(x))$ for all $x \in X(\mathbb{O}(S))$. Invoking Hermite's theorem [Lang 1994, p. 122], we now conclude that there are only finitely many possibilities for $L(f^{-1}(x))$ as x ranges in $X(\mathbb{O}(S))$, and therefore there exists a finite extension L_1/L such that $f^{-1}(X(\mathbb{O}(S))) \subset Y(L_1)$. Enlarging L, we can actually assume that $L = L_1$ and L/K is a Galois extension. Also, expanding S if necessary, we can make sure that if $v \notin S$ splits completely in L (i.e., $L \subset K_v$) then

$$X(\mathbb{O}(S)) \subset f_{K_v}(Y(\mathbb{O}_v)). \tag{2}$$

On the other hand, for almost all nonarchimedean places w of L, the reductions $\underline{X}^{(w)}$ and $\underline{Y}^{(w)}$ modulo w are smooth irreducible varieties over the residue field ℓ_w, and the reduction $\underline{f}^{(w)} : \underline{Y}^{(w)} \to \underline{X}^{(w)}$ is an n-sheeted Galois cover. It follows that

$$\left| \underline{f}_{\ell_w}^{(w)}(\underline{Y}^{(w)}(\ell_w)) \right| = \frac{|\underline{Y}^{(w)}(\ell_w)|}{n}. \tag{3}$$

Since $\underline{X}^{(w)}$ and $\underline{Y}^{(w)}$ are irreducible, by the Lang–Weil theorem [1954], the cardinalities $|\underline{X}^{(w)}(\ell_w)|$ and $|\underline{Y}^{(w)}(\ell_w)|$ are both "approximately equal" to q_w^d, where $q_w = |\ell_w|$ and d is the common dimension of $\underline{X}^{(w)}$ and $\underline{Y}^{(w)}$. Comparing this with (3), we see that for almost all w, the cardinality $\left| \underline{f}_{\ell_w}^{(w)}(\underline{Y}^{(w)}(\ell_w)) \right|$ is only a fraction of $|\underline{X}^{(w)}(\ell_w)|$; in particular, $\underline{f}_{\ell_w}^{(w)}(\underline{Y}^{(w)}(\ell_w)) \neq \underline{X}^{(w)}(\ell_w)$. Since by Hensel's lemma, the reduction map $X(\mathbb{O}_w) \to \underline{X}^{(w)}(\ell_w)$ is surjective, we obtain that

$$f_{L_w}(Y(\mathbb{O}_w)) \neq X(\mathbb{O}_w).$$

(in fact, our argument shows that $f_{L_w}(Y(\mathbb{O}_w))$ is "much smaller" than — in some sense, a "fraction" of — $X(\mathbb{O}_w)$).

This discussion, in conjunction with (2) implies that for almost all v that split completely in L, the set $X(\mathbb{O}(S))$ is not dense in $X(\mathbb{O}_v)$. Since by Chebotarev's density theorem [Lang 1994, p. 169], there are infinitely many v's that split completely in L, we obtain that X does not have strong approximation with respect to S (and in fact that the closure of $X(\mathbb{O}(S))$ in $\prod_{v \notin S} X(\mathbb{O}_v)$ is very "thin"). □

(We note that Minchev [1989] points out another necessary condition for strong approximation in a K-variety X (which is much easier to prove): X needs to be (absolutely) irreducible.)

Remark. It was pointed out to us by Joël Bellaïche that using the version of the Chevalley–Weil theorem given in [Serre 1997, Section 4.2], one can get rid of the normality assumption in Proposition 2.2.

While the proof of Proposition 2.2 for general varieties requires some facts from arithmetic algebraic geometry, there is a much simpler argument in the case of algebraic groups (see [Platonov and Rapinchuk 1994, Section 7.4]). Since most readers are likely to be particularly interested in this case, we will explain the idea using the following example. Consider the canonical isogeny

$$\widetilde{G} = \mathrm{SL}_2 \xrightarrow{\pi} \mathrm{PGL}_2 = G$$

of algebraic groups over a number field K. By the Skolem–Noether theorem, one can think of G as the automorphism group $\mathrm{Aut}(M_2)$ of the degree two matrix algebra. Then for any field extension F/K, again by the Skolem–Noether theorem, we have

$$G(F) = \mathrm{Aut}_F(M_2(F)) = \mathrm{PGL}_2(F).$$

Then there is an exact sequence

$$\widetilde{G}(F) \xrightarrow{\pi_F} G(F) \xrightarrow{\theta_F} F^\times/F^{\times 2} \to 1, \tag{4}$$

where θ_F is induced by the determinant, viz. $gF^\times \mapsto (\det g)F^{\times 2}$. (Alternatively, one can think of G as the special orthogonal group $\mathrm{SO}_3(q)$ of the Killing form q on the Lie algebra $\mathfrak{sl}_2$ — recall that $q = 2x^2 + yz$ in the Chevalley basis; then $\widetilde{G}$ can be identified with $\mathrm{Spin}_3(q)$, and θ_F becomes simply the spinor norm map on $\mathrm{SO}_3(q)(F)$.)

The point is that given *any* finitely generated subgroup $\Gamma \subset G(K)$, its image $\Delta := \theta_K(\Gamma)$ is a *finite group*. If K is a number field, it follows from Chebotarev's density theorem that there are infinitely many nonarchimedean places v of K such that the image of Δ under the natural map $K^\times/K^{\times 2} \to K_v^\times/K_v^{\times 2}$ is trivial. From the exactness of (4) for $F = K_v$, we conclude that for these v we have

$$\Gamma \subset \pi_{K_v}(\widetilde{G}(K_v)) \neq G(K_v).$$

Applying this to $\Gamma = G(\mathbb{O}(S))$ (which is finitely generated), we obtain that for almost all such v,

$$G(\mathbb{O}(S)) \subset \pi_{K_v}(\widetilde{G}(\mathbb{O}_v)) \neq G(\mathbb{O}_v).$$

The latter implies that the closure of $G(\mathbb{O}(S))$ in $\prod_{v \notin S} G(\mathbb{O}_v)$ is of infinite index, for any finite set S of places of K, and hence G fails to have strong approximation.

This type of argument easily generalizes to prove that if a connected algebraic group G over a number field K is not simply connected, then G fails to have strong approximation for any finite set S of places of K (see [Platonov and Rapinchuk 1994, Section 7.4] for the details).

Example. Let $G = \mathrm{GL}_2$. Set $\widetilde{G} = G \times \mathbb{G}_m$. Then the product map $\widetilde{G} \to G$ is an isogeny of degree 2. Composing it with the map

$$\widetilde{G} \to \widetilde{G}, (g,t) \mapsto (g, t^\ell), \qquad \text{for } \ell \geqslant 1,$$

we obtain an isogeny $\widetilde{G} \to G$ of an arbitrary even degree 2ℓ. On the other hand, the map $G \to G$, $g \mapsto (\det g)^\ell g$ for $\ell \geqslant 1$, is an isogeny of an arbitrary odd degree $(2\ell + 1)$. Thus, G has finite-sheeted connected coverings of any degree, in particular, it is not simply connected. In view of the results discussed above, this explains why G does not have strong approximation with respect to any finite S.

2D. ***Strong approximation theorem.*** So far, we have identified two necessary conditions for strong approximation in a connected algebraic group G over a number field K with respect to a finite set S of places of K that contains all archimedean places: the S-arithmetic subgroups (i.e., subgroups commensurable with $G(\mathbb{O}(S))$) must be Zariski-dense, and G must be simply connected. It turns out that for semisimple groups, these conditions are also sufficient. Since the general case easily reduces to absolutely almost simple groups (cf. [Platonov and Rapinchuk 1994, Section 7.4]), we will give a precise statement of the strong approximation theorem only for this case (however, we will include global fields of positive characteristic).

Theorem 2.3 ([Kneser 1965; Platonov 1969] in characteristic zero; [Margulis 1977; Margulis 1991; Prasad 1977] in positive characteristic). *Let G be a connected absolutely almost simple algebraic group over a global field K, and let S be a finite set of places of K. Then G has strong approximation with respect to S (i.e., $G(K)$ is dense in $G(\mathbb{A}_S)$) if and only if*

(1) *G is simply connected;*

(2) *$G_S := \prod_{v\in S} G(K_v)$ is noncompact.*

(We note that for an absolutely almost simple group G, condition (2) is equivalent to $G(\mathbb{O}(S))$ being infinite, and hence Zariski-dense in G; see [Platonov and Rapinchuk 1994, Theorem 4.10]. It should also be mentioned that in the statement of the theorem we included only the names of the main contributors; the interested reader will find more historical remarks at the end of Section 7.4 in [Platonov and Rapinchuk 1994], and also at the end of the current section.)

Remarks. 1. The condition that G is simply connected is used in the proof of sufficiency in Theorem 2.3 in a very peculiar way that is totally unrelated to the above considerations showing that simply connectedness is necessary for strong approximation. More precisely, what we need is the fact that for all $v \notin S$ such that G is K_v-isotropic (i.e., has positive rank over K_v), the group $G(K_v)$ does not have proper (abstract) subgroups of finite index (see Section 2F). It turns out that in the situation at hand, for G simply connected, the group $G(K_v)$ does not, in fact, have *any* proper noncentral normal subgroups. To put this result in perspective, we recall the result of Tits [1964] asserting that given an absolutely almost simple isotropic algebraic group G over a field P with $\geqslant 4$ elements, the subgroup $G(P)^+$ of $G(P)$ generated by the P-rational points of P-defined parabolics, does not have any proper noncentral normal subgroups. In the same paper, Tits proposed a conjecture, which later became known as the *Kneser–Tits conjecture*, that actually $G(P)^+ = G(P)$ *if* G is simply connected. While over general fields this conjecture turns out to be false [Platonov 1980], it holds over nonarchimedean local fields of characteristic zero (finite extensions of the p-adic field $\mathbb{Q}_p$), as shown in [Platonov 1969] (over $\mathbb{R}$ this fact was established much earlier by E. Cartan; see [Platonov and Rapinchuk 1994, Proposition 7.6]). This connection between strong approximation and the Kneser–Tits conjecture was the centerpiece of [Platonov 1969]. We will see another manifestation of this connection in the analysis of strong approximation for arbitrary Zariski-dense subgroups (Section 3), although in a different setting (viz., over finite fields). On the other hand, over a local or a finite field P, we have $G(P)^+ \neq G(P)$ if G is not simply connected, and hence in this case $G(P)$ does have proper noncentral normal subgroups (of finite index). This is where the proof of Theorem 2.3 and the corresponding argument in Section 3 breaks down if one drops the assumption that G is simply connected. Finally, we remark that the Kneser–Tits conjecture has generated a lot of research not associated with strong approximation; see [Gille 2009] for a recent survey.

2. The effect of nonsimply connectedness on strong approximation with respect to a finite set S is different for tori and semisimple groups: for a K-torus T, the quotient $T(\mathbb{A}_S)/\overline{T(K)}$ by the closure of the group of rational points has infinite exponent (Proposition 2.1), while, as follows from Theorem 2.3, for a connected absolutely almost simple nonsimply connected K-group G with a universal K-defined cover $\pi : \widetilde{G} \to G$ such that the group $\widetilde{G}_S$ is not compact, the closure $\overline{G(K)} \subset G(\mathbb{A}_S)$ is a normal subgroup with the infinite quotient $G(\mathbb{A}_S)/\overline{G(K)}$ having finite exponent. (This distinction, of course, reflects the fact that the (algebraic) fundamental group of G is finite, while that of T is infinite.)

3. A connected K-group G may have strong approximation with respect to certain *infinite* sets S of places of K *without* being simply connected. For example, in [Prasad and Rapinchuk 2001], we examined in this context strong approximation in tori (which can never be valid for finite S; see Proposition 2.1). To avoid technical definitions, we will just indicate what our results give in the case of the 1-dimensional split torus $T = \mathbb{G}_m$ over $K = \mathbb{Q}$: *If S is an infinite set of places of K that contains the p-adic places for almost all primes p in a certain arithmetic progression, then the closure $\overline{T(\mathbb{Q})}$ of $T(\mathbb{Q})$ in the group of S-adeles $T(\mathbb{A}_S)$ is of finite index.* The result for general tori is basically the same but contains one important exclusion that has to do with how the arithmetic progression interacts with the splitting field of the torus. This fact is instrumental for the analysis of the congruence subgroup problem: it implies, in particular, that if G is an absolutely almost simple simply connected algebraic group over a number field K, which is an inner form, and S is a set of places of K that contains all archimedean places and also almost all places in a certain generalized arithmetic progression, then the corresponding congruence kernel $C^S(G)$ is trivial, that is, every subgroup of finite index in $G(\mathbb{O}(S))$ contains a suitable congruence subgroup (provided that $G(K)$ has a standard description of

normal subgroups). See [Prasad and Rapinchuk $\geq$ 2012].

4. For general affine varieties, the analogs of conditions (1) and (2) in Theorem 2.3 may not be sufficient for strong approximation, even in homogeneous spaces.

Example. Let $f(x, y, z) = ax^2 + by^2 + cz^2$ be the nondegenerate ternary quadratic form over a number field K, and let $X \subset \mathbb{A}^3$ be a quadric given by $f(x, y, z) = a$. Set $g(x, y) = by^2 + cz^2$. Let S be a finite set of places of K such that $X_S = \prod_{v \in S} X(K_v)$ is noncompact (equivalently, for some $v \in S$ the form f is K_v-isotropic). Then (see Section 2E) X has strong approximation with respect to S if and only if one of the following two conditions holds:

(a) g is K-isotropic.

(b) g is K-anisotropic and there exists $v \in S$ such that g remains anisotropic over K_v and either v is nonarchimedean or f is K_v-isotropic.

It follows that a rational quadric X defined by $x_1^2 + x_2^2 - 2x_3^2 = 1$ (which is simply connected) does not have strong approximation with respect to $S = \{\infty\}$.

2E. ***Strong approximation in homogeneous spaces.*** The fact quoted in the above example is a consequence of the analysis of strong approximation in (affine) homogeneous spaces of algebraic groups. Since these results (found in [Borovoi 1989] and [Rapinchuk 1988]; a detailed exposition of the latter paper was given in [Rapinchuk 1990]) are not as widely known as Theorem 2.3, we briefly mention some of them here for the sake of completeness. The fact that only connected simply connected varieties have a chance to possess strong approximation, by and large, forces us to focus our attention on homogeneous spaces of the form $X = G/H$, where G is a semisimple simply connected algebraic K-group, and H is a K-defined connected reductive subgroup (any such variety is affine and simply connected). Furthermore, given a set S of places of K, it is not difficult to show that for such X, the space X_S is noncompact if and only if G_S is noncompact. Assuming now that G is actually absolutely almost simple, we conclude from Theorem 2.3 that G has strong approximation with respect to S (for a general semisimple group G we need to consider its simple components). Then using Galois cohomology one investigates when strong approximation in G implies strong approximation in $X = G/H$. Here is one easy result in this direction.

Proposition 2.4 [Rapinchuk 1988]. *Let $X = G/H$ be the quotient of a connected absolutely almost simple simply connected algebraic group G defined over a number field K by a connected semisimple simply connected K-subgroup H. Then X has strong approximation with respect to a finite set S of places of K if and only if the space $X_S = \prod_{v \in S} X(K_v)$ is noncompact.*

Now, let $q = q(x_1, \dots, x_n)$ be a nondegenerate quadratic form in $n \geqslant 3$ variables. Consider the quadric $X \subset \mathbb{A}^n$ given by the equation $q(x_1, \dots, x_n) = a$ for some $a \in K^\times$. Assuming that $X(K) \neq \varnothing$, pick $x \in X(K)$. Then $X = G/H$, where $G = \mathrm{Spin}_n(q)$ and $H = G(x)$ (the stabilizer of x); note that $H \simeq \mathrm{Spin}_{n-1}(q')$, where q' is the restriction of q to the orthogonal complement of x. So, it follows from Proposition 2.4 that for $n \geqslant 5$, the quadric X has strong approximation with respect to X if and only if there exists $v \in S$ such that q is K_v-isotropic. The same result remains valid for $n = 4$ even though in this case G is not absolutely almost simple. (Incidentally, this result applies to the defining equation of SL_2 given on page 271, yielding thereby another proof of strong approximations for this group; compare Lemma 1.2.)

The case $n = 3$ is different, as here H is a torus. This case can also be treated in a rather explicit form using the results of Nakayama–Tate on the Galois cohomology of tori. More precisely, let T be a K-torus, and let L be the splitting field of T. As usual, given a module M over the Galois group $\mathrm{Gal}(L/K)$, we let $H^i(L/K, M)$ denote the Galois cohomology group $H^i(\mathrm{Gal}(L/K), M)$. Given a finite set S of places of K, we let $\bar{S}$ denote the set of all extensions of places in S to L, and let $\mathbb{A}_L$ and $\mathbb{A}_{L,\bar{S}}$ denote the rings of adeles and $\bar{S}$-adeles of L. Finally, let $c_L(T) = T(\mathbb{A}_L)/T(L)$ be the adele class group of T over L, and let

$$\delta : H^1(L/K, T(\mathbb{A}_L)) \longrightarrow H^1(L/K, c_L(T))$$

be the corresponding map on cohomology. Then, viewing $T_{\bar{S}}$ and $T(\mathbb{A}_{L,\bar{S}})$ as subgroups of $T(\mathbb{A}_L)$, we have the following statement.

Proposition 2.5 [Rapinchuk 1988]. *Let $X = G/T$, where G is an absolutely almost simple simply connected K-group and $T \subset G$ is a K-torus. Then X has strong approximation with respect to a finite set S of places of K if and only if X_S is noncompact and*

$$\delta(H^1(L/K, T(\mathbb{A}_{L,\bar{S}}))) \subset \delta(\mathrm{Ker}(H^1(L/K, T_{\bar{S}}) \to H^1(L/K, G_{\bar{S}}))),$$

where L is the splitting field of T and $\bar{S}$ consists of all extensions of places in S to L.

This proposition yields the criterion for strong approximation for the quadrics defined by ternary forms we used in Remark 4 of Section 2D. It also implies that for $X = G/T$, one can find a finite set of places S_0 (depending on T) such that X has strong approximation with respect to S whenever $S \supset S_0$. It turns out that this qualitative statement remains valid for quotients by arbitrary connected reductive subgroups. More precisely, using some ideas that eventually led him to theorems of the Nakayama–Tate type for Galois cohomology of arbitrary connected groups, Borovoi proved the following.

Proposition 2.6 [Borovoi 1989]. *Let $X = G/H$ be the quotient of a connected absolutely almost simple algebraic group G over a number field K by its connected reductive K-defined subgroup H. There exists a finite set S_0 of places of K such that X has strong approximation with respect to S_0 (and then, of course, it also has strong approximation with respect to any $S \supset S_0$).*

We remark in passing that the results on strong approximation in homogeneous spaces were used to extend Kneser's method for proving the centrality of the congruence kernel for spinor groups to groups of other classical types as well as G_2 [Rapinchuk 1988; 1989; 1990] (see also [Tomanov 1989a; 1989b]), to establish bounded generation of some S-arithmetic subgroups in orthogonal groups [Erovenko and Rapinchuk 2006], and to study some Diophantine questions involving quadratic forms [Colliot-Thélène and Xu 2009]. (We should mention that the results of the latter work were recently generalized in [Demarche 2011] where the deviation from strong approximation in a connected K-group G has been expressed in terms of a certain subquotient of the Brauer group of G.)

2F. ***On the proof of sufficiency in Theorem 2.3.*** We begin with the following statement that applies to arbitrary Zariski-dense subgroups.

Lemma 2.7. *Let G be an absolutely almost simple algebraic $\mathbb{Q}$-group, and let $\Gamma \subset G(\mathbb{Z})$ be a Zariski-dense subgroup of G. Then for any prime p, the closure $\overline{\Gamma}^{(p)} \subset G(\mathbb{Z}_p)$ is open.*

Proof. Let $\mathfrak{g}$ be the Lie algebra of G as an algebraic group, so that $\mathfrak{g}_{\mathbb{Q}_p}$ is the Lie algebra of $G(\mathbb{Z}_p)$ as a p-adic analytic group. By a theorem of Cartan [Platonov and Rapinchuk 1994, Theorem 3.4], $\Delta := \overline{\Gamma}^{(p)}$ is a p-adic Lie group, of positive dimension as Γ is nondiscrete in $G(\mathbb{Z}_p)$ (the discreteness would force it to be finite, and therefore prevent it from being Zariski-dense). So, the Lie algebra $\mathfrak{h}$ of Δ as a p-adic analytic group is a nonzero $\mathbb{Q}_p$-subalgebra of $\mathfrak{g}_{\mathbb{Q}_p}$. Clearly, $\mathfrak{h}$ is invariant under $\operatorname{Ad}\Gamma$, so the Zariski density of Γ in G implies that $\mathfrak{h} \otimes_{\mathbb{Q}_p} \overline{\mathbb{Q}_p}$ is invariant under $\operatorname{Ad} G$. Since the adjoint representation of G on $\mathfrak{g}$ is irreducible, we conclude that $\mathfrak{h} = \mathfrak{g}_{\mathbb{Q}_p}$, and therefore Δ is open in $G(\mathbb{Z}_p)$ by the implicit function theorem. □

As we will discuss at the beginning of Section 3, this lemma, though useful, falls short of proving any form of strong approximation. We will now indicate additional considerations needed to prove the sufficiency in Theorem 2.3 in characteristic zero, following Platonov's original argument [Platonov 1969]. Let us assume that S contains all archimedean valuations of K. In this case, it is easy to see from the definition of the topology on $G(\mathbb{A}_S)$ that strong approximation is equivalent to the following statement:

For any finite set of places S_1 of K which is disjoint from S, the group $G(\mathcal{O}(S \cup S_1))$ is dense in $G_{S_1} := \prod_{v \in S_1} G(K_v)$.

To showcase the idea, we will now prove this statement in the case where $K = \mathbb{Q}$ and $S_1 = \{p\}$, a single p-adic place such that G is $\mathbb{Q}_p$-isotropic; see [Platonov and Rapinchuk 1994, Section 7.4] for the general case. First, by reduction theory for S-arithmetic groups, $G(\mathcal{O}(S \cup S_1))$ is a *lattice* (i.e., a discrete subgroup of finite covolume) in $G_{S \cup S_1}$; see [Platonov and Rapinchuk 1994, Theorem 5.7]. Since by assumption the group G_S is noncompact, it is not difficult to show (cf. [Platonov and Rapinchuk 1994, Lemma 3.17]) that $G(\mathcal{O}(S \cup S_1)) \subset G(\mathbb{Q}_p)$ is nondiscrete (in particular, infinite), and if Δ denotes the p-adic closure $\overline{G(\mathcal{O}(S \cup S_1))}^{(p)}$, then $G(\mathbb{Q}_p)/\Delta$ carries a finite invariant measure. Next, the fact that $G(\mathcal{O}(S \cup S_1))$ is infinite implies that it is actually Zariski-dense in G (Borel's density theorem; see, for example, [Platonov and Rapinchuk 1994, Theorem 4.10]). Taking into account the nondiscreteness of $G(\mathcal{O}(S \cup S_1))$ in $G(\mathbb{Q}_p)$ and repeating the proof of Lemma 2.7, we conclude that Δ is open in $G(\mathbb{Q}_p)$. Then the existence of a finite invariant measure on $G(\mathbb{Q}_p)/\Delta$ implies that $\Delta \subset G(\mathbb{Q}_p)$ is a subgroup of finite index. On the other hand, since the group G is connected, absolutely almost simple, simply connected and $\mathbb{Q}_p$-isotropic, by the Kneser–Tits conjecture over p-adic fields we have $G(\mathbb{Q}_p) = G(\mathbb{Q}_p)^+$, and therefore the group $G(\mathbb{Q}_p)$ does not have any proper noncentral normal subgroup. In particular, it does not contain any proper subgroups of finite index, and we obtain that $\Delta = G(\mathbb{Q}_p)$, as required.

This argument breaks down in positive characteristic, first and foremost, because Cartan's theorem, which is at the heart of the proof of Lemma 2.7, is valid only in characteristic zero. It should be mentioned that eventually Pink [1998] proved a result which in some sense can be viewed as an analog (or replacement) of Cartan's theorem. The precise general statement is too technical for us to discuss here, so we will only indicate what it yields in one particular case (see Theorem 0.7 in [Pink 1998]): *Let G be an absolutely simple connected adjoint group over a local field F, and assume that the adjoint representation of G is irreducible. If $\Gamma \subset G(F)$ is a compact Zariski-dense subgroup, then there exists a closed subfield $E \subset F$ and a model H of G over E such that Γ is open in $H(E)$.* This sort of result can be used to prove Theorem 2.3 in positive characteristic, but the original argument given virtually simultaneously by Margulis [1977] and Prasad [1977], was different. They derived strong approximation (arguing along the lines indicated above) from the following statement:

Let G be a connected semisimple algebraic group over a local field F, and let $H \subset G(F)$ be a nondiscrete closed subgroup such that $G(F)/H$ carries a finite invariant Borel measure. Then $H \supset G(F)^+$.

Their argument used ergodic considerations and representation theory. More than 25 years later, Pink [2004] used his results from [1998] to give a purely algebraic proof of this theorem, and hence of strong approximation.

3. Strong approximation for Zariski-dense subgroups

3A. ***Overview.*** The strong approximation Theorem 2.3 gives us *precise* information about the adelic closure of S-arithmetic subgroups: for example, if G is an algebraic $\mathbb{Q}$-group that has strong approximation with respect to $S = \{\infty\}$ then for any matrix realization of G, the group $G(\mathbb{Z})$ is dense in $G(\widehat{\mathbb{Z}}) = \prod_p G(\mathbb{Z}_p)$; see Section 2. At the same time, as we explained in Section 1, one can expect a general *qualitative* openness result for the adelic closure of an arbitrary Zariski-dense subgroup. The goal of this section is to discuss some results in this direction.

First, we note one consequence of Lemma 2.7. Let G be a connected absolutely almost simple algebraic $\mathbb{Q}$-group, and let $\Gamma \subset G(\mathbb{Z})$ be a Zariski-dense subgroup of G. Then using the fact that $G(\mathbb{Z}_p)$ is a virtually pro-p group, one easily deduces from Lemma 2.7 that given a *finite* set S of distinct primes, the closure

$$\bar{\Gamma}^{(S)} \subset \prod_{p \in S} G(\mathbb{Z}_p)$$

is open. This statement is already sufficient for some applications; for example, it was used in [Prasad and Rapinchuk 2003] to prove the existence of generic elements in arbitrary finitely generated Zariski-dense subgroups $\Gamma \subset G(K)$, where G is a semisimple algebraic group over a finitely generated field K of characteristic zero; see [Gorodnik and Nevo 2011; Jouve et al. 2013; Lubotzky and Rosenzweig 2012] for more recent work in this direction.[1] On the other hand, if we take S to be the set of *all* primes, the best we can get from Lemma 2.7 is the following:

> *The closure $\widehat{\Gamma}$ of Γ in $G(\widehat{\mathbb{Z}}) = \prod_p G(\mathbb{Z}_p)$ contains $\prod_p W_p$, where $W_p \subset G(\mathbb{Z}_p)$ is open for each p.*

Of course, this does *not* imply that $\widehat{\Gamma}$ is open in $G(\widehat{\mathbb{Z}})$ — for this we need to show that actually $W_p = G(\mathbb{Z}_p)$ for almost all p. The first general result in this direction was the following.

Theorem 3.1 [Matthews et al. 1984]. *Let G be a connected absolutely almost simple simply connected algebraic group over $\mathbb{Q}$.*

---

[1] The article [Prasad and Rapinchuk 2014] in this volume surveys applications of generic elements to the analysis of isospectral locally symmetric spaces; see also [Prasad and Rapinchuk 2009; 2010].

(1) *If* $\Gamma \subset G(\mathbb{Z})$ *is a Zariski-dense subgroup, then the closure* $\widehat{\Gamma} \subset G(\widehat{\mathbb{Z}})$ *is open.*

(2) *If* $\Gamma \subset G(\mathbb{Q})$ *is a finitely generated Zariski-dense subgroup, then for some finite set* S *of places of* $\mathbb{Q}$ *containing* ∞, *the closure of* Γ *in the group of* S*-adeles* $G(\mathbb{A}_S)$ *is open.*

The paper [Matthews et al. 1984] appeared in 1984, but the interest in these sorts of results arose at least 20 years earlier in connection with the study of Galois representations on torsion points of elliptic curves. In fact, in his book on ℓ-adic representations, Serre [1968] pretty much had this theorem for $G = \mathrm{SL}_2$ (at least, all the ingredients of the proof were there).

Parts (1) and (2) are proved in the same way, so let us focus our discussion on the proof of (1) as this will allow us to keep our notations simple. First, it is enough to prove that for almost all primes p, the closure $\bar{\Gamma}^{(p)} \subset G(\mathbb{Z}_p)$ coincides with $G(\mathbb{Z}_p)$. This reduction step is achieved using $(*)$ in conjunction with the fact that for almost all primes p, the group G has a smooth reduction $\underline{G}^{(p)}$ modulo p and the groups $\underline{G}^{(p)}(\mathbb{F}_p)$ are pairwise nonisomorphic almost simple groups (for the reader who is interested only in the case $G = \mathrm{SL}_n$, we will indicate that here, of course, $\underline{G}^{(p)} = \mathrm{SL}_n / \mathbb{F}_p$, and the structural facts quoted above are well-known). Next, it turns out that for almost all p, proving that $\bar{\Gamma}^{(p)} = G(\mathbb{Z}_p)$ reduces to showing that the reduction map $\rho_p : G(\mathbb{Z}_p) \to \underline{G}^{(p)}(\mathbb{F}_p)$ has the property $\rho_p(\Gamma) = \underline{G}^{(p)}(\mathbb{F}_p)$.

Proposition 3.2 (compare [Matthews et al. 1984, 7.3]). *For almost all* p, *if* $\Delta \subset G(\mathbb{Z}_p)$ *is a closed subgroup such that* $\rho_p(\Delta) = \underline{G}^{(p)}(\mathbb{F}_p)$ *then* $\Delta = G(\mathbb{Z}_p)$.

The proof for $G = \mathrm{SL}_2$ was given by Serre [1968, Chapter IV, 3.4].

Lemma 3.3. *Let* $\Delta \subset \mathrm{SL}_2(\mathbb{Z}_p)$, *where* $p > 3$, *be a closed subgroup such that for the reduction map* $\rho_p : \mathrm{SL}_2(\mathbb{Z}_p) \to \mathrm{SL}_2(\mathbb{F}_p)$ *we have* $\rho_p(\Delta) = \mathrm{SL}_2(\mathbb{F}_p)$. *Then* $\Delta = \mathrm{SL}_2(\mathbb{Z}_p)$.

Proof. By assumption, there exists $g \in \Delta$ such that

$$g = \left(\begin{smallmatrix} 1 & 0 \\ 1 & 1 \end{smallmatrix}\right) + ps, \quad \text{with } s \in M_2(\mathbb{Z}_p).$$

We claim that

$$g^p = \left(\begin{smallmatrix} 1 & 0 \\ p & 1 \end{smallmatrix}\right) + p^2 t, \quad \text{with } t \in M_2(\mathbb{Z}_p). \tag{5}$$

Indeed,

$$g^p = \left(I_2 + \left(\left(\begin{smallmatrix} 0 & 0 \\ 1 & 0 \end{smallmatrix}\right) + ps\right)\right)^p = \\ I_2 + p\left(\left(\begin{smallmatrix} 0 & 0 \\ 1 & 0 \end{smallmatrix}\right) + ps\right) + \binom{p}{2}\left(\left(\begin{smallmatrix} 0 & 0 \\ 1 & 0 \end{smallmatrix}\right) + ps\right)^2 + \cdots + \left(\left(\begin{smallmatrix} 0 & 0 \\ 1 & 0 \end{smallmatrix}\right) + ps\right)^p.$$

But clearly

$$\left(\left(\begin{smallmatrix}0&0\\1&0\end{smallmatrix}\right)+ps\right)^k \equiv O_2 (\mathrm{mod}\, p), \quad \text{for any } k \geqslant 2,$$

and in fact

$$\left(\left(\begin{smallmatrix}0&0\\1&0\end{smallmatrix}\right)+ps\right)^k \equiv O_2 (\mathrm{mod}\, p^2) \quad \text{for any } k \geqslant 4,$$

as $\left(\left(\begin{smallmatrix}0&0\\1&0\end{smallmatrix}\right)\right)^2 = O_2$ (the zero matrix). So, since $p > 3$, Equation (5) follows.

As $g^p \in \Delta$, we conclude from (5) that the image Φ of the intersection $\Delta \cap \mathrm{SL}_2(\mathbb{Z}_p, p)$ with the congruence subgroup modulo p in

$$\mathrm{SL}_2(\mathbb{Z}_p, p)/\mathrm{SL}_2(\mathbb{Z}_p, p^2) \simeq \mathfrak{sl}_2(\mathbb{F}_p),$$

where $\mathfrak{sl}_2$ is the Lie algebra of SL_2 (i.e., 2×2-matrices with trace zero), is *nontrivial*. On the other hand, Φ is obviously invariant under Δ, and as $\rho_p(\Delta) = \mathrm{SL}_2(\mathbb{F}_p)$, it is actually invariant under $\mathrm{SL}_2(\mathbb{F}_p)$. But since $p \neq 2$, the group $\mathrm{SL}_2(\mathbb{F}_p)$ acts on $\mathfrak{sl}_2(\mathbb{F}_p)$ irreducibly, implying that $\Delta \cap \mathrm{SL}_2(\mathbb{Z}_p, p)$ surjects onto $\mathrm{SL}_2(\mathbb{Z}_p, p)/\mathrm{SL}_2(\mathbb{Z}_p, p^2)$. However, $\mathrm{SL}_2(\mathbb{Z}_p, p)$ is in fact the Frattini subgroup of the pro-p group $\mathrm{SL}_2(\mathbb{Z}_p, p)$, so the latter fact implies that $\Delta \cap \mathrm{SL}_2(\mathbb{Z}_p, p) = \mathrm{SL}_2(\mathbb{Z}_p, p)$, and our claim follows. □

The general case in Proposition 3.2 is obtained by reduction to the case of SL_2. For this one observes that the group G is quasisplit, and therefore $G(\mathbb{Z}_p)$ contains $H = \mathrm{SL}_2(\mathbb{Z}_p)$, for almost all p. We refer the reader to [Matthews et al. 1984] for further details. (Note that one needs to argue a bit more carefully on p. 529 in [Matthews et al. 1984] to make sure that $\Delta \cap H$ maps onto $\mathrm{SL}_2(\mathbb{F}_p)$ surjectively; this can be achieved by choosing a special H.)

So, to complete the proof of (both parts of) Theorem 3.1, one needs to prove the following.

Theorem 3.4. *Let G be a connected absolutely almost simple simply connected algebraic group over $\mathbb{Q}$, and let $\Gamma \subset G(\mathbb{Q})$ be a finitely generated Zariski-dense subgroup. Then there exists a finite set of primes $\Pi = \{p_1, \ldots, p_r\}$ such that*

(1) *$\Gamma \subset G(\mathbb{Z}_\Pi)$, where $\mathbb{Z}_\Pi = \mathbb{Z}[1/p_1, \ldots, 1/p_r]$;*

(2) *for $p \notin \Pi$ there exists a smooth reduction $\underline{G}^{(p)}$;*

(3) *if $p \notin \Pi$ and $\rho_p : G(\mathbb{Z}_p) \to \underline{G}^{(p)}(\mathbb{F}_p)$ is the corresponding reduction map then $\rho_p(\Gamma) = \underline{G}^{(p)}(\mathbb{F}_p)$.*

Conditions (1) and (2) are routine (in fact, (1) holds automatically if $\Gamma \subset G(\mathbb{Z})$), so the main point is to ensure condition (3). The general idea is the following. Let $\mathfrak{g}$ and $\underline{\mathfrak{g}}^{(p)}$ be the Lie algebras of G and $\underline{G}^{(p)}$. Since Γ is Zariski-dense in G, we conclude that $\mathrm{Ad}\,\Gamma$ acts on $\mathfrak{g}_{\mathbb{Q}}$ absolutely irreducibly. By Burnside's theorem this means that $\mathrm{Ad}\,\Gamma$ spans $\mathrm{End}_{\mathbb{Q}}\, \mathfrak{g}_{\mathbb{Q}}$ as a $\mathbb{Q}$-vector space. Excluding finitely many primes, we can achieve that for any of the remaining primes p, the group

$\operatorname{Ad}\rho_p(\Gamma)$ acts on $\underline{\mathfrak{g}}^{(p)}_{\mathbb{F}_p}$ absolutely irreducibly. This eventually implies that for almost all p we have $\rho_p(\Gamma) = \underline{G}^{(p)}(\mathbb{F}_p)$. This implication would be obvious if we could say that $\rho_p(\Gamma)$ is necessarily of the form $H(\mathbb{F}_p)$, where $H \subset \underline{G}^{(p)}$ is some connected algebraic $\mathbb{F}_p$-subgroup. (Indeed, then the Lie algebra $\mathfrak{h}$ of H would be a nonzero $\rho_p(\Gamma)$-invariant subspace of $\underline{\mathfrak{g}}^{(p)}$, so $\mathfrak{h} = \underline{\mathfrak{g}}^{(p)}$ and $H = \underline{G}^{(p)}$, as $\underline{G}^{(p)}$ is connected for almost all p, yielding the required fact.) Of course, such an a priori description of $\rho_p(\Gamma)$ would be too much to hope for, but important information along these lines, which is sufficient for the proof of Theorem 3.4, is contained in a theorem of Nori [1987].

3B. ***Nori's Theorem.*** Let H be an *arbitrary subgroup* of $\mathrm{GL}_n(\mathbb{F}_p)$. Set

$$X = \{x \in H \mid x^p = 1\}$$

(we will write 1 in place of I_n to simplify notations). Note that if we assume that $p > n$ (which we will throughout this subsection), then the condition $x^p = 1$ characterizes precisely the unipotent elements, i.e., it is equivalent to the condition $(x-1)^n = 0$. For $x \in X$, we can define

$$\log x := -\sum_{i=1}^{p-1} \frac{(1-x)^i}{i}.$$

Furthermore, observing that $(\log x)^n = 0$, we see that for any t in the algebraic closure of $\mathbb{F}_p$, we can define

$$x(t) := \exp(t \cdot \log x), \quad \text{where } \exp z = \sum_{i=0}^{p-1} \frac{z^i}{i!}.$$

(Note that $x(1) = x$.) We regard $x(t)$ as a one-parameter subgroup $\mathbb{G}_a \to \mathrm{GL}_n$. Set

$$H^+ = \langle X \rangle \subset H,$$

and let $\widetilde{H}$ denote the connected $\mathbb{F}_p$-subgroup of GL_n generated by the 1-parameter subgroups $x(t)$ for all $x \in X$.

Theorem 3.5 [Nori 1987]. *If p is large enough (for a given n), then H^+ coincides with $\widetilde{H}(\mathbb{F}_p)^+$, the subgroup of $\widetilde{H}(\mathbb{F}_p)$ generated by all unipotents contained in it.*

Thus, Nori's theorem asserts that if p is large enough compared to n, then any subgroup of $\mathrm{GL}_n(\mathbb{F}_p)$ generated by p-elements is essentially the group of $\mathbb{F}_p$-points of some connected $\mathbb{F}_p$-defined algebraic subgroup of GL_n. Actually, Nori [1987] proves a stronger result stating that for a field F which either has characteristic zero or positive characteristic p that is large enough compared to n, the maps log and exp yield bijective correspondences between nilpotently generated Lie subalgebras of $M_n(F)$ and exponentially generated subgroups of $\mathrm{GL}_n(F)$ (we refer the reader to Nori's paper for precise definitions and detailed statements of these results). Nori's argument was based on algebrogeometric ideas; a different proof was given in [Hrushovski and Pillay 1995] using model-theoretic techniques (the idea of their argument is explained in [Lubotzky and Segal 2003, pp. 399–400]). A very strong result of Larsen and Pink [2011] describing the structure of finite linear groups over fields of positive characteristic gives yet another way of saying that a "typical" subgroup of $\mathrm{GL}_n(\mathbb{F}_p)$ is algebraic.

Given the nature of this article, we will not be able to discuss any details of Nori's argument. All we can offer as compensation is a proof of Nori's results for $\mathrm{GL}_2(\mathbb{F}_p)$.

Lemma 3.6. *Let $H \subset \mathrm{GL}_2(\mathbb{F}_p)$ be a subgroup of order divisible by p, and let $H_p \subset H$ be a Sylow p-subgroup. Then either $H_p \lhd H$ or $H \supset \mathrm{SL}_2(\mathbb{F}_p)$.*

Proof. We may assume that H_p coincides with

$$U := \left\{ \begin{pmatrix} 1 & a \\ 0 & 1 \end{pmatrix} \mid a \in \mathbb{F}_p \right\}.$$

It is well-known that the normalizer of U in $\mathrm{GL}_2(\mathbb{F}_p)$ coincides with $B = TU$ where

$$T := \left\{ \begin{pmatrix} a & 0 \\ 0 & b \end{pmatrix} \mid a, b \in \mathbb{F}_p^\times \right\}.$$

Furthermore, we have the Bruhat decomposition

$$\mathrm{GL}_2(\mathbb{F}_p) = B \cup BwB, \quad \text{where } w = \begin{pmatrix} 0 & 1 \\ -1 & 0 \end{pmatrix}$$

(recall that w normalizes T). Now, if H_p is not normal in H, then it follows from the Bruhat decomposition that H contains an element of the form tw with $t \in T$. Consequently, it also contains

$$U^- := (tw)^{-1} U (tw) = \left\{ \begin{pmatrix} 1 & 0 \\ a & 1 \end{pmatrix} \mid a \in \mathbb{F}_p \right\}.$$

But $\langle U, U^- \rangle = \mathrm{SL}_2(\mathbb{F}_p)$, and our assertion follows. □

So, for any subgroup $H \subset \mathrm{GL}_2(\mathbb{F}_p)$, we have only the following three possibilities:

(1) $H^+ = \{1\}$;

(2) H^+ is conjugate to U;

(3) $H^+ = \mathrm{SL}_2(\mathbb{F}_p)$.

In either case, the assertion of Nori's Theorem is valid.

3C. ***Proof of Theorem 3.4.*** Recall the famous theorem of Jordan:

> *There exists a function* $j(n)$ *on positive integers such that if* $\mathcal{G} \subset \mathrm{GL}_n(\mathcal{K})$ *is a finite linear group over a field* $\mathcal{K}$ *of characteristic zero, then* $\mathcal{G}$ *contains an abelian normal subgroup* $\mathcal{N}$ *such the index* $[\mathcal{G} : \mathcal{N}]$ *divides* $j(n)$.

(In a more common form, Jordan's theorem provides a function $j_0(n)$ for which $\mathcal{G}, \mathcal{N}$ as above satisfy $[\mathcal{G} : \mathcal{N}] \leqslant j_0(n)$; note that given such a function $j_0(n)$, the above statement holds with $j(n) = (j_0(n))!.)$[2] What we need to observe for the proof of Theorem 3.4 is that the assertion of Jordan's theorem remains valid (with the same $j(n)$) for any subgroup $\mathcal{G} \subset \mathrm{GL}_n(\mathbb{F}_p)$ of order not divisible by p.

Indeed, consider the reduction modulo p map $\rho : \mathrm{GL}_n(\mathbb{Z}_p) \to \mathrm{GL}_n(\mathbb{F}_p)$. The kernel $\mathrm{Ker}\,\rho = \mathrm{GL}_n(\mathbb{Z}_p, p)$ is a pro-p group, so since the order of $\mathcal{G} \subset \mathrm{GL}_n(\mathbb{F}_p)$ is prime to p there is a section $\sigma : \mathcal{G} \to \mathrm{GL}_n(\mathbb{Z}_p)$ for ρ over $\mathcal{G}$. Applying the standard Jordan theorem for characteristic zero to $\tilde{\mathcal{G}} := \sigma(\mathcal{G})$, we obtain the corresponding assertion for $\mathcal{G}$. (For the sake of completeness, we would like to indicate that there are various "modular" forms of Jordan's theorem that treat finite subgroups $\mathcal{G} \subset \mathrm{GL}_n(\mathcal{K})$ of order divisible by p where $p = \mathrm{char}\,\mathcal{K}$, starting with [Brauer and Feit 1966]; see [Collins 2008; Weisfeiler 1984a] for subsequent results (we also note that [Bass 1983] provides a generalization to algebraic groups). As we have already mentioned, the most general results about finite linear groups in positive characteristic are contained in [Larsen and Pink 2011].)

Now, suppose that $G \subset \mathrm{GL}_n$. Let $j = j(n)$ be the value of the Jordan function for this n. Set

$$\Gamma^{(j)} = \langle \gamma^j \mid \gamma \in \Gamma \rangle,$$

and $\Phi = [\Gamma^{(j)}, \Gamma^{(j)}]$. Since the regular map $G \to G$, $x \mapsto x^j$, is dominant, and $G = [G, G]$, we conclude that Φ is Zariski-dense in G, in particular, it is nontrivial. Then, by expanding Π, which initially needs to be chosen to satisfy conditions (1) and (2) of the theorem, we may assume that for all $p \notin \Pi$ we have $\rho_p(\Phi) \neq \{1\}$ where $\rho_p : G(\mathbb{Z}_p) \to \underline{G}^{(p)}(\mathbb{F}_p)$ is the reduction modulo p map. In addition, as we explained earlier, by expanding Π further, we may assume for $p \notin \Pi$, the group $\mathrm{Ad}\,\rho_p(\Gamma)$ acts on $\underline{\mathfrak{g}}^{(p)}$ (= the Lie algebra of $\underline{G}^{(p)}$) absolutely irreducibly, and also Nori's theorem is applicable to $\mathrm{GL}_n(\mathbb{F}_p)$. We will now show that the resulting Π is as required.

Let $p \notin \Pi$, and set $H = \rho_p(\Gamma) \subset \mathrm{GL}_n(\mathbb{F}_p)$. First, we observe that p divides the order of H. Indeed, otherwise by the version of Jordan's theorem mentioned

---

[2] Various sources give different expressions for a Jordan function $j_0(n)$; the optimal function is known to be $j_0(n) = (n+1)!$ for $n \geqslant 71$; see [Collins 2007].

above, there would exist an abelian normal subgroup $N \subset H$ of index dividing j. Then $\rho_p(\Gamma^{(j)}) \subset N$, and therefore $\rho_p(\Phi) = \{1\}$, a contradiction. This means that if we define H^+ and $\widetilde{H}$ as in the Nori's theorem, then $\widetilde{H} \neq \{1\}$, and hence the Lie algebra $\tilde{\mathfrak{h}}$ of $\widetilde{H}$ is a nonzero subspace of $\underline{\mathfrak{g}}^{(p)}$. On the other hand, by our construction, $\widetilde{H}$ is normalized by $\rho_p(\Gamma)$, so the space $\tilde{\mathfrak{h}}$ is $\mathrm{Ad}\rho_p(\Gamma)$-invariant. Combining this with the absolute irreducibility of the latter, we obtain that $\tilde{\mathfrak{h}} = \underline{\mathfrak{g}}^{(p)}$, that is, $\widetilde{H} = \underline{G}^{(p)}$. Furthermore, since G is simply connected, so is $\underline{G}^{(p)}$, and therefore by the affirmative answer to the Kneser–Tits conjecture over finite fields, we have $\underline{G}^{(p)}(\mathbb{F}_p) = \underline{G}^{(p)}(\mathbb{F}_p)^+$. Invoking Nori's theorem, we obtain

$$H = \widetilde{H}(\mathbb{F}_p)^+ = \underline{G}^{(p)}(\mathbb{F}_p)^+ = \underline{G}^{(p)}(\mathbb{F}_p),$$

as required. □

Remarks. 1. The proof of Theorem 3.4 sketched above is based on Nori's paper [Nori 1987], and is different from the original argument in [Matthews et al. 1984]. The interested reader can find an outline of this argument (which relied on the classification of finite simple groups) in [Lubotzky and Segal 2003, pp. 397–398].

2. Combining Lemmas 3.3, 3.6 with the above argument, we obtain a virtually complete proof of Theorem 3.1 for $G = \mathrm{SL}_2$, which, as we have pointed out earlier, was essentially present already in [Serre 1968].

3. We stress that the simply connectedness of G was used again to conclude that the group $\underline{G}^{(p)}(\mathbb{F}_p)$ is generated by unipotent elements. This is yet another manifestation of the connection between strong approximation and the Kneser–Tits conjecture that was first pointed out by Platonov [1969].

4. During the workshop, I. Rivin asked if one can give an explicit bound $N = N(\Gamma)$ such that for any $p > N$ we have $\rho_p(\Gamma) = \underline{G}^{(p)}(\mathbb{F}_p)$. In ongoing work with my student A. Morgan, we have been able to produce some bounds of this kind. More precisely, for $g = (g_{ij}) \in \mathrm{SL}_n(\mathbb{Z})$, set

$$\|g\| = \max_{i,j} |g_{ij}|.$$

Furthermore, given a Zariski-dense subgroup $\Gamma = \langle g_1, \ldots, g_d \rangle \subset \mathrm{SL}_n(\mathbb{Z})$, set

$$m = \max_{k=1,\ldots,d} \|g_k\|.$$

Then there exists $N = N(d, m, n)$ such that for any prime $p > N$ we have $\rho_p(\Gamma) = \mathrm{SL}_n(\mathbb{F}_p)$. However, at the time of this writing our bounds are too large to be of practical use.

5. (Due to the referee.) It is worth pointing out two additional results. First, Guralnick [1999, Theorem B] using the classification of finite simple groups proved the following: *Let G be a finite subgroup of* $\mathrm{GL}_n(k)$ *where k is a field of characteristic p with* $p \geqslant \max(n-3, 12)$. *Assume that G has no normal p-subgroups and that G is generated by its elements of order p. Then G is a central product of finite quasisimple Chevalley groups in characteristic p.* This gives a very strong quantitative version of Nori's theorem (under the assumption that G has no normal p-subgroups). Second, it is proved in [Salehi Golsefidy and Varjú 2012, Appendix] that the lower bound on characteristic in Nori's theorem is recursively defined; that is, there is a recursively defined function f such that if $p > f(n)$ then for any subgroup $H \subset \mathrm{GL}_n(\mathbb{F}_p)$ there is an algebraic $\mathbb{F}_p$-subgroup $\tilde{H}$ of GL_n such that $H^+ = \tilde{H}(\mathbb{F}_p)^+$.

3D. *Weisfeiler's theorem.* A far-reaching generalization of Theorem 3.1 was given by B. Weisfeiler. We will state it using the original notation (which is somewhat different from the notation used in the rest of our article).

Theorem 3.7 [Weisfeiler 1984b]. *Let k be an algebraically closed field of characteristic different from 2 and 3, and let G be an almost simple, connected and simply connected algebraic group defined over k. Let Γ be a Zariski-dense finitely generated subgroup of $G(k)$, and let A be the subring of k generated by the traces* $\operatorname{tr}\operatorname{Ad}\gamma$, $\gamma \in \Gamma$. *Then there exists $b \in A$, a subgroup $\Gamma' \subset \Gamma$, and a structure G_{A_b} of a group scheme over A_b on G such that $\Gamma' \subseteq G_{A_b}(A_b)$ and Γ' is dense in $G_{A_b}(\hat{A}_b)$.*

(Here A_b denotes the localization of A with respect to b, and $\hat{A}_b$ the profinite completion of the ring A_b, i.e., the completion with respect to the topology given by all ideals of finite index. To connect this with our previous results, we note that for $A = \mathbb{Z}$, the ring A_b coincides with $\mathbb{Z}[1/p_1, \ldots, 1/p_r]$, where $p_1, \ldots, p_r$ are the primes dividing b, and the completion $\hat{A}_b$ is precisely $\prod_{p \notin \{p_1,\ldots,p_r\}} \mathbb{Z}_p$, that is, the ring of integral S-adeles for $S = \{\infty, p_1, \ldots, p_r\}$.)

In characteristic 2 and 3, one encounters additional problems due to the existence of so-called nonstandard isogenies. We will not get into these technical details here, but roughly speaking one of the problems is that in these exceptional cases the "right" ring or field of definition of Γ may not be the trace ring or field (the subring or subfield of the algebraically closed field k generated by the traces $\operatorname{tr}\operatorname{Ad}\gamma$ for $\gamma \in \Gamma$). The adequate definitions were given by Pink using the notion of so-called minimal triples (which we will not discuss here). In fact [Pink 2000] proves an appropriate version of the openness statement for the adelic closures of Zariski-dense subgroups in all characteristics, was really the final word in the strong approximation saga.

3E. *Applications to group theory*: *Lubotzky's alternative.* One of the most notable applications of strong approximation is the so-called *Lubotzky alternative* for linear groups. It is discussed in detail in [Klopsch et al. 2011, Chapter II] and [Lubotzky and Segal 2003, Window 9], so here we will only state it for linear groups over fields of characteristic zero.

Theorem 3.8 [Lubotzky and Mann 1991]. *Let Γ be a finitely generated linear group over a field of characteristic zero. Then one of the following holds*:

(a) *Γ is virtually solvable*;

(b) *there exists a connected absolutely almost simple simply connected algebraic $\mathbb{Q}$-group G, a finite set $\Pi = \{p_1, \ldots, p_r\}$ of primes such that the group $G(\mathbb{Z}_\Pi)$, where $\mathbb{Z}_\Pi = \mathbb{Z}\left[1/p_1, \ldots, 1/p_r\right]$, is infinite, and a subgroup $\Gamma_1 \subset \Gamma$ of finite index for which the profinite completion $\widehat{\Gamma}_1$ admits a continuous epimorphism onto $G(\widehat{\mathbb{Z}_\Pi})$.*

This statement was applied in [Lubotzky and Mann 1991] to study the *subgroup growth* (= number of subgroups of a given index n) of linear groups; in particular, it leads to the following dichotomy (which we will state here only in characteristic zero, referring the reader to [Lubotzky and Segal 2003] for some minor distinctions that can occur in the case of positive characteristic): if a linear group has polynomial subgroup growth, then it is virtually solvable, but if the growth is not polynomial (hence the group is not virtually solvable), then it is at least $n^{\log n}$.

The interested reader will find more group-theoretic applications of strong approximation in [Klopsch et al. 2011; Lubotzky and Segal 2003] and references therein, and, of course, in other articles contained in this volume.

Acknowledgements

The author is grateful to Joël Bellaïche, Alex Lubotzky and Gopal Prasad for their comments that helped to improve the exposition. Thanks are also due to the anonymous referee for his/her detailed comments and suggestions that were incorporated in the final version. The author was partially supported by NSF grant DMS-0965758, BSF grant 2010149 and the Humboldt Foundation. He is grateful to the MSRI for the hospitality during the workshop "Hot Topics: Thin Groups and Superstrong Approximation" (February 6–10, 2012). During the preparation of the final version, the author was visiting the Department of Mathematics at Yale University whose support is thankfully acknowledged.

References

[Bass 1983] H. Bass, "Theorems of Jordan and Burnside for algebraic groups", *J. Algebra* **82**:1 (1983), 245–254.

[Borovoi 1989] M. V. Borovoi, "Strong approximation for homogeneous spaces", *Dokl. Akad. Nauk BSSR* **33**:4 (1989), 293–296, 380. In Russian.

[Brauer and Feit 1966] R. Brauer and W. Feit, "An analogue of Jordan's theorem in characteristic p", *Ann. of Math.* (2) **84** (1966), 119–131.

[Collins 2007] M. J. Collins, "On Jordan's theorem for complex linear groups", *J. Group Theory* **10**:4 (2007), 411–423.

[Collins 2008] M. J. Collins, "Modular analogues of Jordan's theorem for finite linear groups", *J. Reine Angew. Math.* **624** (2008), 143–171.

[Colliot-Thélène and Xu 2009] J.-L. Colliot-Thélène and F. Xu, "Brauer–Manin obstruction for integral points of homogeneous spaces and representation by integral quadratic forms", *Compos. Math.* **145**:2 (2009), 309–363.

[Demarche 2011] C. Demarche, "Le défaut d'approximation forte dans les groupes linéaires connexes", *Proc. Lond. Math. Soc.* (3) **102**:3 (2011), 563–597.

[Erovenko and Rapinchuk 2006] I. V. Erovenko and A. S. Rapinchuk, "Bounded generation of S-arithmetic subgroups of isotropic orthogonal groups over number fields", *J. Number Theory* **119**:1 (2006), 28–48.

[Gille 2009] P. Gille, "Le problème de Kneser–Tits", pp. 39–81 in *Séminaire Bourbaki* 2007/2008 (Exposé 983), Astérisque **326**, Société Mathématique de France, Paris, 2009.

[Gorodnik and Nevo 2011] A. Gorodnik and A. Nevo, "Splitting fields of elements in arithmetic groups", *Math. Res. Lett.* **18**:6 (2011), 1281–1288.

[Guralnick 1999] R. M. Guralnick, "Small representations are completely reducible", *J. Algebra* **220**:2 (1999), 531–541.

[de la Harpe 2000] P. de la Harpe, *Topics in geometric group theory*, University of Chicago Press, Chicago, IL, 2000.

[Hrushovski and Pillay 1995] E. Hrushovski and A. Pillay, "Definable subgroups of algebraic groups over finite fields", *J. Reine Angew. Math.* **462** (1995), 69–91.

[Jouve et al. 2013] F. Jouve, E. Kowalski, and D. Zywina, "Splitting fields of characteristic polynomials of random elements in arithmetic groups", *Israel J. Math.* **193**:1 (2013), 263–307.

[Klopsch et al. 2011] B. Klopsch, N. Nikolov, and C. Voll, *Lectures on profinite topics in group theory*, London Mathematical Society Student Texts **77**, Cambridge University Press, 2011.

[Kneser 1965] M. Kneser, "Starke Approximation in algebraischen Gruppen, I", *J. Reine Angew. Math.* **218** (1965), 190–203.

[Lang 1983] S. Lang, *Fundamentals of Diophantine geometry*, Springer, New York, 1983.

[Lang 1994] S. Lang, *Algebraic number theory*, 2nd ed., Graduate Texts in Mathematics **110**, Springer, New York, 1994.

[Lang and Weil 1954] S. Lang and A. Weil, "Number of points of varieties in finite fields", *Amer. J. Math.* **76** (1954), 819–827.

[Larsen and Pink 2011] M. J. Larsen and R. Pink, "Finite subgroups of algebraic groups", *J. Amer. Math. Soc.* **24**:4 (2011), 1105–1158.

[Lubotzky and Mann 1991] A. Lubotzky and A. Mann, "On groups of polynomial subgroup growth", *Invent. Math.* **104**:3 (1991), 521–533.

[Lubotzky and Rosenzweig 2012] A. Lubotzky and L. Rosenzweig, "The Galois group of random elements of linear groups", preprint, 2012. To appear in *Amer. J. Math.* arXiv 1205.5290

[Lubotzky and Segal 2003] A. Lubotzky and D. Segal, *Subgroup growth*, Progress in Mathematics **212**, Birkhäuser, Basel, 2003.

[Margulis 1977] G. A. Margulis, "Cobounded subgroups in algebraic groups over local fields", *Funkcional. Anal. i Priložen.* **11**:2 (1977), 45–57. In Russian; translated in *Funct. Anal. Appl.* **11**:2 (1977), 119–122.

[Margulis 1991] G. A. Margulis, *Discrete subgroups of semisimple Lie groups*, Ergebnisse der Mathematik und ihrer Grenzgebiete (3) **17**, Springer, Berlin, 1991.

[Matthews et al. 1984] C. R. Matthews, L. N. Vaserstein, and B. Weisfeiler, "Congruence properties of Zariski-dense subgroups, I", *Proc. London Math. Soc.* (3) **48**:3 (1984), 514–532.

[Minchev 1989] K. P. Minchev, "Strong approximation for varieties over an algebraic number field", *Dokl. Akad. Nauk BSSR* **33**:1 (1989), 5–8. In Russian.

[Nori 1987] M. V. Nori, "On subgroups of $\mathrm{GL}_n(\mathbf{F}_p)$", *Invent. Math.* **88**:2 (1987), 257–275.

[Pink 1998] R. Pink, "Compact subgroups of linear algebraic groups", *J. Algebra* **206**:2 (1998), 438–504.

[Pink 2000] R. Pink, "Strong approximation for Zariski dense subgroups over arbitrary global fields", *Comment. Math. Helv.* **75**:4 (2000), 608–643.

[Pink 2004] R. Pink, "On Weil restriction of reductive groups and a theorem of Prasad", *Math. Z.* **248**:3 (2004), 449–457.

[Platonov 1969] V. P. Platonov, "The problem of strong approximation and the Kneser–Tits hypothesis for algebraic groups", *Izv. Akad. Nauk SSSR Ser. Mat.* **33** (1969), 1211–1219. In Russian; translated in *Math. USSR Izv.* **3**:3 (1969), 1139–1147.

[Platonov 1980] V. P. Platonov, "Algebraic groups and reduced K–theory", pp. 311–317 in *Proceedings of the International Congress of Mathematicians* (Helsinki, 1978), edited by O. Lehto, Acad. Sci. Fennica, Helsinki, 1980.

[Platonov and Rapinchuk 1994] V. Platonov and A. Rapinchuk, *Algebraic groups and number theory*, Pure and Applied Mathematics **139**, Academic Press, Boston, 1994.

[Prasad 1977] G. Prasad, "Strong approximation for semi-simple groups over function fields", *Ann. of Math.* (2) **105**:3 (1977), 553–572.

[Prasad and Rapinchuk 2001] G. Prasad and A. S. Rapinchuk, "Irreducible tori in semisimple groups", *Int. Math. Res. Not.* **2001**:23 (2001), 1229–1242.

[Prasad and Rapinchuk 2003] G. Prasad and A. S. Rapinchuk, "Existence of irreducible $\mathbb{R}$–regular elements in Zariski-dense subgroups", *Math. Res. Lett.* **10**:1 (2003), 21–32.

[Prasad and Rapinchuk 2009] G. Prasad and A. S. Rapinchuk, "Weakly commensurable arithmetic groups and isospectral locally symmetric spaces", *Publ. Math. Inst. Hautes Études Sci.* 109 (2009), 113–184.

[Prasad and Rapinchuk 2010] G. Prasad and A. S. Rapinchuk, "Number-theoretic techniques in the theory of Lie groups and differential geometry", pp. 231–250 in *Fourth International Congress of Chinese Mathematicians* (Hangzhou, 2007), edited by L. Ji et al., AMS/IP Stud. Adv. Math. **48**, Amer. Math. Soc., Providence, RI, 2010. arXiv 0809.2401

[Prasad and Rapinchuk 2014] G. Prasad and A. S. Rapinchuk, "Generic elements in Zariski-dense subgroups and isospectral locally symmetric space", pp. 211–252 in *Thin groups and superstrong approximation*, edited by H. Oh and E. Breuillard, Math. Sci. Res. Inst. Publ. **61**, Cambridge, New York, 2014.

[Prasad and Rapinchuk $\geq$ 2012] G. Prasad and A. S. Rapinchuk, "On centrality of the congruence kernel", In preparation.

[Rapinchuk 1988] A. S. Rapinchuk, "The congruence subgroup problem for algebraic groups and strong approximation in affine varieties", *Dokl. Akad. Nauk BSSR* **32**:7 (1988), 581–584. In Russian.

[Rapinchuk 1989] A. S. Rapinchuk, "On the congruence subgroup problem for algebraic groups", *Dokl. Akad. Nauk SSSR* **306**:6 (1989), 1304–1307. In Russian; translated in *Sov. Math. Dokl.* **39**:3 (1989), 618–621.

[Rapinchuk 1990] A. S. Rapinchuk, *The congruence subgroup problem for algebraic groups*, Habilitationsschrift, Institute of Mathematics of the Academy of Sciences of the BSSR, Minsk, 1990.

[Salehi Golsefidy and Varjú 2012] A. Salehi Golsefidy and P. P. Varjú, "Expansion in perfect groups", *Geom. Funct. Anal.* **22**:6 (2012), 1832–1891.

[Serre 1968] J.-P. Serre, *Abelian l-adic representations and elliptic curves*, W. A. Benjamin, New York, 1968.

[Serre 1971] J.-P. Serre, "Cohomologie des groupes discrets", pp. 77–169 in *Prospects in mathematics* (Princeton, NJ, 1970), Ann. of Math. Studies **70**, Princeton Univ. Press, 1971.

[Serre 1997] J.-P. Serre, *Lectures on the Mordell–Weil theorem*, 3rd ed., Aspects of Mathematics **15**, Vieweg, Braunschweig, 1997.

[Tits 1964] J. Tits, "Algebraic and abstract simple groups", *Ann. of Math.* (2) **80** (1964), 313–329.

[Tits 1966] J. Tits, "Classification of algebraic semisimple groups", pp. 33–62 in *Algebraic groups and discontinuous subgroups* (Boulder, CO, 1965), edited by A. Borel and G. D. Mostow, Amer. Math. Soc., Providence, R.I., 1966.

[Tomanov 1989a] G. Tomanov, "On the congruence-subgroup problem for some anisotropic algebraic groups over number fields", *J. Reine Angew. Math.* **402** (1989), 138–152.

[Tomanov 1989b] G. M. Tomanov, "Congruence subgroup problem for groups of type G_2", *C. R. Acad. Bulgare Sci.* **42**:6 (1989), 9–11.

[Weisfeiler 1984a] B. Weisfeiler, "Post-classification version of Jordan's theorem on finite linear groups", *Proc. Nat. Acad. Sci. U.S.A.* **81**:16 (1984), 5278–5279.

[Weisfeiler 1984b] B. Weisfeiler, "Strong approximation for Zariski-dense subgroups of semisimple algebraic groups", *Ann. of Math.* (2) **120**:2 (1984), 271–315.

asr3x@virginia.edu

Department of Mathematics, University of Virginia, Charlottesville, VA 22904, United States

Thin groups and superstrong approximation
MSRI Publications
Volume **61**, 2013

Generic phenomena in groups: some answers and many questions

IGOR RIVIN

To the memory of Bill Thurston, with gratitude

We give a survey of some known results and of the many open questions in the study of generic phenomena in geometrically interesting groups.

1. Introduction

In this paper we will discuss a number of loosely related questions, which emanate from Thurston's geometrization program for three-dimensional manifolds, and the general Thurston "yoga" that most everything is hyperbolic. We venture quite far afield from three-dimensional geometry and topology — to the geometry of higher-rank symmetric spaces, to number theory, and probability theory, and to the theory of finite groups. In Section 2 we describe the underpinnings from the theory of three-dimensional manifolds as envisaged by W. Thurston. In Section 3 we will describe one natural approach to describing randomness in groups. In Section 5 we describe an approach to actually producing random matrices in lattices in semisimple Lie groups using the philosophy in Section 3. In Section 6 we describe a different approach to randomness, and the questions it raises.

MSC2010: primary 20G25, 20H25, 20P05, 05C81, 20G30, 20F28, 57M50, 20E05, 60F05; secondary 60B15, 60G50, 57M07, 37E30, 20H10, 37A50, 15A36, 11F06.
Keywords: groups, lattices, mapping class group, modular group, random matrix products, three-dimensional manifolds, surfaces, genericity, Zariski-density.

2. Thurston geometry

In so far as this paper is concerned, history begins with Bill Thurston's geometrization program of three-dimensional manifolds. We will begin with the fibered version. The setup is as follows: we have a surface M (a two-dimensional manifold, homeomorphic to a compact surface with a finite number of punctures) and a homeomorphism $\phi : M \to M$. Given this information we construct the *mapping torus* $T_\phi(M)$ of ϕ, by first constructing the product $\Pi = M \times [0, 1]$, and then defining $T_\phi(M)$ to be the quotient space of Π by the equivalence relation which is trivial outside $M \times \{0, 1\}$, where $(x, 0) \sim (\phi(x), 1)$.

One of Thurston's early achievements was the complete understanding of geometric structures on such fibered manifolds. To state the next results we will need to give a very short introduction to the *mapping class group* $\mathrm{Mod}(M)$, which is the group of homeomorphisms of our surface modulo the normal subgroup of homeomorphisms isotopic to the identity — for a longer introduction, see the recently published (but already standard) reference [Farb and Margalit 2011].

In low genus, the mapping class group is easy to understand. For $M \simeq \mathbb{S}^2$, $|\mathrm{Mod}(M)| = 2$; every automorphism of the sphere is isotopic to either the identity map or the antipodal map.

The next easiest case is that of the torus: $M \simeq \mathbb{T}^2$. Then, $\mathrm{Mod}(M) \simeq \mathrm{GL}(2, \mathbb{Z})$. Looking at this case in more detail, we note that the elements of $\mathrm{GL}(2, \mathbb{Z})$ fall into three classes: elliptic (those with a fixed point in the upper halfplane), parabolic (those with a single fixed point p/q on the real axis in $\mathbb{C}$) and the rest (these are hyperbolic, and have two quadratic irrational fixed points on the real axis). Elliptic elements are *periodic*. Parabolic elements leave the (p, q) curve on the torus invariant (they correspond to a *Dehn twist* about this curve). Hyperbolic elements leave no curve invariant. Further, one of their fixed points is attracting, while the other one is repelling. These two fixed points correspond to two orthogonal curves of *irrational slope* on the torus.

In the case where M is the torus with one puncture, Nielsen had proved that $\mathrm{Mod}(M)$ is the same as for $M \simeq \mathbb{T}^2$. After that, things were rather mysterious, until Thurston discovered his classification of surface homeomorphisms, which parallels closely the toral characterization. Thurston's result is that every surface homeomorphism falls into three classes: it is either periodic, or leaves invariant a *multicurve* γ (a collection of simple closed curves on M) — in this case the map is allowed to permute the components of γ, or *pseudo-Anosov*, in which case the map has a pair of orthogonal measured foliations, one of which is expanded by ϕ and the other is contracted. This is a highly nontrivial result which is the beginning of the modern two-dimensional geometry, topology, and dynamics. For a discussion in considerably more depth, see the standard references [Thurston 1988; Poénaru et al. 1979; Casson and Bleiler 1988; Farb and Margalit 2011].

The next theorem ties the above discussion into Thurston's geometrization program for 3-dimensional manifolds (the special case of fibered manifolds was probably the first case of geometrization finished — see J. P. Otal's excellent exposition [2001].) For an in-depth discussion of the various geometries of three-dimensional manifolds, see [Scott 1983].

Theorem 2.1 (Thurston's geometrization theorem for fibered manifolds). *Let $T_\phi(M)$ be as above. We have the following possibilities for the geometry of M.*

(1) *If $M \simeq \mathbb{S}^2$, then $T_\phi(M)$ is modeled on $\mathbb{S}^2 \times \mathbb{R}$.*

(2) *If $M \simeq \mathbb{T}^2$, we have the following possibilities:*
   - (a) *If ϕ is elliptic, $T_\phi(M)$ is modeled on $\mathbb{E}^3$.*
   - (b) *If ϕ is parabolic, $T_\phi(M)$ is a nil-manifold.*
   - (c) *If ϕ is hyperbolic, $T_\phi(M)$ is a solv-manifold.*

(3) *If M is a hyperbolic surface, then:*
   - (a) *If ϕ is periodic, $T_\phi(M)$ is modeled on $\mathbb{H}^2 \times \mathbb{R}$.*
   - (b) *If ϕ is reducible, $T_\phi(M)$ is a graph manifold.*
   - (c) *If ϕ is pseudo-Anosov, $T_\phi(M)$ is hyperbolic.*

An attentive reader will note there are seven special cases, and six out of the eight three-manifold geometries make an appearance. Six out of the seven special cases of the theorem are easy, while the proof of the last case (c) occupies most of the book [Otal 2001]. Thurston's philosophy, moreover, is that "most" fibered (or otherwise) three-manifolds are hyperbolic — the first appearance of this phenomenon in Thurston's work is probably the Dehn surgery theorem [Thurston and Milnor 1979], which states that most Dehn fillings on a cusped hyperbolic manifold yield hyperbolic manifolds, and the last appears in [Dunfield and Thurston 2003; 2006], where it is conjectured that a random three manifold of fixed Heegard genus is hyperbolic. The actual statement that a random *fibered* manifold is hyperbolic seems to have not been published by Thurston, and the honor of first publication of an equivalent question goes to Benson Farb [2006]. We will discuss Farb's precise question below, but first, let's talk about what it means for some property P to be generic for some (possibly) infinite (but countable) set S.

3. An idealist approach to randomness

First, define a measure of size v on the elements of S. This should satisfy some simple axioms:

(1) $v(x) \geq 0$ for all $x \in S$.

(2) The set $S_k = \{x \in S \mid v(x) \leq k\}$ is finite for every k.

Now let P be a predicate on the elements of S —think of a predicate as just a function from S to $\{0, 1\}$. Let $\mathcal{P} \subset S$ be defined as $\mathcal{P} = \{x \in S \mid P(x) = 1\}$, and define $P_k = \{x \in \mathcal{P} \mid v(x) \leq k\}$. We say that the property P is *generic* for S with respect to the valuation v if

$$\lim_{k\to\infty} \frac{|P_k|}{|S_k|} = 1. \tag{1}$$

We say that P is *negligible* with respect to v if

$$\lim_{k\to\infty} \frac{|P_k|}{|S_k|} = 0. \tag{2}$$

Sometimes the above two definitions are not enough, and we say that P has asymptotic density p with respect to v if

$$\lim_{k\to\infty} \frac{|P_k|}{|S_k|} = p. \tag{3}$$

These definitions work well when they work. Here are some examples:

Example 3.1. The set S is the set $\mathbb{N}$ of natural numbers, and the predicate P is $P(x) =$ "x is prime". The valuation v is just the usual archimedean valuation on $\mathbb{N}$, and, as is well known, the set of primes is *negligible*. One can make a more precise statement (which is the content of the prime number theorem; see [Davenport 2000; Newman 1980]).

With definitions as above,

$$\frac{P_k}{S_k} = \Theta\left(\frac{1}{\log k}\right).$$

Example 3.2. Let S be the set of integer lattice points $(x, y) \in \mathbb{Z}^2$, let $\Omega \subset \mathbb{R}^2$ be a Jordan domain, and define the valuation on S as follows:

$$v(x) = \inf\{t \mid x \in t\Omega\}.$$

Further, define the predicate P by $P(x, y) =$ "x is relatively prime to y"; such points are called *visible*, since one can see them from the origin $(0, 0)$. Then, the asymptotic density of P is $1/\zeta(2) = 6/\pi^2$.

The proof of this for Ω being the unit square is classical, and can be found (for example) in [Hardy and Wright 2008] or in the less classical reference [Rivin 2001]. To get the general statement, we first note that the special linear group $\mathrm{SL}(2, \mathbb{Z})$ acts ergodically on the plane $\mathbb{R}^2$ (see [Zimmer 1984]). Now, define a measure μ_t by

$$\mu_t(\Omega) = \frac{1}{t^2} \text{ (the number of points such that } P(x, y) = 1 \text{ in } t\Omega.)$$

Each μ_t is clearly a measure, dominated by the Lebesgue measure, and invariant under the $\mathrm{SL}(2,\mathbb{Z})$ action on $\mathbb{R}^2$. By Helly's theorem [Lax 2002, Section 10.3] It follows that the set $\{mu_t\}$ has a convergent subsequence σ, and by $\mathrm{SL}(2,\mathbb{Z})$ invariance, the limit μ_σ is a constant multiple of the Lebesgue measure, and the constant can be evaluated for some specific Ω, such as the square (more details of the argument can be found in [Kapovich et al. 2007]). Notice that the constant does not depend on σ, so all the convergent subsequences of the set $\{\mu_t\}$ have the same limit, which must, therefore, be the unique limit point of the set.

Example 3.3. Consider the free group on two generators $F_2 = \langle a, b\rangle$. We define the valuation $v(x)$ to be the reduced word length of x. Define the predicate $P(x) =$ "the abelianization $a(x) \in \mathbb{Z}^2$ is a visible point". Then P does *not* have an asymptotic density (see [Kapovich et al. 2007] for a discussion). It does, however, have a an asymptotic *annular* density, defined as follows: Let $X \subset S$, where S, as usual, has a valuation satisfying our axioms. We define $S_k = \{x \in S \mid v(x) = k\}$, and similarly for X_k. Then, the k-th annular density of T is defined by

$$\rho_k(X) = \frac{1}{2}\left(\frac{X_{k-1}}{S_{k-1}} + \frac{X_k}{S_k}\right). \tag{4}$$

We define the *strict annular density* of X to be $\rho_A(X) = \lim_{k\to\infty} \rho_k(X)$, if the limit exists. The general result is this:

Theorem 3.4 [Kapovich et al. 2007]. *Let S be an $\mathrm{SL}(n,\mathbb{Z})$ invariant subset of $\mathbb{Z}^k$, and let $\tilde{S} = a^{-1}S$, where a, as before, is the abelianization map from the free group on k generators F_k to $\mathbb{Z}^k$. Then $\tilde{S}$ has a strict annular density whenever S has an asymptotic density. Moreover, the two densities are equal.*

The proof of Theorem 3.4 uses the ergodicity of the $\mathrm{SL}(n,\mathbb{Z})$ action on $\mathbb{R}^n$, as described in Example 3.2, and the central and local limit theorems of [Rivin 1999] (see also [Rivin 2010a; Sharp 2001]).

The examples above show that the cases where the groups are reasonably simple to describe, the idealistic valuation-based approach is quite successful. However, once the groups are more complicated, this approach often bogs down in at least some ways, the principal of which is that when one talks of negligibility, genericity, or density, one is making a statement about properties of *random* elements of the set S. However, this raises the question of how to generate such random elements (note that the generation method will often hold the keys to our ability to approach asymptotic statements.)

Example 3.5. Let $S = \mathrm{SL}(n,\mathbb{Z})$. There is a natural family of valuations on S — the archimedean valuations associated to the various Banach space norms on the space of $n \times n$ matrices $M^{n\times n}$. Since all these are known to be equivalent, we might as well choose the Frobenius norm (the L^2 norm of a matrix $x \in \mathrm{SL}(n, \mathbb{Z})$ viewed as a vector in $\mathbb{Z}^{n \times n}$.). In other words, in our previous language,

$$v(x) = \sqrt{\sum_{i=1}^{n} \sum_{j=1}^{n} a_{ij}^2}.$$

Let $S_{\leq k}$ be the set of those x in S with $v(x) \leq k$. It is not at all obvious how to find the cardinality of $S_{\leq k}$, though this has been done (relatively recently) for $\mathrm{SL}(2, \mathbb{Z})$ (by Morris Newman [1988][1] and in general by W. Duke, Z. Rudnick, and P. Sarnak [1993], and by A. Eskin and C. McMullen [1993]. The result is that the number of points is asymptotic to a constant times k^{n^2-n} – the constant for $n = 2$ is 6. In any case, enumeration in and of itself is difficult, and enumerating subsets seems more difficult still. An example of this is the (simple to state) question of finding a *uniformly distributed random element* of bounded norm (see Section 5 for more).

Nevertheless, one can try to show some results on linear groups using the archimedean valuation as above, sometimes using the very deep results of A. Nevo and P. Sarnak [2010], which are a major advance on the results of [Duke et al. 1993].

Example 3.6 Rivin 2008; 2009. A generic element of $\mathrm{SL}(n, \mathbb{Z})$ has irreducible characteristic polynomial; furthermore, the Galois group of the characteristic polynomial is generically the full symmetric group. The generic element of $\mathrm{Sp}(n, \mathbb{Z})$ has irreducible characteristic polynomial.

Example 3.7. Let G be a lattice in a semisimple linear group $\mathfrak{G}$, pick a (non-central) element $h \in G$ and for an element $x \in G$ consider the group $H_x = \langle x, h \rangle$. Consider the predicate $P(x) =$ "H_x is Zariski-dense in $\mathfrak{G}$". In [Rivin 2010b] I showed that P is generic in G. Similarly, if $H = G \times G$, the property $P(x, y) =$ "H_x is Zariski-dense in $\mathfrak{G}$" is generic in $G \times G$. The results are based on strong approximation theory, as developed in [Matthews et al. 1984]; see also [Platonov and Rapinchuk 1994].

Question 3.8. How do we tell if a subgroup G of (say) $\mathrm{SL}(n, \mathbb{Z})$ given by its matrix generators is Zariski-dense?

There are two avenues by which to attack Question 3.8. The first is via strong approximation techniques: by [Matthews et al. 1984], almost all modular projections are surjective for any Zariski-dense subgroup G, and further, by [Weigel 1996], the group is Zariski-dense if *any* projection modulo some $p > 3$ is surjective. So, we need only check a finite number of possible bad projections,

[1] P. Sarnak remarks that this result follows immediately from Delsarte's solution of the circle problem in hyperbolic case, but Newman's argument is completely different.

which *is* bounded by [Rapinchuk 2012], but the bound is not what one would call practical, so this approach, while aesthetically pleasing, takes a lot of work to make work.

A completely different approach is the brute force attack: take the group, compute several elements, compute their (matrix) logarithm, and see if the resulting elements generate the Lie algebra of $\mathrm{SL}(n, \mathbb{C})$ as a vector space. This is a much more computationally promising approach, but it requires a lot of work to produce provable results (the logarithm can usually be computed only approximately, so one needs to find the measure of one's confidence in one's results, etc). The method is particularly effective when there are a lot of unipotent elements in the subgroup, since the logarithm of a unipotent integral matrix is a matrix with *rational* entries, so no approximation techniques are necessary (this was pointed out to me by A. Eskin, who, in turn, credits Forni, Matheus, and Zorich).

Added in proof. Since the first version of this paper was written, the author has discovered a third method (much superior to the other two), based on the following observation: if a subgroup of $\mathrm{SL}(n, \mathbb{Z})$ contains two noncommuting matrices of infinite order such that the Galois group of at least one of them is the full symmetric group, then the subgroup is Zariski-dense. Since, as shown in [Rivin 2008; 2009; 2010b], the probability that a random word of length N of a Zariski dense subgroup does *not* have the full Galois group has probability decreasing exponentially in N, and since testing for Galois group equaling S_n is both theoretically and practically fast, this is an excellent practical and theoretical method. The validity of this method is shown, and the result is extended to other algebraic groups, in [Prasad and Rapinchuk 2014, Theorem 9.10], and the general algorithmic aspects are discussed in an upcoming paper of the author.

Example 3.9 [Fuchs and Rivin $\geq$ 2012]. If G is a lattice in $\mathrm{SL}(2, \mathbb{C})$, and $H = G \times G$, then, for any $\alpha > 0$, the property

$$P(x, y) = \text{“the Hausdorff dimension of } \langle x, y \rangle \text{ is at least } \alpha\text{”}$$

is negligible in H.

The argument uses the ergodicity of the action of $\mathrm{SL}(2, \mathbb{Z})$ on the plane (which shows that the attractive and repelling fixed points of elements are equidistributed), and a ping-pong argument, together with bounds on the Hausdorff dimension as in [McMullen 1998].

One can also show (using an idea of Sarnak) that the same result holds in the parabolic setting:

Theorem 3.10. *Consider a pair of elements* one of which is parabolic. *The property*

$$P(x, y) = \text{“The Hausdorff dimension of the limit set of } \langle x, y\rangle \text{ is at least } \alpha\text{”}$$

is negligible for any $\alpha > \frac{1}{2}$.

It is a theorem of Beardon [1968] that any such group *does* have Hausdorff dimension at least $\frac{1}{2}$, so the constant $\frac{1}{2}$ in Theorem 3.10 is best possible.

It is hoped that the techniques used to attack Example 3.9 can be extended to attack Theorem 3.10.

Example 3.11 [Capdeboscq and Rivin ≥ 2012]. If G is a lattice in a semisimple Lie group $\mathfrak{G}$ of rank at least two, and $H = G \times G$, and if we define $P(x, y) = \langle x, y\rangle$ is profinitely dense, then $P(x, y)$ has asymptotic density bounded below.

The idea of the argument is as follows (in the case of $\mathrm{SL}(n, \mathbb{Z})$):

The first observation is that $\mathrm{SL}(n, \mathbb{Z}/NM\mathbb{Z}) = \mathrm{SL}(n, \mathbb{Z}/N\mathbb{Z}) \times \mathrm{SL}(n, \mathbb{Z}/M\mathbb{Z})$, for N, M relatively prime.

The second observation is that random elements (either in the random walk model or in the “archimedean height” model) are eventually equidistributed in modular projections, e.g modulo the product P of the first k primes (this is one of the results of [Rivin 2008]). By the first observation, the behaviors modulo different primes are independent, and so by [Kantor and Lubotzky 1990; Liebeck and Shalev 1995] the probability that the projections onto the first k primes are surjective is bounded below by

$$B_k = \prod_{i=1}^{k} (1 - C(n)/p^{n-1}),$$

where $C(n)$ is a rank-dependent constant (we are using the congruence subgroup property). Since the series

$$\sum_{i=1}^{\infty} \frac{1}{p^{n-1}}$$

converges for $n > 2$, it follows that the products B_k converge to some constant B.

Now, for any ϵ we can pick k in such a way that $|B_k - B| < \epsilon$, while

$$R_k = \sum_{i=k+1}^{\infty} \frac{C(n)}{p^{n-1}} < \epsilon.$$

By the union bound, the probability that some projection is not surjective is bounded above by $(1 - B_k) + R_k \leq B + 2\epsilon$. So, as long as $\epsilon \ll (1 - B)/2$, we get the probability that at least one projection does not surject is bounded above by $(1 - B)/2$. In reality, of course, if the walks get very very long, the true probability of profinite density is bounded below by B.

The observations above show that the modular projections are surjective for all *prime* moduli with positive probability. Then, using some group-theoretic arguments we can show that there is a positive probability of surjection for all moduli. The estimates on the probabilities are completely effective.

Since the first example of a profinitely dense (free) subgroup of $\mathrm{SL}(n, \mathbb{Z})$ was constructed by Steve Humphries in his beautiful paper [1988], we call the groups described in Example 3.11 *Humphries groups*. Now for the question:

Question 3.12. Suppose we are given a subgroup G of $\mathrm{SL}(n, \mathbb{Z})$ given by matrix generators. How hard it is to decide whether it is a Humphries group?

Question 3.12 is quite difficult. Notice that the full lattice $\mathrm{SL}(n, \mathbb{Z})$ is a Humphries group (by the definition above), so there is a natural dichotomy: Either a Humphries group G is the whole $\mathrm{SL}(n, \mathbb{Z})$ or, by the congruence subgroup property, it is infinite index, and further, by [Lubotzky 1999] (see [Weigel 1996] for a sharper result), G is Zariski-dense in $\mathrm{SL}(n, \mathbb{Z})$. This dichotomy is not really relevant from the algorithmic standpoint—we know from the work of Matthews, Vaserstein and Weisfeiler [Matthews et al. 1984] that for any Zariski-dense subgroup G only a finite number of projections is not surjective, and if we could bound the number, we could just check surjectivity for every possible exceptional modulus—such a check can be performed efficiently using, for example, the algorithm of Neumann and Praeger [1992]. Unfortunately, getting a bound using the generators is not so easy. The first advance came (after the author raised the question at an MSRI Hot Topics conference) very recently, in the work by A. Rapinchuk [2012], but the bounds there, while explicit, are not really computationally useful, as Rapinchuk prominently states in the paper.

On the other hand, the beginning of the discussion in the paragraph above begs the question:

Question 3.13. Given a collection of matrices in $\mathrm{SL}(n, \mathbb{Z})$ do they generate $\mathrm{SL}(n, \mathbb{Z})$?

One attempt is as follows: compute the fundamental domain of the span of the matrices on the homogeneous space of $\mathrm{SL}(n, \mathbb{R})$. If we are lucky, and that domain is finite-volume, we can answer the question (the author has conducted a number of experiments along these lines). No general attack seems to be available, and it is not even clear whether the question is decidable! Similar sounding questions (like the membership problem) are undecidable in $\mathrm{SL}(n, \mathbb{Z})$ for $n \geq 4$, but the techniques which show this seem unapplicable here. In special cases (which are central to the study of mirror symmetry; see [Singh and Venkataramana 2012;

Brav and Thomas 2012]) the question can be decided by a rather diverse set of approaches (arithmeticity is proved using T. N. Venkataramana's theorem that if a Zariski-dense subgroup has "opposing unipotents", then it is a lattice; nonarithmeticity use proved using ping-pong — a presentation is obtained for the group, and then it can be checked that for homological reasons the group cannot be a lattice).

4. Punctured (or not) torus

Consider the modular group $\mathrm{SL}(2,\mathbb{Z})$. Our first set of results will use ordering by Frobenius norm of the matrix.

Definition 4.1. The *Frobenius norm* of the matrix $x = \left(\begin{smallmatrix} a & b \\ c & d \end{smallmatrix}\right)$ is

$$\|x\| = \sqrt{a^2+b^2+c^2+d^2}.$$

The first question is:

Question 4.2. How many elements $x \in \mathrm{SL}(2,\mathbb{Z})$ have $\|x\| \le N$?

It is surprising that answer to this question was first written down by Morris Newman a quarter-century ago!

Theorem 4.3 [Newman 1988]. *The number $\mathcal{N}_k$ of elements $x \in \mathrm{SL}(2,\mathbb{Z})$ with $\|x\| \le k$ is asymptotic to $6k^2$.*

Newman's proof begins by reparametrizing $\mathrm{SL}(2,\mathbb{Z})$, as follows. First define the variables

$$A = a+d, \quad B = b+c, \quad C = b-c, \quad D = a-d.$$

We see that $A^2+B^2+C^2+D^2 = 2(a^2+b^2+c^2+d^2)$. Further note that

$$4 = 4(ad-bc) = A^2+C^2-B^2-D^2, \tag{5}$$

while

$$A = \operatorname{tr}\begin{pmatrix} a & b \\ c & d \end{pmatrix}. \tag{6}$$

Since the difference between A and D is $2d$, we know that

$$A \equiv D \mod 2, \tag{7}$$

and for the same reason

$$B \equiv C \mod 2. \tag{8}$$

Then Newman writes down a generating function for the number of matrices in $\mathrm{SL}(2,\mathbb{Z})$ with prescribed Frobenius norm in terms of the theta function

$$\theta(x) = \sum_{n=-\infty}^{\infty} x^{n^2}, \tag{9}$$

and uses classical estimates on the coefficients of products of theta functions to obtain the asymptotic result of Theorem 4.3. Since an exposition of this method would take us too far afield, let's use the parametrization above to count those elements with trace equal to 2 (which is to say, the parabolic elements). Equations (5) and (6) tell us that the number of such matrices with Frobenius norm bounded by k is exactly equal to the number of *Pythagorean triples* of norm bounded by k.

Now, as is well known, pythagorean triples (A, B, C) with $A^2 = B^2 + C^2$ are rationally parametrized by

$$A = u^2 + v^2, \quad B = u^2 - v^2, \quad C = 2uv. \tag{10}$$

With this parametrization, the 2-norm of (A, B, C) equals $\sqrt{2}(u^2 + v^2)$, so the number of pythagorean triples with L^2 norm bounded above by X equals the number of pairs (u, v) with L^2 norm bounded above by $2^{1/4}\sqrt{X}$, which, in turn, is asymptotic to $\sqrt{2}\pi X$. Note that the congruences (7) and (8) tell us that $2uv = a - d$. This overcounts by a factor of two (since (u, v) and $(-u, -v)$ give the same Pythagorean triple, but on the other hand, parabolic matrices are allowed to have trace equal to ± 2, so when the smoke clears, we have:

Fact 4.4. *The number of parabolic matrices in* $\mathrm{SL}(2,\mathbb{Z})$ *and Frobenius norm bounded above by* k *is asymptotic to* $\sqrt{2}\pi k$.

Observation 4.5. The characteristic polynomial of a matrix in $\mathrm{SL}(2,\mathbb{Z})$ factors over $\mathbb{Z}$ if and only if the matrix is parabolic (so has trace ± 2.)

Proof. If $M \in \mathrm{SL}(2,\mathbb{Z})$, the roots of the characteristic polynomial $\chi(M)$ are

$$\frac{\operatorname{tr} M \pm \sqrt{\operatorname{tr}^2 M - 4}}{2}. \tag{11}$$

For $\chi(M)$ to factor, $\operatorname{tr}^2 M - 4$ must be a perfect square, which obviously happens only when $|\operatorname{tr} M| = 2$. □

We thus have:

Theorem 4.6. *The probability of a matrix in* $\mathrm{SL}(2,\mathbb{Z})$ *of Frobenius norm at most* x *having* reducible *characteristic polynomial is asymptotic to* $3\sqrt{2}/(\pi x)$.

5. Looking for random integer matrices

Consider a simple question: Let v be an archimedean valuation in a lattice (below we will be discussing $SL(n, \mathbb{Z})$, but the question is just as interesting for any other lattice in a noncompact Lie Group, not necessarily semisimple).

Question 5.1. Given k, how do we choose a random element x *uniformly* with $v(x) \leq k$?

Even for $SL(2, \mathbb{Z})$, Question 5.1 seems not to have been studied, but here is an idea.

***A line of attack for* $\mathbf{SL(2, \mathbb{Z})}$.** To get an *approximately* uniform element, based on the fact that the homogeneous space of $SL(2, \mathbb{R})$ is the hyperbolic plane $\mathbb{H}^2$: Take a basepoint in $\mathbb{H}^2$ (since we will be eventually interested in $SL(2, \mathbb{Z})$, the Poincaré halfspace model is popular, and there the point $i = \sqrt{-1}$ is a popular choice of basepoint). Now, the matrices in $SL(2, \mathbb{R})$ with Frobenius norm bounded by N translate i by hyperbolic distance at most some $f(N)$, so pick a disk D of radius $g(N)$ in $\mathbb{H}^2$, and pick a point x uniformly at random. x will lie in some fundamental domain of the $SL(2, \mathbb{Z})$ action. The point x corresponds to a lattice $L \subset \mathbb{R}^2$, which can be *reduced* (using Legendre's algorithm — basically continued fractions); this corresponds to finding a matrix $m(x)$ in $SL(2, \mathbb{Z})$ that maps the fundamental domain of x to the "standard" fundamental domain of the modular group $SL(2, \mathbb{Z})$. The matrix $m(x)$ is our candidate for the uniformly random element of $SL(2, \mathbb{Z})$ we seek. (That equidistribution on $\mathbb{H}^2$ leads to equidistribution on $SL(2, \mathbb{R})$ is standard; see [Eskin and McMullen 1993], for example.)

There are two problems with this approach. Firstly, $m(x)$ might not satisfy the norm constraint. If that is the case, we throw it away, and try again — if $g(N)$ is not too big, this process will terminate reasonably quickly. The other problem is that the area in the disk D is only approximately equidistributed amongst fundamental domains (more precisely, the intersection of D with the union of the translates of the basic fundamental domain by matrices satisfying the norm inequality is only approximately equidistributed among these translates). There is a trade-off between the two problems: the larger disk we take, the more uniform the distribution is, but the less likely we are to get a point satisfying the norm inequality, so some care is required in designing this algorithm properly.

***A line of attack for* $\mathbf{SL(n, \mathbb{Z})}$ *and* $\mathbf{Sp(2n, \mathbb{Z})}$.** The method just described can be extended to higher dimensions, for at least the special linear and symplectic groups. The homogeneous space for $SL(n, \mathbb{Z})$ (symmetric positive definite matrices with determinant 1,) and for $Sp(2n, \mathbb{Z})$ (the Siegel halfspace) have been known for several decades, and sampling uniformly from the ball in that space is easy, using what Lie theorists call the KAK decomposition, and most other people call the singular value decomposition. The Haar measure (induced by that on the Lie group) is easy to compute, and a random element in the ball of the homogeneous space is easy to sample.

What is *not* so easy is the lattice reduction step. Lattice reduction is a much studied problem, since the groundbreaking work of Lovasz (as embodied in the LLL algorithm); a good survey is [Nguyen 2011]. (Interestingly, the symplectic version of the problem had not been considered until quite recently; see [Gama et al. 2006]). The problem is that the complexity of *exact* lattice reduction is, at present, exponential (in the dimension n of the lattice). The LLL algorithms, and the various improvements run in polynomial time, but they don't necessarily get us to the canonical fundamental domain. They *do* get us near the canonical fundamental domain, which brings up a fundamental question:

Question 5.2. What are the statistical properties of the currently used lattice reduction algorithms?

In other words, if we run the algorithm we sketched not with a precise lattice reducer, but with an approximate one (like LLL), will the matrices we get be uniformly distributed in the ball in $\mathrm{SL}(n, \mathbb{Z})$ or $\mathrm{Sp}(2n, \mathbb{Z})$?

The basic principle of the method described works for lattices over number field, and not just over $\mathbb{Z}$. In fact, a version of lattice reduction for such is described in [Fieker and Stehlé 2010]. The version over $\mathrm{SL}(2, \mathbb{Z})$ can be easily made to work to generate a random matrix in an *arithmetic* Kleinian group— the continued fraction algorithm analogue is described in [Page 2012]. It would be interesting to analyze the *non*arithmetic case.

6. A nonidealist approach to randomness

A much more tractable, from the computational standpoint, approach to generating random elements of fairly arbitrary (finitely generated) groups is the following:

Take a symmetric generating set $S = \{g_1, \dots, g_k\}$ of our favorite group G (where "symmetric" means that the set is invariant under the map $x \mapsto x^{-1}$), and look at the set W_k words of length k in the elements of S. The statement that the group G is finitely generated means that

$$\bigcup_{k \geq 0} W_k = G.$$

Now, the trick is to use the definitions in Section 3, but apply them not to the group G itself but to the *free monoid* M_S on S, with

$$v(x) = \text{the word length of } x.$$

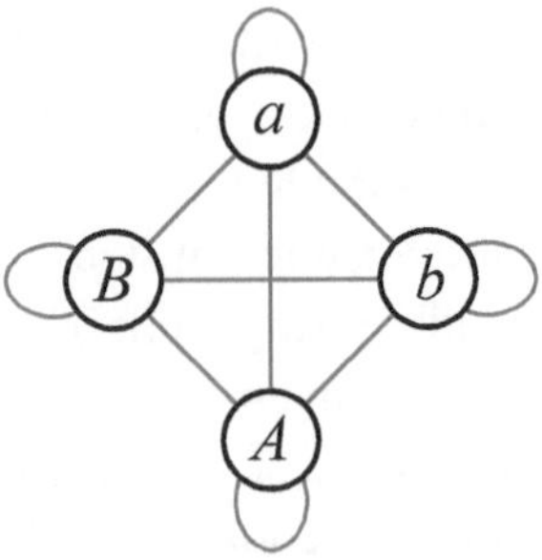

Figure 1. The recognizing automaton for the free monoid on two generators is the (very) complete graph: $bB = aA = 1$.

This approach has many advantages: it is trivial to produce a random element (just multiply elements at random), there are different techniques for proving results, and, in the linear group context this model is closely related to the study of random matrix products, which has a long history and a considerable record of success (a standard reference is [Bougerol and Lacroix 1985]). The (fairly obvious) disadvantage is that the structure of M_S has nothing to do with the structure of G, and it is very difficult to estimate the relationship between the number of occurrences of a given group element in a v-ball of some given radius R. Consequently, doing probability on M_S instead of G is has a certain air of capitulation to it. On the other hand, the free monoid model can be refined, as follows: Elements in the free monoid can be identified with walks in the *complete graph* $\mathcal{K}_k$ on $k = |S|$ vertices (we use the expression in a nonstandard way: every vertex of K_k is connected *to itself* in addition to all the other vertices; see Figure 1.[2])

However, it turns out that this is too broad a context to be able to demonstrate sharp results, and so much of the author's work (see [Rivin 2008, 2009; 2010b] so far has centered on a smaller set of automatic structures: namely, we consider only *undirected* graphs, which, in addition, have the *Perron–Frobenius* property: there is a unique eigenvalue of the adjacency matrix of maximal modulus. This has the immediate benefit of bringing the *free monoid* model closer to reproducing structures of actual interest. For example, Figure 2 shows the graph that generates reduced words (for two generator groups, the general case is similar):

[2]If we remove the requirement that a graph be the (very) complete graph, and, indeed, a directed, as opposed to undirected graph, we find ourselves in the world of *regular languages*, and it was a major discovery of Jim Cannon's, expanded upon by a number of people, including David Epstein and Bill Thurston (see the classic book [Epstein et al. 1992]) that such regular languages are a good way of describing a large class of groups, called *automatic groups*. For such groups, the length of a walk in the defining automaton is a very good valuation — in particular, it coincides with the distance in the Cayley graph from the identity element.

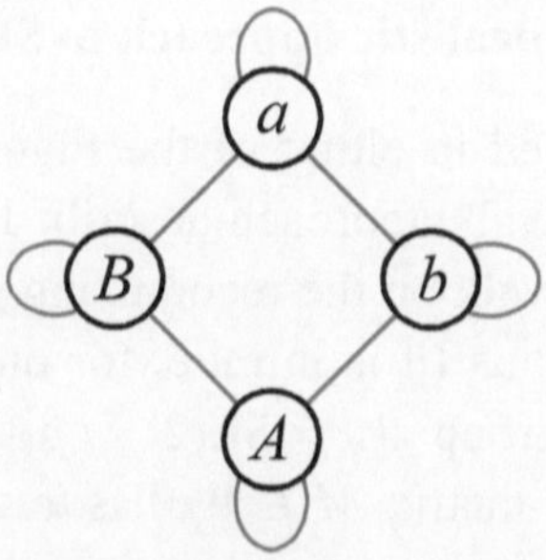

Figure 2. The recognizing automaton for the free group on two generators F_2: $bB = aA = 1$.

This brings up the relatively obvious question:

Question 6.1. Which groups have an automatic structure where the accepting automaton has the properties of being (1) undirected and (2) Perron–Frobenius?

Let's call the combination of properties 1 and 2 *property R*.

Conjecture 6.2. Every word hyperbolic group has property R, with respect to some generating set.

It is not clear that property R is generating-set invariant. While the answer to Question 6.1 is obvious of interest, and the only groups known to have property R are free groups, there is a "cheap" way to extend the techniques to a bigger class of groups, as described in the next section.

What if your group is not free? As a simple example of a nonfree group, we take the *modular group* $\mathcal{M} = \mathrm{SL}(2, \mathbb{Z})$. As is well known, this group is almost, but not quite free. More precisely,

$$\mathcal{M} = \langle S, T \mid S^2, T^3 \rangle = C_2 \star C_2,$$

where C_p is the cyclic group of order p and $\star$ denotes the free product. The obvious symmetric automaton which accepts $C_p \star C_q$ has $p + q - 2$ vertices, corresponding to $T, T^2, \dots, T^{p-1}, S, \dots, S^{q-1}$. Every vertex corresponding to T^i is connected to all vertices of the form S^j. This works wonderfully, except for the minor matter of not representing the identity element and the not-so-minor matter of being bipartite, hence not having property R. However, this can seemingly be fixed by making a new graph with $(p-1)(q-1)$ vertices (corresponding to the products $S^i T^j$) and $q - 1$ start states (corresponding to $T, \dots, q-1$. I believe that this technique will allow the methods used in [Rivin 2008; 2009; 2010b] to be extended to this class of groups.

7. A nonidealistic approach to SL(2, $\mathbb{Z}$)

Using the automata described in either of the Figures 1 or 2, we can study SL(2, $\mathbb{Z}$) using the random walk approach described in Section 6. The idea is simple: consider an n-step walk on the recognizing graph G. Since the group SL(2, $\mathbb{Z}$) is a bit too big for us (it is infinite, for one thing), let's do a quick warm-up, and consider the group $\mathcal{M}_p = \mathrm{SL}(2, \mathbb{Z}/p\mathbb{Z})$. How many elements of $\mathcal{M}_p$ have trace equal to 2? A matrix $M \in \mathcal{M}_p$ has trace 2 if it has the form

$$M = \begin{pmatrix} a & b \\ c & 2-a \end{pmatrix},$$

where $a(2-a) - bc = 1$, which we rewrite as

$$bc + (a-1)^2 = 0. \tag{12}$$

Now, if $a = 1$, Equation (12) has $2p - 1$ solutions ($p - 1$ solutions with b, but not c equal to 0, $p - 1$ solutions with c, but not b, equal to zero, and $(0, 0)$.) If $a \neq 1$, (12) has $p - 1$ solutions of the form $(b, -(a-1)^2/b)$. This gives a total of $2p - 1 + (p-1)^2 = p^2$ matrices with trace equal to 2. On the other hand, the order of $\mathcal{M}_p$ is equal to $p(p^2 - 1)$, so for large p there is a probability of around $1/p$ that M picked uniformly from $\mathcal{M}_p$ picked uniformly at random has trace equal to 2.

What does this have to do with the problem at hand? Note that a matrix which has trace equal to 2 has trace equal to 2 for *every* prime p. This means that if the walks on our graph G are equidistributed in $\mathcal{M}_p$ for some p, the asymptotic probability that an element is parabolic is *at most* $1/p$. The key is this:

Theorem 7.1 [Rivin 2008]. *Let G be a graph with property R, with vertices labeled by generators $\gamma_1, \ldots, \gamma_v$ of a finite group Γ. Then, the walks of length k become equidistributed in Γ, exponentially quickly as a function ofk*, unless *all of the γ_i are sent to the same complex number by some nontrivial one-dimensional irreducible representation of* Γ.

Since $\mathcal{M}_p$ has no irreducible one-dimensional representations, we see that the asymptotic probability is smaller than any $1/p$, and hence the asymptotic density is 0. In fact, using the fact that SL(2, $\mathbb{Z}$), while not enjoying property T, does enjoy property τ for congruence representations (see [Lubotzky 2005]), we can show that the probability of being parabolic decreases *exponentially* in the length of the walk (see [Rivin 2009]).

For specific generating sets, explicit growth rates have been computed in [Takasawa 2001] and [Atalan and Korkmaz 2010]. In the first of these papers, SL(2, $\mathbb{Z}$) is viewed as the mapping class group of the torus. In the second, the authors study the mapping class group of the four-punctured sphere, but the two objects are, in fact, the same (though the generating sets are different). This follows from the fact that for any hyperbolic structure on the punctured torus there is the *elliptic involution*, the quotient by which is an orbifold of signature $(0; 2, 2, 2, \infty)$, while each quadruply punctured sphere admits an order four symmetry group (the Klein four-group) of involutions (this can be seen in many ways, one of which being that each complete finite-area structure on the four times punctured sphere can be realized uniquely as the induced metric on an ideal symplex in $\mathbb{H}^3$; see [Rivin 1994]), the quotient by which is the self-same orbifold. The reader wishing an algebraic proof of this fact can consult [Thompson 1995].

Polynomial versus exponential. The attentive and inquisitive reader may have noticed that in Section 4 we showed that for the matrices of norm bounded above by x, there was a probability of order $1/x$ of finding a parabolic, while this probability decreases *exponentially* fast for the graph walk model. The simplest way of explaining this is the following: the number of distinct elements of length k grows exponentially in k (the order of growth depends on the generating set), so the probability of being parabolic is decaying only polynomially in the size of the sample space.

The second simplest explanation (closely related to the first) is that at least in the simplest possible walk model, the expected norm of the products grows exponentially in the length of the walk (this follows from the classical theory of random matrix products; see [Bougerol and Lacroix 1985]).

The case of $\mathrm{SL}(2, \mathbb{Z})$ is particularly interesting (as it always is). A particularly popular generating set for $\mathrm{SL}(2, \mathbb{Z})$ is the set $\{L, U\}$ where $L = \left(\begin{smallmatrix}1 & 0\\ 1 & 1\end{smallmatrix}\right)$ and $U = \left(\begin{smallmatrix}1 & 1\\ 0 & 1\end{smallmatrix}\right)$. Any matrix $M\left(\begin{smallmatrix}a & b\\ c & d\end{smallmatrix}\right)$, where we assume that $b > a > 0$, $d > c \geq 0$ (the other cases are similar), can be written as $M = U^{a_0} L^{a_1} \ldots U^{a_r}$, where $b/d = [a_0, a_1, \ldots, a_r]$. In this case, the word length of M in terms of the generators L, U is simply the sum of the continued fraction coefficients of b/d — the word length with respect to other generating sets is within a multiple (obtained by writing the "new" generators S, T as words in U, L and vice versa). It turns out that the sum of continued fraction coefficients is not so easy to analyze, though not for lack of trying. The first reference seems to be the paper of Andy Yao and Don Knuth [1975], which showed that the *average* value of the sum $S(p/q)$ of the continued fraction coefficients of p/q taken over all $0 < p < q$ satisfies

$$\frac{1}{q} \sum_{0<p<q} S(p, q) \asymp \frac{6}{\pi^2} (\log q)^2.$$

The $(\log q)^2$ growth is a little deceptive, since the distribution has "fat tails". Indeed, I. Vardi [1993] had shown that for any α satisfying $1 > \alpha > \frac{1}{2}$, we have

$$S(p/q) \leq (\log q)^{1+\alpha}, \quad p < q < n,$$

with at most $O_\epsilon(N^2(\log N)^{1/2-\alpha+\epsilon})$ exceptions.

Vardi's paper is mostly concerned with Dedekind sums, which can be interpreted as the *alternating* sum of continued fraction coefficients. The reader is referred to Vardi's very nice preprint [2009] for more on continued fractions and related mathematics.

8. Higher mapping class groups

The techniques which work for the torus at first appear to fail resoundingly for the mapping class group of more complicated surfaces, since the mapping class group in that setting is not linear. Luckily, there is a workaround: The *Torelli homomorphism* $\mathcal{T}$ is a homomorphism from the mapping class group $\mathcal{M}(S)$ of a (closed, for simplicity, and oriented) surface S of genus g to $\mathrm{Sp}(2g, \mathbb{Z})$. A mapping class ϕ is mapped by $\mathcal{T}$ to its action on the first homology group of S. This action preserves the intersection pairing, and so the image of $\mathcal{T}$ is contained in the symplectic group. In addition, the following fact is standard:

Fact 8.1 [Farb and Margalit 2011]. *The image of $\mathcal{T}$ is all of the symplectic group* $\mathrm{Sp}(2g, \mathbb{Z})$.

Except for the cases of the torus and the four-times punctured sphere, the map $\mathcal{T}$ has a nontrivial kernel, known as the *Torelli group* $\mathfrak{T}(S)$ of the surface S. The Torelli group contains pseudo-Anosov elements, so the Torelli homomorphism does lose a lot of information. Nonetheless, we have: .

Theorem 8.2 [Casson and Bleiler 1988]. *Suppose the characteristic polynomial $\chi(M)$ of a matrix $M = \mathcal{T}(\phi)$ has the following properties*:

(1) *$\chi(M)$ is irreducible.*

(2) *$\chi(M)$ is not cyclotomic.*

(3) *$\chi(M)$ is not of the form $f(x^k)$, for some $k > 1$.*

Then ϕ is pseudo-Anosov.

In view of Fact 8.1, the question of showing that a generic element of $\mathcal{M}(S)$ is pseudo-Anosov reduces to showing that a generic (in the sense of Section 6) element of the symplectic group $\mathrm{Sp}(2g, \mathbb{Z})$ satisfies the conditions of Theorem 8.2. This is a done by what is, philosophically (in a sense that was later made precise by E. Kowalski [2008]), a sieving argument. We show that the properties desired by Theorem 8.2 are enjoyed with a probability *independent of the prime p* by elements of the quotient group $\mathrm{Sp}(2g, \mathbb{Z}/p\mathbb{Z})$. Since, by strong approximation (see [Platonov and Rapinchuk 1994], or, for a more elementary approach, in [Newman 1972]), the reductions modulo different primes are independendent, and the properties described in Theorem 8.2 are assymmetric, in the sense that, for example, in order to conclude that a polynomial is irreducible it is enough to find a *single* prime for which the reduction mod p is irreducible[3], brings us close to the end. The end is achieved thanks to the fundamental equidistribution result Theorem 7.1, together with property T for the groups we are studying for the groups we are interested in to assure exponential convergence [Rivin 2009].

The good news. The argument sketched above (see [Rivin 2008; 2009] for all the details) has many virtues.

Firstly, it is very general — just how general was outlined in [Lubotzky and Meiri 2012a]. In particular, it can be used to show that a generic element in the outer automorphism group of a free group is *irreducible with irreducible powers*, which is the analogue in that setting of being pseudo-Anosov (an element ψ of the automorphism group of a free group is *irreducible* if it does not preserve any splitting of the free group F as a free product $F = G \star H$; it is irreducible with irreducible powers (iwip) if all powers ψ^k are, likewise, irreducible). The importance of iwip automorphisms was first noted in the foundational paper of M. Bestvina and M. Handel [1992]. For automorphisms of free groups there is the analogue of the Torelli homomorphism, which sends an automorphism ψ of a free group F_n to its action on the abelianization $\mathbb{Z}^n$ of F_n. It is easy to show that the Torelli homomorphism $\mathfrak{T}$ is surjective onto the automorphism group of $\mathbb{Z}^n$ — $\mathrm{GL}(n, \mathbb{Z})$, and it is easy to see that an automorphism ψ is irreducible if (of course, *not* only if) the characteristic polynomial of $\mathfrak{T}(\psi)$ is irreducible. Initially, it seems a little frightening to check that ψ is iwip by checking the characteristic polynomials of $(\mathfrak{T}(\psi))^k$ for *every* k for irreducibility, but it turns out (see [Rivin 2008]) that it is enough to check that the Galois group of the characteristic polynomial $\chi(\mathfrak{T}(\psi))$ is the full symmetric group. This can be proved by combining the previous ideas with the idea (going back to van der Waerden) of characterizing Galois groups *via* the factorization patterns of polynomials modulo various primes. This result was later extended in [Jouve et al. 2013] to show that characteristic polynomials of matrices in lattices in semisimple Lie group usually have as big a Galois group as possible (which is to say, the Weyl group of the ambient Lie group).

[3] Note that the argument we used for $\mathrm{SL}(2, \mathbb{Z})$ is much simpler, since there only one (large) prime sufficed, and no strong approximation argument was necessary — it turns out that this kind of argument works for $\mathrm{SL}(n, \mathbb{Z})$ to show that the characteristic polynomial of a generic matrix is irreducible, since it can be shown that the set of irreducible polynomials is a relatively sparse set of subvarieties of the set of coefficients

The results are effective (and explicit) in that they give an exponential rate of convergence of the densities to 0.

The results can be extended without any work to finite index subgroups of $\mathrm{SL}(n,\mathbb{Z})$ and $\mathrm{Sp}(2g,\mathbb{Z})$, and their preimages in the mapping class groups, and, with some work, and much use of the results of [Breuillard et al. 2011], to *thin Zariski-dense subgroups* of such groups, and *their* preimages (these results are still effective, but considerably less explicit than for lattices. It can also be shown, using the results of [Rivin 2010b] that for a *generic* subgroup H of the mapping class group, a generic element of H is pseudo-Anosov.

Bad news. In the context of mapping class groups, our results are not useful for groups which have very small image under the Torelli homomorphism. In particular, the Torelli group itself is completely "orphaned" — Theorem 8.2 is vacuous for elements in the Torelli group. This is all the more galling, since more geometric approaches (see Section 9) show that a generic element of a subgroup of the mapping class group which contains at least two noncommuting pseudo-Anosov elements (below we will call such subgroups *nonelementary*) is pseudo-Anosov. The problem with these approaches is that the convergence rates are completely ineffective, and also they do not apply in the less-geometric situations like the automorphism group of a free group.

Better news. Recently, at least some of the news became less bad, since two groups — A. Lubotzky and C. Meiri [2012b] and J. Malestein and J. Souto [2013] — have extended the results sketched above to the Torelli group, using very similar methods. Lubotzky and Meiri have also extended their results to the Torelli subgroup of the automorphism group of the free group in [Lubotzky and Meiri 2012c]. (The results on the Torelli group make sense only when the genus of the surface in question is at least three — genericity is hard to define for infinitely generated groups).

The first idea of these results goes to the beautiful paper of E. Looijenga [1997], later developed in a more algebraic direction by F. Grunewald and A. Lubotzky [2009]:

Consider a surface S and a double cover $\tilde{S}$. Any homeomorphism the Torelli group lifts to a homeomorphism of $\tilde{S}$. In addition, the image of the lift of the Torelli group under the Torelli homomorphism of $\tilde{S}$ is of finite index in $P\,\mathrm{Sp}(2g-2,\mathbb{Z})$, where $P\,\mathrm{Sp}(\cdot\,,\mathbb{Z})$ denotes $\mathrm{Sp}(\cdot\,,\mathbb{Z})/\{\pm I\}$.

This gives us an indication that we might be able to use linear methods to study the Torelli group. The next ingredient goes back to N. V. Ivanov:

Theorem 8.3 [Ivanov 1992]. *Any non-pseudo-Anosov element of the Torelli subgroup of a surface S leaves invariant an essential simple curve γ on S.*

A much stronger result was shown by B. Farb, C. Leininger and D. Margalit:

Theorem 8.4 [Farb et al. 2008, Proposition 1.4]. *Let γ be a curve and f an element in the Torelli subgroup. Then $i(\gamma, f(\gamma)) \geq 4$ if γ is nonseparating, then $i(\gamma, f^j(\gamma)) \geq 2$, for $j = 1$ or $j = 2$, where $i(x, y)$ denotes the geometric intersection number of curves x and y.*

Finally, it is noted that γ can be used to construct a cover such that the element g in the corresponding $P\,\mathrm{Sp}(2g-2, \mathbb{Z})$ leaves invariant a line in $\mathbb{Z}2g-2$, from which the genericity of pseudo-Anosovs in the *whole* Torelli follows.

In fact, one can combine the above with the methods and results of [Rivin 2010b] it can be shown that a for a generic subgroup of Torelli, a generic element is pseudo-Anosov. However, the silver bullet would be the following:

Conjecture 8.5. For any nonelementary subgroup H of the mapping class group there is a cover $\tilde{S}$ to which H lifts, and such that the image of H under the Torelli homomorphism is not solvable, with the degree of the cover at most polynomial in the sums of the word-lengths of the generators of H.

It should be noted that we are *very* far from being able to resolve Conjecture 8.5. For example, while it is known that for every pseudo-Anosov mapping class ψ there exists *some* cover to which ψ lifts, and such that the ψ is not in the Torelli subgroup for that cover [Koberda 2012], it is *not* known that the image $\mathfrak{T}(\psi)$ is of infinite order! The degree of the cover is effective — the bounds in the paper follow essentially from the results of Edna Grossman [1974] — but, just as in her paper, they are easily doubly exponential in the word length of the element.

9. The geometric approach

Above we have alluded to the "geometric" approach to the mapping class, which uses the curve complex. This approach was used by Joseph Maher [2011] (where he also shows other remarkable results). Maher's results are very general, but not effective. A somewhat more conceptual approach was undertaken in a very nice paper by A. Malyutin, using the followin approach.

First, he uses the central limit theorem of M. Björklund and T. Hartnick [2011] (which is a vast generalization, at the cost of losing effectiveness completely of some of the results of [Rivin 1999]) to show this:

Theorem 9.1 [Malyutin 2011]. *Let G be a countable group and let $\Phi : G \mapsto \mathbb{R}^d$ is a nondegenerate $\mathbb{R}^d$-quasimorphism. Then, for each nondegenerate probability measure μ on G and for every bounded subset $Q \subset \mathbb{R}^d$, there is a constant $C = C(G, \Phi, \mu, Q)$ such that for any $k \in \mathbb{N}$ and $\boldsymbol{x} \in \mathbb{R}^d$ we have*

$$\mu^{*k}(\Phi^{-1}(\boldsymbol{x} + Q)) < Ck^{-d/2},$$

where μ^{*k} denotes the k-fold convolution of μ with itself.

Recall that a *quasimorphism* from a group G to $\mathbb{R}$ is a map $\phi : G \mapsto \mathbb{R}$ such that the following condition holds:

$$\sum_{h,g \in G} \left|\phi(gh) - \phi(g) - \phi(h)\right| < \infty. \tag{13}$$

An $\mathbb{R}^d$ quasimorphism is a map $\Phi : G \to \mathbb{R}^d$ which satisfies the inequality (13) (where $|\cdot|$ now denotes some Banach norm) — a map is an $\mathbb{R}^d$ quasimorphism if and only if its coordinates are garden-variety quasimorphisms. Such a Φ is called *nondegenerate* if its image is not contained in a proper hyperplane.

An immediate corollary of Theorem 9.1 is:

Corollary 9.2. *If a subset S of a countable group G has bounded image under a nondegenerate $\mathbb{R}^d$-quasimorphism $G \mapsto \mathbb{R}^d$, then for every nondegenerate probability measure μ on G there exists a constant $C = C(\mu)$, such that for each $k \in \mathbb{N}$ we have*

$$\mu^{*k}(S) < Ck^{-d/2}.$$

The other ingredient is the result of M. Bestvina and K. Fujiwara [2002], which states that if H is a nonelementary subgroup of the mapping class group of a surface, than there are infinitely many linearly independent quasimorphisms which all map the non-pseudo-Anosov mapping classes to 0.

The Bestvina–Fujiwara theorem and Corollary 9.2 together show that the probability of being non-pseudo-Anosov decreases faster than any polynomial (but does not quite show exponential decay). All constants in the argument are completely ineffective.

Acknowledgements

The author would like to thank Ilan Vardi, Alex Eskin, Inna Capdeboscq, Peter Sarnak, Tania Smirnova-Nagnibeda and Tobias Hartnick for enlightening conversations, and the editors for their patience.

References

[Atalan and Korkmaz 2010] F. Atalan and M. Korkmaz, "Number of pseudo-Anosov elements in the mapping class group of a four-holed sphere", *Turkish J. Math.* **34**:4 (2010), 585–592.

[Beardon 1968] A. F. Beardon, "The exponent of convergence of Poincaré series", *Proc. London Math. Soc.* (3) **18**:3 (1968), 461–483.

[Bestvina and Fujiwara 2002] M. Bestvina and K. Fujiwara, "Bounded cohomology of subgroups of mapping class groups", *Geom. Topol.* **6** (2002), 69–89.

[Bestvina and Handel 1992] M. Bestvina and M. Handel, "Train tracks and automorphisms of free groups", *Ann. of Math.* (2) **135**:1 (1992), 1–51.

[Björklund and Hartnick 2011] M. Björklund and T. Hartnick, "Biharmonic functions on groups and limit theorems for quasimorphisms along random walks", *Geom. Topol.* **15**:1 (2011), 123–143.

[Bougerol and Lacroix 1985] P. Bougerol and J. Lacroix, *Products of random matrices with applications to Schrödinger operators*, Progress in Probability and Statistics **8**, Birkhäuser, Boston, 1985.

[Brav and Thomas 2012] C. Brav and H. Thomas, "Thin monodromy in Sp(4)", preprint, 2012. arXiv 1210.0523

[Breuillard et al. 2011] E. Breuillard, B. Green, and T. Tao, "Approximate subgroups of linear groups", *Geom. Funct. Anal.* **21**:4 (2011), 774–819.

[Capdeboscq and Rivin ≥ 2012] I. Capdeboscq and I. Rivin, "The density of profinite density". In preparation.

[Casson and Bleiler 1988] A. J. Casson and S. A. Bleiler, *Automorphisms of surfaces after Nielsen and Thurston*, London Math. Soc. Student Texts **9**, Cambridge University Press, 1988.

[Davenport 2000] H. Davenport, *Multiplicative number theory*, 3rd ed., Graduate Texts in Mathematics **74**, Springer, New York, 2000.

[Duke et al. 1993] W. Duke, Z. Rudnick, and P. Sarnak, "Density of integer points on affine homogeneous varieties", *Duke Math. J.* **71**:1 (1993), 143–179.

[Dunfield and Thurston 2003] N. M. Dunfield and W. P. Thurston, "The virtual Haken conjecture: Experiments and examples", *Geom. Topol.* **7** (2003), 399–441.

[Dunfield and Thurston 2006] N. M. Dunfield and W. P. Thurston, "Finite covers of random 3-manifolds", *Invent. Math.* **166**:3 (2006), 457–521.

[Epstein et al. 1992] D. B. A. Epstein, J. W. Cannon, D. F. Holt, S. V. F. Levy, M. S. Paterson, and W. P. Thurston, *Word processing in groups*, Jones and Bartlett, Boston, 1992.

[Eskin and McMullen 1993] A. Eskin and C. McMullen, "Mixing, counting, and equidistribution in Lie groups", *Duke Math. J.* **71**:1 (1993), 181–209.

[Farb 2006] B. Farb, "Some problems on mapping class groups and moduli space", pp. 11–55 in *Problems on mapping class groups and related topics*, edited by B. Farb, Proc. Sympos. Pure Math. **74**, Amer. Math. Soc., Providence, RI, 2006.

[Farb and Margalit 2011] B. Farb and D. Margalit, *A primer on mapping class groups*, Princeton Mathematical Series **49**, Princeton University Press, 2011.

[Farb et al. 2008] B. Farb, C. J. Leininger, and D. Margalit, "The lower central series and pseudo-Anosov dilatations", *Amer. J. Math.* **130**:3 (2008), 799–827.

[Fieker and Stehlé 2010] C. Fieker and D. Stehlé, "Short bases of lattices over number fields", pp. 157–173 in *Algorithmic number theory* (Nancy, 2010), edited by G. Hanrot et al., Lecture Notes in Comput. Sci. **6197**, Springer, Berlin, 2010.

[Fuchs and Rivin ≥ 2012] E. Fuchs and I. Rivin, "How thin is thin?". In preparation.

[Gama et al. 2006] N. Gama, N. Howgrave-Graham, and P. Q. Nguyen, "Symplectic lattice reduction and NTRU", pp. 233–253 in *Advances in cryptology—EUROCRYPT 2006* (St. Petersburg, 2006), edited by S. Vaudenay, Lecture Notes in Comput. Sci. **4004**, Springer, Berlin, 2006.

[Grossman 1974] E. K. Grossman, "On the residual finiteness of certain mapping class groups", *J. London Math. Soc.* (2) **9**:1 (1974), 160–164.

[Grunewald and Lubotzky 2009] F. Grunewald and A. Lubotzky, "Linear representations of the automorphism group of a free group", *Geom. Funct. Anal.* **18**:5 (2009), 1564–1608.

[Hardy and Wright 2008] G. H. Hardy and E. M. Wright, *An introduction to the theory of numbers*, 6th ed., Oxford University Press, Oxford, 2008.

[Humphries 1988] S. P. Humphries, "Free subgroups of $SL(n, \mathbf{Z})$, $n > 2$, generated by transvections", *J. Algebra* **116**:1 (1988), 155–162.

[Ivanov 1992] N. V. Ivanov, *Subgroups of Teichmüller modular groups*, Translations of Mathematical Monographs **115**, Amer. Math. Soc., Providence, RI, 1992.

[Jouve et al. 2013] F. Jouve, E. Kowalski, and D. Zywina, "Splitting fields of characteristic polynomials of random elements in arithmetic groups", *Israel J. Math.* **193**:1 (2013), 263–307.

[Kantor and Lubotzky 1990] W. M. Kantor and A. Lubotzky, "The probability of generating a finite classical group", *Geom. Dedicata* **36**:1 (1990), 67–87.

[Kapovich et al. 2007] I. Kapovich, I. Rivin, P. Schupp, and V. Shpilrain, "Densities in free groups and $\mathbb{Z}^k$, visible points and test elements", *Math. Res. Lett.* **14**:2 (2007), 263–284.

[Koberda 2012] T. Koberda, "Asymptotic linearity of the mapping class group and a homological version of the Nielsen–Thurston classification", *Geom. Dedicata* **156** (2012), 13–30.

[Kowalski 2008] E. Kowalski, *The large sieve and its applications: Arithmetic geometry, random walks and discrete groups*, Cambridge Tracts in Mathematics **175**, Cambridge University Press, 2008.

[Lax 2002] P. D. Lax, *Functional analysis*, Wiley, New York, 2002.

[Liebeck and Shalev 1995] M. W. Liebeck and A. Shalev, "The probability of generating a finite simple group", *Geom. Dedicata* **56**:1 (1995), 103–113.

[Looijenga 1997] E. Looijenga, "Prym representations of mapping class groups", *Geom. Dedicata* **64**:1 (1997), 69–83.

[Lubotzky 1999] A. Lubotzky, "One for almost all: Generation of $SL(n, p)$ by subsets of $SL(n, \mathbf{Z})$", pp. 125–128 in *Algebra, K-theory, groups, and education* (New York, 1997), edited by T. Y. Lam and A. R. Magid, Contemp. Math. **243**, Amer. Math. Soc., Providence, RI, 1999.

[Lubotzky 2005] A. Lubotzky, "What is... property (τ)?", *Notices Amer. Math. Soc.* **52**:6 (2005), 626–627.

[Lubotzky and Meiri 2012a] A. Lubotzky and C. Meiri, "Sieve methods in group theory I: Powers in linear groups", *J. Amer. Math. Soc.* **25**:4 (2012), 1119–1148.

[Lubotzky and Meiri 2012b] A. Lubotzky and C. Meiri, "Sieve methods in group theory II: The mapping class group", *Geom. Dedicata* **159** (2012), 327–336.

[Lubotzky and Meiri 2012c] A. Lubotzky and C. Meiri, "Sieve methods in group theory, III: $\mathrm{Aut}(F_n)$", *Internat. J. Algebra Comput.* **22**:7 (2012), 1250062.

[Maher 2011] J. Maher, "Random walks on the mapping class group", *Duke Math. J.* **156**:3 (2011), 429–468.

[Malestein and Souto 2013] J. Malestein and J. Souto, "On genericity of pseudo-Anosovs in the Torelli group", *Int. Math. Res. Not.* **2013**:6 (2013), 1434–1449.

[Malyutin 2011] A. V. Malyutin, "Quasimorphisms, random walks, and transient subsets in countable groups", *Zap. Nauchn. Sem. (POMI)* **390** (2011), 210–236. Republished in *J. Math. Sci. New York* **181**:6 (2012), 871–885.

[Matthews et al. 1984] C. R. Matthews, L. N. Vaserstein, and B. Weisfeiler, "Congruence properties of Zariski-dense subgroups, I", *Proc. London Math. Soc.* (3) **48**:3 (1984), 514–532.

[McMullen 1998] C. T. McMullen, "Hausdorff dimension and conformal dynamics, III: Computation of dimension", *Amer. J. Math.* **120**:4 (1998), 691–721.

[Neumann and Praeger 1992] P. M. Neumann and C. E. Praeger, "A recognition algorithm for special linear groups", *Proc. London Math. Soc.* (3) **65**:3 (1992), 555–603.

[Nevo and Sarnak 2010] A. Nevo and P. Sarnak, "Prime and almost prime integral points on principal homogeneous spaces", *Acta Math.* **205**:2 (2010), 361–402.

[Newman 1972] M. Newman, *Integral matrices*, Pure and Applied Mathematics **45**, Academic Press, New York, 1972.

[Newman 1980] D. J. Newman, "Simple analytic proof of the prime number theorem", *Amer. Math. Monthly* **87**:9 (1980), 693–696.

[Newman 1988] M. Newman, "Counting modular matrices with specified Euclidean norm", *J. Combin. Theory Ser. A* **47**:1 (1988), 145–149.

[Nguyen 2011] P. Q. Nguyen, "Lattice reduction algorithms: Theory and practice", pp. 2–6 in *Advances in cryptology—EUROCRYPT 2011* (Tallinn, Estonia, 2011), edited by K. G. Paterson, Lecture Notes in Comput. Sci. **6632**, Springer, Berlin, 2011.

[Otal 2001] J.-P. Otal, *The hyperbolization theorem for fibered 3–manifolds*, SMF/AMS Texts and Monographs **7**, Amer. Math. Soc., Providence, RI, 2001.

[Page 2012] A. Page, "Computing arithmetic Kleinian groups", preprint, 2012. arXiv 1206.0087

[Platonov and Rapinchuk 1994] V. Platonov and A. Rapinchuk, *Algebraic groups and number theory*, Pure and Applied Mathematics **139**, Academic Press, Boston, 1994.

[Poénaru et al. 1979] *Travaux de Thurston sur les surfaces*, Astérisque **66**, Société Mathématique de France, Paris, 1979.

[Prasad and Rapinchuk 2014] G. Prasad and A. S. Rapinchuk, "Generic elements in Zariski-dense subgroups and isospectral locally symmetric space", pp. 211–252 in *Thin groups and superstrong approximation*, edited by H. Oh and E. Breuillard, Math. Sci. Res. Inst. Publ. **61**, Cambridge, New York, 2014.

[Rapinchuk 2012] A. S. Rapinchuk, "Strong approximation for algebraic groups", preprint, 2012. arXiv 1207.4425

[Rivin 1994] I. Rivin, "Intrinsic geometry of convex ideal polyhedra in hyperbolic 3-space", pp. 275–291 in *Analysis, algebra, and computers in mathematical research* (Luleå, 1992), edited by M. Gyllenberg, Lecture Notes in Pure and Appl. Math. **156**, Dekker, New York, 1994.

[Rivin 1999] I. Rivin, "Growth in free groups (and other stories)", preprint, 1999. arXiv math/9911076

[Rivin 2001] I. Rivin, "Simple curves on surfaces", *Geom. Dedicata* **87**:1-3 (2001), 345–360.

[Rivin 2008] I. Rivin, "Walks on groups, counting reducible matrices, polynomials, and surface and free group automorphisms", *Duke Math. J.* **142**:2 (2008), 353–379.

[Rivin 2009] I. Rivin, "Walks on graphs and lattices—effective bounds and applications", *Forum Math.* **21**:4 (2009), 673–685.

[Rivin 2010a] I. Rivin, "Growth in free groups (and other stories)—twelve years later", *Illinois J. Math.* **54**:1 (2010), 327–370.

[Rivin 2010b] I. Rivin, "Zariski density and genericity", *Int. Math. Res. Not.* **2010**:19 (2010), 3649–3657.

[Scott 1983] P. Scott, "The geometries of 3–manifolds", *Bull. London Math. Soc.* **15**:5 (1983), 401–487.

[Sharp 2001] R. Sharp, "Local limit theorems for free groups", *Math. Ann.* **321**:4 (2001), 889–904.

[Singh and Venkataramana 2012] S. Singh and T. N. Venkataramana, "Arithmeticity of certain symplectic hypergeometric groups", preprint, 2012. To appear in *Duke Math. J.* arXiv 1208.6460

[Takasawa 2001] M. Takasawa, "Enumeration of mapping classes for the torus", *Geom. Dedicata* **85**:1-3 (2001), 11–19.

[Thompson 1995] J. G. Thompson, "4-punctured spheres", *J. Algebra* **171**:2 (1995), 587–605.

[Thurston 1988] W. P. Thurston, "On the geometry and dynamics of diffeomorphisms of surfaces", *Bull. Amer. Math. Soc.* (*N.S.*) **19**:2 (1988), 417–431.

[Thurston and Milnor 1979] W. P. Thurston and J. W. Milnor, "The geometry and topology of three-manifolds", Lecure notes, Princeton University, 1979, http://library.msri.org/books/gt3m/.

[Vardi 1993] I. Vardi, "Dedekind sums have a limiting distribution", *Int. Math. Res. Not.* **1993**:1 (1993), 1–12.

[Vardi 2009] I. Vardi, "Continued fractions from Euclid to the present day", preprint, 2009, http://www.chronomaitre.org/ContinuedFractions.pdf.

[Weigel 1996] T. Weigel, "On the profinite completion of arithmetic groups of split type", pp. 79–101 in *Lois d'algèbres et variétés algébriques* (Colmar, 1991), edited by M. Goze, Travaux en Cours **50**, Hermann, Paris, 1996.

[Yao and Knuth 1975] A. C. Yao and D. E. Knuth, "Analysis of the subtractive algorithm for greatest common divisors", *Proc. Nat. Acad. Sci. U.S.A.* **72**:12 (1975), 4720–4722.

[Zimmer 1984] R. J. Zimmer, *Ergodic theory and semisimple groups*, Monographs in Mathematics **81**, Birkhäuser, Basel, 1984.

rivin@temple.edu

Department of Mathematics, Temple University, Philadelphia, PA 19122, United States

Thin groups and superstrong approximation
MSRI Publications
Volume **61**, 2013

Affine sieve and expanders

ALIREZA SALEHI GOLSEFIDY

The goals of this article are to describe the affine sieve in its most general setting and the fundamental theorem on the subject; to convey the main ideas behind the proof of that theorem and its connection with expansion in linear groups; to survey some of the results on the expansion in linear groups and briefly explain what goes into their proofs; and to collect some questions and conjectures.

1. The affine sieve

1.1. ***What is the affine sieve?*** Lots of problems and theorems in number theory concern the existence of *infinitely many* primes in a subset of integers.

(1) Dirichlet's theorem: *For any integers $a \neq 0$ and b, there are infinitely many integers x such that $ax + b$ is prime if there are no local obstructions — that is, if* $\gcd(a, b) = 1$.

Instead of restricting ourselves to coprime pairs, we can work with $\mathbb{Z}[1/\gcd(a, b)]$ and say for any integers $a \neq 0$ and b there are infinitely many integers x such that $ax + b$ is prime in $\mathbb{Z}[1/\gcd(a, b)]$.

(2) Twin prime conjecture: *There are infinitely many positive integers x such that $x(x + 2)$ has at most two prime factors.*

This is a well-known long standing open problem. If we relax it and ask for infinitely many *almost primes* instead, then we have an affirmative result. Brun developed a combinatorial sieve and proved that there are infinitely many x such that $x(x + 2)$ has at most 20 prime factors. Later more sophisticated sieve methods were developed. As a result Chen proved that there are infinitely many x such that $x(x + 2)$ has at most *three* prime factors.

As we see in this example, in many problems, sieve methods can help us to see what should be expected and prove the existence of infinitely many *almost primes* instead of primes [Halberstam and Richert 1974].

(3) Mersenne prime conjecture: *There are infinitely many positive integers x such that $2^x - 1$ is prime.*

Partially supported by NSF grant DMS-1160472 and an Alfred P. Sloan Research Fellowship.

This set is too sparse and so sieve methods do not give us anything. In fact, it is unknown even whether there are integers r such that $2^x - 1$ has at most r prime factors for infinitely many positive integers x.

In all these examples, we are dealing with a one-parameter subset of integers. One can ask what happens in a multiparameter setting. What should be the "right" question in this setting? In some sense it is not a good idea to just ask about existence of infinitely many desired values, as one can just restrict to a one-parameter subset for that purpose. Bourgain, Gamburd and Sarnak [Bourgain et al. 2010] nicely suggest replacing "infiniteness" with "Zariski density".[1] They also give the following reformulation of the Hardy–Littlewood conjecture, which further convinces us that Zariski density should be the right notion to seek in a multiparameter setting.

Hardy–Littlewood conjecture. *Let $\vec{b} = (b_1, \dots, b_n) \in \mathbb{Z}^n$ and let Λ be a subgroup of $\mathbb{Z}^n$. Let*

$$\Lambda_{\vec{b}} := \left\{ \vec{\lambda} = (\lambda_1, \dots, \lambda_n) \in \Lambda \;\middle|\; \prod_{i=1}^{n} (\lambda_i + b_i) \text{ has at most } n \text{ prime factors.} \right\}.$$

Then the Zariski closure of $\Lambda_{\vec{b}}$ is equal to the Zariski closure of Λ if there are no local obstructions, i.e., if for any square-free integer q there is $\vec{\lambda} \in \Lambda$ such that $\gcd(f_{\vec{b}}(\vec{\lambda}), q) = 1$, where $f_{\vec{b}}(\vec{\lambda}) = \prod_{i=1}^{n} (\lambda_i + b_i)$.[2]

In this formulation, we are looking at the action of the n-th power of the additive group scheme, $\mathbb{G}_a^n$, on $\mathbb{A}^n$ and investigating points in Λ where the value of $f_{\vec{b}} \in \mathbb{Q}[\mathbb{A}^n]$ has at most n prime factors. This point of view makes us wonder what one should expect for an arbitrary algebraic $\mathbb{Q}$-group $\mathbb{G}$, equipped with an algebraic action on a $\mathbb{Q}$-variety $\mathbb{V}$ and a regular function f on $\mathbb{V}$.[3]

General setting of the affine sieve, I. *Let $\Gamma \subseteq \mathbb{G}(\mathbb{Q})$ be a Zariski-dense subgroup of $\mathbb{G}$, where $\mathbb{G}$ is a linear algebraic group defined over $\mathbb{Q}$. Assume that $\mathbb{G}$ acts on a $\mathbb{Q}$-variety $\mathbb{V}$ and that the action is also defined over $\mathbb{Q}$. Let f be a regular function on $\mathbb{V}$ that is defined over $\mathbb{Q}$ and let $x_0 \in \mathbb{V}(\mathbb{Q})$. Under what conditions can we find a positive integer r and a finite set of primes S such that*

$$\{\gamma \in \Gamma \mid f(\gamma \cdot x_0) \text{ has at most } r \text{ prime factors in } \mathbb{Z}_S\}$$

is Zariski-dense in $\mathbb{G}$?

[1] Of course, a subset of the affine line is Zariski-dense if and only if it is infinite.

[2] If $\operatorname{rank}(\Lambda)$ is at least two, this conjecture is proved as a result of [Green and Tao 2010; 2012; Green et al. 2012].

[3] We work with rational numbers instead of integers for simplicity. As a result we end up working with S-integers instead of integers. And similarly to the above formulation of Dirichlet's theorem, we can avoid local obstructions.

Since the action is algebraic, $f \in \mathbb{Q}[\mathbb{V}]$ and $x_0 \in \mathbb{V}(\mathbb{Q})$, we have that $f(g \cdot x_0)$ defines a regular function on $\mathbb{G}$ that is also defined over $\mathbb{Q}$. So without loss of generality, we can directly work with $\mathbb{G}$ and avoid introducing $\mathbb{V}$.

General setting of the affine sieve, II. *Let $\Gamma \subseteq \mathrm{GL}_n(\mathbb{Q})$ and let $\mathbb{G}$ be its Zariski closure. Let $f \in \mathbb{Q}[\mathbb{GL}_n]$. For a positive integer r and a finite set of primes S, let*

$$\Gamma_{r,S}(f) := \{\gamma \in \Gamma \mid f(\gamma) \text{ has at most } r \text{ prime factors in } \mathbb{Z}_S\}.$$

Under what conditions can we find r and S such that $\Gamma_{r,S}(f)$ is Zariski-dense in $\mathbb{G}$?

Our goal here is to describe a general framework. In a given problem, not only it is important to show the existence of r and S, but also to find the best possible r and the right conditions on f and Γ which guarantee that S is empty. (S will be called the set of ramified primes.) It should be added that this general setting was formulated in [Bourgain et al. 2010] (slightly different notations are used in [Bourgain et al. 2010] and the best possible r is called the saturation number.)

Before formulating and justifying the needed conditions, let us quickly reformulate the mentioned results and problems in terms of the above setting. We denote by $\mathbb{G}_a$ and $\mathbb{G}_m$ the additive and multiplicative group schemes.

(1) Brun's fundamental theorem of sieves: let $\Gamma = \mathbb{Z} \subseteq \mathbb{G}_a(\mathbb{Q})$ and $f(x) \in \mathbb{Z}[x]$. Then $\Gamma_{r,\varnothing}(f)$ is Zariski-dense in $\mathbb{G}_a$ for some positive integer r.

(2) Dirichlet's theorem: let $\Gamma = \mathbb{Z} \subseteq \mathbb{G}_a(\mathbb{Q})$ and $0 \neq a, b \in \mathbb{Z}$. Let S be the set of prime factors of $\gcd(a, b)$. Then $\Gamma_{1,S}(ax + b)$ is Zariski-dense in $\mathbb{G}_a$.

(3) Twin prime conjecture: let $\Gamma = \mathbb{Z} \subseteq \mathbb{G}_a(\mathbb{Q})$ and $f(x) = x(x + 2)$. Then $\Gamma_{2,\varnothing}(f)$ should be Zariski-dense in $\mathbb{G}_a$. Chen proved $\Gamma_{3,\varnothing}(f)$ is Zariski-dense in $\mathbb{G}_a$.

(4) Mersenne prime conjecture: let $\Gamma = \langle 2 \rangle \subseteq \mathbb{G}_m(\mathbb{Q}) = \mathbb{Q}^\times$ and $f(x) = x - 1$. Then $\Gamma_{1,\varnothing}(f)$ should be Zariski-dense in $\mathbb{G}_m$.

(5) Hardy–Littlewood conjecture: let $\Gamma \subseteq \mathbb{Z}^n$ and $f(\vec{x}) = \prod_{i=1}^n (x_i + b_i)$ where $b_i \in \mathbb{Z}$. Then $\Gamma_{n,\varnothing}(f)$ should be Zariski-dense in the Zariski closure of Γ if there are no local obstructions.

(6) Bourgain–Gamburd–Sarnak's result: let $\Gamma = \langle \Omega \rangle \subseteq \mathrm{GL}_n(\mathbb{Z})$, $\mathcal{H}$ be its Zariski closure in $(\mathcal{GL}_n)_\mathbb{Z}$ and $f \in \mathbb{Z}[\mathcal{H}]$. If $\mathcal{H}_\mathbb{Q}$ is isomorphic to $\mathbb{SL}_2$ and f is absolutely irreducible and primitive[4], then $\Gamma_{r,\varnothing}(f)$ is Zariski-dense in $\mathcal{H}$ for some positive integer r.

[4] We refer the reader to [Bourgain et al. 2010] for the definition. This condition essentially takes care of local obstructions.

In fact, a much stronger result is proved in [Bourgain et al. 2010]. In the above setting, assume that the family of Cayley graphs $\mathrm{Cay}(\pi_q(\Gamma), \pi_q(\Omega))$ form a family of expanders as q runs through square-free integers. Then if f is absolutely irreducible and primitive, then $\Gamma_{r,\varnothing}(f)$ is Zariski-dense in $\mathcal{H}$ for some positive integer r.

1.2. ***What are the needed conditions? Statement of the main result.*** As we mentioned earlier, if $\Gamma = \langle 2 \rangle \subseteq \mathbb{G}_m(\mathbb{Q})$ and $f(x) = x - 1$, then we do not know whether $\Gamma_{r,S}(f)$ is Zariski-dense in $\mathbb{G}_m$ for some r and S. Sieve methods do not give us anything for this problem. In fact, the heuristics in [Salehi Golsefidy and Sarnak 2013, Appendix] suggest that it is not just the weakness of the method. And as Peter Sarnak says,[5] "*torus is the enemy*!" Here we present two examples where conjecturally the answer to the general setting of the affine sieve should be negative.

Isotropic torus.[6] Heuristics suggest that the number of prime factors of

$$(2^n - 1)(2^{n-1} - 1)$$

should go to infinity as n tends to infinity. This implies that if $\Gamma = \langle 2 \rangle \subseteq \mathbb{G}_m(\mathbb{Q})$ and $f(x) = (x-1)(x-2)$, then, for any r and S, $\Gamma_{r,S}(f)$ is not Zariski-dense in $\mathbb{G}_m$. See [Hardy and Wright 1979, p. 15] or the Appendix to [Salehi Golsefidy and Sarnak 2013] for heuristic considerations of this kind.

Anisotropic torus. Let $\gamma = \left[\begin{smallmatrix} 0 & 1 \\ 1 & 1 \end{smallmatrix}\right]$. Then for any integer n

$$\gamma^n = \begin{bmatrix} f_{n-1} & f_n \\ f_n & f_{n+1} \end{bmatrix},$$

where f_n is the n-th Fibonacci number. Let $\Gamma = \langle \gamma^2 \rangle$. Note that the Zariski closure $\mathbb{G}$ of Γ is isomorphic to the $\mathbb{Q}$-anisotropic torus $R^{(1)}_{\mathbb{Q}[\sqrt{5}]/\mathbb{Q}}(\mathbb{G}_m)$. Let $f(X_{ij}) = X_{12}$. Conjecturally, the number of prime factors of $f(\gamma^{2n})$ goes to infinity [Bugeaud et al. 2005]. Hence again $\Gamma_{r,S}(f)$ cannot be Zariski-dense in $\mathbb{G}$ for any r and S.

In general, I believe the following question should have an affirmative answer (see [Salehi Golsefidy and Sarnak 2013, Appendix] for a heuristic consideration).

---

[5]Personal communication, 2007.

[6]Recall that if A is any commutative algebra, $\mathbb{G}_m(A)$ is the group of units of A. A torus $\mathbb{T}$ defined over a field k is called k-*isotropic* if there is a nontrivial k-homomorphism from $\mathbb{T}$ to $\mathbb{G}_m$. It is called k-*anisotropic* otherwise. Let l be a quadratic extension of k and let $\mathbb{T} = R^{(1)}_{l/k}(\mathbb{G}_m)$ be the kernel of the norm map, for instance. Then $\mathbb{T}$ is k-anisotropic and l-isotropic; e.g., $R^{(1)}_{\mathbb{C}/\mathbb{R}}(\mathbb{G}_m)(\mathbb{R}) \simeq S^1$ and $R^{(1)}_{\mathbb{C}/\mathbb{R}}(\mathbb{G}_m)(\mathbb{C}) \simeq \mathbb{C}^\times$.

Question 1. Let k be a Galois number field and H be a finitely generated subgroup of $k^{\times}$. Is there a polynomial $p_H(x) = p(x) \in k[x]$ such that for any positive integer r

$$\left|\{h \in H \mid N_{k/\mathbb{Q}}(p(h)) \text{ has at most } r \text{ prime factors}\}\right| < \infty?$$

If the answer to Question 1 is affirmative, then one can easily prove the following.

Proposition 2. *Assume Question 1 has an affirmative answer. Let* $\Gamma \subseteq \mathrm{GL}_n(\mathbb{Q})$*, let* $\mathbb{G}$ *be its Zariski closure and* $\mathbb{G}^\circ$ *its (Zariski) connected component of identity. If* $X(\mathbb{G}^\circ) := \mathrm{Hom}(\mathbb{G}^\circ, \mathbb{G}_m)$ *is nontrivial, then there is* $f \in \mathbb{Q}[\mathbb{G}]$ *(which is not constantly zero on a connected component of* $\mathbb{G}$*) such that* $\Gamma_{r,S}(f)$ *is not Zariski-dense in* $\mathbb{G}$ *for any* r *and* S.

The above discussion suggests that one has to assume that $X(\mathbb{G}^\circ) = \{1\}$. This condition is in fact sufficient:

Theorem 3 (fundamental theorem of the affine sieve [Salehi Golsefidy and Sarnak 2013]). *Let* $\Gamma \subseteq \mathrm{GL}_n(\mathbb{Q})$, $\mathbb{G}$ *be the Zariski closure of* Γ*, and* $f \in \mathbb{Q}[\mathbb{G}]$*. If* f *is not constantly zero on a connected component of* $\mathbb{G}$ *and* $X(\mathbb{G}^\circ) = \{1\}$*, then* $\Gamma_{r,S}(f)$ *is Zariski-dense in* $\mathbb{G}$ *for some positive integer* r *and a finite set of primes* S.

It is worth mentioning that the following conditions are equivalent.

(1) $X(\mathbb{G}^\circ) = \{1\}$.
(2) No torus is a homomorphic image of $\mathbb{G}^\circ$.
(3) $X(R(\mathbb{G})) = \{1\}$ where $R(\mathbb{G})$ is the radical of $\mathbb{G}$.
(4) $\mathbb{G}/R_u(\mathbb{G})$ is semisimple where $R_u(\mathbb{G})$ is the unipotent radical of $\mathbb{G}$.
(5) $\mathbb{G} \simeq \mathbb{G}_{ss} \ltimes \mathbb{U}$, where $\mathbb{G}_{ss}$ is a semisimple group and $\mathbb{U}$ is a unipotent group.
(6) The Levi subgroup of $\mathbb{G}$ is semisimple.

A group is called Levi-semisimple if it satisfies these conditions.

1.3. ***Outline of the proof of Theorem 3.*** From this point on, we work in the setting of Theorem 3.

Let us first remark that any unipotent group is a Levi-semisimple group. But the Cayley graphs of finite quotients of a unipotent group cannot form a family of expanders. So one cannot directly appeal to [Bourgain et al. 2010] (see item (6) in page 327). In order to handle this difficulty, stronger results for unipotent groups and perfect groups[7] are proved.

[7] A group $\mathbb{G}$ is called perfect if $[\mathbb{G}, \mathbb{G}] = \mathbb{G}$.

In what follows let us also assume that $\mathbb{G}$ is Zariski-connected. So its derived subgroups $D^i(\mathbb{G})$ are also Zariski-connected. Let us recall that $D^0(\mathbb{G}) = \mathbb{G}$ and

$$D^{i+1}(\mathbb{G}) = [D^i(\mathbb{G}), D^i(\mathbb{G})].$$

Hence after $\dim \mathbb{G}$ steps, we get a perfect group $\mathbb{H} = D^{\dim \mathbb{G}}(\mathbb{G})$. We call it *the perfect core* of $\mathbb{G}$. Since $\mathbb{G}$ is Levi-semisimple, $\mathbb{G}/\mathbb{H}$ is a unipotent group $\mathbb{U}$ and we get the following diagram where each row is an exact sequence

$$\begin{array}{ccccccccc} 1 \to & \mathbb{H} & \to & \mathbb{G} & \xrightarrow{\pi} & \mathbb{U} & \to 1 \\ & \uparrow & & \uparrow & & \uparrow & \\ 1 \to & \mathbb{H} \cap \Gamma & \to & \Gamma & \xrightarrow{\pi} & \pi(\Gamma) & \to 1. \end{array} \tag{1}$$

We also notice that $\Gamma_H := \Gamma \cap \mathbb{H}$ is Zariski-dense in $\mathbb{H}$, and $\pi(\Gamma)$ is Zariski-dense in $\mathbb{U}$. Furthermore since $\mathbb{U}$ is a unipotent $\mathbb{Q}$-group, there is a $\mathbb{Q}$-section $s : \mathbb{U} \to \mathbb{G}$ (alternatively $\mathbb{G}$ is isomorphic to $\mathbb{H} \times \mathbb{U}$ as a $\mathbb{Q}$-variety (and not as a $\mathbb{Q}$-group)). This way we view $\mathbb{G}$ as a fiber bundle over $\mathbb{U}$ and each fiber is a shifted copy of the perfect group $\mathbb{H}$.

The general idea is that in order to find "lots" of desirable points in Γ. First we find "lots" of desirable base points in $\pi(\Gamma)$ and then above each one of them in the fiber we find "lots" of desirable points. It is clear that for the above scheme to work we need to prove certain "uniformity" for the number of prime factors r and ramified primes S for the base points and the fibers.

Let us make these more precise. Since $\mathbb{G}$ as a $\mathbb{Q}$-variety is isomorphic to $\mathbb{H} \times \mathbb{U}$, there are $f_i \in \mathbb{Q}[\mathbb{H}]$ and $p_i \in \mathbb{Q}[\mathbb{U}]$ such that f_i are linearly independent over $\mathbb{Q}$ and $f = \sum f_i \otimes q_i$, i.e.,

$$f(g) = \sum_i q_i(\pi(g)) f_i(s(\pi(g))^{-1} g),$$

for any $g \in \mathbb{G}$. Since $\mathbb{U}$ is unipotent, $\mathbb{Q}[\mathbb{U}]$ is isomorphic to the ring of polynomials in $\dim \mathbb{U}$ many variables. Let $p := \gcd q_i$ and $p_i = q_i / p$. So, for any $\gamma \in \Gamma$,

$$f(\gamma) = p(\pi(\gamma)) \sum_i p_i(\pi(\gamma)) f_i(s(\pi(\gamma))^{-1} \gamma). \tag{2}$$

So if $f(\gamma)$ has few prime factors, then $p(\pi(\gamma))$ and $\gcd_i(p_i(\pi(\gamma)))$ also have few prime factors. Thus, in the unipotent case, we need a stronger result:

Theorem 4 [Salehi Golsefidy and Sarnak 2013, Theorem 4]. *Let Λ be a finitely generated Zariski-dense subgroup of $\mathbb{U}(\mathbb{Q})$ where $\mathbb{U}$ is a unipotent $\mathbb{Q}$-group. Let $p, p_1, \ldots, p_k \in \mathbb{Q}[\mathbb{U}]$. Assume* $\gcd(p_1, \ldots, p_k) = 1$. *Then there are a positive* *integer r and a finite set of primes S such that*

$$\Lambda_{r,S}(p;p_1,\ldots,p_k) := \left\{\lambda\in\Lambda \;\middle|\; \begin{array}{r} p(\lambda)\ \textit{has at most } r \textit{ prime factors in } \mathbb{Z}_S \\ \textit{and } \gcd_i(p_i(\lambda)) = 1 \textit{ in } \mathbb{Z}_S \end{array}\right\}$$

is Zariski-dense in $\mathbb{U}$.

The main tools in the proof of Theorem 4 are Malcev theory of lattices in unipotent Lie groups [Raghunathan 1972] and Brun's combinatorial sieve.

First we notice that Λ is a discrete subgroup of $\mathbb{U}(\mathbb{R})$. Then, by Malcev theorem, since Λ is discrete and Zariski-dense in $\mathbb{U}(\mathbb{R})$, it is a lattice in $\mathbb{U}(\mathbb{R})$. Thus $\log(\Lambda)$ contains a lattice of $\mathrm{Lie}(\mathbb{U})(\mathbb{Q})$. Since $\mathbb{U}$ is a unipotent group, the logarithm is a polynomial map. Hence it is enough to handle the vector group case. We handle this using Brun's combinatorial sieve and a careful induction on the dimension.

By Theorem 4, $X = \pi(\Gamma)_{r,S}(p;(p_i)_i)$ is Zariski-dense in $\mathbb{U}$. We treat any $x\in X$ as a base point and look at the fiber above it. For any $x \in X$, fix $\gamma_x \in \Gamma$ such that $\pi(\gamma_x) = x$. By (2), for any $\gamma_H \in \Gamma \cap \mathbb{H}$, we have

$$\begin{aligned} f(\gamma_x\gamma_H) &= p(x)\sum_i p_i(x) f_i(s(x)^{-1}\gamma_x\gamma_H) \\ &= p(x)L_{\gamma_x s(x)^{-1}}\Big(\sum_i p_i(x) f_i\Big)(\gamma_H). \end{aligned} \tag{3}$$

From (3) and Theorem 4, one can easily prove Theorem 3 using the following.

Theorem 5 [Salehi Golsefidy and Sarnak 2013, Theorem 6]. *Let* Γ *be a finitely generated, Zariski-dense subgroup of a perfect, Zariski-connected* $\mathbb{Q}$*-group* $\mathbb{G}\subseteq \mathrm{GL}_n$. *Let* S_0 *be a finite set of primes and* $f_1,\ldots,f_m \in \mathbb{Q}[\mathbb{G}]$ *be linearly independent over* $\mathbb{Q}$. *Then there are a positive integer r and a finite set of primes S such that* $\Gamma_{r,S}(L_g(\sum_i v_i f_i))$ *is Zariski-dense in* $\mathbb{G}$ *for any* $g \in \mathbb{G}\cap \mathrm{GL}_n(\mathbb{Z}_{S_0})$ *and primitive integer vector* $(v_1,\ldots,v_n)$.

To prove Theorem 5, we start with a single regular function and describe how r and S depend on f and Γ. And then using our description, we uniformly control these parameters for $L_g(\sum_i v_i f_i)$.

We must carefully define a set of ramified primes $S_{\Gamma,f}$ with respect to Γ and f. For given q, let π_q denote the reduction map modulo q, and $\mathfrak{f}_q$ the finite field of order q. The set $S_{\Gamma,f}$ will essentially consists of primes where either $\pi_p(\Gamma)$ is "small" or $V(f)(\mathfrak{f}_p)$[8] is "large". By strong approximation [Nori 1987], one knows that $S_{\Gamma,f}$ is finite.

Using the Bourgain–Gamburd–Sarnak sieve method, we prove the following.

[8]Here we are abusing notation: $V(f)(\mathfrak{f}_p)$ denotes the set of solutions of f in $\mathbb{H}(\mathfrak{f}_p)$.

Theorem 6 [Salehi Golsefidy and Sarnak 2013, Theorem 5]. *In the above setting if $\mathbb{H}$ is perfect and Zariski-connected and $f \in \mathbb{Q}[\mathbb{H}]$, then $\Gamma_{r,S_{\Gamma,f}}(f)$ is Zariski-dense for some positive integer r which depends on the spectral gap of the congruence quotients of Γ, the degree of $V(f)$ and $|S_{\Gamma,f}|$.*

To execute the Bourgain–Gamburd–Sarnak sieve method, one needs to estimate the number of elements of $V(f)(\mathfrak{f}_p)$. This can be done using [Lang and Weil 1954] and Chebotarev density theorem (this is needed as $V(f)$ is not necessarily geometrically irreducible).

After proving Theorem 6, we can finish proof of Theorem 5 using this:

Proposition 7 [Salehi Golsefidy and Sarnak 2013, Proposition 29]. *In the above setting,*

$$\bigcup_{\substack{g \in \mathbb{H} \cap \mathrm{GL}_n(\mathbb{Z}_{S_0}) \\ \gcd_i v_i = 1}} S_{\Gamma, L_g(\sum_i v_i f_i)}$$

is finite.

Theorem 8 ([Salehi Golsefidy and Varjú 2012, Theorem 1]). *Let Ω be a finite symmetric subset of $\mathrm{GL}_n(\mathbb{Z}_S)$ and $\Gamma = \langle \Omega \rangle$. Let $\mathbb{G}$ be its Zariski closure and $\mathbb{G}^\circ$ be its Zariski-component of the identity. Then the Cayley graphs $\mathrm{Cay}(\pi_q(\Gamma), \pi_q(\Omega))$ form a family of expanders as q runs through square-free S-integers if and only if $\mathbb{G}^\circ$ is perfect.*

Theorem 8 is the main analytic tool in the proof of the fundamental theorem of the affine sieve. In the next section, I explain very briefly the outline of proof of Theorem 8 and the groundbreaking results which are behind its proof.

2. Expansion properties of linear groups

2.1. ***Expanders, "thin" subgroups and triple-product: formulation and recent results.*** In lots of problems in communication, one needs high connectivity and low cost. In other words, arbitrarily large highly connected sparse graphs. Such a family of finite graphs is called a family of expanders. Expanders have various interesting applications in computer science and number theory. I refer the reader to the beautiful surveys by A. Lubotzky [2012] and E. Kowalski [2010]. Here I mostly discuss the recent breakthroughs related to Theorem 8.

The first explicit construction of expanders is due to Margulis. He made a remarkable observation that the Cayley graphs of finite quotients of a discrete group with property (T) form expanders. The same ideas show that using Selberg's theorem one can deduce that the Cayley graphs

$$\mathrm{Cay}\left(\mathrm{SL}_2(\mathbb{Z}/n\mathbb{Z}), \left\{\left[\begin{smallmatrix} 1 & \pm 1 \\ 0 & 1 \end{smallmatrix}\right], \left[\begin{smallmatrix} 1 & 0 \\ \pm 1 & 1 \end{smallmatrix}\right]\right\}\right)$$

form expanders though $\mathrm{SL}_2(\mathbb{Z})$ does not have property (T) (in fact it is virtually free). As a result of works of many mathematicians (to name a few Kazhdan, Selberg, Margulis, Burger, Sarnak and Clozel) using automorphic forms and representation theory, the following is proved [Kazhdan 1967; Margulis 1973; Selberg 1965; Sarnak and Xue 1991; Burger and Sarnak 1991; Clozel and Ullmo 2004; Clozel 2003].

Theorem 9. *Let $\mathbb{G} \subseteq \mathrm{GL}_n$ be a semisimple simply connected $\mathbb{Q}$-group, and assume $\Gamma = \mathbb{G} \cap \mathrm{GL}_n(\mathbb{Z}_S) = \langle \Omega \rangle$ is an infinite group. Then the Cayley graphs $\mathrm{Cay}(\pi_m(\Gamma), \pi_m(\Omega))$ form a family of expanders as m runs through positive integers.*

Lubotzky was the first to ask if a *thin group*, that is, a Zariski-dense subgroup of infinite index in an arithmetic lattice, has the same property. He asked if

$$\mathrm{Cay}\left(\mathrm{SL}_2(\mathfrak{f}_p), \left\{\begin{bmatrix} 1 & \pm 3 \\ 0 & 1 \end{bmatrix}, \begin{bmatrix} 1 & 0 \\ \pm 3 & 1 \end{bmatrix}\right\}\right)$$

form expanders or not.[9] Y. Shalom [1997; 1999] constructed the first thin group with *certain* finite quotients whose Cayley graphs form expanders (not congruence quotients). A. Gamburd [2002] is the first to prove Lubotzky's question for "large" thin subgroups of $\mathrm{SL}_2(\mathbb{Z})$. He proved that if the Hausdorff dimension of the limit set of a subgroup of $\mathrm{SL}_2(\mathbb{Z})$ is larger than $5/6$, then the Cayley graphs of its congruence quotients modulo primes form a family of expanders.

Bourgain and Gamburd [Bourgain and Gamburd 2008b] completely answered Lubotzky's question in groundbreaking work. They proved that, if $\Gamma = \langle \Omega \rangle \subseteq \mathrm{SL}_2(\mathbb{Q})$ is Zariski-dense in $\mathbb{SL}_2$, then $\mathrm{Cay}(\pi_p(\Gamma), \pi_p(\Omega))$ form expanders as p runs through large primes. The steps and the ideas of their proof also gave the general framework for all recent work on this area. One of the main tools in their proof is a breakthrough by Helfgott [2008], who proved that if a symmetric generating set A of $\mathrm{SL}_2(\mathfrak{f}_p)$ is not very large — $|A| \le |\mathrm{SL}_2(\mathfrak{f}_p)|^{1-\varepsilon}$ — then its triple-product gets exponentially larger — $|A.A.A| \ge |A|^{1+\delta}$. Using this result coupled with Tao's noncommutative version [2008] of Balog–Szemerédi–Gowers, Bourgain and Gamburd proved a measure-theoretic version of the triple-product theorem ("l^2-flattening phenomena"). (In the next section, I elaborate on this.) Then using Kesten's bound concerning the random walk on a free group and the fact that any proper algebraic subgroup of $\mathbb{SL}_2$ is virtually solvable, they proved that the probability of being in a proper subgroup of $\mathrm{SL}_2(\mathfrak{f}_p)$ after an $l \sim \log(p)$-step random walk is small ("Escape from proper subgroups"). They finished the proof using a lower bound on the dimension of a nontrivial complex irreducible representation of $\mathrm{SL}_2(\mathfrak{f}_p)$ (this idea goes back to [Sarnak and Xue 1991]).

---

[9] Now it is called *Lubotzky's 1-2-3 problem.*

To execute the affine sieve method, Bourgain, Gamburd and Sarnak [2010] considered square-free congruences. They proved that $\mathrm{Cay}(\pi_q(\Gamma), \pi_q(\Omega))$ form expanders as q runs through square-free integers if Γ is a Zariski-dense subgroup of $\mathrm{SL}_2(\mathbb{Z})$. They also conjectured that the "if" part of Theorem 8 should hold if $\mathbb{G}$ is a semisimple group and $\Gamma \subseteq \mathbb{G} \cap \mathrm{GL}_n(\mathbb{Z})$. First they proved a sum-product theorem for $\mathbb{Z}/q\mathbb{Z}$, where q is a square-free integer. Then following Helfgott's argument they proved a triple-product theorem for $\mathrm{SL}_2(\mathbb{Z}/q\mathbb{Z})$ for a square-free integer q. They continued similar to [Bourgain and Gamburd 2008b]. Later P. Varjú [2012] gave an elegant proof that if a family of finite quasisimple groups satisfy a Helfgott-type triple-product property and some additional technical conditions, then any product of them also satisfies the triple-product property. He further showed how to use Tits's kind of argument to escape from proper subgroups when $\mathbb{G} \simeq R_{k/\mathbb{Q}}(\mathbb{SL}_n)$ (k is a number field) and $\Gamma \subseteq \mathrm{SL}_n(\mathcal{O}_k)$.

The next groundbreaking result is the generalization of Helfgott's result to any quasisimple finite group due to independent works of Breuillard, Green and Tao [Breuillard et al. 2011] and Pyber and Szabó [2010]. The main tools in their proofs are Helfgott's ideas (specially the ideas presented in [Helfgott 2011], where he proved that $\mathrm{SL}_3(\mathfrak{f}_p)$ has the triple-product property) and Larsen–Pink inequality [2011].

Theorem 8 mostly relies on [Bourgain and Gamburd 2008b; Breuillard et al. 2011; Pyber and Szabó 2010; Varjú 2012].

2.2. ***Outline of the proof of Theorem 8.*** For the "only if" part, it is enough to know that $\mathrm{Cay}(\pi_{q_i}(\Gamma), \pi_{q_i}(\Omega))$ form expanders for an infinite sequence of positive integers q_i. One can easily prove this using the following.

(1) $\Gamma \cap \mathbb{G}^\circ$ is a congruence subgroup of Γ. And so without loss of generality one can assume that $\mathbb{G}$ is Zariski-connected.

(2) There is a uniform upper bound for the order $|\pi_{q_i}(\Gamma)/[\pi_{q_i}(\Gamma), \pi_{q_i}(\Gamma)]|$ of the abelianization of $\pi_{q_i}(\Gamma)$.

(3) $\Gamma/[\Gamma, \Gamma]$ is a Zariski-dense finitely generated subgroup of $(\mathbb{G}/[\mathbb{G}, \mathbb{G}])(\mathbb{Q})$.

(4) π_{q_i} commutes with $\iota : \mathbb{G} \to \mathbb{G}/[\mathbb{G}, \mathbb{G}]$ when q_i has large enough prime factors.

To prove the "if" part, similar to all the recent works on this subject [Bourgain and Gamburd 2008b; 2008a; 2009; Bourgain and Varjú 2012], we prove *escape from proper subgroups* and *l^2-flattening*.

The general picture of a random walk on the Cayley graph of $\pi_q(\Gamma)$. Let us first remark that $\mathcal{G}_{q,\Omega} := \mathrm{Cay}(\pi_q(\Gamma), \pi_q(\Omega))$ form a family of expanders if and only if the random walk on $\pi_q(\Gamma)$ with the probability law $\pi_q[\mathcal{P}_\Omega]$ gets arbitrarily close to the equidistribution in $l \sim \log q$-steps. So we essentially study this random walk and the idea is this:

(1) If Ω is chosen carefully, then after $O(\log q)$-steps not only the probability law is not concentrated on any point but even it is not concentrated on any coset of a proper subgroup. This is called *escape from proper subgroups.*

(2) If we start with a probability law which is not concentrated on a coset of a proper subgroup, then either already the probability of being at the identity is pretty close to $1/|\pi_q(\Gamma)|$ or the probability of returning to identity gets much closer in the next step of the random walk (we get a power-saving). This is called *l^2-flattening*.

(3) At this stage, we can appeal to [Sarnak and Xue 1991] and use representation theory to say that in finitely many steps we get a flat probability law.

The precise formulation of "escape from proper subgroups" and the ideas behind its proof. First for simplicity, let us assume that Ω freely generates Γ. So $\mathrm{Cay}(\Gamma, \Omega)$ is a regular tree which is the covering space of all the finite graphs $\mathcal{G}_{q,\Omega}$. And the random walk on $\pi_q(\Gamma)$ in $O(\log q)$-steps can be completely understood by the random walk on the tree. This means in order to understand the behavior of the random walk on $\pi_q(\Gamma)$ in $O(\log q)$-steps, one can focus on the behavior of the random on Γ and study "small" lifts of elements of $\pi_q(\Gamma)$ (here we view $\mathbb{Z}_S$ as a discrete subgroup of $\mathbb{R} \cdot \prod_{p \in S} \mathbb{Q}_p$ and use the S-norm).

On the other hand, notice that the weight of a proper subgroup H with respect to the normalized counting (probability) measure on $\pi_q(\Gamma)$ is equal to $[\pi_q(\Gamma) : H]^{-1}$. If we want to get arbitrarily close to this probability law in $l \sim \log q$-steps, we should be able to get $\pi_q[\mathcal{P}_\Omega^{(l)}](H) \ll [\pi_q(\Gamma) : H]^{-\delta}$ in $l = O(\log q)$-steps.

Proposition 10. *Let $\Omega \subseteq \mathrm{GL}_n(\mathbb{Q})$ be a finite set and $\Gamma = \langle \Omega \rangle$. Assume the Zariski closure $\mathbb{G}$ of Γ is Zariski-connected and perfect. Then there are $\delta > 0$ and a finite symmetric subset $\Omega' \subseteq \Gamma$ such that*

$$\pi_q[\mathcal{P}_{\Omega'}^{(l)}](H) \ll [\pi_q(\Gamma) : H]^{-\delta},$$

for any proper subgroup H of $\pi_q(\Gamma)$ and any even integer $l \sim \log q$.

Here $\mu^{(l)}$ is the l-fold convolution of μ with itself and for any S and q, π_q is either the quotient map from $\mathbb{Z}_S \to \mathbb{Z}_S/q\mathbb{Z}_S$ or any other similar map.

To prove Proposition 10, by the above discussion, we have to look at small lifts of elements of H. Using Nori's theorems [1987], we prove that small lifts of a large subgroup of H are in a proper algebraic subgroup of $\mathbb{G}$. So we have to prove that the weight of any proper algebraic subgroup of $\mathbb{G}$ in the random walk on Γ with respect to the probability law $\mathcal{P}_{\Omega'}$ exponentially decays (for some choice of $\Omega' \subseteq \Gamma$).

In the spirit of Chevalley's theorem, we look for projective representations such that a proper algebraic subgroup fixes a point in one of them. If we had *finitely* many *irreducible* representations $\rho_i : \mathbb{G} \to \mathrm{GL}(\mathbb{V}_i)$ such that

(1) for any i, $\rho_i(\Gamma)$ contains *proximal* elements, and

(2) any proper connected algebraic subgroup fixes a point in $\mathbb{P}(\mathbb{V}_i)$ for some i,

then we could have used Tits's method to find "ping-pong players" which move around any projective point in all of these representations. And then finish the proof using Kesten's bounds for random walks on a tree.

In fact, if $\Gamma \subseteq \mathbb{G} \cap \mathrm{GL}_n(\mathbb{Z})$ and $\mathbb{G}$ is semisimple, then it is relatively easy to construct such representations (using [Goldsheid and Margulis 1989]). However, if $\mathbb{G}$ is not semisimple, then the unipotent radical is in the kernel of any irreducible representation. So one needs another technique to detect proper subgroups of $\mathbb{G}$ which map onto the semisimple part of $\mathbb{G}$. Even when $\mathbb{G}$ is semisimple and we are in S-arithmetic setting, finding these representations would be still challenging (as [Goldsheid and Margulis 1989] does not work over nonarchimedean fields).

To overcome these difficulties, we construct finitely many irreducible representations $\rho_i : \mathbb{G} \to \mathrm{GL}(\mathbb{V}_i)$ (which factor through the semisimple part of $\mathbb{G}$) and algebraic families $\{\phi_{i,w} : \mathbb{G} \to \mathrm{Aff}(\mathbb{V}_i)\}_{w \in \mathbb{W}_i}$ of affine representations defined over local fields K_i (the base parameter w changes in a vector group $\mathbb{W}_i$) such that:

(1) The linear part of $\phi_{i,w}$ is ρ_i and $\mathbb{G}(K_i)$ does not fix any point in $\mathbb{V}_i(K_i)$ via $\phi_{i,w}$ for any $w \neq 0$ (the representations above 0 take care of proper subgroups which do not surject onto the semisimple part of $\mathbb{G}$).

(2) For any i, $\rho_i(\Gamma)$ is unbounded in $\mathrm{GL}(\mathbb{V}_i(K_i))$.

(3) Any proper connected algebraic subgroup $\mathbb{H}$ of $\mathbb{G}$ either
   (a) fixes a projective point in $\mathbb{P}(\mathbb{V}_i(K_i))$ via ρ_i for some i, or
   (b) fixes a point v in $\mathbb{V}_i(K_i)$ via $\phi_{i,w}$ for some i and norm one vector $w \in \mathbb{W}_i(K_i)$.

We also give a somewhat new technique for constructing "ping-pong players" which does not rely on the existence of proximal elements.

We finish the proof of Proposition 10 studying random walks in affine spaces and proving that the probability of staying in a bounded set decays exponentially.

The precise statement of l^2-flattening. I have already given its formulation in item (2). Let us see the precise statement.

Proposition 11. *Let Γ and $\mathbb{G}$ be as in Proposition 10. Then for any $\varepsilon > 0$, there is $\delta > 0$ such that the following holds:*

Let μ be a probability measure on $\pi_q(\Gamma)$. Assume that

$$\|\mu\|_2 > |\pi_q(\Gamma)|^{-1/2+\varepsilon} \quad \text{and} \quad \mu(gH) < [\pi_q(\Gamma) : H]^{-\varepsilon},$$

for any $g \in \pi_q(\Gamma)$ and any proper subgroup $H < \pi_q(\Gamma)$. Then

$$\|\mu * \mu\|_2 < \|\mu\|_2^{1+\delta},$$

for any square-free integer q.

As I said earlier, [Bourgain and Gamburd 2008b] (also see [Varjú 2012]) used Tao's noncommutative version of Balog–Szemerédi–Gowers theorem to prove that, for a symmetric probability measure μ on a group, $\mu * \mu$ is not substantially flatter than μ only when $\mu * \mu$ is concentrated on an almost subgroup. So to prove Proposition 11, one has to understand almost subgroups of $\pi_q(\Gamma)$; or alternatively prove a triple-product theorem.

When $\mathbb{G}$ is semisimple, one can get such a result using works of [Breuillard et al. 2011] or [Pyber and Szabó 2010] (to get prime modulus for simple groups) and [Varjú 2012] (to extend it to square-free modulus for semisimple groups). (I refer the reader to a nice survey by B. Green [2010]).

To extend it to perfect groups, we prove a kind of bounded generation result and the general idea has some similarities with [Alon et al. 2001].

3. Final remarks and questions

Finding the best possible r (in [Bourgain et al. 2010], it is called the saturation number) and S for a given Γ and f in the setting of Theorem 3 is an extremely hard task. I have already mentioned the connection of this question with twin prime and Hardy–Littlewood conjectures. In [Bourgain et al. 2010], more interesting connections are mentioned, for example, divisibility of area of Pythagorean triangles and integral Apollonian circle packings (ACP). And since then, there have been lots of works on integral ACP, for example: [Bourgain and Fuchs 2011; Kontorovich and Oh 2011; Bourgain and Kontorovich 2012]. The main tool in the study of an integral ACP is its group $\mathcal{A}$ of symmetries. It is observed that $\mathcal{A}$ is a Kleinian group which is generated by a set of Möbius inversions $S = \{s_1, s_2, s_3, s_4\}$ (see [Graham et al. 2003] or [Sarnak 2007]). For instance, already in [Sarnak 2007], it is showed that in any primitive ACP there are infinitely many pairs of tangent circles with prime curvatures. In fact, much stronger result is proved. Sarnak considered the nerve $N(\mathcal{P})$ of a given ACP $\mathcal{P}$, that is, a graph whose vertices are circles in $\mathcal{P}$ and, for $C_1, C_2 \in \mathcal{P}$, $\{C_1, C_2\}$ is an edge if and only if C_1 and C_2 are tangent. Then he considered the subgraph generated by vertices $N^P(\mathcal{P})$ with prime curvature and proved that this subgraph is a union of trees all of whose vertices are of infinite degree (in particular, there are arbitrarily large chain of circles).

Now I would like to add a bit more structure to the nerve $N(\mathcal{P})$ of $\mathcal{P}$. Let us attach 2-cells and 3-cells to $N(\mathcal{P})$ to get a contractible space and call it the simplicial complex $\mathcal{C}(\mathcal{P})$ of $\mathcal{P}$. For a given positive integer r, let $\mathcal{C}_r^P(\mathcal{P})$ be the contractible subcomplex generated by the vertices whose curvature has at most r prime factors, for example, $\mathcal{C}_1^P(\mathcal{P}) = N^P(\mathcal{P})$. A corollary of Fuchs' result [2010] implies that $\mathcal{C}_{28}^P(\mathcal{P})$ has infinitely many 3-cells (her result even implies that the boundary of the nerve of the 3-cells in $\mathcal{C}_{28}^P(\mathcal{P})$ is also infinite). Conjecturally the same result should be true for $\mathcal{C}_2^P(\mathcal{P})$.

In light of the recent advancements, it seems interesting to study these complexes.

Question 12. In the above setting:

(1) [Sarnak 2007] Study the densities and the distributions of the connected components of $N^P(\mathcal{P})$.

(2) What can we say about the 2-cells and the 3-cells of $\mathcal{C}_r^P(\mathcal{P})$?

(3) Is there any r such that $\mathcal{C}_r^P(\mathcal{P})$ contains arbitrarily large chains of 3-cells?

For each 3-cell C in $\mathcal{C}(\mathcal{P})$, let $p(C)$ be the product of the curvatures of its vertices. For any path $w = (1 = \gamma_1, \dots, \gamma_k)$ of length k in the Cayley graph $\mathrm{Cay}(\mathcal{A}, S)$ which starts from the identity and any 3-cell C, we can consider $f_w(C) := \prod_i p(\gamma_i \cdot C)$. So f_w is a polynomial of degree $4k$ which is a product of $4k$ linear functions. Then Theorem 3 (together with the bound on r given in the proof!) says that there is a positive integer r_0 such that $\Gamma_{r_0 k^2, \varnothing}(f_w)$ is Zariski-dense in Γ; in particular, there are infinitely many chains of length k in $\mathcal{C}^P_{r_0 k^2}(\mathcal{P})$. The third part of Question 12 (in average) asks if $\Gamma_{r_0 k, \varnothing}(f_w)$ is infinite for some w.

Question 12 can be a test to see how much we can push affine sieve methods and get better bounds for the saturation number. When Γ is a lattice in a semisimple Lie group, using best bounds toward Ramanujan conjecture, Nevo and Sarnak [2010] gave sharp bounds on the saturation number which are similar to the bounds known for the classical case of one variable. I believe the next place to look for such bounds is where Γ is a thin group which contains a lattice in a subgroup (similar to the group of isometries of the ACP).

The general philosophy behind Lubotzky's 1-2-3 problem is that the Zariski-topology of Γ not only dictates the congruence topology on Γ (by strong approximation [Weisfeiler 1984], when $\mathbb{G}$ is simply connected semisimple) but also tells us about the analytical behavior of the congruence quotients. Theorem 8 says that indeed this way of thinking is completely true if $\Gamma \subseteq \mathrm{GL}_n(\mathbb{Q})$. To be precise, if Ω_1 and Ω_2 generate two Zariski-dense subgroups of $\mathbb{G} \cap \mathrm{GL}_n(\mathbb{Z}_S)$, then either both of the families $\{\mathcal{G}_{q,\Omega_1}\}$ and $\{\mathcal{G}_{q,\Omega_2}\}$ as q runs through square-free S-integers are expanders or neither of them are. However as soon as we enlarge $\mathbb{Q}$, Zariski-topology might not detect some of the properties of Γ (see [Salehi Golsefidy and Varjú 2012, Example 5]):[10]

Example 13. There are finite subsets Ω_1 and Ω_2 of $\mathrm{GL}_n(\mathbb{Z}[i])$ such that $\{\mathcal{G}_{\mathfrak{q},\Omega_1}\}$ is a family of expanders as $\mathfrak{q}$ runs through square-free Gaussian integers and $\{\mathcal{G}_{\mathfrak{q},\Omega_2}\}$ is NOT a family of expanders as $\mathfrak{q}$ runs through square-free Gaussian integers.

Let $\mathcal{H}$ be the Heisenberg group scheme over $\mathbb{Z}$ and $\mathcal{C}$ be its scheme-theoretic center. Then it is well-known that the symplectic group scheme $\mathrm{Sp}(\mathcal{V})$ acts on $\mathcal{H}$ (where $\dim \mathcal{V}_{\mathbb{Q}} = \dim \mathcal{H}_{\mathbb{Q}} - 1$). Let $\mathcal{L} = \mathrm{Sp}(\mathcal{V}) \ltimes \mathcal{H}$. And let Γ_1 be the group generated by $\mathcal{L}(\mathbb{Z})$ and $\mathcal{C}(\mathbb{Z}[i])$ in $\Gamma_2 = \mathcal{L}(\mathbb{Z}[i])$. Then one can show that for any generating sets Ω_1 and Ω_2 of Γ_1 and Γ_2, respectively, we have:

(1) Γ_1 and Γ_2 are both Zariski-dense in $\mathcal{L}_{\mathbb{Q}[i]}$;

(2) $\{\mathrm{Cay}(\pi_{\mathfrak{q}}(\Gamma_1), \pi_{\mathfrak{q}}(\Omega_1))\}$ is not a family of expanders as $\mathfrak{q}$ runs through square-free Gaussian integers;

(3) $\{\mathrm{Cay}(\pi_{\mathfrak{q}}(\Gamma_2), \pi_{\mathfrak{q}}(\Omega_2))\}$ is a family of expanders as $\mathfrak{q}$ runs through square-free Gaussian integers.

Though Example 13 says that in general even over a number field one should be cautious, it should be said that if the Zariski closure is semisimple we are in good shape [Salehi Golsefidy and Varjú 2012, Corollary 6]:

Corollary 14. *If a finite set $\Omega \subseteq \mathrm{GL}_n(\overline{\mathbb{Q}})$ generates a Zariski-dense subgroup of an adjoint form semisimple group, then $\{\mathcal{G}_{\mathfrak{q},\Omega}\}$ is a family of expanders as $\mathfrak{q}$ runs through square-free ideals of $\mathcal{O}_k(S)$ for some S, where k is the trace-field of $\langle\Omega\rangle$.*

Now one can ask if Corollary 14 is true for a linear group over $\mathbb{C}$ (or any other field) and arbitrary finite index ideals. (This is a form of [Lubotzky 2012, Conjecture 2.25]; also, see [Salehi Golsefidy and Varjú 2012, Question 4]).

Question 15. If a finite set $\Omega \subseteq \mathrm{GL}_n(\mathbb{C})$ generates a Zariski-dense subgroup of an adjoint form semisimple group, then is $\{\mathcal{G}_{\mathfrak{a},\Omega}\}$ a family of expanders as $\mathfrak{a}$ runs through finite index ideals of the trace ring of $\langle\Omega\rangle$?

If one just wants to relax the square-free condition, then it should be true in the generality of Theorem 8:

---

[10] This shows [Lubotzky 2012, Conjecture 2.25] as written is not correct.

Conjecture 16. *If a finite subset $\Omega \subseteq \mathrm{GL}_n(\mathbb{Q})$ generates a Zariski-dense subgroup of a perfect Zariski-connected group, then $\{\mathcal{G}_{m,\Omega}\}$ is a family of expanders as m runs through all the positive integers.*

Bourgain and Varjú [2012] proved Conjecture 16 when $\Gamma \subseteq \mathrm{SL}_n(\mathbb{Z})$ is Zariski-dense in $\mathbb{SL}_n$ (earlier a similar result for powers of primes was proved by Bourgain and Gamburd [2008a; 2009]).

Another interesting question is if the positive characteristic analogue of Theorem 8 (or its generalizations to arbitrary modulus) holds [Salehi Golsefidy and Varjú 2012, Question 3].

Question 17. If a finite subset $\Omega \subseteq \mathrm{GL}_n(\mathfrak{f}_l(t))$ generates a Zariski-dense subgroup of a perfect Zariski-connected group, then is $\{\mathcal{G}_{q(t),\Omega}\}$ a family of expanders as $q(t)$ runs through square-free polynomials with large degree prime factors?

Since Nori's theorems are extensively used in [Salehi Golsefidy and Varjú 2012] and they do not hold over $\mathfrak{f}_l(t)$, one needs new ideas to handle Question 17. An affirmative answer to Question 17 have immediate applications to arithmetic over global function fields and sieve methods in group theory in the sense of [Lubotzky and Meiri 2012].

Acknowledgments

I am grateful to Peter Sarnak and Peter Varjú for their collaborations. I am in debt to Peter Sarnak for pointing out his description of the "prime subgraph" of the nerve of an integral ACP. I would like to thank the anonymous referee for comments and suggestions.

References

[Alon et al. 2001] N. Alon, A. Lubotzky, and A. Wigderson, "Semi-direct product in groups and zig-zag product in graphs: connections and applications (extended abstract)", pp. 630–637 in *42nd IEEE Symposium on Foundations of Computer Science* (Las Vegas, NV, 2001), IEEE Computer Soc., Los Alamitos, CA, 2001.

[Bourgain and Fuchs 2011] J. Bourgain and E. Fuchs, "A proof of the positive density conjecture for integer Apollonian circle packings", *J. Amer. Math. Soc.* **24**:4 (2011), 945–967.

[Bourgain and Gamburd 2008a] J. Bourgain and A. Gamburd, "Expansion and random walks in $\mathrm{SL}_d(\mathbb{Z}/p^n\mathbb{Z})$, I", *J. Eur. Math. Soc. (JEMS)* **10**:4 (2008), 987–1011.

[Bourgain and Gamburd 2008b] J. Bourgain and A. Gamburd, "Uniform expansion bounds for Cayley graphs of $\mathrm{SL}_2(\mathbb{F}_p)$", *Ann. of Math.* (2) **167**:2 (2008), 625–642.

[Bourgain and Gamburd 2009] J. Bourgain and A. Gamburd, "Expansion and random walks in $\mathrm{SL}_d(\mathbb{Z}/p^n\mathbb{Z})$, II", *J. Eur. Math. Soc. (JEMS)* **11**:5 (2009), 1057–1103.

[Bourgain and Kontorovich 2012] J. Bourgain and A. Kontorovich, "On the strong density conjecture for integral Apollonian circle packings", preprint, 2012. arXiv 1205.4416

[Bourgain and Varjú 2012] J. Bourgain and P. P. Varjú, "Expansion in $SL_d(\mathbf{Z}/q\mathbf{Z})$, q arbitrary", *Invent. Math.* **188**:1 (2012), 151–173.

[Bourgain et al. 2010] J. Bourgain, A. Gamburd, and P. Sarnak, "Affine linear sieve, expanders, and sum-product", *Invent. Math.* **179**:3 (2010), 559–644.

[Breuillard et al. 2011] E. Breuillard, B. Green, and T. Tao, "Approximate subgroups of linear groups", *Geom. Funct. Anal.* **21**:4 (2011), 774–819.

[Bugeaud et al. 2005] Y. Bugeaud, F. Luca, M. Mignotte, and S. Siksek, "On Fibonacci numbers with few prime divisors", *Proc. Japan Acad. Ser. A Math. Sci.* **81**:2 (2005), 17–20.

[Burger and Sarnak 1991] M. Burger and P. Sarnak, "Ramanujan duals, II", *Invent. Math.* **106**:1 (1991), 1–11.

[Clozel 2003] L. Clozel, "Démonstration de la conjecture τ", *Invent. Math.* **151**:2 (2003), 297–328.

[Clozel and Ullmo 2004] L. Clozel and E. Ullmo, "Équidistribution des points de Hecke", pp. 193–254 in *Contributions to automorphic forms, geometry, and number theory*, edited by H. Hida et al., Johns Hopkins Univ. Press, Baltimore, 2004.

[Fuchs 2010] E. Fuchs, *Arithmetic properties of Apollonian circle packings*, Ph.D. thesis, Princeton University, 2010, http://math.berkeley.edu/~efuchs/efuchsthesis.pdf.

[Gamburd 2002] A. Gamburd, "On the spectral gap for infinite index "congruence" subgroups of $SL_2(\mathbb{Z})$", *Israel J. Math.* **127** (2002), 157–200.

[Goldsheid and Margulis 1989] I. Y. Goldsheid and G. A. Margulis, "Lyapunov exponents of a product of random matrices", *Uspekhi Mat. Nauk* **44**:5(269) (1989), 13–60. In Russian. Translation in *Russian Math. Surveys* **44**:5 (1989), 11–71.

[Graham et al. 2003] R. L. Graham, J. C. Lagarias, C. L. Mallows, A. R. Wilks, and C. H. Yan, "Apollonian circle packings: number theory", *J. Number Theory* **100**:1 (2003), 1–45.

[Green 2010] B. Green, "Approximate groups and their applications: work of Bourgain, Gamburd, Helfgott and Sarnak", 2010. Lecture at the Joint Mathematics Meetings of the American Math. Soc., San Francisco, January 15, 2010).

[Green and Tao 2010] B. Green and T. Tao, "Linear equations in primes", *Ann. of Math.* (2) **171**:3 (2010), 1753–1850.

[Green and Tao 2012] B. Green and T. Tao, "The Möbius function is strongly orthogonal to nilsequences", *Ann. of Math.* (2) **175**:2 (2012), 541–566.

[Green et al. 2012] B. Green, T. Tao, and T. Ziegler, "An inverse theorem for the Gowers $U^{s+1}[N]$-norm", *Ann. of Math.* (2) **176**:2 (2012), 1231–1372.

[Halberstam and Richert 1974] H. Halberstam and H. Richert, *Sieve method*, Academic Press, New York, 1974.

[Hardy and Wright 1979] G. H. Hardy and E. M. Wright, *An introduction to the theory of numbers*, 5th ed., Oxford University Press, New York, 1979.

[Helfgott 2008] H. A. Helfgott, "Growth and generation in $SL_2(\mathbb{Z}/p\mathbb{Z})$", *Ann. of Math.* (2) **167**:2 (2008), 601–623.

[Helfgott 2011] H. A. Helfgott, "Growth in $SL_3(\mathbb{Z}/p\mathbb{Z})$", *J. Eur. Math. Soc. (JEMS)* **13**:3 (2011), 761–851.

[Kazhdan 1967] D. A. Kazhdan, "On the connection of the dual space of a group with the structure of its closed subgroups", *Funkcional. Anal. i Prilozhen.* **1** (1967), 71–74. In Russian. Translation in *Functional Analysis and its Applications*, **1** (1967) (1) 63–65.

[Kontorovich and Oh 2011] A. Kontorovich and H. Oh, "Apollonian circle packings and closed horospheres on hyperbolic 3-manifolds", *J. Amer. Math. Soc.* **24**:3 (2011), 603–648.

[Kowalski 2010] E. Kowalski, "Sieve in expansion", Preprint, 2010. arXiv 1012.2793v1

[Lang and Weil 1954] S. Lang and A. Weil, "Number of points of varieties in finite fields", *Amer. J. Math.* **76** (1954), 819–827.

[Larsen and Pink 2011] M. J. Larsen and R. Pink, "Finite subgroups of algebraic groups", *J. Amer. Math. Soc.* **24**:4 (2011), 1105–1158.

[Lubotzky 2012] A. Lubotzky, "Expander graphs in pure and applied mathematics", *Bull. Amer. Math. Soc.* (*N.S.*) **49**:1 (2012), 113–162.

[Lubotzky and Meiri 2012] A. Lubotzky and C. Meiri, "Sieve methods in group theory I: Powers in linear groups", *J. Amer. Math. Soc.* **25**:4 (2012), 1119–1148.

[Margulis 1973] G. A. Margulis, "Explicit constructions of expanders", *Problemy Peredači Informacii* **9**:4 (1973), 71–80. In Russian. Translated in *Problems of Information Transmission*, **9**:4 (1973), 325–332.

[Nevo and Sarnak 2010] A. Nevo and P. Sarnak, "Prime and almost prime integral points on principal homogeneous spaces", *Acta Math.* **205**:2 (2010), 361–402.

[Nori 1987] M. V. Nori, "On subgroups of $\mathrm{GL}_n(\mathbf{F}_p)$", *Invent. Math.* **88**:2 (1987), 257–275.

[Pyber and Szabó 2010] L. Pyber and E. Szabó, "Growth in finite simple groups of Lie type of bounded rank", preprint, 2010. arXiv 1005.1858

[Raghunathan 1972] M. S. Raghunathan, *Discrete subgroups of Lie groups*, Ergebnisse der Math. **68**, Springer, New York, 1972.

[Salehi Golsefidy and Sarnak 2013] A. Salehi Golsefidy and P. Sarnak, "Affine sieve", *J. Amer. Math. Soc.* **26**:4 (2013), 1085–1105.

[Salehi Golsefidy and Varjú 2012] A. Salehi Golsefidy and P. Varjú, "Expansion in perfect groups", *Geom. Funct. Anal.* **22**:6 (2012), 1832–1891.

[Sarnak 2007] P. Sarnak, Letter to Lagarias about integral Apollonian packings, 2007, http://web.math.princeton.edu/sarnak/AppolonianPackings.pdf.

[Sarnak and Xue 1991] P. Sarnak and X. X. Xue, "Bounds for multiplicities of automorphic representations", *Duke Math. J.* **64**:1 (1991), 207–227.

[Selberg 1965] A. Selberg, "On the estimation of Fourier coefficients of modular forms", pp. 1–15 in *Proc. Sympos. Pure Math., Vol. VIII*, Amer. Math. Soc., Providence, R.I., 1965.

[Shalom 1997] Y. Shalom, "Expanding graphs and invariant means", *Combinatorica* **17**:4 (1997), 555–575.

[Shalom 1999] Y. Shalom, "Expander graphs and amenable quotients", pp. 571–581 in *Emerging applications of number theory* (Minneapolis, MN, 1996), edited by D. A. Hejhal et al., IMA Vol. Math. Appl. **109**, Springer, New York, 1999.

[Tao 2008] T. Tao, "Product set estimates for non-commutative groups", *Combinatorica* **28**:5 (2008), 547–594.

[Varjú 2012] P. P. Varjú, "Expansion in $\mathrm{SL}_d(\mathcal{O}_K/I)$, I square-free", *J. Eur. Math. Soc.* (*JEMS*) **14**:1 (2012), 273–305.

[Weisfeiler 1984] B. Weisfeiler, "Strong approximation for Zariski-dense subgroups of semisimple algebraic groups", *Ann. of Math.* (2) **120**:2 (1984), 271–315.

golsefidy@ucsd.edu *Mathematics Department, University of California, San Diego, San Diego, CA 92093-0112, United States*

Thin groups and superstrong approximation
MSRI Publications
Volume **61**, 2013

Notes on thin matrix groups

PETER SARNAK

Dedicated to the memory of Frances Wroblewski

We give a brief overview of the developments in the theory, especially the fundamental expansion theorem. Applications to diophantine problems on orbits of integer matrix groups, the affine sieve, group theory, gonality of curves and Heegaard genus of hyperbolic three manifolds, are given. We also discuss the ubiquity of thin matrix groups in various contexts, and in particular that of monodromy groups.

1. The fundamental expansion theorem

The Chinese remainder theorem for $\mathrm{SL}_n(\mathbb{Z})$ asserts, among other things, that for $q \geq 1$, the reduction $\pi_q : \mathrm{SL}_n(\mathbb{Z}) \to \mathrm{SL}_n(\mathbb{Z}/q\mathbb{Z})$ is onto. Far less elementary is the extension of this feature to $G(\mathbb{Z})$ where G is a suitable matrix algebraic group defined over $\mathbb{Q}$. The general form of this phenomenon for arithmetic groups is known as strong approximation and it is well understood [Platonov and Rapinchuk 1994].

There is a quantification of the above that is not as well known as it should be, as it turns out to be very powerful in many contexts. We call this "superstrong" approximation and it asserts that if we choose a finite symmetric ($s \in S$ if $s^{-1} \in S$) generating set S of $\mathrm{SL}_n(\mathbb{Z})$, then the congruence Cayley graphs (X_q, S) form an expander family as q goes to infinity (see [Hoory et al. 2006] for the definition and properties of expanders). Here the vertices x of the $|S|$-regular connected graph (X_q, S) are the elements of $\mathrm{SL}_n(\mathbb{Z}/q\mathbb{Z})$ and the edges run from x to sx, $s \in S$. The proof of this expansion property for $\mathrm{SL}_2(\mathbb{Z})$ has its roots in Selberg's lower bound of $\frac{3}{16}$ for the first eigenvalue λ_1 of the Laplacian on the hyperbolic surface $\Gamma\backslash\mathbb{H}$, Γ a congruence subgroup of $\mathrm{SL}_2(\mathbb{Z})$ [Selberg 1965]. This bound is an approximation to the Ramanujan/Selberg conjecture for automorphic forms on $\mathrm{GL}_2/\mathbb{Q}$. The generalizations of the expansion property to $G(\mathbb{Z})$ where G is say a semisimple matrix group defined over $\mathbb{Q}$ is also known thanks to developments towards the general Ramanujan conjectures that have been established [Burger and Sarnak 1991; Clozel 2003; Sarnak 2005]. This general expansion for these $G(\mathbb{Z})$ also goes by the name "property τ" for congruence subgroups [Lubotzky 2005].

Let Γ be a finitely generated subgroup of $\mathrm{GL}_n(\mathbb{Z})$ (more generally later on we allow it to be in $\mathrm{GL}_n(K)$ where K a number field) and denote its Zariski closure: $\mathrm{Zcl}(\Gamma)$, by G. If Γ is of finite index in $G(\mathbb{Z})$, then the discussion above of strong and superstrong approximation can be applied. However, if Γ is of infinite index in $G(\mathbb{Z})$, then $\mathrm{Vol}(\Gamma\backslash G(\mathbb{R})) = \infty$ and the techniques used to prove both of these properties don't apply. In this case we call Γ "thin". It is remarkable that under suitable natural hypotheses, strong approximation continues to hold in this thin context. The first result in this direction is [Matthews et al. 1984], and Weisfeiler extended it much further. More recent and effective treatments of this can be found in [Nori 1987] and [Larsen and Pink 2011]. An example of the statement of strong approximation in this context is: suppose that $\mathrm{Zcl}(\Gamma) = \mathrm{SL}_n$, then there is a $q_0 = q_0(\Gamma)$ such that $\pi_q : \Gamma \to \mathrm{SL}_n(\mathbb{Z}/q\mathbb{Z})$ is onto whenever $(q, q_0) = 1$. That the expansion property might continue to hold for thin groups was first suggested by Lubotzky [1993]. Thanks to a number of major developments by many people [Sarnak and Xue 1991; Gamburd 2002; Helfgott 2008; Bourgain and Gamburd 2008b; Bourgain et al. 2010; Pyber and Szabó 2010; Breuillard et al. 2011; Varjú 2012], the general expansion property is now known. The almost final version (almost because of the restriction that q be squarefree) is due to Salehi and Varjú [2012].

Theorem (the fundamental expansion). *Let $\Gamma \le \mathrm{SL}_n(\mathbb{Q})$ be a finitely generated group with a symmetric generating set S. Then the congruence graphs $(\pi_q(\Gamma), S)$, for q squarefree and coprime to a finite set of primes* (*which depend on Γ*), *are an expander family if and only if G^0, the identity component of $G := \mathrm{Zcl}(\Gamma)$, is perfect* (*i.e.*, $[G^0, G^0] = G^0$). *Moreover the determination of the expansion constant is in principle effective, if not feasible.*[1]

I will not review the techniques leading to the proof of this theorem (they have been discussed in many places including Kowalski and Tao's blogs) other than to point out that it involves three steps, the opening, the middlegame and the endgame. The endgame establishes the expansion by combining sufficiently strong (but still quite crude) upper bounds for the number of closed circuits in these graphs with largeness properties of the dimensions of the irreducible representations of the finite groups $G(\mathbb{Z}/q\mathbb{Z})$. In some cases (indeed all for which reasonable bounds for the expansion are known) the proof involves the endgame only [Sarnak and Xue 1991; Gamburd 2002]. In the general case, the upper bounds for the number of closed circuits is derived combinatorially. The opening and middlegame involve showing that smaller subsets of $G(\mathbb{Z}/p\mathbb{Z})$ grow substantially when multiplied by themselves at least three times (see [Helfgott

[1] It remains an open problem how to match to some extent in this general setting the quality of expansion that is known when Γ is arithmetic.

2008] and the extensions [Pyber and Szabó 2010] and [Breuillard et al. 2011]). A critical ingredient in the early treatments was the "sum-product" theorem [Bourgain et al. 2004] in finite fields. The middlegame is concerned with moderately large sets and is further handled by the crucial "flattening lemma" [Bourgain and Gamburd 2008b]. The latter also has its roots in combinatorics appealing to the Balog–Szemeredi theorem [Balog and Szemerédi 1994; Gowers 2001]. When q is not prime, the analysis and combinatorics is far more complicated and difficult due to the many subgroups of $G(\mathbb{Z}/q\mathbb{Z})$. It is handled in [Bourgain et al. 2010] for SL_2 and in [Varjú 2012] in general.

2. Applications

2.1. ***The affine sieve and diophantine analysis.*** The impetus for developing the expansion property for thin groups arose in connection with diophantine problems (in particular sieve problems for values of polynomials) on orbits of such thin groups [Bourgain et al. 2010]. Both strong approximation and superstrong approximation are crucial ingredients in executing a Brun combinatorial sieve in this setting. The theory is by now quite advanced and in particular the basic theorem of the affine sieve has been established in all cases where it is expected to hold [Salehi Golsefidy and Sarnak 2011].

For various special examples, such as for integral Apollonian packings, which has turned out to be one of the gems of the theory [Sarnak 2011], much more can be said thanks to special features. Firstly, in this case one can develop an archimedean count for the number of points in an orbit in a large region. This is done by combining spectral methods (using techniques which when Γ is a geometrically finite subgroup of $O(n-1,1)(\mathbb{R})$ go back to [Patterson 1976; Sullivan 1979; Lax and Phillips 1982]) with ergodic theoretic methods [Kontorovich and Oh 2011; Oh and Shah 2013; Lee and Oh 2013; Vinogradov 2012]. For the diophantine applications, one needs an archimedean spectral gap for the induced congruence groups, rather than the combinatorial expansion. [Bourgain et al. 2011] establishes the transfer of this information from the combinatorial to archimedean setting in this infinite volume case.

Two recent highlights of these developments are the "almost all" local-to-global results of [Bourgain and Kontorovich 2013a; 2013b]. The first concerns integral Apollonian packings and the question is which numbers are curvatures? The expected local-to-global conjecture [Graham et al. 2003; Fuchs and Sanden 2011] is proven for all but a zero density set of integers (the conjecture asserts that there are only a finite number of exceptions). Prior to that Bourgain and Fuchs [2011] had shown that the number of integers that are achieved is of positive density. The second development concerns the Zaremba problem, which asserts that if $A \geq 5$, the set of integers $q \geq 1$ for which there is a $1 \leq b \leq q$, $(b, q) = 1$ and for which the coefficients of the continued fraction of b/q are bounded by A, consists of all of $\mathbb{N}$. In [Bourgain and Kontorovich 2013b] the theory of thin subgroups of $\mathrm{SL}_2(\mathbb{Z})$ is extended to thin semi(sub)groups (one has to abandon direct spectral methods and replace them by dynamical ones [Lalley 1989; Bourgain et al. 2011]). In [Bourgain and Kontorovich 2013b] it is shown that for $A \geq 50$, the set of exceptions to the Zaremba conjecture is of zero density in $\mathbb{N}$.

2.2. *Random elements in Γ.* It is well known that for any reasonable notion of randomness, the random $f \in \mathbb{Q}[x]$ is irreducible and has Galois group the full symmetric group on the degree of f symbols. In [Rivin 2008] the study of such questions for the characteristic polynomial f_γ of a random element γ in $\mathrm{Sp}(2g, \mathbb{Z})$ and more general Γs, was initiated. The random element in $\mathrm{Sp}(2g, \mathbb{Z})$ is generated by running a symmetric random walk with respect to a measure μ whose support generates $\mathrm{Sp}(2g, \mathbb{Z})$. The expansion property is used via a sieving argument to show that the probability that f_γ is reducible is exponentially small. This and some generalizations are then coupled with the theory of the mapping class group M to show that the random element in M is pseudo-Anosov. These irreducibility questions and much more,[2] are extended and refined, especially in terms of the sieves that are applied, in [Kowalski 2008; Jouve et al. 2013; Lubotzky and Rosenzweig 2012]. Again, strong and superstrong approximation plays a central role.

In a different direction, Lubotzky and Meiri [2012] examine some group theoretic questions for linear groups using a random walk and a sieve. An example of what they show is: Let Γ be a finitely generated subgroup of $\mathrm{GL}_n(\mathbb{C})$ which is not virtually solvable, then the set of proper powers $P := \bigcup_{m=2}^{\infty}\{\gamma^m : \gamma \in \Gamma\}$, is exponentially small (in terms of hitting P in a long random walk). In particular, this resolved an open question as to whether finitely many translates of P can cover Γ, the answer being no.

2.3. *Gonality and Heegaard genus.* A compact Riemann surface of genus g can be realized as a covering of the plane of degree at most $g+1$ (Riemann–Roch). The gonality $d(X)$ of X is the minimal degree of such a realization. Unlike $g(X)$, $d(X)$ is a subtle conformal invariant. In [Zograf 1984] (see also [Yau 1996; Abramovich 1996]) the differential geometric inequality of [Yang and Yau 1980] is extended to the setting of $X = \Gamma\backslash\mathbb{H}$, a finite area quotient (orbifold) of the hyperbolic plane. If $A(X)$ is its area and $\lambda_1(X)$ its first Laplace eigenvalue,

[2] For example to showing that it is very unlikely that a random three dimensional manifold in the Dunfield–Thurston model [2006], has a positive first Betti number.

then

$$d(X) \geq \frac{\lambda_1(X)A(X)}{8\pi}. \tag{2-1}$$

This together with the known bounds towards the Ramanujan/Selberg conjectures for congruence (arithmetic) Xs (see [Blomer and Brumley 2011] for the best bounds for GL_2/K, where K is a number field which is what is relevant here) imply that for these Xs, the ratio of any two of $d(X)$, $A(X)$ and $(g(X)+1)$ is bounded universally from above and below.

There is a generalization of (2-1) to finite volume quotients $X = \Gamma\backslash\mathbb{H}^m$ (orbifolds) of hyperbolic m-space [Agol et al. 2008]. This is stated in terms of Li and Yau's notion [1982] of conformal volume. It gives an inequality between $\mathrm{Vol}(X)$, $\lambda_1(X)$ and the conformal volume of a piecewise conformal map of X into S^n. Again, this, together with the known universal lower bounds for $\lambda_1(X)$ when X is congruence arithmetic [Burger and Sarnak 1991; Clozel 2003], gives a linear in the volume, lower bound for the conformal volume of a conformal map of X to S^m. This has a nice application to reflection groups. A discrete group of motions of $\mathbb{H}^m$ is called a reflection group if it is generated by reflections (a reflection of $\mathbb{H}^m$ is a nontrivial isometry which fixes an $m-1$ dimensional hyperplane). Using the inequalities mentioned above, one shows (see [Long et al. 2006] for $m=2$ and [Agol et al. 2008] for $m>2$) that the set of maximal arithmetic reflection groups is finite for each m. Now Vinberg [1984] and Prokhorov [1986] have shown that for $m \geq 1000$, a reflection group can never be a lattice. Thus the totality of all maximal arithmetic reflection groups is finite.

Equation (2-1) has interesting applications to diophantine equations. As observed in [Abramovich 1991; Frey 1994], Faltings' finiteness theorem [1991] for rational points on subvarieties of abelian varieties can be used to prove finiteness of rational points on curves, whose coordinates lie in the union of all number fields of a bounded degree, as long as one can show the gonality of the curve is large enough. For example, if $X_0(N)/\mathbb{Q}$ is the familiar modular curve of level N and if D is given, then for $N \geq 230D$ (this value following from (2-1) and explicit Ramanujan bounds), the set of points on $X_0(N)$ with coordinates in the union of all number of fields of degree at most D, is finite! Recently Ellenberg, Hall and Kowalski [Ellenberg et al. 2012] have applied similar reasoning to a diophantine problem on a tower of curves. It arises from questions of reducibility and symmetry of specializations of members of a 1-parameter family of varieties. The curves that arise (as the parameter) are determined by the monodromy group Γ of the family (see below), and it lies in $\mathrm{Sp}(2g, \mathbb{Z})$ and is assumed to be Zariski dense in $\mathrm{Sp}(2g)$. In order to show that the gonalities of the curves in question increase quickly enough, they use the combinatorial expansion that is provided by the fundamental expansion theorem. Typically it is not known if Γ is thin or not in this context (see Section 3), but the beauty of the fundamental theorem is that one does not need to know!

There is an inequality similar to (2-1) for the Heegaard genus of a hyperbolic 3-manifold X. It is known that such an X can be decomposed into two handle bodies with common boundary a surface of genus h (called a Heegaard splitting). The minimal genus of such a surface in a splitting is called the Heegaard genus of X which we denote by $g(X)$. Like the gonality, it is a much more subtle (this time topological) invariant of X than its volume. In [Lackenby 2006] (see Theorem 4.1 and [Buser 1982]) it is shown that for complete X of finite volume

$$g(X) \geq \frac{\min[\lambda_1(X), 1] \cdot \mathrm{Vol}(X)}{32\pi}. \tag{2-2}$$

Applying this together with the universal lower bounds for λ_1 for congruence arithmetic X's shows that the Heegaard genus of a congruence hyperbolic three manifold is in order of magnitude, a linear function of its volume. In particular, any arithmetic 3-manifold has an infinite tower (by congruence subgroups) of coverings whose Heegaard genus grows linearly with the volume. One can ask if the same is true for any hyperbolic 3-manifold and the answer is yes as was shown in [Long et al. 2008]. Using local rigidity of lattices in $\mathrm{SL}_2(\mathbb{C})$ one can realize Γ where $X = \Gamma\backslash\mathbb{H}^3$, as a finitely generated subgroup of $\mathrm{SL}_2(K)$, where K is some number field. If Γ is not arithmetic then Γ is thin (in $\mathrm{SL}_2(O_K)$ perhaps allowing denominators at finitely many places), since its projection on the identity embedding of K into $\mathbb{C}$ is discrete. Using the fundamental expansion theorem gives a lower bound on λ_1 for a "congruence tower" of Γ and one then applies (2-2).

A related application of the expansion is to some questions in knot theory. Answering a question of Gromov, Pardon [2011] recently showed that there are isotopy classes of knots in S^3 which have arbitrary large distortion. In fact he shows that torus knots have this property. In [Gromov and Guth 2012] a large family of knots with large distortion is constructed using hyperbolic 3-manifolds X. Such an X can be realized as a degree 3 cover of S^3 branched over a knot K [Hilden 1976; Montesinos 1976]. Gromov and Guth [2012] show that the distortion $\delta(K)$ of this K satisfies $\delta(K) \gg \mathrm{Vol}(X)\lambda_1(X)$, (the implied constant being universal). From this and the lower bound for λ_1 when X varies over congruence arithmetic 3-manifolds (or a congruence thin tower and using the fundamental expansion theorem) one concludes that K and all knots isotopic to it has arbitrarily large distortion by choosing such X of large volume.

2.4. ***Rotation groups.*** Let $\Gamma = \langle \sigma_1, \sigma_2, \ldots, \sigma_t \rangle$ be a finitely generated subgroup of the group $\mathrm{SO}_3(\mathbb{R})$. There is an archimedean analogue of the expander property for the congruence graphs in this setting and which likewise has many applications [Lubotzky 1994; Sarnak 1990]. Define T_σ to be the averaging operator on functions on the two sphere S^2 by:

$$T_\sigma f(x) = \sum_{j=1}^{t} \big[f(\sigma_j x) + f(\sigma_j^{-1} x) \big]. \tag{2-3}$$

T_σ is self adjoint on $L^2(S^2, dA)$, where dA is the rotation invariant area element on S^2, and its spectrum is contained in $[-2t, 2t]$. The spectral gap property is that $2t$ (which is an eigenvalue with eigenvector the constant function) is a simple and isolated point of the spectrum. It is not hard to see that this property depends only on Γ and not on the generators. It is conjectured that Γ has such a spectral gap if and only if $\mathrm{Zcl}(\Gamma) = \mathrm{SO}_3$ (which in this case is equivalent to the topological closure of Γ being $\mathrm{SO}_3(\mathbb{R})$). A lot is known towards this conjecture. The first example of a Γ with a spectral gap was given by Drinfeld [1984] and this provided the final step in the solution of the Ruziewicz problem; that the only finitely additive rotationally invariant measure defined on Lebesgue measurable subsets of S^2, is a multiple of dA. His proof of the spectral gap makes use of an arithmetic such Γ together with the full force of automorphic forms and the solution of the Ramanujan conjectures for holomorphic cusp forms on the upper half plane. In [Gamburd et al. 1999] many thin Γs are shown to have a spectral gap. The best result known is the analogue of the fundamental expansion theorem in this context [Bourgain and Gamburd 2008a; 2010], and it suffices for most applications. It asserts that if the matrix elements of members of Γ are algebraic, then the conjecture is true for Γ. Like the very thin cases of the fundamental expansion theorem, part of the proof here relies on additive combinatorics. This time one needs the full force of the proof of the local Erdős–Volkmann ring conjecture [Edgar and Miller 2003; Bourgain 2003] — that a subset of $\mathbb{R}$ which is closed under addition and multiplication has Hausdorff dimension zero or one. As far as some concrete applications of the spectral gap for these groups, we mention the speed of equidistribution of directions associated with general quaquaversal tilings of three dimensional space [Draco et al. 2000; Radin and Sadun 1998] and constructions of quantum gates in the theory of quantum computation (the Solovay–Kitaev theorem; see [Harrow et al. 2002]).

3. Ubiquity of thin groups

Given a finitely generated group Γ in $\mathrm{GL}_n(\mathbb{Z})$, one can usually compute $G = \mathrm{Zcl}(\Gamma)$ without too much difficulty. On the other hand, deciding if Γ is thin can be formidable. In fact one is flirting here with questions that have no decision procedures (I thank Rivin for alerting me to these pitfalls that are close by). For example, if $\Gamma = \mathrm{SL}_2(\mathbb{Z}) \times \mathrm{SL}_2(\mathbb{Z})$ then there is no decision procedure to determine if an element $A \in \Gamma$ is in the group generated by a general set of say seven elements [Mikhailova 1958]. Even for Gromov hyperbolic groups, the question of whether a finitely generated subgroup generates a finite index subgroup, has no decision procedure [Rips 1982; Baumslag et al. 1994]. Mercifully strong and superstrong approximation only ask about $\mathrm{Zcl}(\Gamma)$. Still one is curious about thinness when applying these theorems and sometimes for good reason. For example, in the affine sieve setting, the quality of the expansion impacts the results dramatically (see [Nevo and Sarnak 2010] for the cases when Γ is a lattice) while the diophantine orbit problems become more standard ones of integer points on homogeneous varieties, when Γ is a lattice. Whether the typical Γ is thin or not is not so clear, and may depend on how Γ arises.

3.1. ***Schottky, ping-pong.*** Schottky groups in which the generators play ping-pong [Tits 1972; Breuillard and Gelander 2003] are one of the few classes of discrete groups whose group theoretic structure is very simple. If one chooses $A_1, A_2, \dots, A_\ell$ independently and at random in $\mathrm{SL}_n(\mathbb{Z})(n \geq 2)$, then with high probability $\Gamma = \langle A_1, \dots, A_\ell \rangle$ will be free on these generators, Zariski dense in SL_n and thin. If the A_j are chosen at the m-th step of a μ-random walk $(m \to \infty)$ and support (μ) generates $\mathrm{SL}_n(\mathbb{Z})$, then this was proved in [Aoun 2011]. A more geometric version is proven in [Fuchs and Rivin $\geq$ 2012] where the As are chosen independently and uniformly by taking them from the set of Bs with $\max(\|B\|, \|B^{-1}\|)$ less than X. Here $\| \ \|$ is any euclidean norm on the space of matrices and $X \to \infty$. Not only is Γ thin but it is very thin in the sense that the Hausdorff dimension of the limit set of Γ acting on $\mathbb{P}^{n-1}(\mathbb{R})$ is arbitrarily small.

3.2. ***Nonarithmetic lattices.*** If $\Gamma \leq G$ with $G \neq \mathrm{SL}_2(\mathbb{R})$, is an irreducible nonarithmetic lattice in a semisimple real group G, then Γ is naturally thin in the appropriate product by its conjugates. The argument is the same as the one in Section 2.3 using local rigidity. The certificate of being thin is that Γ is discrete in the factor corresponding to G. Examples of this kind which come from monodromy of hypergeometric differential equations in several variables are given in [Deligne and Mostow 1986] and in one variable in [Cohen and Wolfart 1990]. It appears that these were the first examples of thin monodromy groups (see Section 3.5 below). Other examples of thin monodromy groups in products of SL_2s are given in [Nori 1986] and these examples aren't even finitely presented. Teichmüller curves in the moduli space M_2 of curves of genus 2, give via Abel–Jacobi, curves in A_2 whose monodromies (inclusion of fundamental groups) are thin [McMullen 2003]. Here too the Zariski closure is a nontrivial product in Sp_4 and the thinness follows from having a discrete projection.

3.3. ***Reflection groups in hyperbolic space.*** Let f be an integral quadratic form in n-variables and of signature $(n-1, 1)$. For $n \geq 3$, $O_f(\mathbb{Z})$ the group of integral automorphs of f is a lattice in $G = O_f(\mathbb{R})$. The reflective subgroup R_f is the subgroup of $O_f(\mathbb{Z})$ which is generated by all the hyperbolic reflections which are in $O_f(\mathbb{Z})$. R_f is a normal subgroup of $O_f(\mathbb{Z})$ and if it is nontrivial, then $\mathrm{Zcl}(R_f) = O_f$. Vinberg [1984] and Nikulin [1987] have examined the question of when R_f is of finite index in $O_f(\mathbb{Z})$ (they call such an f reflective). In particular, in [Nikulin 1987] it is shown that there are only finitely many fs (up to integral equivalence) which are reflective. Thus for all but finitely many fs, R_f, if it is nontrivial, is a thin group in $\mathrm{GL}_n(\mathbb{Z})$ (albeit infinitely generated). Note that Nikulin's theorem fails for $n = 2$. If f is a binary quadratic form, then f is reflective if and only if it is ambiguous in the sense of Gauss (see [Sarnak 2007]) and Gauss determined the ambiguous forms in his study of genus theory.

3.4. ***Rotation groups.*** An interesting family of rotation groups are the groups $\Gamma(m, n)$, $m \geq 3$, $n \geq 3$ generated by σ_m and τ_n where

$$\sigma_m = \begin{bmatrix} \cos 2\pi/m & \sin 2\pi/m & 0 \\ -\sin 2\pi/m & \cos 2\pi/m & 0 \\ 0 & 0 & 1 \end{bmatrix}, \quad \tau_n = \begin{bmatrix} 1 & 0 & 0 \\ 0 & \cos 2\pi/n & \sin 2\pi/n \\ 0 & -\sin 2\pi/n & \cos 2\pi/n \end{bmatrix}.$$

That is, $\Gamma(m, n)$ is a subgroup of $\mathrm{SO}_f(\mathbb{R})$, $f(x_1, x_2, x_3) = x_1^2 + x_2^2 + x_3^2$, generated by two rotations about orthogonal axes and of orders m and n respectively. These arise in the theory of quaquaversal tilings of 3-space and their generalizations [Conway and Radin 1998; Radin and Sadun 1998].

As abstract groups, these are free products of two cyclic (or dihedral) groups amalgamated over a similar such group (except for $\Gamma(4, 4)$ which is finite and which we avoid); see [Radin and Sadun 1999]. This description can be used to decide the question of whether $\Gamma(m, n)$ is thin or not and also to show that thin is the rule rather than the exception. If

$$K = \mathbb{Q}\big(\cos 2\pi/m, \sin 2\pi/m, \cos 2\pi/n, \sin 2\pi/n\big),$$

then K is a totally real Galois extension of $\mathbb{Q}$ with abelian Galois group $G_{m,n}$. It is plain that $\Gamma(m, n)$ is a subgroup of $\mathrm{SO}_f(O[\frac{1}{2}])$, where O is the ring of integers of K. Moreover, since $\Gamma(m, n)$ is infinite, the powers of 2 in the denominators of the matrix entries of $\Gamma(m, n)$ must be unbounded (otherwise $\Gamma(m, n)$ would be a discrete subgroup of the compact group $\prod_{v|\infty} \mathrm{SO}_f(K_v)$). Hence the smallest S-arithmetic group to contain a subgroup commensurable with $\Gamma(m, n)$ is $\mathrm{SO}_f(O_S)$ where O_S are the S-integers of K, and S consists of the places of K dividing

2. Our thinness question is whether $\Gamma(m, n)$ is of finite or infinite index in the latter. If $|S| \geq 2$, then any finite index subgroup of $\mathrm{SO}_f(O_S)$ is a lattice in the higher rank group, $\prod_{v|(2)} \mathrm{SO}_f(K_v)$. By well known rigidity properties of such lattices [Margulis 1991] (or one can argue with vanishing of first cohomology groups) and the description of $\Gamma(m, n)$ mentioned above, it follows that $\Gamma(m, n)$ cannot be such a lattice. That is if $|S| \geq 2$, then $\Gamma(m, n)$ is thin and the former holds most of the time (for example if $G_{m,n}$ is not cyclic then $|S| \geq 2$). If $|S| = 1$, then $\Gamma(m, n)$ may be arithmetic and it is so in some special cases.[3] Perhaps the most interesting cases where $|S| = 1$ are when $m = 4$ and $n = 2^\nu$, $\nu \geq 3$, for which 2 is totally ramified. These have been investigated in [Robinson 2006; Serre 2009]. Serre shows that for $\nu = 3$ and 4, $\Gamma(4, 2^\nu)$ is arithmetic (in fact, $\Gamma(4, 2^\nu) = \mathrm{SO}_f(O[\frac{1}{2}])$) while for $\nu \geq 5$, it is thin. The thinness is proven by comparing the Euler characteristics $\chi(\mathrm{SO}_f(O[\frac{1}{2}]))$ and $\chi(\Gamma(4, 2^\nu))$, the first using a Tamagawa number computation and the second from the abstract group description of $\Gamma(4, 2^\nu)$.

3.5. *Monodromy groups.* The oldest and perhaps most natural source of finitely generated linear groups comes from monodromy in all of its guises. These include the very classical case of monodromy of the hypergeometric differential equation which we discuss further below, as well as that of a family of varieties varying over a base with its monodromy action on cohomology. For large families, and in cases where the monodromy has been computed, it appears almost always to be arithmetic. The question as to whether such monodromy groups are arithmetic was first raised in [Griffiths and Schmid 1975]. For example for the universal family of smooth projective hypersurfaces of degree d and dimension n in projective space, the monodromy representation on $H^n(X_0, \mathbb{Z})$, X_0 a base hypersurface, is an arithmetic subgroup of $\mathrm{GL}(H^n(X_0))(\mathbb{Z})$; see [Beauville 1986] where the exact level in $G(\mathbb{Z})$ is determined. For smaller families such as cyclic covers of $\mathbb{P}^1$, which have recently been studied in [McMullen 2013] in connection with the thinness question, the story is similar. More precisely, consider the family of curves (in affine coordinates) given by

$$C_a : y^d = (x - a_1)(x - a_2) \cdots (x - a_{n+1}), \tag{3-1}$$

where the parameters a vary so that $a_i \neq a_j$, for $i \neq j$. The fundamental group of the space of as is the pure braid group and it has a monodromy representation on $H_1(C, \mathbb{Z}) \cong \mathbb{Z}^{2g}$, g the genus of C_a, and again C is a fixed base curve. Answering a question in [McMullen 2013], Venkataramana [2012] shows that if $n \geq 2d$, then

[3] The quaquaversal tiling [Conway and Radin 1998] has symmetry group $\Gamma(3, 6)$, which is arithmetic [Serre 2009], while the dite/kart tiling [Radin and Sadun 1998] has symmetry $\Gamma(10, 4)$, for which $K = \mathbb{Q}(\cos \pi/10)$, $G_{10,4} = \mathbb{Z}/2\mathbb{Z} \times \mathbb{Z}/2\mathbb{Z}$ and $|S| = 2$; hence the latter is thin.

the image of the monodromy representation of the braid group in $\mathrm{GL}(H_1(C))(\mathbb{Z})$ is arithmetic. This generalizes a result of [A'Campo 1979] for $d = 2$. The proof is based on another result of Venkataramana [1994] which asserts that for $\mathbb{Q}$ rank two or higher arithmetic groups, a Zariski dense subgroup which contains enough elements from opposite horospherical subgroups is necessarily arithmetic. If $n < 2d$, then as observed in [McMullen 2013], there are examples based on the nonarithmetic lattices of [Deligne and Mostow 1986] in $\mathrm{SU}(2, 1)$ which are thin (one such is $n = 3$ and $d = 18$).

The thinness story for monodromy groups of one parameter families is less clear. We discuss in some detail the very rich examples of the classical hypergeometric equation. Let $\alpha, \beta \in \mathbb{Q}^n$ and consider the ${}_nF_{n-1}$ algebraic hypergeometric equation:

$$Du = 0, \tag{3-2}$$

where $D = (\theta + \beta_1 - 1)(\theta + \beta_2 - 1)\cdots(\theta - \beta_n - 1) - z(\theta + \alpha_1)\cdots(\theta + \alpha_n)$ and $\theta = z\,d/dz$.

The equation is regular outside $\{0, 1, \infty\}$ and the fundamental group

$$\pi_1(\mathbb{P}^1 - \{0, 1, \infty\})$$

has a representation in GL_n gotten by analytic continuation of a basis of solutions to (3-2) along curves in the thrice punctured sphere. Its image in GL_n is denoted by $H(\alpha, \beta)$ and is the monodromy group in question (defined up to conjugation in GL_n). $H(\alpha, \beta)$ is generated by the local monodromies A, B, C ($C = A^{-1}B$) gotten from loops about 0, ∞ and 1, respectively; see [Beukers and Heckman 1989] for a detailed description. We restrict to Hs which can be conjugated into $\mathrm{GL}_n(\mathbb{Z})$, which is equivalent to the characteristic polynomials of A and B being products of cyclotomic polynomials.[4] Such $H(\alpha, \beta)$ are self-dual and according to [Beukers and Heckman 1989], their Zariski closures $G(\alpha, \beta)$ are either finite, O_n or Sp_n, and they determine which it is explicitly in terms of α and β. Our interest is whether $H(\alpha, \beta)$ is of finite or infinite index in $G(\alpha, \beta)(\mathbb{Z})$. Other than the cases where $H(\alpha, \beta)$ (or equivalently $G(\alpha, \beta)$) are finite, all of which are listed in [Beukers and Heckman 1989], there are few cases where $H(\alpha, \beta)$ itself is known.

Recently Venkatamarana [2012] has shown that for n even and

$$\begin{aligned} \alpha &= \left(\frac{1}{2} + \frac{1}{n+1}, \frac{1}{2} + \frac{2}{n+1}, \dots, \frac{1}{2} + \frac{n}{n+1}\right), \\ \beta &= \left(0, \frac{1}{2} + \frac{1}{n}, \frac{1}{2} + \frac{2}{n}, \dots, \frac{1}{2} + \frac{n-1}{n}\right), \end{aligned} \tag{3-3}$$

[4] We assume further that (α, β) are primitive in the sense of [Beukers and Heckman 1989].

$H(\alpha, \beta)$ is arithmetic (here $G(\alpha, \beta) = \mathrm{Sp}(n)$). He deduces this by showing that for these exact parameters, the monodromy representation of $\pi_1(\mathbb{P}^1 - \{0, 1, \infty\})$ factors through a representation of the braid group on (3-1) with $d = 2$. In particular the arithmeticity follows from the arithmeticity of the latter.

The very fruitful Dwork family (see [Katz 2009; Harris et al. 2010]) $n \geq 4$ even, and

$$\begin{aligned} \alpha &= (0, 0, \ldots, 0), \\ \beta &= \left(\frac{1}{n+1}, \frac{2}{n+1}, \ldots, \frac{n}{n+1}\right), \end{aligned} \tag{3-4}$$

is apparently different. Again $G(\alpha, \beta) = \mathrm{Sp}(n)$ and for $n = 4$, the local monodromies are

$$A = \begin{bmatrix} 0 & 0 & 0 & -1 \\ 1 & 0 & 0 & -1 \\ 0 & 1 & 0 & -1 \\ 0 & 0 & 1 & -1 \end{bmatrix} \quad \text{and} \quad C = \begin{bmatrix} 1 & 0 & 0 & 5 \\ 0 & 1 & 0 & -5 \\ 0 & 0 & 1 & 5 \\ 0 & 0 & 0 & 1 \end{bmatrix}. \tag{3-5}$$

Very recently, Brav and Thomas [2012] have shown that A and C in (3-5) play generalized ping-pong on certain subsets of $\mathbb{P}^3$, from which it follows that $H(\alpha, \beta) \cong \mathbb{Z}/5\mathbb{Z} * \mathbb{Z}$. From rigidity, or the first cohomology properties of finite index subgroups of $\mathrm{Sp}(4, \mathbb{Z})$, it follows that $H(\alpha, \beta)$ must be thin. It seems likely that $H(\alpha, \beta)$ is thin for the whole Dwork family, that is, $n \geq 4$, but other than showing that the corresponding A and Cs play ping-pong, there appear to be no known means of proving this and no infinite family of thin $H(\alpha, \beta)$s with $G(\alpha, \beta)$ symplectic is known. For $n = 4$ there are 112 such $H(\alpha, \beta)$s in $\mathrm{Sp}(4, \mathbb{Z})$ [Singh and Venkataramana 2012].[5] Using extensions of the technique in [Venkataramana 2012] it is shown in [Singh and Venkataramana 2012] that of these, 63 are arithmetic. Of these, three[6] correspond to the 14 hypergeometrics associated with certain Calabi–Yau three folds (see [Doran and Morgan 2006; Chen et al. 2008]). Of the other 11, seven are shown to be thin in [Brav and Thomas 2012], again by finding ping-pong sets in $\mathbb{P}^3$. This leaves four of these Calabi–Yau's for which the thinness question is open. It would be interesting to understand the geometric significance, if there is one, for $H(\alpha, \beta)$ being thin or not in these families.

What is lacking above is a certificate for $H(\alpha, \beta)$ being thin that can be applied for example to families (i.e., $n \to \infty$). A robust such certificate has been provided in the case that $G(\alpha, \beta)(\mathbb{R})$ is of rank one and $n > 3$ [Fuchs et al. 2013]. In these cases $G(\alpha, \beta)$, as a group defined over $\mathbb{Q}$ is O_f, where f is a rational

[5]Remarkably, all of these are realized geometrically [Doran and Morgan 2006, Theorem 2.12].
[6]$((0,0,0,0), (\frac{1}{6}, \frac{1}{6}, \frac{5}{6}, \frac{5}{6}))$, $((0,0,0,0), (\frac{1}{6}, \frac{1}{4}, \frac{3}{4}, \frac{5}{6}))$, $((0,0,0,0), (\frac{1}{10}, \frac{3}{10}, \frac{7}{10}, \frac{9}{10}))$.

quadratic form in an odd number of variables and of signature $(n-1, 1)$ (over $\mathbb{R}$). We call these (α, β)s hyperbolic hypergeometrics and besides a (long) list of sporadic examples, they come in seven infinite parametric families [Fuchs et al. 2013]. Our conjecture for these is that thin rules, that is for all but finitely many of the hyperbolic hypergeometrics, $H(\alpha, \beta)$ is thin. This is proved in [Fuchs et al. 2013] for a number (but not all) of the seven families. For example for n odd consider the two families

$$\alpha = \left(0, \frac{1}{n+1}, \frac{2}{n+1}, \ldots, \frac{n-1}{2(n+1)}, \frac{n+3}{2(n+1)}, \ldots, \frac{n}{n+1}\right), \quad \beta = \left(\frac{1}{2}, \frac{1}{n}, \frac{2}{n}, \ldots, \frac{n-1}{n}\right), \tag{3-6}$$

and

$$\alpha = \left(\frac{1}{2}, \frac{1}{2n-2}, \frac{3}{2n-2}, \ldots, \frac{2n-3}{2n-2}\right), \quad \beta = \left(0, 0, 0, \frac{1}{n-2}, \frac{2}{n-2}, \ldots, \frac{n-3}{n-2}\right). \tag{3-7}$$

Both of these families are hyperbolic hypergeometrics and for both $H(\alpha, \beta)$ is thin for $n \geq 5$ and is arithmetic for $n = 3$.

The proof is based on the following principle: if $\psi : G(\mathbb{Z}) \to K$ is a morphism onto a group K for which $|\psi(H(\alpha, \beta))\backslash K| = \infty$, then certainly $H(\alpha, \beta)$ is of infinite index in $G(\mathbb{Z})$. Now in the higher rank cases there are no useful such ψs (by the Margulis normal subgroup theorem [1991] in these cases if K is infinite then $ker(\psi)$ is finite), however, in the rank one case such ψs may exist and yield a certificate of thinness. Indeed in this hyperbolic case if R_f is the Vinberg reflection subgroup described in Section 3.3, then as mentioned there, except for finitely many fs, $K_f := O_f(\mathbb{Z})/R_f$ is infinite. To use this one needs to analyze the image of $H(\alpha, \beta)$ in K_f. The key observation is that up to the finite index the hyperbolic hypergeometrics are generated by Cartan involutions.[7] These are linear reflections of $\mathbb{Q}^n$ which induce isometries on hyperbolic space given by geodesic inversions in a point [the hyperbolic reflections are generated by root vectors v in $\mathbb{Z}^n$ outside the light cone ($f(v) > 0$) while the Cartan involutions by root vectors w in $\mathbb{Z}^n$ inside the light cone, in fact $f(w) = -2$]. In order to examine the image of a group generated by such Cartan involutions in K_f, consider the "minimum distance graph," X_f. Its vertices are the integral Cartan root vectors $V_{-2}(\mathbb{Z}) = \{v \in \mathbb{Z}^n : f(v) = -2\}$, and v and w are joined if $f(v, w) = -3$. One can show that the components of X_f consist of finitely many isomorphism types and each is the Cayley graph of a finitely generated

[7] The local monodromy C about 1 is always a pseudoreflection and in these cases yields a Cartan involution.

Coxeter group. The main lemma [Fuchs et al. 2013] asserts that if $\Sigma \subset V_{-2}(\mathbb{Z})$ is a connected component of X_f then the image of the group generated by the Cartan involutions r_v with roots $v \in \Sigma$, is a finite subgroup of K_f.[8] This together with Vinberg and Nikulin's theorems gives a robust certificate for the thinness of these hyperbolic hypergeometric monodromies. As far as I know (3-6) and (3-7) give the first family of thin monodromy groups in high dimensions for which G is simple.

We end with some comments about the arithmetic Ramanujan conjectures. The gonality of a congruence arithmetic surface being linear in its genus and the Heegaard genus of a congruence hyperbolic three manifold being linear in its volume, as well as the proof that there are only finitely many maximal arithmetic reflection groups, all appeal to the uniform lower bounds for λ_1 for all such manifolds. This follows from what is known towards the Ramanujan conjectures but it does not follow from the fundamental expansion theorem since the latter only applies to one tower at a time. As far as the general Ramanujan conjectures, some progress has been made since [Sarnak 2005]. Namely in [Arthur 2013] a precise formulation of the Ramanujan conjectures for these groups is given, and moreover it is shown (assuming forms of the fundamental lemma which themselves should be theorems before too long) that these conjectures will follow if one can prove the Ramanujan conjectures for GL_m.

Acknowledgement

These brief notes cover a lot of ground. I thank my collaborators, the people whose work is quoted and the many mathematicians with whom I have discussed aspects of the theory connected with these thin groups. Thanks to the referee for pointing me to the relevant applications in [Gromov and Guth 2012].

References

[Abramovich 1991] D. Abramovich, *Subvarieties of abelian varieties and of jacobians of curves*, Ph.D. thesis, Harvard University, 1991, http://search.proquest.com/docview/303906247.

[Abramovich 1996] D. Abramovich, "A linear lower bound on the gonality of modular curves", *Int. Math. Res. Not.* **1996**:20 (1996), 1005–1011.

[A'Campo 1979] N. A'Campo, "Tresses, monodromie et le groupe symplectique", *Comment. Math. Helv.* **54**:2 (1979), 318–327.

[Agol et al. 2008] I. Agol, M. Belolipetsky, P. Storm, and K. Whyte, "Finiteness of arithmetic hyperbolic reflection groups", *Groups Geom. Dyn.* **2**:4 (2008), 481–498.

[Aoun 2011] R. Aoun, "Random subgroups of linear groups are free", *Duke Math. J.* **160**:1 (2011), 117–173.

[8] The proof makes use of the quite special feature of the binary form $g = x^2 + 3xy + y^2$ of being integrally equivalent to $-g$ (called reciprocal in [Sarnak 2007]).

[Arthur 2013] J. Arthur, *The endoscopic classification of representations: Orthogonal and symplectic groups*, Colloquium publications **61**, Amer. Math. Soc., Providence, RI, 2013.

[Balog and Szemerédi 1994] A. Balog and E. Szemerédi, "A statistical theorem of set addition", *Combinatorica* **14**:3 (1994), 263–268.

[Baumslag et al. 1994] G. Baumslag, C. F. Miller, III, and H. Short, "Unsolvable problems about small cancellation and word hyperbolic groups", *Bull. London Math. Soc.* **26**:1 (1994), 97–101.

[Beauville 1986] A. Beauville, "Le groupe de monodromie des familles universelles d'hypersurfaces et d'intersections complètes", pp. 8–18 in *Complex analysis and algebraic geometry* (Göttingen, 1985), edited by H. Grauert, Lecture Notes in Math. **1194**, Springer, Berlin, 1986.

[Beukers and Heckman 1989] F. Beukers and G. Heckman, "Monodromy for the hypergeometric function ${}_nF_{n-1}$", *Invent. Math.* **95**:2 (1989), 325–354.

[Blomer and Brumley 2011] V. Blomer and F. Brumley, "On the Ramanujan conjecture over number fields", *Ann. of Math.* (2) **174**:1 (2011), 581–605.

[Bourgain 2003] J. Bourgain, "On the Erdős–Volkmann and Katz–Tao ring conjectures", *Geom. Funct. Anal.* **13**:2 (2003), 334–365.

[Bourgain and Fuchs 2011] J. Bourgain and E. Fuchs, "A proof of the positive density conjecture for integer Apollonian circle packings", *J. Amer. Math. Soc.* **24**:4 (2011), 945–967.

[Bourgain and Gamburd 2008a] J. Bourgain and A. Gamburd, "On the spectral gap for finitely-generated subgroups of SU(2)", *Invent. Math.* **171**:1 (2008), 83–121.

[Bourgain and Gamburd 2008b] J. Bourgain and A. Gamburd, "Uniform expansion bounds for Cayley graphs of $\mathrm{SL}_2(\mathbb{F}_p)$", *Ann. of Math.* (2) **167**:2 (2008), 625–642.

[Bourgain and Gamburd 2010] J. Bourgain and A. Gamburd, "Spectral gaps in $\mathrm{SU}(d)$", *C. R. Math. Acad. Sci. Paris* **348**:11-12 (2010), 609–611.

[Bourgain and Kontorovich 2013a] J. Bourgain and A. Kontorovich, "On the strong density conjecture for integral circle packings", preprint, 2013. arXiv 1205.4416

[Bourgain and Kontorovich 2013b] J. Bourgain and A. Kontorovich, "On Zaremba's conjecture", preprint, 2013. arXiv 1107.3776

[Bourgain et al. 2004] J. Bourgain, N. Katz, and T. Tao, "A sum-product estimate in finite fields, and applications", *Geom. Funct. Anal.* **14**:1 (2004), 27–57.

[Bourgain et al. 2010] J. Bourgain, A. Gamburd, and P. Sarnak, "Affine linear sieve, expanders, and sum-product", *Invent. Math.* **179**:3 (2010), 559–644.

[Bourgain et al. 2011] J. Bourgain, A. Gamburd, and P. Sarnak, "Generalization of Selberg's $\frac{3}{16}$ theorem and affine sieve", *Acta Math.* **207**:2 (2011), 255–290.

[Brav and Thomas 2012] C. Brav and H. Thomas, "Classical and derived monodromy of the quintic threefold", preprint, 2012, http://people.maths.ox.ac.uk/brav/monodromy.pdf.

[Breuillard and Gelander 2003] E. Breuillard and T. Gelander, "On dense free subgroups of Lie groups", *J. Algebra* **261**:2 (2003), 448–467.

[Breuillard et al. 2011] E. Breuillard, B. Green, and T. Tao, "Approximate subgroups of linear groups", *Geom. Funct. Anal.* **21**:4 (2011), 774–819.

[Burger and Sarnak 1991] M. Burger and P. Sarnak, "Ramanujan duals, II", *Invent. Math.* **106**:1 (1991), 1–11.

[Buser 1982] P. Buser, "A note on the isoperimetric constant", *Ann. Sci. École Norm. Sup.* (4) **15**:2 (1982), 213–230.

[Chen et al. 2008] Y.-H. Chen, Y. Yang, and N. Yui, "Monodromy of Picard–Fuchs differential equations for Calabi–Yau threefolds", *J. Reine Angew. Math.* **616** (2008), 167–203.

[Clozel 2003] L. Clozel, "Démonstration de la conjecture τ", *Invent. Math.* **151**:2 (2003), 297–328.

[Cohen and Wolfart 1990] P. Cohen and J. Wolfart, "Modular embeddings for some nonarithmetic Fuchsian groups", *Acta Arith.* **56**:2 (1990), 93–110.

[Conway and Radin 1998] J. H. Conway and C. Radin, "Quaquaversal tilings and rotations", *Invent. Math.* **132**:1 (1998), 179–188.

[Deligne and Mostow 1986] P. Deligne and G. D. Mostow, "Monodromy of hypergeometric functions and nonlattice integral monodromy", *Inst. Hautes Études Sci. Publ. Math.* 63 (1986), 5–89.

[Doran and Morgan 2006] C. F. Doran and J. W. Morgan, "Mirror symmetry and integral variations of Hodge structure underlying one-parameter families of Calabi–Yau threefolds", pp. 517–537 in *Mirror symmetry, V*, edited by N. Yui et al., AMS/IP Stud. Adv. Math. **38**, Amer. Math. Soc., Providence, RI, 2006.

[Draco et al. 2000] B. Draco, L. Sadun, and D. Van Wieren, "Growth rates in the quaquaversal tiling", *Discrete Comput. Geom.* **23**:3 (2000), 419–435.

[Drinfeld 1984] V. G. Drinfeld, "Finitely-additive measures on S^2 and S^3, invariant with respect to rotations", *Funct. Anal. and Appl.* **18**:3 (1984), 245–246.

[Dunfield and Thurston 2006] N. M. Dunfield and W. P. Thurston, "Finite covers of random 3-manifolds", *Invent. Math.* **166**:3 (2006), 457–521.

[Edgar and Miller 2003] G. A. Edgar and C. Miller, "Borel subrings of the reals", *Proc. Amer. Math. Soc.* **131**:4 (2003), 1121–1129.

[Ellenberg et al. 2012] J. S. Ellenberg, C. Hall, and E. Kowalski, "Expander graphs, gonality, and variation of Galois representations", *Duke Math. J.* **161**:7 (2012), 1233–1275.

[Faltings 1991] G. Faltings, "Diophantine approximation on abelian varieties", *Ann. of Math.* (2) **133**:3 (1991), 549–576.

[Frey 1994] G. Frey, "Curves with infinitely many points of fixed degree", *Israel J. Math.* **85**:1-3 (1994), 79–83.

[Fuchs and Rivin $\geq$ 2012] E. Fuchs and I. Rivin, "Finitely generated subgroups of $\mathrm{SL}_n(\mathbb{Z})$ are generically thin".

[Fuchs and Sanden 2011] E. Fuchs and K. Sanden, "Some experiments with integral Apollonian circle packings", *Exp. Math.* **20**:4 (2011), 380–399.

[Fuchs et al. 2013] E. Fuchs, C. Meiri, and P. Sarnak, "Hyperbolic monodromy groups for the hypergeometric equation and Cartan involutions", preprint, 2013. arXiv 1305.0729

[Gamburd 2002] A. Gamburd, "On the spectral gap for infinite index "congruence" subgroups of $\mathrm{SL}_2(\mathbf{Z})$", *Israel J. Math.* **127** (2002), 157–200.

[Gamburd et al. 1999] A. Gamburd, D. Jakobson, and P. Sarnak, "Spectra of elements in the group ring of SU(2)", *J. Eur. Math. Soc. (JEMS)* **1**:1 (1999), 51–85.

[Gowers 2001] W. T. Gowers, "A new proof of Szemerédi's theorem", *Geom. Funct. Anal.* **11**:3 (2001), 465–588.

[Graham et al. 2003] R. L. Graham, J. C. Lagarias, C. L. Mallows, A. R. Wilks, and C. H. Yan, "Apollonian circle packings: Number theory", *J. Number Theory* **100**:1 (2003), 1–45.

[Griffiths and Schmid 1975] P. Griffiths and W. Schmid, "Recent developments in Hodge theory: A discussion of techniques and results", pp. 31–127 in *Discrete subgroups of Lie groups and applicatons to moduli* (Bombay, 1973), edited by W. L. Baily, Oxford Univ. Press, Bombay, 1975.

[Gromov and Guth 2012] M. Gromov and L. Guth, "Generalizations of the Kolmogorov–Barzdin embedding estimates", *Duke Math. J.* **161**:13 (2012), 2549–2603.

[Harris et al. 2010] M. Harris, N. Shepherd-Barron, and R. Taylor, "A family of Calabi–Yau varieties and potential automorphy", *Ann. of Math.* (2) **171**:2 (2010), 779–813.

[Harrow et al. 2002] A. W. Harrow, B. Recht, and I. L. Chuang, "Efficient discrete approximations of quantum gates", *J. Math. Phys.* **43**:9 (2002), 4445–4451.

[Helfgott 2008] H. A. Helfgott, "Growth and generation in $\mathrm{SL}_2(\mathbb{Z}/p\mathbb{Z})$", *Ann. of Math.* (2) **167**:2 (2008), 601–623.

[Hilden 1976] H. M. Hilden, "Three-fold branched coverings of S^3", *Amer. J. Math.* **98**:4 (1976), 989–997.

[Hoory et al. 2006] S. Hoory, N. Linial, and A. Wigderson, "Expander graphs and their applications", *Bull. Amer. Math. Soc.* (*N.S.*) **43**:4 (2006), 439–561.

[Jouve et al. 2013] F. Jouve, E. Kowalski, and D. Zywina, "Splitting fields of characteristic polynomials of random elements in arithmetic groups", *Israel J. of Math.* **193**:1 (2013), 263–307.

[Katz 2009] N. M. Katz, "Another look at the Dwork family", pp. 89–126 in *Algebra, arithmetic, and geometry: In honor of Yu. I. Manin, II*, edited by Y. Tschinkel and Y. Zarhin, Progr. Math. **270**, Birkhäuser, Boston, MA, 2009.

[Kontorovich and Oh 2011] A. Kontorovich and H. Oh, "Apollonian circle packings and closed horospheres on hyperbolic 3-manifolds", *J. Amer. Math. Soc.* **24**:3 (2011), 603–648.

[Kowalski 2008] E. Kowalski, *The large sieve and its applications: Arithmetic geometry, random walks and discrete groups*, Cambridge Tracts in Mathematics **175**, Cambridge University Press, 2008.

[Lackenby 2006] M. Lackenby, "Heegaard splittings, the virtually Haken conjecture and property (τ)", *Invent. Math.* **164**:2 (2006), 317–359.

[Lalley 1989] S. P. Lalley, "Renewal theorems in symbolic dynamics, with applications to geodesic flows, non–Euclidean tessellations and their fractal limits", *Acta Math.* **163**:1-2 (1989), 1–55.

[Larsen and Pink 2011] M. J. Larsen and R. Pink, "Finite subgroups of algebraic groups", *J. Amer. Math. Soc.* **24**:4 (2011), 1105–1158.

[Lax and Phillips 1982] P. D. Lax and R. S. Phillips, "The asymptotic distribution of lattice points in Euclidean and non–Euclidean spaces", *J. Funct. Anal.* **46**:3 (1982), 280–350.

[Lee and Oh 2013] M. Lee and H. Oh, "Effective circle count for Apollonian packings and closed horospheres", *Geom. Funct. Anal.* **23**:2 (2013), 580–621.

[Li and Yau 1982] P. Li and S. T. Yau, "A new conformal invariant and its applications to the Willmore conjecture and the first eigenvalue of compact surfaces", *Invent. Math.* **69**:2 (1982), 269–291.

[Long et al. 2006] D. D. Long, C. Maclachlan, and A. W. Reid, "Arithmetic Fuchsian groups of genus zero", *Pure Appl. Math. Q.* **2**:2 (2006), 569–599.

[Long et al. 2008] D. D. Long, A. Lubotzky, and A. W. Reid, "Heegaard genus and property τ for hyperbolic 3-manifolds", *J. Topol.* **1**:1 (2008), 152–158.

[Lubotzky 1994] A. Lubotzky, *Discrete groups, expanding graphs and invariant measures*, Progress in Mathematics **125**, Birkhäuser, Basel, 1994.

[Lubotzky 2005] A. Lubotzky, "What is... property (τ)?", *Notices Amer. Math. Soc.* **52**:6 (2005), 626–627.

[Lubotzky and Meiri 2012] A. Lubotzky and C. Meiri, "Sieve methods in group theory, I: Powers in linear groups", *J. Amer. Math. Soc.* **25**:4 (2012), 1119–1148.

[Lubotzky and Rosenzweig 2012] A. Lubotzky and L. Rosenzweig, "The Galois group of random elements of linear groups", preprint, 2012. arXiv 1205.5290

[Lubotzky and Weiss 1993] A. Lubotzky and B. Weiss, "Groups and expanders", pp. 95–109 in *Expanding graphs* (Princeton, NJ, 1992), edited by J. Friedman, DIMACS Ser. Discrete Math. Theoret. Comput. Sci. **10**, Amer. Math. Soc., Providence, RI, 1993.

[Margulis 1991] G. A. Margulis, *Discrete subgroups of semisimple Lie groups*, Ergebnisse der Math. (3) **17**, Springer, Berlin, 1991.

[Matthews et al. 1984] C. R. Matthews, L. N. Vaserstein, and B. Weisfeiler, "Congruence properties of Zariski-dense subgroups, I", *Proc. London Math. Soc.* (3) **48**:3 (1984), 514–532.

[McMullen 2003] C. T. McMullen, "Billiards and Teichmüller curves on Hilbert modular surfaces", *J. Amer. Math. Soc.* **16**:4 (2003), 857–885.

[McMullen 2013] C. T. McMullen, "Braid groups and Hodge theory", *Math. Ann.* **355**:3 (2013), 893–946.

[Mikhailova 1958] K. A. Mikhailova, "The occurrence problem for direct products of groups", *Dokl. Akad. Nauk SSSR* **119** (1958), 1103–1105. In Russian.

[Montesinos 1976] J. M. Montesinos, "Three-manifolds as 3-fold branched covers of S^3", *Quart. J. Math. Oxford Ser.* (2) **27**:105 (1976), 85–94.

[Nevo and Sarnak 2010] A. Nevo and P. Sarnak, "Prime and almost prime integral points on principal homogeneous spaces", *Acta Math.* **205**:2 (2010), 361–402.

[Nikulin 1987] V. V. Nikulin, "Discrete reflection groups in Lobachevsky spaces and algebraic surfaces", pp. 654–671 in *Proceedings of the International Congress of Mathematicians, I* (Berkeley, CA, 1986), edited by A. M. Gleason, Amer. Math. Soc., Providence, RI, 1987.

[Nori 1986] M. V. Nori, "A nonarithmetic monodromy group", *C. R. Acad. Sci. Paris Sér. I Math.* **302**:2 (1986), 71–72.

[Nori 1987] M. V. Nori, "On subgroups of $\mathrm{GL}_n(\mathbf{F}_p)$", *Invent. Math.* **88**:2 (1987), 257–275.

[Oh and Shah 2013] H. Oh and N. A. Shah, "Equidistribution and counting for orbits of geometrically finite hyperbolic groups", *J. Amer. Math. Soc.* **26**:2 (2013), 511–562.

[Pardon 2011] J. Pardon, "On the distortion of knots on embedded surfaces", *Ann. of Math.* (2) **174**:1 (2011), 637–646.

[Patterson 1976] S. J. Patterson, "The limit set of a Fuchsian group", *Acta Math.* **136**:3-4 (1976), 241–273.

[Platonov and Rapinchuk 1994] V. Platonov and A. Rapinchuk, *Algebraic groups and number theory*, vol. 139, Pure and Applied Mathematics, Academic Press, Boston, MA, 1994.

[Prokhorov 1986] M. N. Prokhorov, "Absence of discrete groups of reflections with a noncompact fundamental polyhedron of finite volume in a Lobachevskiĭ space of high dimension", *Izv. Akad. Nauk SSSR Ser. Mat.* **50**:2 (1986), 413–424. In Russian; translated in *Math. USSR Izv.* **28**:2 (1987), 401–411.

[Pyber and Szabó 2010] L. Pyber and Szabó, "Growth in finite simple groups of Lie type", preprint, 2010. arXiv 1001.4556

[Radin and Sadun 1998] C. Radin and L. Sadun, "Subgroups of SO(3) associated with tilings", *J. Algebra* **202**:2 (1998), 611–633.

[Radin and Sadun 1999] C. Radin and L. Sadun, "On 2-generator subgroups of SO(3)", *Trans. Amer. Math. Soc.* **351**:11 (1999), 4469–4480.

[Rips 1982] E. Rips, "Subgroups of small cancellation groups", *Bull. London Math. Soc.* **14**:1 (1982), 45–47.

[Rivin 2008] I. Rivin, "Walks on groups, counting reducible matrices, polynomials, and surface and free group automorphisms", *Duke Math. J.* **142**:2 (2008), 353–379.

[Robinson 2006] G. R. Robinson, "A subgroup of SO(3, $\mathbb{R}$) generated by rotations of orders 4 and 8", *J. Algebra* **306**:1 (2006), 201–207.

[Salehi Golsefidy and Sarnak 2011] A. Salehi Golsefidy and P. Sarnak, "Affine sieve", preprint, 2011. arXiv 1109.6432

[Salehi Golsefidy and Varjú 2012] A. Salehi Golsefidy and P. P. Varjú, "Expansion in perfect groups", *Geom. Funct. Anal.* **22**:6 (2012), 1832–1891.

[Sarnak 1990] P. Sarnak, *Some applications of modular forms*, Cambridge Tracts in Mathematics **99**, Cambridge University Press, 1990.

[Sarnak 2005] P. Sarnak, "Notes on the generalized Ramanujan conjectures", pp. 659–685 in *Harmonic analysis, the trace formula, and Shimura varieties*, edited by J. Arthur et al., Clay Math. Proc. **4**, Amer. Math. Soc., Providence, RI, 2005.

[Sarnak 2007] P. Sarnak, "Reciprocal geodesics", pp. 217–237 in *Analytic number theory*, edited by W. Duke and Y. Tschinkel, Clay Math. Proc. **7**, Amer. Math. Soc., Providence, RI, 2007.

[Sarnak 2011] P. Sarnak, "Integral Apollonian packings", *Amer. Math. Monthly* **118**:4 (2011), 291–306.

[Sarnak and Xue 1991] P. Sarnak and X. X. Xue, "Bounds for multiplicities of automorphic representations", *Duke Math. J.* **64**:1 (1991), 207–227.

[Selberg 1965] A. Selberg, "On the estimation of Fourier coefficients of modular forms", pp. 1–15 in *Proc. Sympos. Pure Math., VIII*, edited by A. L. Whiteman, Amer. Math. Soc., Providence, R.I., 1965.

[Serre 2009] J. P. Serre, "Le groupe quaquaversal, vu comme groupe S-arithmétique", *Oberwolfach Rep.* **6**:2 (2009), 1421–1426.

[Singh and Venkataramana 2012] S. Singh and T. Venkataramana, "Arithmeticity of certain symplectic hypergeometric groups", preprint, 2012. arXiv 1208.6460

[Sullivan 1979] D. Sullivan, "The density at infinity of a discrete group of hyperbolic motions", *Inst. Hautes Études Sci. Publ. Math.* 50 (1979), 171–202.

[Tits 1972] J. Tits, "Free subgroups in linear groups", *J. Algebra* **20** (1972), 250–270.

[Varjú 2012] P. P. Varjú, "Expansion in $\mathrm{SL}_d(\mathbb{O}_K/I)$, I square-free", *J. Eur. Math. Soc. (JEMS)* **14**:1 (2012), 273–305.

[Venkataramana 1994] T. N. Venkataramana, "On systems of generators of arithmetic subgroups of higher rank groups", *Pacific J. Math.* **166**:1 (1994), 193–212.

[Venkataramana 2012] T. Venkataramana, "Arithmeticity of the braid group at roots of unity", preprint, 2012. arXiv 1204.4778V2

[Vinberg 1984] È. B. Vinberg, "Absence of crystallographic groups of reflections in Lobachevskiĭ spaces of large dimension", *Trudy Moskov. Mat. Obshch.* **47** (1984), 68–102. In Russian; translated in *Trans. Mosc. Math. Soc.* (1985), 75–112.

[Vinogradov 2012] I. Vinogradov, "Effective bisector estimate with application to Apollonian circle packings", preprint, 2012. arXiv 1204.5498

[Yang and Yau 1980] P. C. Yang and S. T. Yau, "Eigenvalues of the Laplacian of compact Riemann surfaces and minimal submanifolds", *Ann. Scuola Norm. Sup. Pisa Cl. Sci.* (4) **7**:1 (1980), 55–63.

[Yau 1996] S.-T. Yau, "An application of eigenvalue estimate to algebraic curves defined by congruence subgroups", *Math. Res. Lett.* **3**:2 (1996), 167–172.

[Zograf 1984] P. G. Zograf, “Small eigenvalues of automorphic Laplacians in spaces of parabolic forms”, *Zap. Nauchn. Sem. Leningrad. Otdel. Mat. Inst. Steklov.* (*LOMI*) **134** (1984), 157–168. In Russian; translated in *J. Soviet Math.* **36**:1 (1987), 106–114.

sarnak@math.princeton.edu *Department of Mathematics, Princeton University, Princeton, NJ 08544-1000 United States*